New Essays on Tarski and Philosophy

Edited by
DOUGLAS PATTERSON

OXFORD
UNIVERSITY PRESS

Great Clarendon Street, Oxford OX2 6DP

Oxford University Press is a department of the University of Oxford.
It furthers the University's objective of excellence in research, scholarship,
and education by publishing worldwide in

Oxford New York

Auckland Cape Town Dar es Salaam Hong Kong Karachi
Kuala Lumpur Madrid Melbourne Mexico City Nairobi
New Delhi Shanghai Taipei Toronto

With offices in

Argentina Austria Brazil Chile Czech Republic France Greece
Guatemala Hungary Italy Japan Poland Portugal Singapore
South Korea Switzerland Thailand Turkey Ukraine Vietnam

Oxford is a registered trade mark of Oxford University Press
in the UK and in certain other countries

Published in the United States
by Oxford University Press Inc., New York

© the Several Contributors 2008

The moral rights of the authors have been asserted
Database right Oxford University Press (maker)

First published 2008

All rights reserved. No part of this publication may be reproduced,
stored in a retrieval system, or transmitted, in any form or by any means,
without the prior permission in writing of Oxford University Press,
or as expressly permitted by law, or under terms agreed with the appropriate
reprographics rights organization. Enquiries concerning reproduction
outside the scope of the above should be sent to the Rights Department,
Oxford University Press, at the address above

You must not circulate this book in any other binding or cover
and you must impose this same condition on any acquirer

British Library Cataloguing in Publication Data

Data available

Library of Congress Cataloging in Publication Data

Data available

Typeset by Laserwords Private Ltd, Chennai, India
Printed in Great Britain
on acid-free paper by
CPI Antony Rowe, Chippenham, Wiltshire

ISBN 978–0–19–929630–9

1 3 5 7 9 10 8 6 4 2

Contents

	List of Contributors	vii
1	Introduction *Douglas Patterson*	1
2	Tarski and his Polish Predecessors on Truth *Roman Murawski and Jan Woleński*	21
3	Polish Axiomatics and its Truth: On Tarski's Leśniewskian Background and the Ajdukiewicz Connection *Arianna Betti*	44
4	Tarski's Conceptual Analysis of Semantical Notions *Solomon Feferman*	72
5	Tarski's Theory of Definition *Wilfrid Hodges*	94
6	Tarski's Convention T and the Concept of Truth *Marian David*	133
7	Tarski's Conception of Meaning *Douglas Patterson*	157
8	Tarski, Neurath, and Kokoszyńska on the Semantic Conception of Truth *Paolo Mancosu*	192
9	Tarski's Nominalism *Greg Frost-Arnold*	225
10	Truth, Meaning, and Translation *Panu Raatikainen*	247
11	Reflections on Consequence *John Etchemendy*	263
12	Tarski's Thesis *Gila Sher*	300
13	Are There Model-Theoretic Logical Truths that are not Logically True? *Mario Gómez-Torrente*	340

14	Truth on a Tight Budget: Tarski and Nominalism *Peter Simons*	369
15	Alternative Logics and the Role of Truth in the Interpretation of Languages *Jody Azzouni*	390

Index 431

List of Contributors

Jody Azzouni, Professor of Philosophy, Tufts University

Arianna Betti, Vrije Universiteit Amsterdam

Marian David, Professor of Philosophy, University of Notre Dame

John Etchemendy, Provost, Stanford University

Solomon Feferman, Professor of Mathematics and Philosophy and Patrick Suppes Professor of Humanities and Social Sciences, Stanford University

Greg Frost-Arnold, Assistant Professor of Philosophy, University of Nevada, Las Vegas

Mario Gómez-Torrente, Instituto de Investigaciones Filosóficas, Universidad Nacional Autónoma de México

Wilfrid Hodges, Professorial Fellow, School of Mathematical Sciences, Queen Mary, University of London

Paolo Mancosu, Professor of Philosophy, University of California, Berkeley

Roman Murawski, Faculty of Mathematics and Computer Science, Adam Mickiewicz University

Douglas Patterson, Associate Professor of Philosophy, Kansas State University

Panu Raaikainen, Academy Research Fellow, Academy of Finland, and Docent in Theoretical Philosophy, University of Helsinki

Gila Sher, Professor of Philosophy, University of California, San Diego

Peter Simons, Professor of Philosophy, University of Leeds

Jan Woleński, Jagiellonian University Institute of Philosophy

1
Introduction

Douglas Patterson

If we go by the manner in which his contributions have been received up to the present day, there are three Alfred Tarskis. One is the pure mathematician and preternaturally clear logician and father of model theory, "the most methodologically sophisticated definer of all time" (Belnap 1993, 132). This is the Tarski known to logicians and very formally inclined philosophers. A second Tarski is known in continental Europe as the most eminent member of the Lvov–Warsaw school of philosophers following Kazimerez Twardowski—Stanislaw Leśniewski, Jan Łukasiewicz, Kazimierez Ajdukiewicz, Tadeusz Kotarbiński, and so on (Woleński 1989)—and as a philosopher who was in close contact with movements such as Hilbertian formalism (Sinaceur 2001) and the positivism of the Vienna Circle (Szaniawski 1993). The third Tarski is the one known among mainstream philosophers of language in the English speaking world. This Tarski came from nowhere to propose certain technical means of defining truth, means exploited in one way by Davidson (2001) and criticized in another by Field (1972). This Tarski taught us that sentences such as " 'snow is white' is true if and only if snow is white" are very important to the theory of truth. This third Tarski also proposed a simple approach to the paradox of the liar, one that provides a convenient foil for more sophisticated accounts, and he held a strange view to the effect that natural languages are "inconsistent."

My aim in assembling this collection of essays is to encourage us to rediscover the real Tarski behind these three appearances. My own journey began with acquaintance with the third, as I read mainstream work in the philosophy of language. I also knew something of the first Tarski from my own study of logic and mathematics. I think I did wonder, as many do, who Tarski was and how he had come to exercise such a large influence, as I knew nothing about his background aside from a few familiar anecdotes. Things changed for me—prompted, I recall, by a remark of Lionel Shapiro's to the effect that though many people talk about "The Concept of Truth in Formalized Languages" few people actually read it—when I sat down, more or less on a whim, to make myself one of the exceptions. What I found was vastly more interesting than what I'd been told about. Since then I have been reading and re-reading Tarski, learning what I can about where he came from and how his views developed, and rethinking the significance of what he said. In a way my motives in putting this collection together have been entirely selfish: I wanted to learn more about Tarski, and I have.

More sociably, however, I want to get three mutually isolated groups of scholars talking to one another. I don't believe that Tarski's remarks on formal topics from the seminal pre-war years can be understood without attention both to the philosophical and mathematical background in which he worked. To take one example here, Tarski's presentation of "Tarski's Theorem" on the indefinability of truth—known today as the result that arithmetic truth is not arithmetically definable, or, a bit more philosophically put, that no language sufficient for expressing a weak version of arithmetic can consistently express basic aspects of its own semantics—is intimately bound up with his views on expressibility and type theory, views Tarski attributes to Leśniewski and even Husserl. These make the difference between what is now said, which is that arithmetic truth is not arithmetically definable, and what Tarski said at the time, which is that it isn't definable at all. There are also questions, raised by Gomez-Torrente 2004, as to whether Tarski proves the result usually attributed to the passage, or a different, syntactic result. In the seminal works of the 1930s one finds an emerging body of formal results still embedded in a bygone era. Much has been lost along the way, some well, some not, but we cannot understand the development of modern logic, a development in which Tarski played a central role, without setting its beginning in proper context.

When it comes to the second, genuinely historical Tarski, my hope is to help to rescue him and, even more importantly his teachers, from the obscurity into which they have fallen in the English speaking world. There is no reason why Leśniewski, Łukasiewicz, Ajdukiewicz, Kotarbiński, Twardowski, and their compatriots are not accorded at least the status of Carnap, Reichenbach, Schlick, Neurath, and Hempel. My belief here is that this isn't really anything more than a legacy of the many misfortunes Poland endured in the twentieth century. Lvov and Warsaw before the Second World War seem poised to play roles equivalent to that of Vienna in the history of philosophy before being cut off by fascism and communism. As a result, in the English speaking world undergraduate and even graduate students learn nothing of the Polish school, and many important texts are walled off by archaic or missing translations. I hope that this work will stimulate interest in Tarski's predecessors.

As for the third Tarski, I believe him to be a vastly less able philosopher than the real one. Among other things, I argue here and elsewhere that his approach to the semantic paradoxes is different from, and vastly better than, what is usually attributed to him. He also has things to say about definition that I believe have a good deal to teach us about meaning and analyticity. The arsenal of stock criticisms of Tarski—that his methods of defining truth make contingent truths about the semantics of language into necessary or logical truths, that he was misguided to think that languages can be "inconsistent," that he advocated a simplistic formalization-and-hierarchy approach to the paradoxes—nearly all fail to make contact with Tarski's actual views. In the light of what we can learn about Tarski's background and the development of his thought, it is time for a re-evaluation of his contributions. This has been to some extent ongoing for some time—especially, for example, in the debate following the appearance of Etchemendy's *The Concept of Logical Consequence*—but I hope to accelerate the process with this volume.

In the service of all of these goals I have solicited essays from a genuinely international group of scholars, ranging from those directly knowledgeable about Tarski's Polish background, through scholars familiar with other aspects of his philosophical development, to those more interested in understanding Tarski in the light of contemporary thought. I have arranged the essays roughly in historical order, from essays about his influences and teachers, through direct textual discussions of his work and on to more general evaluations of his ideas in light of what we now know about various topics. It bears mention, in connection with a discussion of what is included here, that since this collection is concerned with Tarski's philosophical views and the evaluation thereof, textually our primary focus is on the more overtly philosophical works surrounding the seminal period of the early 1930s—roughly, then, the contents of the English anthology *Logic, Semantics, Metamathematics*, plus some outliers. This book is not a collection of essays directly concerned with Tarski's purely mathematical and logical work of later decades, though that work does get mentioned in some of the contributions.

By way of introducing the content of the essays that follow, and of knitting them together, more can still be said. Many ways of organizing an introduction would serve, but here I'll focus on the idea that Tarski's work is fundamentally focused on the relation of logical consequence. Many other schemata for an introduction would have done just as well, but this one will serve us well enough. Familiar characterizations of deductive inference of the sort often presented on the first day of an introductory logic class focus on the idea that one sentence "follows from" some others, or "is a consequence" of them when, if the latter are true, the former must be true as well. Logic, broadly construed, is the systematic study of this relationship, and has been from Aristotle's definition of the syllogism onward. As Etchemendy puts it in his contribution:

Among the characteristics claimed for logically valid arguments are the following: If an argument is logically valid, then the truth of its conclusion follows necessarily from the truth of the premises. From our knowledge of the premises we can establish, without further investigation, that the conclusion is true as well. The information expressed by the premises justifies the claim made by the conclusion. And so forth. These may be vague and ill-understood features of valid inference, but they are the characteristics that give logic its *raison d'être*. They are why logicians have studied the consequence relation for over two thousand years.

The essays collected in this volume can be seen as addressing Tarski's treatment of four basic questions about logical consequence.

(1) How are we to understand truth, one of the notions in terms of which logical consequence is explained? *What* is it that is preserved in valid inference, or that such inference allows us to discover new claims to have on the basis of old?

(2) *Among what* does the relation of logical consequence hold? Assertions? Token sentences? Types of sentences? Interpreted sentences? Propositions? Judgments? Several or all of the foregoing?

(3) Given answers to the first two questions, what is involved in the consequence relationship itself? What is the *preservation* at work in "truth preservation"?

(4) Finally, what do the notions of truth and consequence thus explored have to do with what Etchemendy has above as "the information expressed by" a sentence? What do truth and consequence so construed have to do with *meaning*?

Let us look further into each of these four topics, both in their more and less formal guises, and set out what the essays presented here have to say about them. Given the spirit of times an interest in deduction was often an interest in axiomatic systems or "deductive sciences" and so these will be with us throughout the discussion.

TRUTH

Tarski both proposed a criterion of adequacy for the definition of truth and presented a method for formulating definitions that are adequate according to the criterion in some tractable cases. He also gave a famous proof that nothing of the sort was possible in a range of other cases. All of these topics require our attention, as does the topic of what Tarski thought definition and definability were. However, we have to begin with a more basic question: what gave Tarski the idea that a rigorous treatment of these topics was needed?

The matter is actually less clear than it might appear. One might think that there had been a clear need at some point for such definitions, and that the time was right for someone with Tarski's talents to provide them. This appears, however, not to have been the case; indeed, the concept of truth, Tarski himself insists at the outset of his most famous work, is something of which "every reader possess in greater or less degree and intuitive knowledge" (1983, 153). Why, then, is an involved project of definition required? As Solomon Feferman explains in his contribution, Tarski seems to have worked comfortably with the informal notion of truth in a structure from 1924 onwards, just as did contemporaries like Skolem and Gödel. Indeed, as Vaught notes, for everyone in the field, "it had been possible to go even as far as the completeness theorem by treating truth (consciously or unconsciously) essentially as an undefined notion—one with many obvious properties" (1974, 161). What moved Tarski to attempt a more rigorous understanding and ultimately a definition of the notion, if previously the informal understanding had sufficed?

Feferman attributes the impetus for the definition to Tarski's desire to find a clearer way of expressing some of the results he was obtaining, in particular in his oft-discussed Warsaw seminar of the late 1920s, in which the "American Postulate Theorists" Langford, Veblen, and Huntington were studied at length (see Scanlan 1991 on the relation), as was Skolem's work on quantifier elimination. He attributes it as well to Tarski's feeling that mathematicians distrusted the "metamathematical" concepts of truth, definability, and so on both inherently and due to the appearance of the semantic and set-theoretic paradoxes. Tarski hoped that by offering precise definitions of these notions using the mathematical tools at his disposal he might better be able to express his results and set the worries of mathematicians to rest. Arianna Betti adds detail to the historical picture here by noting, in response to Feferman's suggestion that perhaps little more than Tarski's fastidiousness was the impetus for the project of definition, that in fact Tarski seems to have got the idea of rigorous definition of semantic notions from Ajdukeiwicz, who demanded such definition in a yet untranslated work from 1921. Wilfrid Hodges, in turn, emphasizes Leśniewski's influence: Tarski's aim in defining semantic notions, on Hodges's reading, was to work out the project of "intuitionistic formalism" conceived of as setting out the contents of the mind of the proving mathematician; to this end rigorous definitions of semantic notions were required to meet the Leśniewskian requirement of clarity about the meanings of the symbols employed in rigorous thought about any topic, *a fortiori* in "metamathematics." Hodges offers a detailed series of hypotheses as to how Tarski's other work of the time, especially the work on quantifier elimination from the Warsaw seminar, but also the influence of Kotarbiński's *Elementy*, fed into this interest to produce the truth definition as we know it.

Those are the historical antecedents to Tarski's definition of truth. As Tarski's aim was to give a *definition* of truth, our next topic is Tarski's theory of definition. Now there are two schools of thought about definition; these are often conflated, as Hodges notes, crediting Leśniewski with having raised the problem in an ingenious way:

There were two broad views in circulation, which should have been recognized as incompatible but were not. The first view (that of *Principia*) is that the addition of a definition to a deductive theory does *nothing* to the deductive theory; it simply sets up a convention that we can rewrite the expressions of the deductive theory in a shorter way and perhaps more intuitive way. People who took this view were divided about whether the definition is the convention or a piece of text... that expresses the convention. The second view is that adding a definition to a deductive theory creates a new deductive theory which contains a new symbol. In this case too one could think of the definition as the process of adding the new symbol and any attached formulas, or one could single out some particular formula in the new system as the 'definition' of the new symbol.

It is absolutely crucial to keep in mind that Tarski follows Leśniewski in adhering to the second conception: a formal definition is *not* a mere abbreviation, but is rather an otherwise ordinary sentence that plays a specified deductive role in a theory, the role now commonly understood (e.g. Suppes 1956 and Belnap 1993) in terms of eliminability and conservativeness relative to a prior theory and set of contexts. I stress the importance of this in my own contribution, so I will not elaborate further here. Hodges traces in detail the antecedents for Tarski's treatment of definition, finding Tarski to have been influenced in equal measure by Leśniewski and Kotarbiński.

Viewed in these terms a definition of truth will be a sentence that, relative to some theory and a set of contexts—always, in Tarski's case, extensional contexts, in accord with the general Polish distrust of intensional notions—allows one to eliminate "is true" from every sentence in which it appears (*salva veritate* of course) and that allows one to prove nothing free of the expression "true" that one could not prove in the "true"-free sub-theory. Note, then, in particular that a definition is not necessarily intended to be without content relative to this "true"-free sub-theory, and that the philosophical significance of the definition will turn crucially on theses about this prior theory, about the contexts in which and the expressions in favor of which "is true" is eliminated, and on claims about the philosophical significance of the full theory that includes the definition. None of these questions so much as make sense on the more common understanding of a definition as a mere abbreviation of some compound expression of a theory. (A related question is this: in what sense did Tarski intend to offer a "conceptual analysis" of truth and related notions? Tarski's definition is often taken to be a paradigm case of conceptual analysis, and was taken to be such, for instance, by Carnap. For a treatment of Tarski's definition of this sort, see Feferman; for reasons to think that Tarski shouldn't be read this way, see the considerations on p. 113 of Hodges' chapter.)

We may now turn to the best-known aspects of Tarski's treatment of truth: his celebrated condition of adequacy on a definition, Convention T, and his development of the technical means for meeting it in a certain variety of cases. I'll begin with the criterion of adequacy. Tarski proposes that a good definition of truth will imply, for every sentence of the language for which it is offered, a sentence of the form "*s* is true in L if and only if p" where what is substituted for "p" translates *s*. These are commonly known as "T-sentences." He treats these as "partial definitions" of truth and asks of a definition only that it imply them and that all truths are sentences (Tarski 1983, 188). Our first questions about this should be two: First, where does Tarski get the idea that doing *this* will answer to "the classical questions of philosophy" (1983, 152)? Second, what reasons does he himself give for thinking so, and what are we to make of them?

Here Jan Woleński and Roman Murawski trace the history of the idea among Tarski's Polish predecessors, finding antecedents for Tarski's criterion of adequacy in Twardowski's formulation of the familiar Aristotelian dictum "to say of what is, that it is, is true"—a formulation to which, because of Twardowski's influence, all those in his circle adhered. This conception was widely agreed by its proponents to be a sensible way of working out the idea of truth as correspondence, and Tarski says as much in his central publications. Within this tradition a second, more specific tradition also developed. On this second view, a sentence *s* and the claim, of it, that it is true are "equivalent." As Woleński and Murawski discuss, proponents of this second idea generally thought of it as a way of making the first more specific. This second tradition came to its fruition in the idea, which Tarski himself attributes to Leśniewski, that when it comes to truth the T-sentences are the heart of the matter. In Tarski's work the Polish approach to truth comes into full flower. As Tarski sees it, the idea that a definition of truth just needs to imply the T-sentences both expresses a "classical" conception of truth with philosophical credentials running all the way back to Aristotle and, simultaneously, makes the provision of a clear and rigorous definition of truth—and thus an answer to the "classical questions of philosophy"—a purely logical matter of crafting a definition that, when added to a theory, implies each member of a certain clearly specified set of sentences.

The results have been debated ever since. The reader will note that there must be a good deal to dispute, since it is a staple of the contemporary debate that conceptions of truth closely allied to the T-sentences are the antithesis of "correspondence" theories of truth. Nearly all of the contributors to this volume weigh in on the question of the significance of implication of the T-sentences in one way or another. Panu Raatikainen and I respond to the familiar charge that definitions that imply the

T-sentences make what ought to be contingent, empirical truths about the meanings of symbols into necessary truths in two very different ways. Marian David offers a detailed study of all aspects of Convention T, and contrasts his own conception of it with the one he takes to be more common, the common conception being that Convention T and the definitions it certifies allow us to discern nothing in common among the various defined notions of truth. David argues that in fact Tarski uses expressions like "true sentence" in a context sensitive way, so that they inherit their extensions from a unitary concept of truth in concert with the salience, in context, of a particular language. In my own contribution I attempt in a different way to answer the charge that Tarski does nothing to tell us what various truth-definitions constructed in accord with his methods have in common by relating the idea that a good definition should imply the T-sentences back to the Aristotelian conception of truth that Tarski inherited from his teachers.

Jody Azzouni has an entirely different take on the import of Convention T: in requiring translation of the object language into the metalanguage in the service of defining truth for it, Convention T forces us to take the object languages for which we define truth as expressively very similar to our metalanguage. This was fine, Azzouni notes, for the cases in which Tarski was interested, but it ties the application of the strategy for definition to languages that *are* sufficiently similar to the language in which the definition is stated: extensional languages based on classical logic. However, this seriously undermines its empirical applicability on Azzouni's view, since we can't assume of actual languages the sentences of which we might call true that they are logically or expressively similar to our own. Azzouni likewise will reject broadly Quinean and Davidsonian arguments that we cannot make sense of languages that are particularly expressively different from our own as spuriously based on this restriction of truth and interpretation to cases where translation is possible.

Another aspect of the debate about the philosophical import of Convention T and definitions it licenses concerns whether the "material adequacy" for which implying the T-sentences is sufficient is merely extensional adequacy, that is, whether Convention T has any function other than to ensure that an expression defined and intended to be a truth predicate in fact applies to all and only truths. Here Hodges is quite adamant this is what is to be found in Convention T, while I make it my business to argue that Convention T carries, and is intended to carry, the philosophical weight of Tarski's project of making clear the concept of truth, and thereby serves to do far more than merely guarantee extensional adequacy. (Other authors, e.g. Simons and Etchemendy, are with Hodges here, though in offhand remarks.) Though I disagree with Hodges, I nevertheless cannot over-stress the importance of his discussion of "adequacy" and "correctness" an the German and Polish terms they translate: in this respect the English of even the second edition of *Logic, Semantics, Metamathematics* seriously misrepresents the German and Polish, and everyone will do well to pay heed to what Hodges sets out.

Having determined that he had philosophical, logical, and technical reasons to want a definition that implies the T-sentences, Tarski's great technical achievement in "The Concept of Truth in Formalized Languages" was to propose a method of crafting a definition successful by these standards that is applicable in a range of important cases. As is often noted, the result was the first genuine compositional semantics, and the chapter thus stands at the source of linguistics and formal semantics as we know them today. Extensive introductory treatments are available from many authors (e.g. Soames 1999), so I will say just a little here by way of orienting the reader who is relatively new to the topic. The languages in question are languages that allow the multiply quantified statements and inferences involving them so ubiquitous mathematics. These languages allow the formation of infinitely many statements, including infinitely many involving many quantifiers. The question is: how are the truth values of multiply general claims determined?

In the case of sentential connectives like "and" and "not," the truth value of the sentence they form depends on the truth value of its parts. Tarski's insight was that since "for all x, Fx" doesn't *have* a complete sentence as a part, a two step-procedure was required on which first some semantic treatment was given of what really are the immediate parts of quantified sentences, and the truth value of these sentences is determined directly by the value of their immediate parts. Now, Frege's parallel insight was that the immediate "part" of a quantified claim is, as it is put in Fregean terms, a complex predicate, and Tarski realized that he needed to associate semantic properties with complex predicates, or, as he called them, "sentential functions"—open, as opposed to closed, sentences, as we know them today—and then work out how the relevant semantic values of all such items were determined by the semantic values of some finite stock of them.

The method, then, is to focus on the relation of a predicate's being "true of" something, which becomes Tarski's more technical notion of satisfaction. "Is red," for instance, is *true of* some things, and false of others. For a given language amenable to Tarski's treatment, there is some finite stock of primitive, or "lexical," such predicates, about each of which we can simply say *of what* it is true. Tarski then applies the methods of forming complex sentences that were already understood: composition of complex open sentences from simple ones by the application of sentential operators. In addition, quantification is handled in what becomes a straightforward way by exploiting a degenerate case of the "true of" relation: "there is an x such that Fx" is true of a given object just in case there is *some* object of which "Fx" is true. ("All" is defined in the usual way as "it is not the case that there is an x such that not.")

The result, when the final open position in a "sentential function" is closed off by a quantifier, is that "there is an x such that Fx" (where "F" may involve further quantifications) is satisfied by an object just in case some object is F. So put that may not look like news, but remember that the point is not to understand what "there is an x such that Fx" *means*, but to determine in detail, for each of an *infinite* number of such sentences, whether or not they are true based on a finite list of assignments of semantic value to primitive expressions (and, of course, some claims about what there is). Perhaps one insight with which Tarski can be credited is recognizing that the second task is not to be confused with the first. Note that if there *is* an object that is F, then "there is an x such that Fx" will be true of everything. Furthermore, intuitively, in such a case the sentence is true. Tarski thus proposes that we simply understand a *true* sentence as one that is *true of* (in his terms, *satisfied by*) everything;

this is upshot of Definitions 22 and 23 of "The Concept of Truth in Formalized Languages."

There will of course be an inclination here to cry "trivial!", one that I've positively courted by allowing that we might as well understand "satisfied by" as "true of." The method of recursion on satisfaction is no kind of analysis or reduction of truth or "is true" all by itself. As I emphasized just a moment ago, it isn't intended to be: it is, rather, a method of stating for each of an infinite number of sentences in a quantificational language under what conditions it is true. When it comes to understanding or analyzing truth itself, as Field 1972 famously points out, avoiding the triviality charge depends entirely on getting rid of "true of" (or "is satisfied by") in favor of something else. As Field also notes, Tarski does so by exploiting the finitude of the stock of lexical predicates, simply listing, for each, what it is true of. Debate continues to rage over what the virtues and vices are of Tarski's treatment of the satisfaction conditions of primitive predicates and the definition that results. (A few of my favorite contributions, in addition to Field 1972, include Soames 1984, Etchemendy 1988, Davidson 1990, and Heck 1997.) What has to be kept in mind, I'd insist, are what Tarski's demands on a definition were. We need clearly to separate the question of whether what he did was adequate by his standards from whether what he did was adequate by ours. Tarski wants a definition of truth, and thus of satisfaction, that is eliminative and conservative relative to an extensional background language, and the fact of the matter is that his procedure results in as much. What the larger picture is here turns crucially on what the value is of an eliminative definition relative to an extensionally formulated theory, and these issues are part and parcel of our evaluation of the significance of Convention T itself. Does a definition that allows us to eliminate "is true" as applied to sentences and "is true of" as applied to predicates relative to extensional contexts tell us what truth is in any important sense? It is here that Convention T again plays a role, since Tarski doesn't merely show us some way or other to get rid of semantic expressions; the definition in terms of satisfaction, with satisfaction in turn defined enumeratively, allows us to eliminate these expressions in a way that satisfies Convention T and thereby expresses Tarski's "Aristotelian" conception of truth. Our evaluation of his particular attempts at definition has to be bound up at least with an assessment of Convention T, as discussed above, as well as with our assessment of other issues.

A number of our contributors focus on the significance of Tarski's definitions and method, and on the conception of truth on which they are based. Mancosu carefully reviews the debate over Tarskian definition among the logical positivists surrounding his presentation at the First International Congress on the Unity of Science, held in 1935 in Paris. The positivists had been extremely skeptical of the notion of truth as being metaphysically loaded, but, as Mancosu notes, "Tarski's theory of truth seemed to many to give new life to the idea of truth as correspondence between language and reality." Many, but not all: Neurath was vehemently opposed to Tarski's approach. Seeing the immediate reaction to Tarski's work on the part of his contemporaries, and Tarski's reactions thereto (summarized in 1944's "The Semantic Conception of Truth") can help us to understand what Tarski himself intended. Mancosu's study of the ensuing debate, in which Carnap, Kokoszynska and others were also caught up, makes clear that many of the themes in the discussion of the significance of Tarski's definitions developed in this early reception. Is Tarski's "theory" a correspondence theory? Does it rehabilitate the idea that truth involves language-world relations? If so, is this good? Does the theory rehabilitate the notion of truth on some reasonable conception of the standards of clarity or "scientific" accuracy?

The technical details of Tarski's strategy for definition, in the cases where he applies it, are fruitfully studied in terms of some of the mathematics in which he was interested in the late 1920s, in particular Skolem's work on quantifier elimination and related work—now recognizable as very early work in model theory—of the "American Postulate Theorists" such as Huntington and Langford. Solomon Feferman's contribution sets Tarski's work in the context of the successes of set theoretic topology among mathematicians in Warsaw during Tarski's formative years. Hodges also discusses these connections and proposes a detailed timeline for the invention of the strategy in which the interaction of Tarski's work on quantifier elimination with his reading of Kotarbiński on truth, semantics, and definition plays the crucial role. Peter Simons also presents a number of the details in the context of a discussion as to what extent they can be reconciled with the nominalistic view that abstract objects do not exist, a topic to which we will return in the next section. On a different, more critical note, as mentioned above, Azzouni discusses how, from his perspective, Tarski's method for defining truth runs together the interpretation of the object language and the mere provision of a device for attributing truth to its sentences.

There are a good many more issues that are relevant to the assessment of Tarski's project of definition and the way he carries it out in the cases he takes to be tractable; here I will mention a few as a guide to the reader. (1) An additional question about the formal methods of the work of the 1930s concerns the extent to which model-theoretic and semantic notions as we currently know them play a role in Tarski's discussions and results. This topic is discussed to some extent by Feferman, who finds such notions in Tarski's work as early as 1924, and is also relevant to my discussion of whether or not Tarski presents a semantic or rather a purely syntactic form of the indefinability theorem in "The Concept of Truth in Formalized Languages." (2) As for Tarski's famously meticulous insistence on the distinction between object and metalanguage, it is discussed in a number of our essays—Hodges, for instance, finds that it was much more Tarski's concern than definition itself. I try to make clear that we need to be equally attentive to Tarski's insistence that the difference between metalanguage and meta-metalanguage is crucial to studies of the definability and indefinability of truth. (3) Tarski's allegiance to the type theory he inherited from Leśniewski, Husserl, and *Principia*, and his later shift to first-order logic is a rich area of study which is relevant to a number of the foregoing issues. (The reader is invited here to see the remarks on the topic in Mancosu 2005 and de Rouilhan 1998 for more on this topic.) Feferman suggests a contrast between Tarski's model-theoretic way of doing mathematics in many instances and his retention of type theory in some writings. Hodges also discusses the role that Tarski's type theory plays in some of his discussions of definition, while I discuss its role in Tarski's remarks on the indefinability of truth.

I will have to allow the above to suffice by way of introduction to the discussion of Tarski's views on truth. To conclude this section, we can look at Tarski's discussions of the indefinability of truth in cases where he takes his methods to be inapplicable. Tarski is famous for his early statement, following Gödel's incompleteness results, of the theorem that we know today as the result that arithmetic truth is not arithmetically definable. The basic technical point here is that, as Gödel showed, for a theory meeting certain conditions (implying a weak system of arithmetic is sufficient), we will be able to prove that for every predicate F of the language of theory a sentence of the form $S \leftrightarrow F(<S>)$ will be a theorem of the theory, where $<S>$ is a "standard name" of S or a number associated with it. This is usually known as a "diagonal lemma" and proving it is, of course, the hard part of getting the result. (The reader should consult a standard textbook presentation here, e.g. Boolos, Burgess, and Jeffrey 2002.) The application bearing Tarski's name is obtained by noting that if we assume that one of the predicates of the language of the theory means "is true" (or, more basically, that it is true of all and only true sentences of the theory) then we have to consider what follows for sentences involving negation and this predicate, T: By the above result, there will be a sentence S such that $S \leftrightarrow \sim T(<S>)$ is a theorem. But this sentence is the negation of the T-sentence for S; put colorfully, S is relevantly like "this sentence is not true," which appears to be true if and only if it is not true. From here very minimal resources get one an explicit contradiction.

My chapter includes an extended discussion of the textual and interpretive issues surrounding Tarski's discussion of this result as it appears in "The Concept of Truth in Formalized Languages," and I offer an interpretation of the significance of the result, for Tarski, in terms of the overall account of meaning I attribute to him. On my view, in the 1930s Tarski was more interested in exploring the expressibility of the intuitive notion of truth in a mathematically tractable way than he was in defining a set of sentences. He took the intuitive notion of truth to require that the T-sentences be theorems, and therefore refused to countenance the possibility of a language that expressed the notion of truth but in which all such sentences were not treated as theorems. Faced with a result that showed that the intuitive notion couldn't consistently be expressed compatibly with such definition for languages of sufficient richness, Tarski was happier to cleave to his Aristotelian notion of truth. The later history of the topic, and in particular the explosion of approaches to semantic paradox that involve alternative logics and give up on Tarski's requirement that the T-sentences for the object language be theorems of the metalanguage provides a striking contrast to Tarski's own take on the phenomena.

TRUTH BEARERS

Having looked at Tarski's treatment of truth itself, we can turn now to discussion of Tarski's treatment of the bearers of truth. Tarski is rather famous for having been a "tortured nominalist"—philosophically, it seems, he was sympathetic to the nominalism of his teachers, especially Leśniewski and Kotarbiński. Indeed, Tarski thought so well of Kotarbiński and his views that as late as 1955 Tarski's translation (co-produced with David Rynin) of Kotarbiński's "The Fundamental Ideas of Pansomatism," an exposition of Kotarbiński's bracing nominalism, appeared in *Mind*. (Intriguingly, the article, written in 1935, includes an extended defense of the paratactic account of attitude ascriptions as part of the project of identifying everything "psychical" with something material. This material comes complete with an explanation for how such "psychical enunciations" merely appear to give rise to intensional contexts and includes discussions that anticipate adverbial theories of perception and the idea that in quotation and attitude ascription one imitates the subject of the ascription rather than referring to some inner state or abstract entity. Kotarbiński ultimately suggests that the account might allow one to eliminate apparent reference to "inner experience." It seems to me that this article, which appeared in English in a major journal, is not cited anywhere near as often as it should be, given the importance that most of these ideas have had in the relevant literatures in the past few decades.)

Tarski's nominalism notwithstanding, his training in mathematics was thoroughly shot through with the intuitive Platonism of the discipline, and Tarski hardly veered from this in his published work—one exception being a series of remarks in "The Concept of Truth in Formalized Languages" itself. The main point of conflict is the logician's need to treat sentences and relations among them as abstracta, rather than as, say, the concrete sentence tokens a nominalist could countenance—on a strict nominalist view the claim, for instance, that if two claims are true, so is their conjunction, might seem liable to fail for the seemingly extraneous reason that what would be the conjunction hasn't been written down by anyone and hence doesn't exist to even *be* true. Here is a typical passage, following upon Tarski's presentation of a series of axioms about which expressions are part of the object language:

Some of the above axioms have a pronounced existential character and involve further consequences of the same kind. Noteworthy among these consequences is the assertion that the class of all expressions is infinite (to be more exact, denumerable). From the intuitive standpoint this may seem doubtful and hardly evident, and on that account the whole axiom system may be subject to serious criticism... I shall not pursue this difficult matter any further here.*
The consequences mentioned could of course be avoided if the axioms were freed to a sufficient degree from existential assumptions. But the fact must be taken into consideration that the elimination or weakening of these axioms, would considerably increase the difficulties of constructing the metatheory...

*For example, the following truly subtle points are here raised. Normally, expressions are regarded as the products of human activity (or as classes of such products). From this standpoint the assumption that there are infinitely many expressions appears to be obviously nonsensical. But another possible interpretation of the term 'expression' presents itself: we could consider all physical bodies of a particular form and size as expressions... The assertion of the infinity of the number of expressions... [then] forms a consequence of the hypotheses which are normally adopted in physics or geometry.

(1983, 174–5)

In the text one might see some allusion to Leśniewski, but the note, in its reference to "bodies" and "products of human activity" makes clear that the main concern is

Kotarbiński and Twardowski. Kotarbiński's *reism*, on which everything is a concrete material body, is discussed by Woleński and Murawski, as is Twardowski's discussion of human activities and their products. Kotarbiński's philosophical nominalism was, in turn, based on the more technical presentations of Leśniewski's ontology and mereology. As I mentioned above, Leśniewski and his relations with Tarski are discussed at length in Betti's contribution. When it comes to Kotarbiński, Woleński and Murawski discuss reism and its influence on Tarski at length, noting that the primary source of Tarski's inner conflict was indeed the clash between his nominalist scruples and his need for an sufficient supply of expressions over which to quantify in doing logic. They also note Tarski's sympathy for the idea that language considered as a product of human activity is essentially finitistic.

Given these predilections, it isn't surprising that Tarski found himself "tortured" by the assumptions required to go forward in logic and mathematics. One intriguing view into Tarski's usually off-the-record philosophical sympathies comes from a series of Carnap's notes on meetings at Harvard in 1940–1941, as discussed in Greg Frost-Arnold's contribution. Tarski met with Carnap and Quine (and sometimes Goodman and Hempel) to talk about the issues involved in devising a "finitistic" language for science. As Frost-Arnold explains it, for Tarski, the requirement that a language be relevantly "finite" derived from views on the conditions required for the language to be fully *intelligible*. "Finiteness" came to being first-order with a finite vocabulary (by the late 1930s Tarski's move away from type theory to first-order logic was complete), with, furthermore, first-order variables ranging over concrete objects only. Strikingly, on this basis Tarski claims that he doesn't really "understand" classical mathematics, and operates with it only as a "calculus." (Perhaps it is remarkable that seminal contributions to mathematics were made by someone who claimed, off the record, that these contributions couldn't really be understood.) Tarski, Carnap, and Quine had a series of discussions about what could be done within the confines of full intelligibility as they construed it, and Frost-Arnold details in particular the convolutions of their attempts to make a series of physical objects serve as the natural numbers. Frost-Arnold works to unearth the assumed notion of understanding that underwrote these efforts, and finds it in the twin ideas that full intelligibility, in a positivist spirit, ruled out "metaphysical" claims (among with the likes of Quine of course included set theory), and that a proper respect for natural science required one not to prejudge the size of the universe by assuming an infinite number of objects (compare here the note from "The Concept of Truth in Formalized Languages" discussed above).

Tarski's nominalism having been rather severe and very much at odds with his work on truth, the question arises how much of the work on truth really could be made acceptable to a hardened nominalist. This is the question Peter Simons addresses in his contribution. In the service of a fully nominalist take on Tarski's strategy of definition, Simons proposes that we reinstate token sentences as truth bearers and that we construe any apparent quantification over sets or classes as being, rather, over pluralities. Central to the account here are Simons' methods for dealing nominalistically with sentence-forming operators in addition to functors. Simons concludes that one can in fact, though with a good deal of complication that a Platonist can avoid, give a Tarskian truth definition in a fully nominalist manner.

As for the bearers of the consequence relation, then, they are sentences taken as abstracta for Tarski. Given this "tortured" slide into Platonism it is noteworthy that Tarski doesn't countenance anything like propositions, given that he is in the early period always insistent that sentences are to be taken as interpreted and fully meaningful, and that he is obviously happy with the idea that sentences can equivalent in meaning. The absence of such notions from his work—and the more manifest hostility to notions such as "state of affairs" and "fact"—I would attribute again to Kotarbiński's influence, for which the reader can see, in the first instance, the article from *Mind* mentioned above, as well as to Leśniewski's: as both intensional and abstract, propositions and their ilk would be doubly off-limits in the Polish school; sentences were only half that bad. (To the influence of his teachers we can also attribute the greater sympathy shown for the idea that particular mental acts may legitimately be taken as bearers of truth, as discussed by Woleński and Murawski and mentioned in another way by Simons.)

CONSEQUENCE

We have now covered Tarski's treatment of what is preserved in logical consequence, truth, and that among which the consequence relation holds, namely type sentences. We can now turn to his treatment of consequence itself, as presented in "On the Concept of Logical Consequence." As Etchemendy discusses, Tarski reduces the logical truth of a sentence to the truth of an associated generalization, and the logical validity of an inference to the fact that no arguments of a class associated with the inference have true premises but a false conclusion. (As is suggested by Woleński and Murawski's contribution, there is an analogy here to Łukasiewicz's account of probability.) The associated generalizations and sets of arguments are singled out by taking the original sentence or argument and, as Tarski construed it, allowing all *non-logical* constant expressions to be replaced by other expressions available in extensions of the language. Thus, an instance of modus ponens such as

> If there is smoke, then there is fire.
> There is smoke.
> ———————————————
> There is fire.

is valid because no result of uniformly substituting other expressions for anything but "if... then" results in an argument that has true premises but a false conclusion.

As Tarski mentions in a footnote, though he arrived at the analysis independently, it bears a "far-reaching analogy" (1983, 417) to the account Bernard Bolzano proposed a century earlier. Two facts about the account bear immediate mention. First, the account of consequence is entirely extensional: notions of logical truth and validity are reduced to the obtaining of various truths about the actual world. As Etchemendy notes, if successful, this would be a striking and fruitful reduction: the account of one claim's being a consequence of others would be purged of modal and epistemic notions such as necessity, a priority, and so on. Second, the account assumes an account of which vocabulary is logical. Different notions of consequence will derive from different sets of "fixed" vocabulary and thus in order for Tarski's account of consequence to succeed as an explication of the usual notion of deductive "following from" the selection of *logical* constants is crucial.

Etchemendy trenchantly criticizes the account on both points in *The Concept of Logical Consequence*. The book in turn provoked numerous responses from defenders of Tarski or critics of Etchemendy, and in his contribution Etchemendy sets out his criticism in a renewed form, adding an extensive discussion of the positive conception of consequence he prefers to Tarski's. Etchemendy holds that Tarski's analysis is conceptually inadequate, in that if validity were mere truth-preservation in a set of arguments, deductive inference wouldn't be a guide to the formation of new true beliefs from old ones; given a set of premises one held true, and an argument for some conclusion, one could only conclude that either the conclusion was true or the argument was invalid. The epistemic features of deductive inference simply cannot be accounted for, on Etchemendy's view, by any account that reduces consequence to truth-preservation in an associated set of arguments. Etchemendy likewise holds that the epistemically important features of consequence cannot be recovered by a careful account of the logical constants; even invalid arguments couched in terms of paradigm logical constants will having nothing but truth-preserving instances in perversely selected worlds: truth-preservation even with "logical constants" restricted to obvious cases is, on Etchemendy's view, still insufficient for validity. Furthermore, Etchemendy argues, the analysis will not even be extensionally correct "whenever the language, stripped of the meanings of the non-logical constants, remains relatively expressive, or the world is relatively homogenous, or both." This point is related to the previous one: truth-preservation by all instances of an argument form can be guaranteed by features of the world that are intuitively independent of the consequence relation. Against the conception of logic that results from adherence to the Tarskian analysis of consequence, Etchemendy advocates a much broader perspective, from which second-order logic, modal and epistemic logic, and even the study of the informational content of databases, maps, and diagrams is as fully logical as propositional or first-order logic. Etchemendy sets out a characterization of the conception of model theory as representational semantics advocated in the earlier work, and he discusses the way in which the undue influence of Tarski's conception of consequence has hampered our understanding of logic itself, as well as of mathematics and other disciplines.

Gila Sher, by contrast, favors Tarski's analysis and hence takes very seriously the need to provide a proper characterization of the logical constants and of the domain of logic. In her contribution she focuses on the criterion of logicality set out by Tarski in the 1966 lecture "What are Logical Notions?" In the lecture Tarski suggests that "logical notions" are those that are invariant under all 1–1 permutations of the universe—intuitively, then, logic is the general science of structure. Since mathematics is also often taken to be the general science of structure, Sher has a good deal to say about the relationship between logic and mathematics, extending the defense of the conception of logic, mathematics, and their relationship—one she here calls "mathematicism" in opposition to "logicism"—she has defended in previous work. One aspect of Sher's difference from Etchemendy here is that while Etchemendy sees logical consequence as a matter of the functioning of expressive devices in a system of representation, Sher sees it as a matter of structural relationships in what is represented. The difference is in the first instance one of emphasis, since the functioning of expressive devices is connected to the structure of what is represented by the notion of truth, but when claims are added about what structure is, or about how expressive devices function, the criteria of logicality can easily diverge. In Sher's case this comes in the selection of universal isomorphisms as definitive of logical structure: this leaves the propositional connectives plus various first- and higher-order quantifiers as logical constants. By contrast, Etchemendy is happy to treat various topic-specific inferences (as, e.g., in epistemic logic or geometry) as logical and he ultimately finds the question of which constants are the logical constants a red herring; Sher, in turn, disagrees. It is to be noted here that Sher finds her view "Tarskian" in ways that Etchemendy would dispute, due to their disagreement about what Tarski's view was; the reader should see their respective essays for the details of the disagreement. Sher closes her chapter with a response to Solomon Feferman's criticism of her views in Feferman 1999.

Mario Gómez-Torrente, in turn, considers what the correct formulation of Tarski's view of logical truth, as determined by Tarski's conception of consequence, is, and then asks whether it is correct—that is, whether it correctly characterizes logical truths as we know them. He begins by distinguishing Tarski's claim on his reading from other possible readings; Tarski, he argues, couldn't have meant to restrict his criterion of logicality to the fixed list of the usual first-order constants, but he also could not have intended to include too much among the constants. He arrives at the view that Tarski's model-theoretic conception of logical truth is one on which a sentence is logically true if it is true in all models that reinterpret its non-logical constants, where logical constants include the usual first-order constants plus any other extensional constants which have a plausible intuitive claim to logicality. Having isolated a reading of Tarski's conception of logicality, and a rough demarcation of the logical constants, Gómez-Torrente goes on to argue that nevertheless the set of truths singled out as "logical truths" on such a conception of logicality will nevertheless not coincide with a plausible intuitive sense of what is and is not logically true; this is established by consideration of a series of examples intended to show that sentences that are logically true according to the Tarskian criterion Gómez-Torrente adumbrates can fail to be necessary or *a priori* truths. Thus, like Etchemendy, Gómez-Torrente is concerned that Tarski's reduction of logical consequence (and thereby logical truth) fails adequately to capture the modal and epistemic characteristics held important in the usual conceptions of consequence such as "the conclusion must be true if the premises are true" or "the conclusion can be known on the basis of the premises alone." He closes, however, by suggesting that a modified form of the Tarskian criterion will be defensible. For standard reasons, it appears helpful to distinguish the "attitudinal contents" of sentences like "Hesperus = Hesperus" and

"Hesperus = Phosphorus"; the first, on Gómez-Torrente's view, has an attitudinal content under which it is *a priori* knowable, while the second does not. Given this distinction as applied generally, Gómez-Torrente's claim is that a Tarskian criterion of logicality is workable as long as it includes the claim that the attitudinal contents of various expressions are held fixed.

MEANING

If logical consequence is the preservation of truth among its bearers, then a full philosophy of logic will cover truth, the bearers, and the nature of preservation. We have now covered, in outline, Tarski's philosophy of logic from the seminal period of the early 1930s as examined by the essays in this collection. In this final section I will briefly mention a few topics not adequately touched on above, topics that can be brought under the heading of Tarski's conception of meaning and its relation to the truth and logic.

We have just looked at Gómez-Torrente's discussion of the relation between Tarski's conception of logical truth as he understands it, and "intuitive" logical truth. It is a staple of philosophical treatments of language as we know them to take seriously our impressions of what meaning is and can be, since the object of our study is our language, something with which we are intimately familiar—which isn't to say that it is always easy explicitly to know that with which one is familiar. Tarski is no exception here; he speaks often of "intuitive knowledge" and the "intuitive conception of truth," for instance, as well as about related notions such as the adequate usage of expressions. What is the role of these appeals in Tarski's view, and how are his more formal constructions intended to be related to them? On a related note, what are we to make of his appeal to translation as something apparently unproblematic? I discuss these issues in my contribution to a significant extent, arguing that Tarski always took some basic grasp on the concept of truth as fundamental and viewed his discussions of the derivational and semantic aspects of language as beholden to it. This is related to the fact that Tarski always took his languages to be interpreted, something noted by many of our contributors, including Raatikainen and David in their discussions of Convention T. Relevant here, also, are Tarski's early pronouncements of adherence to Leśniewski's "intuitionistic formalism," a topic discussed at some length by Hodges. Another topic here concerns Tarski's relations to Hilbert, and the extent to which he did or did not engage in "formal" as opposed to "contentual" axiomatics. Betti discusses this, as does Feferman, and I myself comment on Tarski's relations with Hilbert. On this topic one should also see the papers of Sinaceur that appear in the references below.

A final topic relevant to the relation of meaning and logic is Tarski's attitude to logicism. At many junctures Tarski seems willing to call what today we would think is unquestionably mathematics "logic" or "a system of mathematical logic." Relevant here is his early adherence to the type theory he inherited from Leśniewski, *Principia* and other sources, but, as Sher makes it her business to argue, there may be deeper connections between logic and mathematics as Tarski sees them than this merely historical point would indicate. Tarski's relation to logicism is discussed in a number of other contributions, including mine and David's.

CONCLUSION

I hope the above is of some use in orienting the reader. As mentioned, I have arranged the essays that follow in a roughly chronological order, beginning with essays on Tarski's background, running through those that primarily concern the more overtly "philosophical" texts of the late 1920s to the early 1940s, on through those that concern later developments, and finishing with essays that are concerned primarily with the evaluation of Tarski's contributions. Following this introduction there is a short set of references; these are suggestions for further reading, including useful introductions to Tarski, important works on the historical background, and some work on topics that I would have liked to include here. I don't pretend that this list is anything close to complete; I simply offer it as a place where those wanting to learn more may begin their research. Those wanting a fully bibliography of Tarski's work can do no better than the nearly complete Givant 1986; this is included in the four volume collected papers Givant and McKenzie 1986. For Tarski's life, readers are urged to consult Feferman and Feferman 2004. The references provided with the individual essays will, in turn, provide more accurate guides for further reading on specific topics.

I thank Éditions Le Fennec for permission to reprint Solomon Feferman's "Tarski's Conceptual Analysis of Semantical Notions," which originally appeared in 2004 in *Sémantique et épistémologie*, edited by A. Benmakhlouf, in the series *Débats Philosophiques*.

I would like to thank all of my contributors for being part of this project. They have provided me with material that I believe makes for a very good collection, and I hope the whole does justice to each of the parts they have contributed. Most of all I thank Sandra Lapointe for encouragement, advice, and support during the long process of putting the volume together.

SUGGESTED READINGS

Adamowicz, Z., Artemov, S., Niwinski, D., Orlowska, E., Romanowska, A., Woleński, J. (2004) *Provinces of Logic Determined*: Essays in the Memory of Alfred Tarski Parts I, II, and III. Annals of Pure and Applied Logic 126, 127.
Awodey, Steve and Reck, Erich H. (2002) Completeness and Categoricity Part I: Nineteenth-Century Axiomatics to Twentieth-Century Metalogic. *History and Philosophy of Logic* 23, 1–30.
Belnap, Nuel (1993) On Rigorous Definitions. *Philosophical Studies* 72, 115–46.
Boolos, George, Burgess, John and Jeffrey, Richard (2002) *Computability and Logic* 4th edn. New York: Cambridge University Press.
Coffa, Alberto (1986) From Geometry to Tolerance: Sources of Conventionalism in Nineteenth-Century Geometry. In R. G. Colodny ed., *From Quarks to Quasars*. Pittsburgh: University of Pittsburgh Press, 3–70.

—— (1991) *The Semantic Tradition from Kant to Carnap: To the Vienna Station.* Cambridge: Cambridge University Press.
Coniglione, F., Poli, R., and Woleński, J. eds. (1993) *Polish Scientific Philosophy: The Lvov-Warsaw School.* Amsterdam: Rodopi.
Davidson, Donald (1990) The Structure and Content of Truth. *The Journal of Philosophy* 87, 279–328.
—— (2001) *Inquiries into Truth and Interpretation*, 2nd edn. New York: Oxford University Press.
de Roulihan, P. (1998) Tarski et l'universalité de la logique. In *Le formalisme en Question. Le tournant des années trente.* F. Nef and D.Vernant, eds. Paris: Vrin, 85–102.
Etchemendy, John (1988) Tarski on Truth and Logical Consequence. *The Journal of Symbolic Logic* 53, 51–79.
—— (1990) *The Concept of Logical Consequence.* Cambridge, MA: Harvard University Press.
Ewald, William (1996) *From Kant to Hilbert: A Source Book in the Foundations of Mathematics.* 2 vols. New York: Oxford University Press.
Feferman, Anita Burdman and Feferman, Solomon (2004) *Alfred Tarski: Life and Logic.* Cambridge: Cambridge University Press.
Feferman, Solomon (1999) Logic, Logics and Logicism. *Notre Dame Journal of Formal Logic* 40: 31–54.
Ferreirós, Jose (1999) *Labyrinth of Thought: A History of Set Theory and its Role in Modern Mathematics.* Basel: Birkhäuser.
Field, Hartry (1972) Tarski's Theory of Truth. *Journal of Philosophy* 69: 347–75
Friedman, Michael (1999) *Reconsidering Logical Positivism.* Cambridge: Cambridge University Press.
Frost-Arnold, Greg (2004) Was Tarski's Theory of Truth Motivated by Physicalism? *History and Philosophy of Logic* 25, 265–80.
García-Carpintero, M. (1996) What is a Tarskian Definition of Truth? *Philosophical Studies* 82: 113–44.
Givant, Steven (1986) Bibliography of Alfred Tarski. *Journal of Symbolic Logic* 51.
—— (1991) A Portrait of Alfred Tarski. *The Mathematical Intelligencer* 13: 16–32.
—— (1999) Unifying Threads in Alfred Tarski's Work. *The Mathematical Intelligencer* 21: 47–58.
Givant, Steven and McKenzie, Ralph (1986) *Alfred Tarski: Collected Papers*, 4 vols. Basel: Birkhäuser.
Gómez-Torrente, Mario (1996) Tarski on Logical Consequence. *Notre Dame Journal of Formal Logic* 37: 125–51.
—— (2004) The Indefinability of Truth in the "Wahrheitsbegriff." *Annals of Pure and Applied Logic* 126: 27–37.
Grattan-Gunness, I. (2000) *The Search for Mathematical Roots: Logics, Set Theories and the Foundations of Mathematics from Cantor through Russell to Gödel.* Princeton: Princeton University Press.
Heck, Richard G. Jr. (1997) Tarski, Truth and Semantics. *The Philosophical Review* 106, 533–54.
Henkin, L., Addison, J., Chang, C. C., Craig, W., Scott D., and Vaught, R. L. eds. (1974) *Proceedings of the Tarski Symposium.* Providence: American Mathematical Society.
Hiż, H. (1966) 'Kotarbiński on Truth.' *Studies in Polish Civilization*, D. Wandycz, ed., New York: Columbia University Press, 426–31.
Hodges, Wilfrid (1985) Truth in a Structure. *Proceedings of Aristotelian Society* 86: 135–51.
Kotarbiński, Tadeusz (1955) The Fundamental Ideas of Pansomatism. A. Tarski and D. Rynin trans. *Mind* 64: 488–500.

Kotarbiński, Tadeusz (1966) *Gnosiology: The Scientific Approach to the Theory of Knowledge*. Oxford: Oxford University Press.
Kuratowski, Kazimierez (1980) *A Half Century of Polish Mathematics*. Oxford: Pergamon.
Langford, Cooper H. (1926) Some Theorems on Deducibility. *Annals of Mathematics* 28, 16–40.
Leśniewski, Stanislaw (1929) Grundzüge eines neuen Systems der Grundlagen der Mathematik. *Fundamenta Mathematicae* 14, 1–81.
McCall, Storrs, ed. (1967) *Polish Logic 1920–1939*. Oxford: Clarendon Press.
Mancosu, Paolo (2005) Harvard 1940–1941: Tarski, Carnap and Quine on a Finitistic Language of Mathematics for Science. *History and Philosophy of Logic* 26, 327–57.
Putnam, Hilary (1985) A Comparison of Something with Something Else. *New Literary History*, 17, 61–79. Reprinted in Putnam 1994, *Words and Life*, J. Conant ed. Cambridge, MA: Harvard University Press.
Ray, Greg (2003) Tarski and the Metalinguistic Liar. *Philosophical Studies* 115: 55–80.
Rojszczak, Artur (2002) 'Philosophical Background and Philosophical Content of the Semantic Definition of Truth,' *Erkenntnis* 56: 29–62.
Scanlan, Michael (1991) Who were the American Postulate Theorists? *Journal of Symbolic Logic* 56: 981–1002.
—— (2003) American Postulate Theorists and Alfred Tarski. *History and Philosophy of Logic* 24: 307–25.
Sinaceur, Hourya (2001) Alfred Tarski: Semantic Shift, Heuristic Shift in Metamathematics. *Synthese* 126: 49–65.
—— (forthcoming) Tarski's Practice and Philosophy: Between Formalism and Pragmatism. In Lindström, Palmgren, Segerberg and Stoltenberg-Hansen, eds. *Logicism, Intuitionism, and Formalism: What Has Become of Them?* Springer Verlag.
Skolimowski, H. (1967) *Polish Analytical Philosophy*. London: Routledge and Kegan Paul.
Soames, Scott (1984) What is a Theory of Truth? *Journal of Philosophy* 81: 411–29.
—— (1999) *Understanding Truth*. New York: Oxford University Press.
Sundholm, Göran (2003) Tarski and Łesniewski on Languages with Meaning versus Languages without Use: A 60th Birthday Provocation for Jan Woleński. In J. Hintikka, T. Czarnecki, K. Kijania-Placek, T. Placek, and A. Rojszczak eds. *Philosophy and Logic. In Search of the Polish Tradition*. Dordrecht: Kluwer Academic Publishers.
Suppes, P. *Introduction to Logic*. Princeton, NJ: Van Nostrand.
—— (1988) Philosophical Implications of Tarski's Work, *Journal of Symbolic Logic* 53: 80–91.
Szaniawski, K., ed. (1989) *The Vienna Circle and the Lvov-Warsaw School*. Dordrecht: Kluwer Academic Publishers.
Twardowski, B. and Woleński, J. eds. (1994) *Sixty Years of Tarski's Definition of Truth*. Warsaw: Philed.
Vaught, R. L. (1974) Model Theory Before 1945. In Henkin et al., 1974.
Woleński, Jan (1989) *Logic and Philosophy in the Lvov-Warsaw School*. Dordrecht: Kluwer Academic Publishers.
—— (1995) Mathematical Logic in Poland 1900–1939: People, Circles, Institutions, Ideas. *Modern Logic* 5: 363–405.
—— (2002) From Intentionality to Formal Semantics (from Twardowski to Tarski). *Erkenntnis* 56: 9–27.
Woleński, Jan and Kohler, E. eds. (1999) *Alfred Tarski and the Vienna Circle*. Dordrecht: Kluwer Academic Publishers.

2
Tarski and his Polish Predecessors on Truth

Roman Murawski and Jan Woleński

> Almost all researchers who pursue the philosophy of exact sciences in Poland are indirectly or directly the disciples of Twardowski, although his own work could hardly be counted within this domain.
>
> (Tarski 1992, 20)

This is Tarski's description of the genesis of Polish investigations in mathematical logic, or more precisely those done inside the Lvov-Warsaw School.[1] Since the semantic theory of truth belongs to the philosophy of the exact sciences, we conclude that Tarski considered himself as a member of Twardowski's heritage.[2] Sociologically it is obvious that he was, as Tarski was a student of Kotarbiński, Leśniewski, and Łukasiewicz, that is, direct disciples of Twardowski. Not very much is known, however, about direct contacts between Tarski and Twardowski. Almost all the information we have comes from Twardowski's *Diary*.[3] On September 7, 1927 Twardowski described Banach's lecture on the concept of limit at the first Polish mathematical congress and says that "there came several of my acquaintances from Warsaw, except Łukasiewicz, Sierpiński, Tarski and others"—a remark that at least let us know that Twardowski took Tarski to be among his acquaintances.

Perhaps the most interesting record in the *Diary*, however, concerns Tarski's chapter on truth, delivered in Polish Philosophical Society in Lvov on December 15, 1930: "Very interesting and also very well construed." Other fragments of Twardowski's *Diaries* about Tarski mention the problem of the latter's candidacy for a professorship in Lvov (Twardowski supported Tarski; see also Feferman and Feferman 2004, 66–8), mutual meetings (Tarski often visited Twardowski in Lvov), exchanges

---

Roman Murawski acknowledges the support of the Foundation for Polish Science during the writing of this article.

[1] See Skolimowski 1967 and Woleński 1989 for detailed presentations of this philosophical formation.

[2] Currently there is a problem with spelling the name 'Lvov'. 'Lwów' is the Polish version, 'Lviv'—Ukrainian. Some Ukrainians say that 'Lvov' is a Russian word. We take the last as the English spelling of 'Lwów.'

[3] See Twardowski 1997, Part I, p. 323, Part II, pp. 110–13, 176, 179, 180, 205, 296, 331, 336, 352, 369, 372.

of letters, and the preparation of the German version of Tarski 1933 for *Studia Philosophica* (it was published in 1936). We have also a letter of Twardowski to Leśniewski (see Feferman and Feferman 2004, 100–2) written in 1935 in which the former supports Tarski's professorship in Warsaw. Although the relations between Twardowski and Tarski had never been particularly close, all accessible evidence allows one to assert that they were good.

Here, however, we are much more interested in the substantial influence of Twardowski and his direct students on Tarski's work on truth than in obvious sociological links. We intend to show that this influence was important. Although the mathematical side of the semantic theory of truth is independent of its philosophical background, the latter cannot be properly understood without taking into account the aletheiology (we propose this word as an equivalent for 'philosophy of truth') developed by Tarski's Polish philosophical ancestors. We will discuss the views of Twardowski, Łukasiewicz, Leśniewski, Zawirski, Czeżowski, and Kotarbiński.[4] The last philosopher will be treated more extensively than the rest, because his influence on Tarski was greater than that of anybody else, save perhaps Leśniewski. However, as Leśniewski's aletheiology is extensively treated by Arianna Betti's chapter in this volume, we restrict our remarks on Leśniewski to a very few.[5]

TWARDOWSKI

Twardowski's main work on truth (1900) concerned the problem of aletheiological relativism. His understanding of truth and its absoluteness or relativity is as follows:

The term "a truth" designates a true judgment. Therefore, all judgments that are true, that possess the characteristic of truthfulness, are truths. Hence, it is always possible to use the expression "a true judgment" instead of the term "a truth". It then follows that expressions "relative truth" and "absolute truth" mean the same as the expressions "relatively true judgment" and "absolutely true judgment."

Those judgments that are unconditionally true, without any reservations, irrespective of any circumstances, are called "absolute truths"—judgments, therefore that are true always and everywhere. On the other hand those judgments that are true only under certain conditions, with some measure of reservation, owing to particular circumstances, are called "relative truths", such judgments are therefore not true always and everywhere.

(1900, 148)[6]

Twardowski, following Bolzano and Brentano, rejected the view that there exist relative truths, though he was interested in categorizing the reasons some had for

---

[4] Although we concentrate on aletheiology, p. 36 below goes beyond the problem of truth. This section is devoted to nominalism, one of the most intriguing of Tarski's views.

[5] This chapter uses some material published earlier in Woleński and Simons 1989, Woleński 1990, Woleński 1993*a* and *b*, Woleński 1994*b*, Woleński 1995, Woleński 1999; see also Vuissoz 1998.

[6] If our bibliography lists a translation or another edition of an original work, page references are to later sources.

accepting some truths as relative. One such reason, according to Twardowki's survey, stems from elliptical formulations of some judgments through the use of occasional words, like 'now', 'here', 'I', etc.—for example the apparent relativity of the truth of "It is raining today" to a time and a place. Other relativist arguments, he notes, point out the relativity of various evaluations (for example, of 'bathing is healthy') to some salient person, or appeal to the view that empirical hypotheses are neither true or false, but always only probable. Twardowski held that all these arguments are erroneous. In particular, on his view one should sharply distinguish sentences from complete propositions. Only the former can appear as relatively true or false. For example, the sentence 'Today it is raining in Lvov' does not express a complete proposition. After eliminating 'today' and inserting a concrete date, we obtain a sentence that does express a complete proposition, for example, 'December 17, 1899 it is raining in Lvov', which is absolutely true or false. The same treatment applies to evaluations, because we should complete 'bathing is healthy' by indicating a person. Hence, though some sentences are relatively true or false, only complete propositions are absolutely true or false. Twardowski also pointed out that the relativity of truth is at odds with principles of excluded middle and non-contradiction.

If we analyse the most typical case of aletheiological relativity, that with respect to time, the view that truth is absolute can be displayed by two sub-theses:

(1) A proposition A is true at t if and only if it is true at every $t' \leq t$;
(2) A proposition A is true at t if and only if it is true at every $t' \geq t$.

The first sub-thesis expresses the principle of sempiternality of truth (A is true at t if and only if it is true at every earlier moment), while the second gives the principle of the eternality of truth (A is true if and only if it is true at every later moment). If (1) and (2) are accepted, truth does not need to be indexed by time. Twardowski, in accord with his distaste for relativism, did in fact accept (1) and (2) and thereby held that truths are, if ever true, always true.

Turning to his other views on truth, Twardowski had some reservations about the concept of correspondence. In this he followed Brentano. Twardowski's own definition of truth was as follows:

(3) An affirmative judgment is true if its object exists, an negative judgment, if its object does not exist. An affirmative judgment is false if its object does not exist; a negative judgment, if its object exists.

(Twardowski 1975, 208)

Twardowski considered (3) to be a version of Aristotle's definition given in *Metaphysics* 1011b (to say of what is that it is not, or of what is not that is, is false, while to say of what is that it is, or of what is not that is not, is true). On the other hand, Twardowski rejected another of Aristotle's formulations, namely that of *Metaphysics* 1051e, which defines truth in terms of thinking the separated to be separated and the combined to be combined. The main argument Twardowski accepted as telling against this second Aristotelian definition is that it is inconsistent with the 'idiogenic' account of judgments as *sui generis* acts, rather than combinations of presentations: as truths were unitary on Twardowski's view, truth could not be defined in terms of combination and separation as Aristotle's 1051e has it. Twardowski directed this same sort of objection against Russell, arguing further the Russellian notion of a fact was unclear (Twardowski 1975). In general, he had doubts as to whether typical wordings of the correspondence theory (the theory of 'transcendent correspondence' as he called it) were satisfactory. He accused them of being based on unclear metaphysical assumptions concerning what propositions were. Although he agreed that correspondence theories do not offer criteria of truth which would allow one to recognize which judgments were true, he did not consider this sufficient ground for objection. Twardowski also criticized various non-classical definitions of truth and in particular he argued against pragmatism and coherentism on the grounds that they violated the metalogical principles of excluded middle and contradiction.

To sum up, many of Twardowski's views became important for the further development of thinking about truth in Poland.[7] First of all, his defense of the absolute concept of truth was accepted by most Polish philosophers, an aspect of thought about truth that became important for the discussions of many-valued logic. Here Twardowski and Leśniewski defended the view that there are only two truth values. By contrast, Kotarbiński at first admitted judgments which are indefinite, at least until Leśniewski convinced him to accept strict bivalence. On the other hand, as is well known, Łukasiewicz agreed with eternality, but rejected sempiternality as leading to fatalism. Secondly, Twardowski was the first to point out that some metalogical laws (of excluded middle, of non-contradiction) are associated with the absolute character of truth. Thirdly, Twardowski's criticism of non-classical truth-definitions (for instance of the 'utilitarian' conception) became standard in Poland. Fourthly, his doubts concerning the usual formulation of the correspondence theory were shared by his students. As we will see later in this chapter, all of these views find expression in Tarski's work.

Finally, although it was not directly related to the problem of truth, Twardowski introduced (see Twardowski 1912) a distinction between actions and products which was applied by Polish philosophers to the analysis of all mental activities, including the use of language. In particular, this distinction allowed a fruitful approach to the meaning of linguistic expressions. A special group of acts, the psycho-physical, included linguistic activities. Every psycho-physical act has its content, which is intuitively apprehended and objectivized as the meaning of a given expression. Moreover, every act has its object, that is, an entity to which the act is directed.[8] These views will be directly relevant to our discussion of Tarski's somewhat fraught remarks on the bearers of truth.

---

[7] See Woleński 1989 for a full account of the development summarized here with bibliographical references.
[8] The same concerns purely mental acts, that is, acts without physical components: they, too, had entities toward which they were directed. However, most of Twardowski's students considered thinking as essentially linked with the use of language. Thus, the distinction between purely mental and psycho-physical acts was widely rejected in the Lvov-Warsaw School.

ŁUKASIEWICZ

Although Łukasiewicz is famous for his contributions to many-valued logic, we will omit all problems related to this topic, as Tarski was not particularly interested in it. Although he did some technical work in the area in the 1920s and 1930s, he had little respect for this line of logical investigations in the later stages of his career. He wrote:

[...] I hope that no creators of many-valued logics are present, so [...] I can speak freely—I should say that the only one of these systems for which there is any hope of survival is that of Birkhoff and von Neumann. [...] This system will survive because it does fulfill a real need.

(Tarski 2000, 25)

(The quotation is likely an allusion to Łukasiewicz and signals rather poor relations among them after the Second World War.) However, there are other points in Łukasiewicz's views on truth which are important in the present context. Although Łukasiewicz considers propositions as proper bearers of truth, he locates them as existing in a language (see Łukasiewicz 1910, passim). Hence, Twardowski's distinction between sentences and propositions became of secondary importance to Łukasiewicz. In his later works, he always regarded sentences as the objects of logical investigation. Łukasiewicz, following Twardowski, sharply distinguished truth and its criteria (see Łukasiewicz 1911). He proposed the following definition as a version of Aristotle's from *Metaphysics* 1011b (see Łukasiewicz 1910, p. 15):

(4) An affirmative proposition is true if it ascribes a property to an object, which is possessed by this object; a negative proposition is true if it rejects a property, which is not possessed by a given object.

Łukasiewicz also gave a version of the Liar paradox which was used by Tarski (see 1915; it is unfortunate that the relevant passages of this chapter are not included into Łukasiewicz 1970). It was as follows:

(5) the sentence printed in the line m on the page n of this book is false,

where m and n refers to the appropriate line of the appropriate book. Łukasiewicz's response was to maintain that (5) is ill-formed and as such cannot be a value of a propositional variable.

Łukasiewicz also worked on the foundations of probability (see Łukasiewicz 1913). In particular, he argued that sentences are true or false, and, thereby, cannot be considered as merely probable. Probability can be ascribed only to indefinite sentences, that is, formulas with free variables. Now if Px is such a formula, $p(Px)$ (= the probability of Px) is its logical value, which is measured by the relation of the number of values satisfying Px to the number of all possible values. In a particular case, Px is true if all values satisfy it, and false if it is satisfied by no value. We can say that truth defined in such a way conforms to the following condition:

(6) Px is true if and only if $\forall x Px$ is true.

The relation of this to certain aspects of Tarski's treatment of truth will be discussed below.

ZAWIRSKI, CZEŻOWSKI

Although as of the present writing no definitive historical link can be established between Zawirski and Czeżowski and Tarski, the contributions of these two authors clearly anticipate Tarski's work and they bear mention here. Zawirski (1914), like Łukasiewicz, construed propositions as items of a language and denied that truth and falsehood could have degrees. Following Twardowski he favored the idiogenic theory of judgments and Aristotle's formulation from *Metaphysics* 1011b. However, he defended (1914, 57–8) the nihilistic account of truth, saying that every attribution of truth or falsehood is either an assertion or denial of that to which truth is apparently attributed, more precisely, the assertion of reality or the rejection of reality. This account of truth was also discussed by Kotarbiński, as we will see below.

Czeżowski (see Czeżowski 1918, p. 7) was the first author in Poland to focus on the formula later called the T-scheme:

Truth is an characteristic attribute of sentences [note 'sentences', not 'propositions'—RM, JW]. [...]. We assert truth or falsehood about every sentence. However, truth is a property of a particular importance. If a certain sentence *A* is true, the sentence *A is true* is also true, if one of them is false, the same simultaneously concerns the second: the sentences *A* and *A is true* are equivalent.

One should perhaps add that the equivalence of *A* and *A is true* occurs in Couturat 1905, p. 84, translated into Polish in 1918.

KOTARBIŃSKI

As Leśniewski (and Łukasiewicz) were his masters in logic, so Kotarbiński was in philosophy. Tarski's main background was in mathematics but he very seriously studied philosophy under Kotarbiński. Tarski really revered Kotarbiński. One of the indications of this can be seen in the dedication of his 1956 collection (Tarski 1956) of fundamental papers *Logic, Semantics, Metamathematics*. The dedication reads: "To his teacher TADEUSZ KOTARBIŃSKI. The author." The dedication for the second edition (1983) which appeared after the death of Kotarbiński was: "To the memory of his teacher TADEUSZ KOTARBIŃSKI. The author." This is remarkable when one takes into account that Tarski had many teachers who influenced his scientific interests, in particular Łukasiewicz, Sierpiński, and Leśniewski, the last of whom was his dissertation advisor. When asked by doctoral students in Berkeley who his teacher was, Tarski replied "Kotarbiński". Leśniewski's name was never mentioned. Add also that Kotarbiński's photo had a privileged position on Tarski's desk. People who observed meetings of Kotarbiński with Tarski were very impressed by their mutual relations and the great respect of the pupil for his treacher.⁹ He translated (together with David Rynin) into English Kotarbiński's chapter 'Zasadnicze myśli pansomatyzmu' (The Fundamental Ideas of Pansomatism). The chapter was originally published in Polish in 1935, the translation appeared in *Mind* in 1955 and has been also included into Tarski's *Collected Works* (1986, vol. 3) on the explicit request of Tarski himself.[10]

Kotarbiński's doctrine of reism (called also pansomatism or concretism) is a form of physicalistic nominalism. The main reistic thesis is that there exist only singular, spatio-temporal, material things, some of them equipped with psyche. Thus there are no abstract entities like properties, relations, or state of affairs.[11] Kotarbiński was very strongly influenced by Leśniewski's logical and philosophical ideas. Leśniewski was also a nominalist. His calculus of names, called ontology (LO, for brevity), was considered as the logical basis of reism. The concept of an object as defined in LO became the central tool for Kotarbiński. According to LO, a is an object if and only if a is something; a exists if and only if something is a. One can prove that LO implies that only individual objects exist. Thus, things in the reistic sense are individuals as defined in LO, and being material is their additional universal attribute. Things are usually mereological complexes, that is, aggregates of material pieces. This idea was formally elaborated in Leśniewski's mereology.

Reism determined some essential features of Kotarbiński's theory of truth (see Hiż 1966, Woleński 1990 for a general account). Since from the point of view of reism there are no propositions (they are abstract objects and rejected by reism), the predicate 'is true' cannot be applied to such entities. Although Kotarbiński did not admit propositions in the psychological sense either (because he also banished abstract contents from the furniture of the world), he recognized the existence of subjects performing mental acts, that is, if somatic bodies with mental acts as their proper parts. Hence, as in the case of later Brentano, truth can be attributed to acts of thinking or speaking of concrete persons, for example, one can think or speak truly or not. This use of 'truly' indicates that Kotarbiński to some extent advocated a kind of adverbial theory of truth (see Pasquerella 1989). However, for Kotarbiński, sentences understood as inscriptions or sounds are the principal bearers of truth on the reistic position. Although he noticed that 'is true' is predicated both of acts (thoughts) as well as of sentences, and considered this situation to be puzzling, in the end he agreed that it was tolerable. Kotarbiński distinguished at least three interpretations of sentences (*Elementy*, 104–5): idealistic (sentences are ideal objects), psychologistic (sentences are psychical entities), and nominalistic. He adopted the last. For Kotarbiński,

---

⁹ Marian Przełęcki told us about the meeting in Bucharest at the International Congress of Logic, Methodology and Philosophy of Science in 1971, in which it was clear that there was a great affection between the two men.

¹⁰ Kotarbiński was the first reviewer of Tarski 1933 (see Kotarbiński 1934).

¹¹ Since we are not interested in reism as such, we do not enter into a more detailed analysis of this view. For assessments of reism, sympathetic as well as critical, see the papers collected in Woleński (ed.)1990. Let us note that reism was accepted by Brentano in his later philosophy. See Woleński 1996 for comparisons of various forms of reism.

inscriptions or sounds are things in the normal sense.[12] This is clearly expressed in *Elementy*, his *opus magnus*:[13]

> [...] in the nominalistic (outward) interpretation, *propositio* [that is, a sentence—RM, JW] [...] means [...] the symbol itself, the inscription, the statement, the linguistic phrase or formulation.
>
> (1929, 104)

> [...] There are no "truths" or "falsehoods," if they should be any so called "ideal objects", some so called "objects from the world of content." There are only persons who are thinking in a true way and persons thinking in a false way as well as true sentences and false sentences. Hence terms "truth" and "falsehood" will be proper names, and they will be non-empty, if by "truth" one will understand "true sentence" and by "falsehood"—"false sentence."
>
> (1929, 109)[14]

Kotarbiński was an advocate of the absolute character of truth and an opponent of the relativist approach; he closely followed Twardowski in this (observe also the similarity between 'truth' and 'true sentence' in both philosophers), although the position of Kotarbiński was perhaps slightly weaker than that of his teacher, perhaps because Kotarbiński in this early stage had some reservations concerning the absoluteness of sentences about the future. (See Woleński 1990 on this problem.) In Kotarbiński 1926 he remarked:

> The controversy between absolutism and relativism has not been sufficiently explained so far [...], but at least in the domain of scientific sentences absolutism is undoubtedly right.
>
> (135)

According to Kotarbiński, being true or false does not depend on who is uttering the given sentence or on the circumstances in which they do so. In *Elementy* he wrote:

> The reader has certainly seen that the position of the relativism is weaker. Hence, though relativism attracts some minds today (see, e.g., writings of pragmatists) as it did in the period of Greek sophists [...], so among good specialists in the domain of logic relativism is not popular.
>
> (113)

Kotarbiński distinguished the real from the verbal understanding of truth.[15] This seems to be his original contribution to the theory of truth.[16] According to him in

---

[12] See Rojszczak 2005 for a detailed account of truth-bearers, including Brentano and other adverbialists.

[13] Since the title of the English translation (Kotarbiński 1966) of his (1929) is very unfortunate we will give—when referring to this work—the first word of the Polish title. However, we quote after the English edition.

[14] The second passage supplements the first one, but also contains a combination of adverbialism and reism.

[15] In Kotarbiński 1926 the terms 'real' and 'nihilistic' were used. See also Kotarbiński 1934.

[16] Unfortunately, Kotarbiński did not point out representatives of these views. Since we do not know whether he had Zawirski (see p. 26 above) in his mind. Brentano could be another possibility, because he anticipated the prosentential acount of truth.

some contexts the predicate 'true' (resp. 'false') is not necessary; it plays exclusively the role of a stylistic ornament and does not add anything to the content of a sentence. One can reformulate such a sentence without using the term 'true' (or 'false'). Hence the statement 'The proposition that Warsaw is the capital of Poland is true' can be replaced by the statement 'Warsaw is the capital of Poland.' In this use, 'true' conforms to the 'nihilistic' theory of Zawirski.

But Kotarbiński notices that this is not always the case. For example, the expressions 'The theory of relativity is true' or 'What has been said by Plato is true' cannot be reformulated in this way. By omitting the predicate 'is true' in these sentences one gets expressions not only of other senses but even of a different grammatical type, namely they become names and not sentences. Hence in various contexts the predicate 'is true' ('false') is necessary and cannot be eliminated. In such cases the adjectives 'true' and 'false' are used in a real, and not merely verbal, sense. The nihilist account of truth in Kotarbiński's sense corresponds to some extent to a variety of views covering the redundancy theory (Ramsey 1927), deflationism (Field 1994), minimalism (Horwich 1999), prosentencialism (Grover 1992), or disquotationalism (Quine 2004). According to nihilism as Kotarbiński understood it, the sentence 'Snow is white is true' (or 'It is true that snow is white') says no more than does the sentence 'Snow is white.' Hence nothing is added to the sentence by adding the suffix *is true*. Hence one can claim that the predicate *is true* is empty and adds nothing. So it does not represent or attribute any particular property to its subject. The fact that in our language there are the predicates 'is true' and 'is false' is of a historical but not of a logical interest. As has been said above, Kotarbiński accepted the nihilistic theory of truth only with respect to verbal (in fact: redundant) uses of the predicate 'is true' ('false') and claimed that those predicates are indispensable in various important contexts. Hence the nihilistic theory does not suffice.

Kotarbiński, in *Elementy* (chapter 3, §17), understood the classical and utilitarian conceptions of truth as the two basic conceptions. According to the first a truth is that which corresponds to or is in agreement with reality, and according to the second, 'true' means 'useful' (in some respect). One of the forms of utilitarian understanding is pragmatism, which claims that truth is just the property of a proposition which makes an action based on it efficient. Having distinguished those two senses Kotarbiński explicitly expressed his preference for the classical understanding. On the other hand he was aware that the phrase 'accordance with reality' is not precise enough and has a rather metaphorical character when understood by analogy to pictures or copies. In *Elementy* he wrote:

> Let us [...] pass to the classical doctrine and ask what is understood by "accordance with reality". The point is not that a true thought should be a copy or simile of the thing of which we are thinking, as a painted copy or a photograph is. A brief reflection suffices to recognize the metaphorical nature of such comparison. A different interpretation of "accordance with reality" is required. We shall confine ourselves to the following: "John thinks truly if and only if John thinks that things are so and so, and things are in fact so and so."

(106–7)

As we see Kotarbiński preferred here unequivocally weak over strong correspondence, that is, he did not invoke such notions as simililarity or isomorphism in order to explain the concept of correspondence (see Woleński 1993*a* for a more detailed account of the distinction of strong and weak correspondence).

TARSKI'S VIEWS RELATED TO THE PREVIOUS SECTIONS

Tarski considered his analysis of the concept of truth as a logical (mathematical) enterprise, as well as a philosophical one, as is explicitly asserted in his treatise on truth (Tarski 1933) in both its opening and closing passages:

The present article is almost wholly devoted to a single problem—*the definition of truth*. Its task is to construct—with reference to a given language—*a materially adequate and formally correct definition of the term 'true sentence'*. This problem [...] belongs to the classical questions of philosophy [...].

(1954)

[...] in its essential parts the present work deviates from the main stream of methodological study [that is, metalogical or metamathematical; the scope of methodological study should be seen here in a wider sense than in the Hilbert school, that is, as not restricted to finitary proof theory—RM, JW]. Its central problem—the construction of the definition of true sentence and establishing the scientific foundations of the theory of truth—belongs to the theory of knowledge and forms one of the chief problems of philosophy. I therefore hope that this work will interest the student of the theory of knowledge and that he will be able to analyse the results contained in it critically and to judge their value for further research in this field, without allowing himself to be discouraged by the apparatus of concepts and methods used here, which in places have been difficult and have not been used in the field in which he works.

(266–7)

Hence, it is quite legitimate to look at Tarski's philosophical background. As far as the matter concerns terminology and a broad philosophical perspective Tarski usually refers to *Elementy*:

A good analysis of various intuitive conceptions concerning the notion of truth is contained in Kotarbiński's book [*Elementy*].

(1932, 615)

[...] in writing the present article I have repeatedly consulted [*Elementy*] and in many points adhered to the terminology there suggested.

(1933, 153, note 1)

A critical discussion of various conceptions of truth can be found in [*Elementy*].

(1944, 695, note 6)

Yet the substantial links between Tarski and other philosophers from the Lvov-Warsaw School are at least as important. We address these issues in this section. A special problem will be discussed later on (p. 38).

Classical, correspondence, etc.

Tarski followed Kotarbiński in understanding the contrast between the classical and utilitarian truth-definitions as the main opposition in aletheiology.[17] He also (see Tarski 1944, p. 698, note 38) referred to Kotarbiński as a person who interpreted the semantic conception of truth as a version of the classical theory. Thus, Tarski's claim that he semantically developed the classical tradition was entirely coherent with Twardowski and his tradition.[18] In fact Tarski adhered to the classical correspondence conception of truth and that in just the formulation given by Kotarbiński. At the very beginning of Tarski 1933 we read:

[A] true sentence is one which says that the state of affairs is so and so, and the state of affairs indeed is so and so.

(1933, 155)[19]

An important note is associated with this passage, which reads:

Very similar formulations are found in Kotarbiński [1929] [. . .] where they are treated as commentaries which explain approximately the classical view of truth.

In several places Tarski stressed that his conception of truth coincides with the intuitive classical Aristotelian one and refers to various authors, a fact stressed by reviews such as Kotarbiński 1934 and Scholz 1937. Commenting on intuitions underlying the semantic definition of truth, he wrote:

We should like our definition to do justice to the intuitions which adhere to the *classical Aristotelian conception of truth*—intuitions which find their expression in the well-known words of Aristotle's *Metaphysics*:

To say of what is that it is not, or of what is it not that it is, is false while to say of what is that is, or of what is not that it is not, is true.

If we wished to adapt ourselves to modern philosophical terminology, we could perhaps to express this conception by means of the familiar formula:

The truth of a sentence consists in its agreement with (or correspondence to) reality.

(For a theory of truth which is to be based on upon the latter formulation the term "correspondence theory" has been suggested).

If, on the other hand, we should decide to extend the popular usage of the term "*designate*" by applying it not only to names, but also to sentences, and if we agreed to speak of the designata of sentences as "states of affairs," we could possibly to use for the same purpose the following phrase:

---

[17] Note, however, that other philosophers from the Lvov-Warsaw School, in particular, Leśniewski considered aletheiological pragmatism as the most important rival of the classical position.

[18] This does not mean that every philosopher from this school accepted the classical definition, but the exceptions were rare, for example, the consensus account was advocated by Poznański and Wundheiler.

[19] A caution is required here. In particular, the phrase 'state of affairs' has no technical meaning, that is, it does not commit us to an ontology of states of affairs. Tarski (or rather Woodger, the translator of Tarski 1933 into English) used it as a substitute for Kotarbiński's (see below p. 32) 'things are so and so.'

A sentence is true if it designates an existing state of affairs.

However, all these formulations can lead to various misunderstandings, for none of them is sufficiently precise and clear (though this applies much less to the original Aristotelian formulation than to either of the others); at any rate, none of them can be considered a satisfactory definition of truth.

(1944, 666–7)

Three points are worthy of note. Firstly, Tarski, like Twardowski, Łukasiewicz, and Kotarbiński, took the quoted passage from Aristotle's *Metaphysics* as the best approximation of the Stagirite intuition about truth. Secondly, the term 'state of affairs' has here a misleading technical ontological connotation (see note 19). Thirdly, and most importantly in the present context, Tarski considers formulations with 'agreement' or 'designating states of affairs' as not quite satisfactory. This appears to be a legacy of Twardowski's and Kotarbiński's skepticism concerning the concept of correspondence.

The Liar paradox, satisfaction, the T-scheme

Tarski (1933, 157) explicitly used Łukasiewicz's version of the Liar paradox (see p. 25 above), but he never said that self-referential sentences are ill-formed. Tarski's view was rather that they should not appear in properly constructed formalized languages.[20] In this sense, he would agree with Łukasiewicz that the Liar sentences and other self-referential constructions could not be values of sentential variables. The defective character of such sentences consists in their role in generating semantic antinomies. In general, we can say that according to the logicians of the Lvov-Warsaw School, good symbolic notations should not lead to contradictions caused by rules of formation.

Tarski defined truth (he also identified truths with true sentences) via satisfaction: a sentence is true if and only if it is satisfied by all sequences of objects. It is difficult to say whether Tarski was influenced by Łukasiewicz in this respect (see p. 25 above), although Łukasiewicz 1913 was among the best-known philosophical papers in Poland. Anyway, (6) is a consequence of Łukasiewicz's account as well as the semantic definition of truth. According to Tarski, the intuitive content of this definition is captured by the T-scheme 'A is true if and only if A' (with additional constraints concerning protection against antinomies), but it is not clear whether Tarski recognized that Czeżowski (see p. 26 above) formulated the equivalence of A and 'A is true' as the rule governing the concept of truth.

Absolutism vs. relativism

In Tarski 1933 (199–200) we find the distinction between the absolute concept of truth and that expressed by the phrase 'true sentence in an individual domain of

---

[20] Concerning the Liar paradox, Tarski was much more influenced by Leśniewski, but we omit this issue.

individuals'. According to Tarski, the former is a special case of the latter. It is unclear whether Tarski himself attributed any philosophical significance to this distinction. On the other hand, the semantic definition of truth was used (see Kokoszyńska 1936, Kokoszyńska 1948, Kokoszyńska 1951) for making precise the distinction in question. In particular, Kokoszyńska, who was a good expert on Tarski's views, considered his theory as absolute and argued that the reference to models or languages does not entail (see also Woleński 1994) the relativity of truth. Although we have no explicit comment from Tarski about this issue, we can say that his ideas are coherent with absolutism.[21]

The verbal and real use of 'is true'

Tarski in various places referred to Kotarbiński's distinction between real and verbal usage of the predicate 'is true' and to the nihilistic theory of truth (see Tarski 1944, Tarski 1969).[22] In particular, Tarski shared Kotarbiński's opinion that the predicate 'is true' is not always eliminable. He generalized Kotarbiński's argument in a very interesting way (Tarski 1944, pp. 682–3):

Some people have [...] urged that the term "*true*" in the semantic sense can always be eliminated, and that for this reason the semantic conception of truth is altogether sterile and useless. And since the same considerations apply to other semantic notions, the conclusion has been drawn that semantics as a whole is purely a verbal game and at best only a harmless hobby.

But the matter is not quite simple [...] The sort of elimination here discussed cannot always be made. It cannot be done in the case of universal statements which express the fact that all sentences of a certain type are true, or that all true sentences have a certain property. For example, we can prove in the theory of truth the following statement:

All consequences of true sentences are true.

However, we cannot get rid here of the word "*true*" in the simple matter contemplated.

Again, even in the case of particular sentences having the form "*X is true*" such a simple elimination cannot always be made. In fact, the elimination is possible only in those cases in which the name of the sentence which is said to be true occurs in a form that enables us to reconstruct the sentence itself.

(1944, 682–3)

The non-eliminability of 'is true' was important for Tarski, because it armed him against the view that "the semantic conception of truth is altogether sterile and useless."[23]

---

[21] Note, however, that Jan Tarski, the son of Alfred, told one of us (Jan Woleński) that his father considered the absoluteness of truth as truth's important feature.

[22] In Tarski 1969 we find an explicit reference to Kotarbiński and the assertion that the name 'nihilistic theory of truth' was suggested by him.

[23] Although we agree with Tarski, we would not like to suggest that this issue is uncontroversial. The sentence 'all consequences of true sentences are true' can be rendered in the context of the redundancy theory as follows: $\forall A \forall B((A \in Cn\{B\} \land B) \Rightarrow A)$. Applying the T-scheme gives: if B is true, so is A. However, this translation is much more complicated than the original and assumes a quite considerable amount of logic, for instance, the rules for quantifiers for propositional

Truth-bearers

Tarski, similarly to Kotarbiński, Leśniewski, and Łukasiewicz claimed that the predicate 'is true' (resp. 'is false') should be applied only to sentences. He did not exclude other bearers of truth, such as thoughts or judgments, but his nominalistic preferences (inherited from Leśniewski and Kotarbiński) determined that he considered linguistic expressions, in particular, sentences as primary bearers of semantic properties.[24] This view, together with the role of items of the sentential syntactic category in logic, led him to construe a language as a set of sentences.[25] Hence, a closer analysis of the concept of a sentence was of the utmost importance, for on Tarski's view the issue was relevant to nominalism. In Tarski 1930 he wrote:

Sentences are most conveniently regarded as inscriptions, and thus as concrete physical bodies.

(1930, 62)

According to this explanation language consists of expressions conceived as tokens. Yet Tarski was fully aware of the fact that this purely nominalistic theory of language created serious difficulties for logic, particularly metalogic and metamathematics. This led him to the idea that linguistic expressions should be considered not as concrete inscriptions but as types; that is, as shapes of tokens (mathematically speaking, types are classes of abstractions from similar tokens). Tarski expressed this new approach in the following way (he refers to Kotarbiński; also to *Principia Mathematica* of Whitehead and Russell):

Statements (sentences) are always treated here as a particular kind of expression, and thus as linguistic entities. Nevertheless, when the terms 'expression', 'statement', etc., are interpreted as names of concrete series or printed signs, various formulations which occur in this work do not appear to be quite correct, and give the appearance of a widespread error which consists in identifying expressions of like shape. [...] In order to avoid both objections of this kind and also the introduction of superfluous complications into discussion, which would be connected among other things with the necessity of using the concept of likeness of shape, it is convenient to stipulate that terms like 'word', 'expression', 'sentence', etc., do not denote concrete series of signs but the whole class of such series which are of like shape with the series given.

(1933, 156)

Tarski considered this new account as more convenient for logic.[26]

---

variables. It is also debatable whether the fundamental limitative theorems hold without a precise truth-definition, but we do not enter into this topic.

[24] See p. 36 for a more detailed account of Tarski's nominalism. Let us add a word about Łukasiewicz in this context. He accepted nominalism with respect to truth-bearers more as a useful practical solution than a theoretically justified standpoint (see also note 26 below).

[25] Adopting this view in logic consisted in a radical departure from the traditional logic for which sentences (or propositions or judgments) and concepts (notions, names) constituted equally important building-blocks of logic.

[26] Tarski commenting on his view about expressions as tokens added the following note in 1956 (see Tarski 1956, p. 62): "This [...] expresses the views of the author when this article was

It is interesting that Tarski carried the analysis of the concept of a sentence beyond that of his predecessors. In Tarski 1933 we find at least four different understandings of this concept:

(a) an expression of a special syntactical category (this interpretation is the most suitable for formalized languages):

[A]mong all possible expressions which can be formed with these signs those called sentences are distinguished by means of purely structural properties.

(166)

(b) a sentential function of a kind (also good for formalized languages):

x is a sentence (or a meaningful sentence)—in symbols $x \in S$—if and only if x is a sentential function and no variable v_k is a free variable of the function x.

(178)

(c) a psycho-physical product (although the second sentence points out an essential defect of this position, the use of the actions/product distinction introduced by Twardowski is remarkable):

Normally, expressions are regarded as the products of human activity (or as classes of such products). From this standpoint the supposition that there are infinitely many expressions appears to be obviously nonsensical.

(174 note 2)

(d) a physical body (we have here also critical comments):

But another possible interpretation of the term 'expression' presents itself: we could consider all physical bodies of a particular form and size as expressions. The kernel of the problem is then transferred to the domain of physics. The assertion of the infinity of the number of expressions is then no longer senseless and even forms a special consequence of the hypotheses which are normally adopted in physics or geometry.

(174 note 2)

Although Tarski had serious reservations with respect to (c) and (d) concerning the number of admissible formulas, he still was sympathetic to considering language as finitistic (observe again the importance of the distinction between act and products):

In the course of our investigation we have repeatedly encountered [...] the impossibility of grasping the simultaneous dependence between objects which belong to infinitely many semantical categories; the lack of terms of 'infinite order'; the impossibility of including in *one* process of definition, infinitely many concepts; and so on [...]. I do not believe that these phenomena can be viewed as a symptom of the formal incompleteness of the actually existing languages—their cause is to be sought rather in the nature of language itself: language, which is a product of human activity, necessarily possesses a 'finitistic' character, and cannot

originally published and does not adequately reflect his present attitude." Although this formulation is slightly cryptic, one can assume that Tarski alludes here to his transition to the view that linguistic expressions are types.

serve as as adequate tool for the investigation of facts, or for the construction of concepts, of an eminently 'infinitistic' character.

(253 note 1)[27]

Nominalism

As we noted at the beginning of this chapter nominalism is one of the most intriguing of Tarski's views. It was clearly stated in Mostowski's chapter on Tarski (see also Suppes 1988 for Tarski's caution in announcing his philosophical views):

> Tarski, in oral discussions, has often indicated his sympathies with nominalism. While he never accepted the "reism" of Tadeusz Kotarbiński, he was certainly attracted to it in the early phase of his work. However, the set theoretical methods that form the basis of his logical and mathematical studies compel him constantly to use the abstract and general notions that a nominalist seeks to avoid. In the absence of more extensive publications by Tarski on philosophical subjects, this conflict appears to have remained unresolved.

(Mostowski 1967, 81)[28]

Some sources clearly confirm Tarski's pro-nominalist position. On April 29–30, 1965, he was chairing the joint meeting (held in Chicago) of the Association for Symbolic Logic and the American Philosophical Association on the philosophical implications of Gödel's incompleteness theorems. Tarski's remarks are preserved on tape. He said:

> I happen to be, you know, a much more extreme anti-Platonist. [...] However, I represent this very [c]rude, naïve kind of anti-Platonism, one thing which I would describe as materialism, or nominalism with some materialistic taint, and it is very difficult for a man to live his whole life with this philosophical attitude, especially if he is a mathematician, especially if for some reasons he has a hobby which is called set theory.

(Feferman and Feferman 2004, 52)

Other similar remarks by or about Tarski are collected in Feferman and Feferman 2004; these quotations are taken from Tarski's speech at the celebration of his seventieth birthday as remembered by Chihara, Chateaubriand, and the Fefermans themselves:

> I am a nominalist. This is a very deep conviction of mine. It is so deep, indeed, that even after my third reincarnation, I will still be a nominalist. [...] People have asked me, 'How can you, a nominalist, do work in set theory and logic, which are theories about things you do not believe in?' ... I believe that there is a value even in fairy tales.

(Feferman and Feferman 2004, 52)

---

[27] It appears that this question was very important for logicians in Warsaw in the interwar period. Łukasiewicz (see Łukasiewicz 1936, p. 240) notes the tension between the fact that we have only a finite number of expressions and our need in logic for infinitely many formulas. Doubtless this question must have been discussed in Warsaw and Tarski was the first who mentioned it in print.

[28] It is interesting that Mostowski himself was also attracted by reism, at least on special occasions, namely when he encounters very abstract constructions in set theory. See Kotarbińska 1984, p. 73. Let us add that Tarski was ready to discuss philosophical matters in conversations and seminars.

[I am] a tortured nominalist.

(Feferman and Feferman, 2004, 52)

Elsewhere Tarski has said more specifically that he subscribed to reism or concretism (a kind of physicalistic nominalism) of his teacher Tadeusz Kotarbiński.

(Feferman and Feferman 2004, 352 note 10)

Note, however, that Tarski, contrary to Kotarbiński, never based his nominalism or reism on Leśniewski's system LO. On the other hand we should note that the Feferman's statement contradicts Mostowski's claims. See also Mycielski 2004, pp. 215–17 about Tarski's nominalistic sympathies.

Fortunately we can now say more about Tarski's sympathies to nominalism. This is possible due to the discovery of Carnap's protocols from the discussions between him, Tarski, Quine and (occasionally) Russell at Harvard in the early 1940s.[29] Carnap recorded the following remarks on nominalism and finitism:

Tarski: At bottom, I only understand a language that fulfills the following conditions:

1. Finite number of individuals.
2. Realistic (Kotarbiński): The individuals are physical things.[30]
3. Non-Platonic: Only variables for individuals (things) occur, not for universals (classes etc.)

(Mancosu, 2005, 342)

The following exchange is also recorded:

I [Carnap]: Should we construct the language of science with or without types?

He [Tarski]: Perhaps something else will emerge. One would hope and perhaps conjecture that the whole general set theory, however beautiful it is, will in the future disappear. With the higher types Platonism begins. The tendencies of Chwistek and others ('Nominalism') of speaking only of what can be named are healthy. The problem is only how to find a good implementation.

(Mancosu 2005, 334)

Mancosu also reports this summary, by Carnap, of views of Tarski's Polish predecessors and Tarski's own shift away from them, as he learned of them from Tarski:

The Warsaw logicians, especially Leśniewski and Kotarbiński saw a system like PM [*Principia Mathematica*—RM, JW] (but with simple type theory) as the obvious system form. This restriction influenced strongly all the disciples; including Tarski until the 'Concept of Truth' (where the finiteness of the levels is implicitly assumed and neither transfinite types nor systems without types are taken into consideration; they are discussed only in the Postscript added later). Then Tarski realized that in set theory one uses with great success a different

---

[29] These protocols are in the Rudolf Carnap Collection in Pittsburgh. We are using here Frost-Arnold 2004 and Mancosu 2005 and quoting after them.

[30] Frost-Arnold adds here (Frost-Arnold 2004, p. 278): "Later, Tarski relaxes this requirement: the number of individuals is allowed to be infinite or finite; neither is assumed." Mancosu (2005, p. 343) writes that this condition should be corrected to read 'reistic' as opposed to 'realistic'. This makes sense on account of the reference to Kotarbiński.

system form. So he eventually came to see this type-free system form as more natural and simpler.

(Mancosu 2005, 333–4)

Although all this indicates Tarski's decisive sympathies towards nominalism, reism, and so on, we should note once again the dissonance in Tarski's views, namely between his logical and mathematical practice and some of his philosophical views; Tarski himself was aware of this situation as the quoted passages show. To understand Tarski's attitude one should take into account the attitude of Polish mathematicians and logicians (see Murawski 2004 for a more extensive treatment of this question). According to it one should study problems using any fruitful methods and making no philosophical presuppositions. There is no need to announce one's philosophical views concerning the investigated problems because this does not belong to scientific duty, this is a 'private' affair. Tarski's attitude was in full accordance with this. To some extent he followed the pattern of doing philosophy in the Lvov-Warsaw School. Twardowski and his students distinguished 'metaphysicism', that is, limiting concrete research by metaphysical assumptions, from genuine scientific work. Although in philosophy this attitude is even more difficult to maintain, if it can be maintained at all, than in mathematics, it had an importance influence on Tarski.

LANGUAGE AND MEANING

We believe that one of the most important of Tarski's philosophical remarks about the background of the semantic theory of truth is this:

It remains perhaps to add that we are not interested here in 'formal' languages in sciences in one special sense of the word 'formal', namely sciences to the signs and expressions of which no material sense is attached. For such sciences the problem here discussed [the problem of truth—RM, JW] has no relevance, it is not even meaningful. We shall always ascribe quite concrete and, for us, intelligible meanings to the signs which occur in the language we shall consider. The expressions which we call sentences still remain sentences after the signs which occur in them have been translated into colloquial language. The sentences which are distinguished as axioms seem to us materially true, and in choosing rules of inference we are always guided by the principle that when such rules are applied to true sentences the sentences obtained by their use should also be true.

(1933, 166–7)

We will not enter into complex issues concerning the concept of meaning, nor do we claim that Tarski defined this notion in the quoted passage or elsewhere. He did not do so, and it is well known that he avoided saying what meaning is. He believed that meanings are in language and that this is enough for a logician. The importance of Tarski's words stems from the fact that he explains what he means by formal language and that he understood that the concept of truth has no application for purely formal (syntactic) systems. Thus, the concept of interpretation is fundamental, but one must grasp meaning in order to know how signs are interpreted. According to Tarski meanings are intuitively grasped.

This view has its roots in Leśniewski. He introduced so-called intuitive formalism (we prefer this label over 'intuitionistic formalism' used in the original) in the following way:

> Having no predilection for 'various mathematical games' that consist in writing out according to one or another conventional rule various more or less picturesque formulae which need not be meaningful or even—as some of the 'mathematical gamers' might prefer—which should necessarily be meaningless, I would not have taken the trouble to systematize and to often check quite scrupulously the directives of my system, had I not imputed to its theses a certain specific and completely determined sense, in virtue of which its axioms, definitions, and final directives [...] have for me an irresistible intuitive validity. I see no contradiction therefore, in saying that I advocate a rather radical 'formalism' in the construction of my system even though I am an obdurate 'intuitionist'. Having endeavoured to express my thoughts on various particular topics by representing them as a series of propositions meaningful in various deductive theories, and to derive one proposition from others in a way that would harmonize with the way I finally considered intuitively binding, I know no method more effective for acquainting the reader with my logical intuitions than the method of formalizing any deductive theory to be set forth. By no means do theories under the influence of such formalizations cease to consist of genuinely meaningful propositions which for me are intuitively valid. But I always view the method of carrying out mathematical deduction on an 'intuitionistic' basis of various logical secrets as considerably less expedient method.
>
> (Leśniewski 1929, 487–8)

However, Leśniewski's view about the role of intuitive grasping of meaning extended views of Twardowski. Let us recall that every mental act is intentional and it has a content which is obvious to the acting mind, whatever is mental. In this respect, there was no difference between mentalism and reism. Since, to repeat once again, the meanings of linguistic expressions are objectivized mental contents, intentionally directed to objects, immediate and intuitive grasping of them (meanings) is comprehensible. Moreover, semantic properties of expressions derive from the intentional character of acts. This means that the essential features of linguistic activities are displayed adequately by properties of the corresponding expressions (see Woleński 2002 for a more detailed account). This theoretical scheme, though incomplete as a theory of meaning, functions well as an explanation of how interpretations come in. In particular, there is no conflict between formalized and interpreted languages.[31]

References

Couturat, L. (1905) *L'Algebre de la logique*, Paris: Gauthier-Villars. Eng. tr. *The Algebra of Logic*, Gottingham Robinson trans. Chicago: The Open Court Publishing Company (1914).

Czeżowski, T. (1918) 'Imiona i zdania. Dwa odczyty' ('Names and Sentences: Two Lectures'). *Przegląd Filozoficzny* 21, 3–22.

---

[31] Benis Sinaceur [forthcoming] introduces the term 'semantic formalism' in order to capture Tarski's position. This term seems to us quite apt.

Feferman, A. and Feferman, S. (2004) *Alfred Tarski. Life and Logic*. Cambridge: Cambridge University Press.
Field, H. (1994) 'Deflationist Views of Meaning and Conent'. *Mind* 103, 249–85.
Frost-Arnold, G. (2004) 'Was Tarski's Theory of Truth Motivated by Physicalism?' *History and Philosophy of Logic* 25, 265–80.
Grover, D. (1992) *A Prosentential Theory of Truth*. Princeton, NJ: Princeton University Press.
Hiż, H. (1966) 'Kotarbiński on Truth'. *Studies in Polish Civilization*, D. Wandycz ed. New York: Columbia University Press, 426–31.
Horwich, P. (1999) *Truth*, 2nd edn. New York: Oxford University Press.
Kokoszyńska M. (1936) 'Über den Absoluten Wahrheitsbegriff und einige andere semantische Begriffe', *Erkenntnis* 6, 143–65. Repr. in *Logische Rationalismus: Philosophische Schriften der Lemberg-Warschauer Schule*. D. Pearce and J. Woleński eds. Frankfurt M: Athenäum, 276–92.
—— (1948) 'What Means a "Relativity" of Truth'. *Studia Philosophica* 3, 167–75.
—— (1951) 'A Refutation of the Relativism of Truth'. *Studia Philosophica* 4, 93–149.
Kotarbińska, J. (1984) 'Głos w dyskusji' ('A Contribution to Discussion'), *Studia Filozoficzne* 5 (222), 69–73.
Kotarbiński, T. (1926) *Elementy logiki formalnej, teorji poznania i metodologji* (*The Elements of Formal Logic, Epistemology and Methodology*), authorized manuscript, D. Steinberżanka ed. Warsaw: Scientific Circle for Philosophy of Students of Warsaw University and Circle for Natural Science of Students of Warsaw University.
—— (1929) *Elementy teorji poznania, logiki formalnej i metodologji nauk* (*The Elements of Epistemology, Formal Logic and Methodology of Sciences*). Lwów: Ossolineum. 2nd enl. edn, Wrocław–Warszawa–Kraków: Ossolineum, 1961. Eng. tr. (from 2nd edn) by O. Wojasiewicz: Kotarbiński 1966.
—— (1934) 'W sprawie pojęcia prawdy' ('Concerning the Notion of Truth'), *Przegląd Filozoficzny* 37, 85–91.
—— (1935) 'Zasadnicze myśli pansomatyzmu' ('The Fundamental Ideas of Pansomatism'). *Przegląd Filozoficzny* 38, 283–94. Eng. tr. by A. Tarski and D. Rynin: Kotarbiński 1955.
—— (1955) 'The Fundamental Ideas of Pansomatism'. *Mind*, 64, 488–500, *Mind* 65, 288. Repr. in Tarski 1986, v. 3, 577–91.
—— (1966) *Gnosiology: The Scientific Approach to the Theory of Knowledge*, Oxford: Pergamon.
Leśniewski, S. (1929) 'Grundzüge eines neuen Systems der Grundlagen der Mathematik'. *Fundamenta Mathematicae* 14, 1–81. Eng. tr. "Fundamentals of a New System of the Foundations of Mathematics", M. P. O'Neill trans., in S. Leśniewski, *Collected Works*, S. J. Surma, J. Strzednicki, and D. I. Barnett eds. Dordrecht: Kluwer Academic Publishers, 410–605.
Łukasiewicz, J. (1910) *O zasadzie sprzeczności u Arystotelesa* (On the Principle of Contradiction in Aristotle), Kraków: Polska Akademia Umiejętności. Germ. tr. *Über den Satz des Widerspruchs bei Aristoteles*, J. Barski trans. Hildesheim: Georg Olms Verlag.
—— (1911) 'Zagadnienie prawdy' ('The Concept of Truth'). *Księga Pamiątkowa XI Zjazdu Lekarzy i Przyrodników Polskich*, Nakładem Towarzystwa Gospodarczego, Kraków, 84–6. Repr. in Łukasiewicz 1998, 55–6.
—— (1913) *Die logischen Grundlagen der Wahrscheinlichkeitsrechnung*, Polska Akademia Umiejętności; partial Eng. tr. by O. Wojtasiewicz in Łukasiewicz 1970, 16–63.

—— (1915) 'O nauce' ('On Science'). *Poradnik dla samouków*, Warszawa: v. I, Heflich i Michalski, 15–34. Repr. in Łukasiewicz 1998, 9–33. Partial Eng tr. by O. Wojtasiewicz, 'On Creative Elements in Science' in Łukasiewicz 1970, 1–15.

—— (1936) 'Logistyka s filozofia' ('Logistic and Philosophy'). *Przegląd Filozoficzny* 39, 115–31. Eng. tr. by O. Wojtasiewicz, in Łukasiewicz 1970, 218–35.

—— (1970) *Selected Works*. L. Borkowski ed. Amsterdam: North-Holland Publishing Company.

—— (1998) *Logika i metafizyka. Miscellanea* (Logic and Metaphysics. Miscellanea), J. J. Jadacki ed. Warszawa: Wydział Filozofii i Socjologii Uniwersytetu Warszawskiego

Mancosu, P. (2005) 'Harvard 1940–1941: Tarski, Carnap and Quine on a Finitistic Language of Mathematics for Science'. *History and Philosophy of Logic* 26, 327–57.

Mostowski, A. (1967) 'Tarski Alfred'. *The Encyclopedia of Philosophy*, v. 8, Edwards ed. New York: The Macmillan Comp. and the Free Press.

Murawski, R. (2004) 'Philosophical Reflection on Mathematics in Poland in the Interwar Period'. *Annals of Pure and Applied Logic* 127, 325–37.

Mycielski, J. (2004) 'On the Tension between Tarski's Nominalism and His Model Theory (Definitions for a Mathematical Model of Knowledge)', *Annals of Pure and Applied Logic* 126, 215–24.

Pasquerella, L. (1989) 'Brentano and Kotarbiński on Truth'. *The Object and Its Identity*. Dordrecht: Kluwer Academic Publishers, 98–106.

Quine, W. V. (2004) *Philosophy of Logic*, 2nd edn. Cambridge, MA: Harvard University Press.

Ramsey, F. P. (1927) 'Facts and Propositions'. *Aristotelian Society Supplementary Volume* 7, 153–70.

Rojszczak, A. (2005) *From the Act of Judging to the Sentence*. Dordrecht: Kluwer Academic Publishers.

Scholz, H. (1937) 'Besprechung: *Studia Philosophica I*'. *Deutsche Literaturzeitung* 58, 1914–17.

Sinaceur, H. (forthcoming) 'Tarski's Practice and Philosophy: Between Formalism and Pragmatism'. *Logicism, Intuitionism, Formalism: What has Become of Them?* S. Lindström, K. Segerberg, and V. Stoltenberg-Hansen eds. Berlin: Springer Verlag.

Skolimowski, H. (1967) *Polish Analytical Philosophy*. London: Routledge and Kegan Paul.

Suppes, P. (1988) 'Philosophical Implications of Tarski's Work', *Journal of Symbolic Logic* 53, 80–91.

Tarski, A. (1930) 'Fundamentale Begriffe der Methodologie der deduktiven Wissenschaften I', *Monatshefte für Mathematik und Physik*, 37, 361–404. Eng. tr., 'Fundamental Concepts of the Methodology of the Deductive Sciences', in Tarski 1956, 60–109.

—— (1932) 'Der Wahrheitsbegriff in den Sprachen der deduktiven Disziplinen'. *Akademie der Wissenschaften in Wien, mathematisch-naturwissenschaftliche Klasse, Akademische Anzeiger* 69, 23–5. Repr. in Tarski 1986, 613–17.

—— (1933) *Pojęcie prawdy w językach nauk dedukcyjnych* (*The Concept of Truth in the Languages of Deductive Sciences*). Warszawa: Towarzystwo Naukowe Warszawskie. Eng. tr. *The Concept of Truth in Formalized Languages* Woodger trans., in Tarski 1956, 152–278.

—— (1944) 'The Semantic Conception of Truth and the Foundations of Semantics'. *Philosophy and Phenomenological Research* 4, 341–75. Repr. in Tarski 1986, v. 2, 665–9.

—— (1956) *Logic, Semantics, Metamathematics: Papers From 1923 To 1938*, Woodger trans. Oxford: Clarendon Press. 2nd edn, J. Corcoran ed., Hackett Publishing Company, Indianapolis (1983).

Tarski, A. (1969) 'Truth and Proof'. *Scientific American* 220, 63–77. Repr. in Tarski 1986, v. 4, 399–423.

—— (1986) *Collected Papers*, vols. 1–4, S. R. Givant and R. N. McKenzie eds. Basel: Birkhäuser.

—— (1992) 'Drei Briefe an Otto Neurath' [25. IV. 1930, 10. VI. 1936, 7. IX. 1936], Haller ed., J. Tarski trans., *Grazer Philosophische Studien* 43, 1–31.

—— (2000) 'Address at the Princeton University Bicennential Conference on Problems of Mathematics (December 17–19, 1946)', H. Sinaceur ed. *The Bulletin of Symbolic Logic* 1, 1–44.

Twardowski, K. (1900) 'O tzw. prawdach względnych'. *Księga Pamiątkowa Uniwersytetu Lwowskiego ku uczczeniu pięćsetnej rocznicy Fundacji Jagiellońskiej*. Lwów: Uniwersytet Lwowski, 64–93. Eng. tr. 'On So-Called elative Truths', A. Szylewicz trans., in Twardowski 1999, 147–69.

—— (1912) 'O czynnościach i wytworach. Kilka uwag z pogranicza psychologii, gramatyki i logiki', in *Księga Pamiątkowa ku uczczeniu 250 rocznicy założenia Uniwersytetu Lwowskiego przez króla Jana Kazimierza*, Lwów: Uniwersytet Lwowski, 1–33; Eng. tr. 'Actions and Products: Some Remarks from the Boderline of Psychology, Grammar and Logic', A. Szylewicz trans., in Twardowski 1999, 103–32.

—— (1975) 'Teoria poznania' (wykłady akademickie w r. a. 1924/25), *Archiwum Historii Filozofii i Myśli Społecznej* 21, 241–99. Eng. tr. 'Theory of Knowledge: A Lecture Course 1924/25', A. Szylewicz trans., in Twardowski 1999, 103–32.

—— (1997) *Dzienniki* (Diaries), Part I: 1915–1927, Part II: 1928–1936, R. Jadczak ed. Toruń: Adam Marszałek.

—— (1999) *On Actions, Products and Other Topics in Philosophy*, in J. Brandl and J. Woleński eds. Amsterdam: Rodopi.

Vuissoz, F. (1998) *La conception sémantique de la vérite. Logique et philosophie chez Alfred Tarski*. Neuchâtel: Centre de Recherches Semiologiques, Université de Neuchâtel.

Woleński, J. (1989) *Logic and Philosophy in the Lvov-Warsaw School*. Dordrecht: Kluwer Academic Publishers.

—— (1990) 'Kotarbiński, Many-Valued Logic and Truth', in Woleński (ed.) 1990, 190–197. Repr. in Woleński 1999, 115–20.

—— (1993*a*) 'Two Concepts of Correspondence', *From the Logical Point of View* 2, 42–55.

—— (1993*b*) 'Tarski as a Philosopher'. *Polish Scientific Philosophy: The Lvov-Warsaw School*, F. Coniglione, R. Poli, and J. Woleński eds. Amsterdam: Rodopi, 319–38.

—— (1994*a*) 'Jan Łukasiewicz on the Liar Paradox, Logical Consequence, Truth and Induction'. *Modern Logic* 4, 392–400. Repr. in Woleński 1999*b*, 121–5.

—— (1994*b*) 'Theories of Truth in Austrian Philosophy'. <http://www.fmag.unict.it/~polphil/Polphil/LvovWarsaw/WolTruth.html>. Repr. in Woleński 1999, 150–75.

—— (1995) 'On Tarski's Background'. *From Dedekind to Gödel*, J. Hintikka ed. Dordrecht: Kluwer Academic Publishers, 331–41. Repr. in Woleński 1999, 126–33.

—— (1996) 'Reism in the Brentanian Tradition'. *The School of Franz Brentano*, L. Albertazzi *et al.* eds. Dordrecht: Kluwer Academic Publishers, 357–375. Repr. in Woleński 1999, 179–90.

—— (1999) *Essays in the History of Logic and Logical Philosophy*. Kraków: Jagiellonian University Press.

—— (2002) 'From Intentionality to Formal Semantics (From Twardowski to Tarski)'. *Erkenntnis* 56, 9–27.

—— (ed.) (1990) *Kotarbiński: Logic, Semantics and Ontology*. Dordrecht: Kluwer Academic Publishers.

Woleński, J. and Simons, P. (1989) 'De Veritate: Austro-Polish Contributions to the Theory of Truth from Brentano to Tarski'. *The Vienna Circle and the Lvov-Warsaw School*, K. Szaniawski ed. Dordrecht: Kluwer Academic Publishers, 391–442.

Zawirski, Z. (1914) *O modalności sądów* (On the Modality of Propositions). Lwów: Nakładem Polskiego Towarzystwa Filozoficznego.

3

Polish Axiomatics and its Truth: On Tarski's Leśniewskian Background and the Ajdukiewicz Connection

Arianna Betti

Leśniewski used to say that he had 100 percent genius doctoral students, his only student being Alfred Tarski (Woleński 1995, 68 note 11). According to unwritten sources,[1] however, when Tarski's *The Concept of Truth in the Languages of Deductive Sciences* came out in 1933, the master did not approve. Why not? In the first chapter of his monograph Tarski credits Leśniewski with crucial results on the semantics of natural language. As I showed in a previous chapter (Betti 2004), Leśniewski's early solution to the Liar reveals that it was indeed he who first avowed the impossibility of giving a satisfactory theory of truth for ordinary language, as well as the necessity of sanitation of the latter for scientific purposes. Of Leśniewskian origin were also Tarski's analysis of quotation marks, the idea that truth is language-relative, the notion of a closed language, and the finding that natural language is such a language. But these are all negative results concerning the semantics of natural language, a diagnosis, if you will. How about the positive results, the medicine? Tarski's own solution to the Liar and the cure he proposes for the illnesses of natural language apparently did not coincide with his master's ultimate remedy—at least, nothing similar to the very idea of Tarski's enterprise can be found in Leśniewski. As Tarski wrote in 1944,

Leśniewski did not anticipate the possibility of a rigorous development of the theory of truth, and still less of a definition of this notion.

(1944, 695 note 7)

Work on this chapter has been funded by the Netherlands Organization for Scientific Research (Project 275-80-001) and by the ELV-AKT project at the Institut d'Histoire et de Philosophie des Sciences et des Techniques (CNRS/Université Paris I/Ecole Normale Supérieure). Many thanks to Hein van den Berg, Anna Brożek, Solomon Feferman, Bjørn Jespersen, Wim de Jong, Paolo Mancosu, Marije Martijn, Douglas Patterson, Göran Sundholm, Richard Zach, and Jan Woleński for discussion (including exchanges on remote ancestors of this chapter), comments on content, language and style, information and help with source material.

[1] Jan Woleński, oral communication.

The reason for this is probably that a Tarski-like theory of truth must have appeared to Leśniewski to offer an insufficiently intuitive solution to the malady of semantic antinomy. But in what sense exactly? A proper answer is still missing. Lack of textual evidence is one reason, but another, equally important reason is that, from a broader point of view, we also do not yet know enough about the specific cultural context in which the answer must be sought. It is the aim of this chapter to address some aspects of this context.

As Mazurkiewicz put it, Leśniewski did not just make contributions, he created a great system of the foundations of mathematics. At Leśniewski's death in 1939, the system comprised three deductive theories, Protothetic, Ontology (the two forming together his logic), and Mereology, an extra-logical theory of collective classes which was the first rigorously formulated formal theory of parts and wholes. These theories aren't exactly the present-day logician's bread and butter. They look weird, they are idiosyncratic and complicated, and, to make matters worse, Leśniewski's writing style was catastrophic. In Betti 2008*a* I argue that the idiosyncrasy—peculiar symbolism included—far from being gratuitous, was the result of a deep epistemological concern, and that Leśniewski's grandiose project of a new, up-to-date, paradox-free logicist *Characteristica Universalis* expressed commitment to a millennia-old model of scientific rationality, the 'Classical Model of Science' (or 'Classical Ideal of Science', as I will say in this chapter).

Was Tarski concerned with the same problems? Was he close to the same spirit that animated Leśniewski?

There is a sense in which the answer to both these questions is simply No. For Tarski did not strive towards the construction of a similar, all-encompassing *Characteristica*.[2] Though his work had been fundamental to Leśniewski's project, Tarski did not take up—with some notable exceptions—the philosophical underpinnings of Leśniewski's formal techniques, despite the fact that in Leśniewski's view philosophical underpinnings and formal techniques were inextricably intertwined and that Leśniewski's influence upon Tarski in matters formal had been strong. Leśniewski's work was the result of a monolithic obsession with building The One Beautiful True Logic (in fact, The One Beautiful True Language of the Deductive Sciences), but Leśniewski's One Beautiful True Logic, notwithstanding its claim to perfection, never became a mainstream focus of research. Tarski, in contrast, produced a constellation of particular results in various mathematical areas, results that eventually enjoyed a high degree of fruition, viability, and impact—and wrote far better. They could not be more different.

Nonetheless, some aspects of Tarski's work seem to share the spirit of Leśniewski's methodological concerns. In particular, one may wonder, in the light of Leśniewski's

---

[2] In this chapter I shall concentrate on the differences between Leśniewski and Tarski rather than on specific similarities, such as extensionality (though a more generally Polish mark), the structural-descriptive method, the theory of definitions and its importance, and more stylistic ones such as rigor and precision, terminology, fussiness about proper attribution, and curious small similarities in Tarski's early work like the way of letting the references precede the text, noticed also by Sundholm 2003, 115–16 (see also the Leśniewski-like "Notations" in Tarski 1924*a*, 69–70). On similarities see also Simons 1987, 19–21, 23–4 and Betti 2004, 278–83.

strict adherence to what I called above The Classical Ideal of Science, whether Tarski's work on the methodology of deductive sciences, or 'metamathematics', couldn't also be understood in terms of this ideal and thus be close, after all, to Leśniewski's ideas. Yet we shall see that, despite some important similarities, in particular as to the weight that metatheoretical thinking had for both, there were crucial differences between the two as to what metatheory and its needs are. In short, I shall claim that the Classical Ideal, though apparently shared by both, changed shape and intent in Tarski's hands to a very significant extent—enough, indeed, that Leśniewski could not possibly have recognized Tarski's contributions as an answer to his own concerns.

I shall deal with these differences later in the chapter, after preparing the ground for this in the next section, which discusses Tarski's scientific background in Poland with particular reference to Leśniewski. This is followed by an account of the breakup in their relationship. Then I shall contend that Tarski's motivation to undertake the clarification of fundamental semantical notions came from a 1921 work by Ajdukiewicz in which the need for analysis of such notions was presented as a goal for the axiomatics of deductive sciences. The Ajdukiewicz connection is important not only because it answers directly the question of the motivation of Tarski's enterprise, raised recently by Solomon Feferman (Feferman, this volume), but also for two other reasons. First, it provides the historical confirmation to Ignacio Jané's recent claim that the "common concept" of consequence that Tarski sought to clarify was the concept in use in axiomatics (Jané 2006); secondly, it settles the old question of the influence of Bolzano upon Tarski's notion of logical consequence, which has been debated since Heinrich Scholz noticed the similarity in the mid-1930s (Scholz 1936–7).

100 PERCENT GENIUSES

Tarski enrolled at the Section of Mathematics and Physics at the Faculty of Philosophy at Warsaw University in the autumn of 1919, attending lectures in philosophy by Tadeusz Kotarbiński, in mathematics (in particular topology and set theory) by Zygmunt Janiszewski, Kazimierz Kuratowski, Stefan Mazurkiewicz, and Wacław Sierpiński, and in logic by Jan Łukasiewicz and Stanisław Leśniewski. Under the latter's supervision Tarski obtained, in the spring of 1924, his Ph.D. (his two advisors were Sierpiński and Łukasiewicz) (Jadacki 2003, 116; trans. 144). Łukasiewicz, Leśniewski, and Kotarbiński belonged to the same generation of philosophers of the Lvov-Warsaw School founded by their teacher, Kazimierz Twardowski.

Tarski focused on mathematics and logic early on.[3] In those years Leśniewski was immersed in the construction of his system. Up to 1920, when he was convinced by Leon Chwistek to employ formal symbols, Leśniewski used a strictly regimented natural language for the formulation of his results. After 1920 he started using a formal language of his own invention and translated all the results he had

---

[3] On Tarski's education see also Givant 1991 and Feferman and Feferman 2004, chapter 2 and Interlude I.

obtained into the new symbolism, presenting them in a course on Mereology. During this period he also worked on the axiomatic foundations of Ontology (Leśniewski 1927–31, 154–60; trans. 364–71). His propositional calculus, Protothetic, was essentially completed in 1922.

Tarski contributed in a decisive way to the development of these systems. In 1921 Tarski and Kuratowski, independently from one another, had obtained some results in Mereology and, between 1921 and 1922, Tarski contributed to the axiomatization of Leśniewski's Ontology, as well as to a simplification of the directives of a protothetical system known as ⊛1, and he gave the simplest axiomatization known at the time of another protothetical system, ⊛3.[4] Most significantly, however, in 1922 Tarski made a discovery whose importance to the entire edifice of Leśniewski's systems is hard to overestimate (Leśniewski 1929, 11; trans. 419). By defining conjunction by means of the biconditional and the universal quantifier, Tarski made it possible for (a system of) Protothetic to be based on the biconditional as a single primitive functor, which Leśniewski used also to formulate definitions. In the light of the strong aesthetic element in the architecture of Leśniewski's systems and its bearings on epistemological issues, this result was of exceptional significance. The result, which earned Tarski his doctorate, was published as Tarski 1923. In 1924 Tarski obtained some new theses (theorems) of Protothetic (Leśniewski 1938, 27; trans. 676), while in 1925 he gave a method for reducing to a single axiom the axiom system of any system of Protothetic with the directives of Leśniewski's ⊛4 and with implication as its single primitive term (Leśniewski 1929, 58; trans. 467). In 1926 Tarski drew Leśniewski's attention to the connection between Mereology and Whitehead's theory of events.[5] In 1929 he built a system of geometry based on Mereology (Tarski 1929). In the meantime, in 1925, Tarski had obtained the *venia legendi* in philosophy of mathematics (Jadacki 2003, 117; trans. 145) and had started lecturing next to his former teachers Leśniewski and Łukasiewicz; and so the formidable *trójka* was born that made Warsaw in those years arguably the most important research centre in the world for formal logic.

Leśniewski acknowledged Tarski's contribution to his systems, in 1927, as follows:

The system of the foundations of mathematics I have constructed owes a series of significant improvements to Mr. Alfred Tarski [...] Regarding the concrete results of Tarski's reflections in connection with my system, I will try to present them *explicite*; because of the nature of things, however, I will not be able to present properly all of Tarski's occasional critical remarks, which undermined this or that link of my theoretical conceptions at the various different stages of the construction of my system, and all the subtle and sympathetic counsel and often

---

[4] Leśniewski mentioned Kuratowski's and Tarski's results in Mereology in chapter 8 of *On the Foundations of Mathematics* ("On certain conditions established by Kuratowski and Tarski which are sufficient and necessary for P to be the class of objects A"), not being able, however, to present the original proofs given by them. For the simplification of the directives of Protothetic, cf. Leśniewski 1929, 39–42; trans. 448–50; for the axiomatization of ⊛3, cf. Leśniewski 1929, 46–7; trans. 456. For Ontology, cf. Leśniewski 1930, 131; trans. 627.

[5] See Leśniewski 1927–31, 286 note 1; trans. 258 fn 84, where Leśniewski reports on the discussion with Tarski on the formalization of Whitehead's theory.

impalpable suggestions, from which I had the opportunity to profit in numerous conversations with Tarski.

(Leśniewski 1927–31, 168–9; trans 180, reproduced here with changes)

Logic wasn't Tarski's only interest, however. As mentioned above, Tarski studied mathematics as well. Set theory in particular became Tarski's mathematical specialty, and his research and interest in this field were soon to mark all his work in logic and algebra. In June 1924, immediately after defending his Ph.D. in philosophy, Tarski had applied to complete his studies in mathematics and physics, which he finished in 1926 (Jadacki 2003, 117; trans. 144). Still in 1924, Tarski published three important papers in set theory, among which a chapter with Banach containing the famous Banach–Tarski 'Paradox' (Tarski 1924*a*, 1924*b*, Banach and Tarski 1924).

With respect to Tarski's mathematical interests, two things are important. The first thing is that Leśniewski had no admiration for the set theoretical research to which Tarski was devoting himself. Leśniewski acknowledged only *collective classes* and was radically opposed to the notion of *distributive class* (a set in the usual sense of set theory). As he used to call Protothetic 'Logistic' up to 1927 (Leśniewski 1927–31, 165 note 1; trans 176 note 3, and Lindenbaum and Tarski 1926, 196), Leśniewski kept calling Mereology 'General Theory of Sets' for a number of years—no doubt the reason for this was that the only set theory Leśniewski could agree to *was* Mereology. His use of terms like 'set' and 'class' at that time related to Mereology: a Leśniewskian *collection* (*zbiór*) or *set* (*mnogość*) was a (concrete) collection of objects *a*—a heap of *a*s—and a *class* (*klasa*) the (concrete) collection of all objects *a*—the heap of all *a*s. In opposition to Leśniewski, one of Tarski's teachers in mathematics, Sierpiński, was the Polish champion of Cantorian set theory. Famously, Sierpiński had a most open stance as to the assumptions and the methods to admit in mathematics; he was, for instance, a major player in the process leading to the widespread acceptance of the Axiom of Choice, which at the time was viewed with suspicion by many mathematicians.[6]

The second important thing is that at the time set theory, though in rapid development, had not yet reached its status of unrivalled foundation for mathematics. For instance, as late as 1926, Leśniewski's systems were put on a par—as far as technical results were concerned—by Tarski and Lindenbaum with both the *Principia Mathematica* and Zermelo's set theory:

These results can be developed *mutatis mutandis* in different deductive systems: thus equally well in that of *Principia Mathematica* of Mr. Russell and Mr. Whitehead or in the Ontology of Mr. Leśniewski [...] or in the Set Theory of Mr. Zermelo.

(Lindenbaum and Tarski, 1926, my translation)[7]

---

[6] On this see Moore 1982, chapter 4.

[7] Note that, to put it roughly, Mereology can be seen as an alternative to set theory from a metaphysical point of view, but in the sense of the technical results mentioned by Lindenbaum and Tarski here it is not Mereology that can be seen as an alternative to set theory, for actually the role played here by set theory can be played by parts of Ontology: it is in (Protothetic and) Ontology that as much classical mathematics can be reconstructed as in either edition of *Principia Mathematica*. Cf. Leśniewski 1930, 113–14; trans. 608.

Thus at the time of his teaching appointment in 1925, Tarski was working in two rival fields, Leśniewski's systems and set theory. Moreover, although he hadn't stopped contributing to his master's systems, Tarski had—in the spirit of the Warsaw mathematical *milieu*—started using a variety of methods that, when applied to logic and the foundations of mathematics, were ultimately to distance him from Leśniewski's whole approach more than did set theory alone.

First, between 1926 and 1929, in his Warsaw seminar,[8] Tarski worked on and obtained important results in what constituted nearly the whole of model theory at that time. For instance, he worked on categoricity, obtaining what is known now as the upward Löwenheim–Skolem theorem, and on results leading to the notion of *elementary equivalence*; in 1931 he developed a decision procedure for elementary algebra and geometry (published only in 1948) which applied the technique of quantifier elimination originally developed by Löwenheim, used by Skolem, and exploited by the American Postulate Theorists, in particular C. H. Langford.[9]

My mention of model theory in the previous paragraph needs qualification. Model theory, a discipline that Tarski himself helped to establish, did not yet exist in name at that time: Tarski was to use the word 'model' only later, in 1935–6; and the semantic notions involved in the works considered by Tarski were used informally (Mancosu, Zach, and Badesa, 2008, chapter 8).[10] Nonetheless, the results just quoted belong to model theory in today's sense, i.e. a formal study of the relationship between a language and its interpretations (Mancosu, Zach, and Badesa 2007, 117–18). Importantly for our comparison, in Leśniewski's work model-theoretical considerations are entirely absent.

Secondly, Tarski's interests in logical calculi went far beyond Leśniewski's systems and did not remain limited to classical systems or to the axiomatic method. Actually, by 1930 Tarski's main interest was not the logical systems he was investigating but the very conceptual framework in which the investigation was carried out (Blok and Pigozzi 1988, 40). His work with Łukasiewicz on sentential calculi had this general character, and it included study of many-valued logics as well as the use of the method of matrices, a *metamathematical*, in particular, algebraic, method for the definition of a logic which provided an alternative to the usual axiomatic method and was of broader application, being applicable, for instance, to logics that are not *a priori* finitely axiomatizable (Blok and Pigozzi 1988, 42).

Metamathematics, as it had emerged in the work of Hilbert, was the investigation of logical or epistemological questions concerning logical or mathematical structures or methods with the aid of mathematical tools.[11] Leśniewski, in contrast to Tarski, never did any metamathematics in this sense. Metatheory (metalogic) he certainly did, but he was a logicist in the fashion of Frege and Russell, and nothing like the use of algebraic methods in metatheory was near to his thinking about logic. Besides, not

---

[8] Feferman and Feferman 2004: 73 refer to "the 'exercise sessions' for the seminar at the University of Warsaw led by Jan Łukasiewicz."

[9] For more on this, see Vaught 1974, 159–63; Vaught 1986, 869–70; Blok and Pigozzi 1988, 43–5; Mancosu, Zach, and Badesa 2008, 132–3.

[10] On the history of model theory see also Badesa 2004.

[11] On Tarskian metamathematics see the illuminating Sinaceur 2001.

only Leśniewski's method remained always strictly axiomatic, but Leśniewski considered Łukasiewicz's many-valued logics useless in science, and the third value unintelligible.[12]

When in 1930 Tarski went to Vienna for the first time, he gave three lectures to Karl Menger's Mathematical Colloquium, all devoted to the very topics that, as we just saw, were, among his interests, the most remote from Leśniewski's ways: set theory, fundamental concepts of the deductive sciences in terms of the consequence relation (metamathematics with a model-theoretical approach), and work on the sentential calculus including Łukasiewicz's three-valued logic (Feferman and Feferman 2004, 81).

It is worth mentioning that Leśniewski's uncompromising stance was rather the exception in the Lvov-Warsaw School. For, generally speaking, the Lvov-Warsaw School at its zenith was marked by a liberal attitude towards the use of all admissible mathematical methods, non-constructive ones included, and it was not committed to any particular philosophical position. In this, the spirit of the school was similar to that which prevailed among the Warsaw mathematicians (Woleński 2003, Duda 2004, 293). Leśniewski's spirit was quite different. In particular, he never approved of the emphasis on ends over means typical of the Warsaw mathematicians, and his commitment to nominalism and to a peculiar form of constructivism remained extreme.

ONE'S PARENTS' CLOTHES[13]

We know that at a certain point Leśniewski and Tarski grew apart. While nothing points to a specific episode as the cause, the break-up was radical and involved their personal relationships as well. It is not easy to ascertain when the problems began, but in a letter to Twardowski from September 1935, Leśniewski wrote:

In connection with a series of facts in recent years [...] I feel a sincere antipathy towards Tarski.[14]

Which "facts in recent years"? The whole matter remains to a considerable extent speculative for lack of sources. There is little doubt, though, that, as to the personal aspect of the story leading to the break-up, a major role was played by Leśniewski's anti-Semitism.[15] As to the intellectual aspect, besides the general circumstance that, as we saw in the previous section, by 1930 Tarski's main interests had

---

[12] See Łukasiewicz, Smolka, and Leśniewski 1938. As to metatheory and metalogic in Leśniewski: there is no difference between the two insofar as Leśniewski's metatheory applies only to logical theories. Mereology, the only extra-logical theory Leśniewski built, though based on a logical basis whose development is ruled by a formal metatheory, requires no additional rules.

[13] After John Bayley: Wordsworth's poems "are like one's parents' clothes—always out of fashion," quoted from Clive James' review of *The Power of Delight*, TLS May 27 2005, 4.

[14] Leśniewski to Twardowski, September 8, 1935; full translation by A. O. V. LeBlanc on the *Polish Philosophy Page*, <http://www.fmag.unict.it/~polphil/PolPhil/Lesnie/LesnieDoc.html#Leśniewski>.

[15] The letter, also quoted in Feferman and Feferman 2004, 103 and Woleński 1995, 68–9 has an openly anti-Semitic content. See also Feferman and Feferman 2004, 41–2.

diverged considerably from Leśniewskian orthodoxy, four specific facts from around 1928–9 could have provoked Leśniewski's "sincere antipathy." Before we review them one by one here below, however, we should note that speaking of an intellectual or philosophical 'break-up' is possible only on the assumption that Tarski had previously been some sort of faithful Leśniewskian. As I shall point out, and as we have already begun to see, there is good reason to doubt this.

Fact number one. In 1928 both Leśniewski and Łukasiewicz resigned from the board of *Fundamenta Mathematicae*, the journal of the Warsaw mathematical group. The journal was quite a novelty for the times as it was devoted exclusively to set theory, the foundations of mathematics, and connected mathematical fields, and represented the rather unique situation of fruitful collaboration between logicians and mathematicians to be found in Warsaw (Kuratowski 1973, 32–9). Sierpiński had been the editor-in-chief since 1920 together with Mazurkiewicz, and Leśniewski and Łukasiewicz belonged to the board as responsible for the development of mathematical logic and the foundations of mathematics. In 1927, in the third chapter of *On the Foundations of Mathematics*, the work in Polish with which he broke an eleven-year silence, Leśniewski voiced his opposition to set theory by attacking the notion of (distributive) class, as "an object 'devised' by logicians for the annoyance of many generations" (Leśniewski 1927–31, 200; trans. 219). Among the mathematicians inventing "objects that do not exist" Leśniewski mentioned Sierpiński. Sierpiński was in good company, as Leśniewski also criticized Hausdorff, Dedekind, Schröder, Zermelo, Fraenkel, Whitehead, and Russell, and the much admired Frege—for Leśniewski was, despite his "best efforts in this direction, unable to understand" what Frege's 'extension of a concept' meant (Leśniewski's 1927–31, 193; trans. 211).[16] But Sierpiński was the only colleague in that company, and it has been conjectured that Leśniewski's words ignited the *Fundamenta* fight. As Woleński tells the story, when Leśniewski submitted his long article on Protothetic in *Fundamenta* (Leśniewski 1929), Sierpiński made "some very critical and sarcastic comments" on it (Woleński 1995, 67). The fight ended with Leśniewski resigning from the board of *Fundamenta* and withdrawing the second part of the 1929 article. Łukasiewicz joined Leśniewski out of support, and the action resulted in a rupture between the Warsaw mathematicians and the Warsaw logicians.[17] Tarski did not take a public stance on the matter, but we can be confident that in spirit he sided with the mathematicians (Feferman and Feferman 2004, 41).

Fact number two. The results contained in Tarski's *opus magnum, The Concept of Truth in the Languages of Deductive Sciences*, which appeared in 1933, were already completed in 1929 "in significative part" and presented in two lectures in 1930 (Tarski 1933, 3, note 2; Tarski 1935, 7 note 3; Tarski 1956, 154 note 1).[18] In

---

[16] On this chapter and the criticism of Frege, see Sinisi 1969.

[17] See also Sundholm 2003, 122, Feferman and Feferman 2004, 41. The continuation of the 1929 article (§12) is Leśniewski's 1939.

[18] The lectures were published in Polish in 1930–1 as Tarski 1930/31. From the footnote just quoted in the text it is apparent that the whole monograph was written before the 1930/31 report appeared. A report of a talk in German containing the main results of the monograph appeared as Tarski 1932.

§4 Tarski wrote that Leśniewski's system—though the only complete system of mathematical logic known to him, formally impeccable and a dream of precision—was "an extremely thankless object for methodological and semantical investigations."[19] Now, since methodological and semantical concerns inform the whole body of his work, Leśniewski must have not liked this statement very much.[20] What Tarski's words reveal is, in fact, a fundamentally different attitude toward the *way* in which methodological and semantical research was to be done. Note that, more generally, Tarski also indicates 1929 (in a footnote of his chapter on definability from 1931) as the year in which he obtained the cluster of results in metamathematics to which *The Concept of Truth* belongs—the (model-theoretical) general method of reconstructing a number of metamathematical notions (including definability, truth and universally valid propositional function) in mathematics (Tarski 1931, 211 note 2).[21] And note, *en passant*, that in that very chapter Tarski no longer mentions Leśniewski's Ontology as a possible foundation on a par with the theory of *Principia Mathematica* or Zermelo's 1908 set theory, as he had done in the 1926 chapter with Lindenbaum; he works with a simplified version of the theory of *Principia* (1931, 213; 1956, 113).

Fact number three. In 1929 Leśniewski attacked Zermelo's set theory again, this time on the basis of the fact that his "architectonically refined construction" lacked intuitive foundation (Leśniewski 1929, 6; trans. 413). In the same year Zermelo gave nine lectures in Warsaw, during which he presented models for his (improved) axiomatization of set theory. In the *Postscript* for the German translation of the *Concept of Truth* (§7, April 13, 1935) Tarski says he is no longer convinced that the Leśniewski-inspired theory of semantical categories he adheres to in the body of the monograph has a privileged link with our intuitions regarding the meaningfulness of a scientific language, and he holds that it makes sense, instead, to see what happens when one takes into account type-free languages. In particular, Tarski calls Zermelo's type-free set theory as a "much more convenient and actually much more frequently applied apparatus for the development of logic and mathematics" (Tarski 1935, 397 note 106, 1956, 271 note 1).[22] This fact is relevant not only because Tarski's passage on semantical categories was directed literally against one of Leśniewski's firmest convictions about the language of logic but also because type theory is an integral part of logicism, Leśniewski's particular brand of the latter aside. As Tarski himself pointed out much later, for mathematics to be reducible to logic, the universe must be that of Russell-Whitehead type theory, with membership as a defined logical notion (or, *mutatis mutandis*, the language equipped with Leśniewski-like semantic categories, we might add on our part); if, by contrast, the universe is that of set theory, with the membership relation as an undefined primitive notion, mathematics does not reduce to logic (Tarski 1986*a*, 152–3).[23]

[19] Tarski 1933, 61 note 56, my translation; Tarski 1935, 328 note 56; in Tarski 1956, 210 note 2 the passage is missing and Leśniewski is not mentioned.
[20] See also Sundholm 2003, 119.
[21] This part of the footnote is missing in the English translation (Tarski 1956, 111 note 1).
[22] See also Sundholm 2003, 121–2.
[23] This in turn has immediate bearing on the notion of logicality. For Leśniewski the question of what counts as logical and what counts as extralogical had a simple answer: a theory belongs to logic

Fact number four. The whole of the first chapter of *The Concept of Truth* comes from Leśniewski's investigations on the semantic richness of natural language. At that time Leśniewski's system was still unpublished, and so was his body of research on semantic antinomies. It is quite possible that Leśniewski found the footnote in which Tarski ascribes these results to him (1933, 4 note 3; 1935, 267 note 3; 1956, 155 note 1) too little to count as proper attribution, and he must have foreseen that Tarski's name, rather than his own, was going to be associated with the results in question. This is, indeed, what happened (for the record, Leśniewski never published his own version of the results). Moreover, when he introduces the notion of metalanguage in chapter 2, Tarski does not even mention that the distinction between metalanguage and object-language had been introduced by Leśniewski.[24]

The four facts just reviewed allow us to date the problems between Leśniewski and Tarski to around 1928–9. What we might wonder now is whether the facts are evidence of any genuine philosophical break-up. As we saw in the previous section, Tarski did not stop working with Leśniewski's systems after his Ph.D. thesis in any way we might consider abrupt.[25] He just increasingly concentrated on other areas, and, at most, if one considers the nature of Tarski's results relevant to Leśniewski's systems after 1924, those systems were for Tarski just one object of study among several. This last point is connected to the second fact above, the emergence of Tarski's metamathematics, which I take to be at the source of the real theoretical clash between Tarski and Leśniewski. I will come back to this in the next section.

If we want to take the four facts to be evidence of a scientific rupture, the salient point is to ascertain whether Tarski was ever a 'Leśniewskian' to begin with. Was he? Some published remarks would appear to indicate that he was, but on closer examination these remarks provide little reason to think anything more than at times Tarski found it in his interest to say something positive about his teacher. One example is a passage from "Fundamental Concepts of the Methodology of the Deductive Sciences" (1930) in which Tarski professes himself a disciple of Leśniewski in adhering to the latter's "intuitionistic formalism":

[. . .] no particular standpoint regarding the foundations of mathematics is presupposed in the present work. Only incidentally, therefore, I may mention that my personal attitude towards this question agrees in principle with that which has found emphatic expression in the writings of S. Leśniewski and which I would call *intuitionistic formalism*.[26]

if the grammar of its language is allowed to grow, that is, if new semantic categories—types—can be added to the language (see Luschei, 1962, 105). Tarski had quite a different view on this issue: at least in 1936, in accord with the shift expressed in the *Postscript*, he was sceptical that objective criteria of logicality could be found. Some thirty years later his view remained that the matter was not solved once for all. See Tarski 1936: 200; Tarski 2002: 188 and Tarski 1986*a*: 152–3.

[24] Tarski makes up for this only later, in 1936; on this and on various issues connected to this fourth fact, see Betti 2004, 280–1.

[25] This is contrary to what Feferman and Feferman (2004, 102) suggest.

[26] Tarski 1956, 62 with a few changes with respect to the German version (see 1930*a*, 363). Tarski had a tendency to edit passages as works went through translation, and his remarks on Leśniewski are particularly prone to this treatment. For example, in Tarski 1956 some passages about Leśniewski's systems are removed or changed, but not all. Some, but not all, removals regarded

Sundholm interprets this as meaning that at that time Tarski was still "true to his Leśniewskian calling" (2003, 116).[27] But both the letter of this passage and the context in which it appeared point elsewhere. Tarski's claim that his investigations in metamathematics are neutral with respect to this or that philosophical position as to the foundations of mathematics does not match Leśniewski's stance in the least. Being a Leśniewskian means being heavily committed to a quite specific position on the foundations of mathematics. Therefore, Tarski could not have been both a sincere Leśniewskian and at the same time have assumed "no particular standpoint regarding the foundations of mathematics." One might speculate that after the Sierpiński affair in 1928–9, the passage above had rather the purpose to show that Tarski sided with Leśniewski after all. Alternatively, we might take the passage to be an example of Tarski's lifelong habit of professing himself to work in an area and with tools at odds with his convictions in philosophy. It is known that on one occasion Tarski called himself a "tortured nominalist," referring to his nominalistic preferences being at odds with his work in set theory (Feferman and Feferman 2004, 52). By this token, we might conclude that Tarski's personal philosophical inclinations were genuinely Leśniewskian after all, although his work was not. Yet a hiatus of this kind between philosophical convictions and practice was unacceptable from Leśniewski's standpoint. So either way, the passage is no evidence of Leśniewskian observance on Tarski's part.

Similar considerations hold for the passage in §4 of the *Concept of Truth* (1933) in which Tarski gives "little but a paraphrase" of a Leśniewskian passage on semantic categories.[28] This is far from being evidence of Tarski's siding with Leśniewski. At least in part it is mere homage to his one-time master, for both the theory and the way in which Tarski uses it have non-Leśniewskian features. First, in contrast to Tarski, for Leśniewski there are no distinct semantic categories for names of individuals and names *of classes* of individuals: in Leśniewski's Ontology, names are allowed plural reference, and singular, common (that is, those having plural reference) and empty names fall into one category, that of names (compare Tarski 1933, 67; 1935, 336; 1956, 217). Secondly, for Leśniewski all expressions except the quantifier, including thus composite expressions and not merely variables, belong to a semantic category (compare by contrast Tarski 1933, 68 note 62; 1935, 336 note 62; 1956, 217 note 1); moreover, since no variable occurs free in Leśniewski's systems, no classification of expressions based on their free variables is possible (compare Tarski 1933, 70; 1935, 339; 1956, 219). Thirdly, Tarski allows 'hypostatizations' of categories, that is, he allows—for *practical purposes*—that Leśniewskian *linguistic* types are turned into Russellian *objectual* types, so that also all *individuals* and not just all *names of individuals* belong to the same semantic category (e.g. Tarski 1933, 70; 1935, 339;

---

passages that had lost their purpose in a context in which Leśniewski himself or logicians working within his systems were not interlocutors anymore; some passages about Leśniewski became instead less laudatory (cf. note 19 above, and Tarski 1933, 69 note 65; Tarski 1935, 338 note 65; the last fragment being deleted in Tarski 1956, 218 note 2). See also note 42 below.

[27] See, however, Sundholm 2003, 125 note 36.

[28] Sundholm 2003: 117–18. The passage in question is Tarski 1933, 67; Tarski 1935, 335; Tarski 1956, 215. The similarity is with Leśniewski 1929, 14, trans. 421.

1956, 219).²⁹ Finally, and most importantly, in Leśniewski there is no place for a hierarchy of languages of different orders, with predicates applying, in the metalanguage, to expressions of a lower-order object-language (compare Tarski 1933, 93; 1935, 336; 1956, 244). There is just *language* (and metalanguage for it). There are no order restrictions, the hierarchy of categories is finite at each stage but constructively unbounded and potentially infinite, and no truth predicate is either needed or defined.

Given the non-Leśniewskian traits of Tarski's views about semantic categories in the body of the *Concept of Truth* just quoted, the remark on Zermelo in the *Postscript* added to the revised German version from 1935 (the third fact above) cannot be taken to be a sign of a sudden change of allegiance from Leśniewski to Zermelo, however profound a trace the latter might have left on Tarski during his visit to Warsaw. For one thing, Tarski's views were already too remote from Leśniewskian credos for a remark of that kind to count as apostasy. Moreover, Tarski cannot be said to have simply moved to Zermelo's side, because, as some have observed, in general Tarski actually seems to have felt more at home in simple type theory than in Zermelo–Fraenkel set theory (Simons 1987, 19–21; Bellotti 2003, 409).

What is likely is that, when republishing his masterpiece into the *lingua franca* of philosophy at the time, German, Tarski, whose international reputation was by then established, no longer cared to show deference to Leśniewski and his systems. We should not forget that a significant difference between the Polish and the German *Concept of Truth* was the context in which they were published. The remarks that Tarski had on Leśniewski had made sense in the 1933 Polish original, but they made far less sense in the 1935 German translation. Leśniewski was a central, formidable figure in Warsaw, but he was barely known internationally. In particular, by the mid-thirties he was one of the few Poles to refuse contact with the other most important European centre of action for scientific philosophy—Vienna. While the 1935 translation to which the *Postscript* was added was in progress, Tarski was already on his second stay in Vienna,³⁰ a stay during which, famously, he explained his theory of truth to Popper and Carnap. In the same year Tarski took part in the Unity of Science Conference in Paris. Whereas many Poles—Ajdukiewicz, Kotarbiński, Chwistek, Kokoszyńska, Zawirski, Hosiasson-Lindenbaum, Jaśkowski, and Lindenbaum—joined Tarski, Leśniewski did not. When Neurath had urged him to submit a chapter for the conference, Leśniewski declined for practical reasons (and his "slow work method").³¹ Likewise, when Neurath invited him to take part in the Organizing Committee for the 1937 Unity of Science Congress, Leśniewski wrote that despite several points of contact between his thought and that of some exponents of the Unity

---

²⁹ This occurs even in Tarski 1929. A similar 'incorrectness' is Tarski's *very* non-Leśniewskian identification of—we would say now—tokens of the same type, cf. Tarski 1933, 5 note 5; Tarski 1935, 269 note 5; Tarski 1956, 156 note 1. For a survey of the background logic used by Tarski, cf. Mancosu 2006: 245 note 10.

³⁰ Kokoszyńska to Twardowski from Paris, July 22, 1935, Kazimierz Twardowski Archives, Instytut IFiS PAN, Warsaw.

³¹ Leśniewski to Neurath, August 18, 1935, Wiener Kreis Archive, Rijksarchief Noord-Holland, Haarlem.

of Science group, he did not "feel by any means close enough to the entire group to be able to belong to its official representative organization."[32] As far as we know, this was Neurath's last attempt to involve Leśniewski in the Unity of Science movement.

Thus in general I do not think that Tarski underwent anything like a conversion from a Leśniewskian past, and in any case I do not think that any abrupt conversion took place between the Polish and the German version of the *Concept of Truth* in particular. It seems to me, rather, that the whole story was more, from Tarski's point of view, a fight for freedom from a 100 percent genius master, one whose commitment to a radical *philosophical* position was, for an extraordinarily gifted and ambitious *mathematician*, very much in the way. And as far as Tarski's career in Poland was concerned, kind remarks on Leśniewski and his systems would—one might speculate—do no harm, especially if they, indeed, conveyed genuine "personal" convictions as well.

Logicism, logicality, and metatheoretical research on the (proper) foundation of mathematics in the most general sense were simply among Tarski's genuine concerns: these topics were, of course, Leśniewski's, and the way in which Leśniewski thought about them was, no doubt, influential upon Tarski. But what Tarski wanted to do, and did with them was to have his own go at them, one that took him in another direction from Leśniewski.

Leśniewski's personal aversion might well have been concretely prompted by one in particular or more of the four facts mentioned above; from the purely theoretical point of view it was Tarski's development, in 1929, of his metamathematical method—a circumstance linked to the second fact—that set Tarski fully at odds with Leśniewski. To see this we need to broaden the perspective a little, by taking into account an important external factor: the *Zeitgeist*.

LEŚNIEWSKI, TARSKI, AND THE CLASSICAL IDEAL OF SCIENCE

The split between Tarski and Leśniewski was not just both personal and theoretical, it was also embedded in larger historical developments. According to Sundholm, in order to understand the relationship between Leśniewski and Tarski properly it is necessary to be aware of a tension between two paradigms: the "logic-in-use tradition of Frege" and "the metamathematical tradition of Hilbert" (Sundholm 2003, 114). There is some truth in this, but I doubt that talk of *two* "paradigms" in this context is helpful. From what I say in the rest of this chapter it follows that there was rather *one* paradigm that underwent modification—if talk of paradigms is in place at all. Suppose now we avoid the talk of paradigms, stick to a quite general and neutral

---

[32] "Nun aber, obwohl ich ziemlich viele Berührungspunkte finde, die zwischen meinen theoretischen Tendenzen und dessen von einzelnen Vertretern der Gruppe der Einheit der Wissenschaft bestehen, fühle ich mich jedoch bisher keineswegs dieser ganzen Gruppe nahe genug um ihrer offiziellen repräsentativen Organisation angehören zu können." Leśniewski to Neurath, July 14, 1937, Wiener Kreis Archive, Rijksarchief Noord-Holland, Haarlem.

formulation and put things this way: Leśniewski belonged to *The Old* and Tarski to *The New*. What is meant by *Zeitgeist* in this context, therefore, is the growing popularity of the New. And what we can ask now is: what, exactly, is the Old? And what, exactly, is the New?

On a previous occasion I proposed to interpret the Old as adherence to what I shall call here the Classical Ideal of Science: Leśniewski belonged to the Old because he adhered to that Ideal (Betti 2008*b*, section 1). As I will conceive of it, adhering to the Classical Ideal of Science, like Leśniewski did, means thinking that a science S worth its name must obey the following cluster of conditions:

(1) All propositions and all concepts (or terms) of S concern a *specific set of objects* or are about a *certain domain of being(s)*.
(2a) There are in S a number of so-called *fundamental concepts* (or terms).
(2b) All other concepts (or terms) occurring in S are *composed of* (or are *definable from*) these fundamental concepts (or terms).
(3a) There are in S a number of so-called *fundamental propositions*.
(3b) All other propositions of S *follow from* or *are grounded in* (or *are provable* or *demonstrable from*) these fundamental propositions.
(4) All propositions of S are *true*.
(5) All propositions of S are *universal* and *necessary* in some sense or another.
(6) All propositions of S are *known to be true*. A non-fundamental proposition is known to be true through its *proof* in S.
(7) All concepts or terms of S are *adequately known*. A non-fundamental concept is adequately known through its composition (or definition).

These seven conditions and their history are discussed at more length in De Jong and Betti 2008, where they are presented as an ideal that informed thinking about science, almost without exception, for more than two thousand years beginning at least with Aristotle's *Analytica Posteriora*.[33] Among those who took the Classical Ideal to be the proper framework one ought to follow in the shaping of logic and of any other deductive theory we find Frege, and, as I mentioned, Leśniewski.

If the Old is represented by adherence to this framework, how about the New? Is it captured by *not* adhering to it, then? The answer is not that simple, at any case not simply Yes. For the historical developments in which Leśniewski and Tarski were bound up can still be quite aptly understood in terms of the Classical Ideal of Science. This is what I will in part endeavor to show in the following two sections.

We can take the Classical Ideal as being built out of two clusters of requirements corresponding roughly to the Leibnizian distinction between *lingua characteristica* and *calculus ratiocinator*: the cluster formed by (2) and (3), concerning the order of terms and the propositions, can be seen as matching the ideal of *calculus ratiocinator*, while the cluster formed by the remaining requirements, (1), (4), (5), (6), and (7),

---

[33] Cf. the Appendix to De Jong and Betti 2008 for an account of (1)–(7) as a more suitable and neutral tool than the previous systematizations of Scholz, Beth, and Dijksterhuis. See also Betti 2008*b*, section 2. Various ancestors of the (1)–(7) framework have been set up and applied by Wim de Jong since 1986, cf. De Jong 1986.

concerning homogeneity, truth, universality, necessity, and knowledge, can be seen as matching the ideal of *lingua characteristica*. Now, as known, there exist already two frameworks in the philosophy of logic aimed at embodying Leibniz's notions of *lingua characteristica* and *calculus ratiocinator*, van Heijenoort's "logic as language *versus* logic as calculus" (van Heijenoort 1967) and Hintikka's "language as calculus *versus* language as universal medium" (Hintikka 1996). Sundholm's distinction between "languages with meaning" and "languages without use" aims at capturing the core of both these frameworks at once (Sundholm 2003, 113).[34] I prefer to understand the distinctions in terms of the Classical Ideal, and I hope some of the reasons for this will emerge in the remainder of the chapter. In brief, my proposal will be to see the New as involving a change in the *lingua characteristica* cluster of requirements and the way it is accounted for. In particular, the New involved a revision of (4), the Truth Requirement, with deep repercussions on the epistemological requirements (6–7). To what extent this amounted to a departure from the Classical Ideal might well depend on one's epistemological convictions, but I do think that, on one important understanding, the change was radical. One thing I wish to make clear in any case is that, according to my account, and in contradistinction to van Heijenoort-like accounts, on the one hand metatheoretical investigations are perfectly compatible with the Old, while, on the other, siding with the New does not necessarily mean eschewing (forms of) foundationalism.

Now, Leśniewski's systems match the Classical Ideal, and in a surprisingly strict way. This is shown primarily by the way in which the systems are actually built, as Leśniewski never addressed the Ideal itself systematically in print. He did discuss his conception of axiomatic science to a certain extent in conversations and during lectures, however, and some of his ideas on these topics can be found in a chapter by one of his students, Bolesław Sobociński, in which aesthetic requirements for well-constructed axiom systems are discussed (Sobociński 1956). The requirements described by Sobociński are informal. Note that 'informal' in the mouth of a Leśniewskian does not mean 'casual', 'easygoing', or anything of the kind, but rather: not directly encoded in Leśniewski's metatheory. The latter consists of *formal(ized)* rules (the 'directives') for adding an expression to a certain stage of development of a system (Leśniewski 1929, 76; trans. 485). The directives contain special metalinguistic terms, and are preceded by the explanations of such terms. For instance, the first directive for Protothetic §5 from 1929 says, briefly put, that you can add a protothetical definition to the system, where a protothetical definition, as stated in explanation XLIV, must meet a full *eighteen* conditions (1929, 70–3; trans. 479–81).

In the beginning of chapter 6 of *Introduction to Logic*, Tarski introduces a cluster of conditions strikingly similar, at first sight, to the Classical Ideal of Science as introduced above. Satisfaction and truth, definability, logical consequence, logical operation, axiomatizability, formalized deductive systems are not just a few concepts in which Tarski happened to take an interest and set out to analyse. They all relate to the

---

[34] Note that contrary to what these frameworks suggest, we should be wary of treating every *calculus* as a *calculus ratiocinator* insofar as the latter is, arguably, the *calculus* aspect of a deductive system formulated in a *lingua characteristica* and inseparable from such a language, cf. Korte 2008.

Classical Ideal. Satisfaction and truth relate to (1) and (4), definability relates to (2b), axiomatizability to (3a), and logical consequence to (3b). Logicality relates both to the general foundational problem of logicism and to the place of logic within a hierarchy of deductive sciences—the problem of the subject-matter of logic—so (1) again. The notion of a deductive system relates to the form of a properly formalized science *S* and thus to the entire set of conditions.

As will become clear at the end of the next section, however, in spite of appearances, Tarski's (take on the) Classical Ideal of Science differed considerably from Leśniewski's. To see this we need first of all to cast some light on the different conceptions of metatheory that Leśniewski and Tarski held. For Tarski metamathematics (the methodology of deductive sciences, or methodology of mathematics) concerns the fundamental principles "to be applied in the construction of logic and mathematics" (Tarski 1941: 117). Doubtless, in this broad sense metamathematics extends to Leśniewski's metatheory. But, as we saw, by 1929 Tarski was doing metamathematics with a model-theoretic approach, set-theoretical tools, and a much broader range of interests in logic than Leśniewski's systems. As I pointed out in the previous sections, all this, and especially the use of *mathematical* tools in metatheory, was extraneous to Leśniewski. Besides this, two fundamental differences between Leśniewski's metatheory and Tarski's metamathematics in particular deserve our attention.

The first difference is that Leśniewski's metatheory concerns only the *calculus ratiocinator* bit of the Classical Ideal, in particular (2b) and (3b) above: it tells what definitions and theorems are, and when and how one can add them to a system. It does not say anything about the *lingua characteristica* bit, in particular nothing about the Truth Requirement (4). The same holds for the informal requirements described by Sobociński, which go considerably beyond the *calculus ratiocinator* cluster (2)–(3), taking care, on one reading, of (6–7) (Betti 2008*a*, section VII). Neither in Leśniewski's practice nor in Sobociński's report on Leśniewski's views on axiomatics do we find any felt need to say what truth in a formal system is: the idea is rather that the rules ought to be formulated in such a way that the system one attains, by inscribing *true* axioms and following *correct* inference rules *is* a system of truths—that is, a system obeying Requirement (4). Thus for Leśniewski all there is to say on truth in his formal systems was summed up by Tarski—I believe—as follows:

We can try to speak like this: a sentence of a certain system is true if and only if it is a thesis of that system.

(1930/31, 4)

One manner to describe Tarski's work in semantics in the light of the Classical Ideal and by way of contrast with Leśniewski is to say that Tarski provided a fully formalized understanding of requirement (4) in metamathematical terms. As will become clearer at the end of the next section, this step was not just a completion of a task that Leśniewski had left unfinished; no, it was a step that went directly against Leśniewski's conception of axiomatics expressed in the Classical Ideal of Science, and one having momentous repercussions on the Ideal itself. Leśniewski's investigations into semantic antinomies and the semantic closure of natural language led him to design systems in which no semantic notions are allowed in the object-language. But Leśniewski did not allow semantic notions in the metalanguage either. In Leśniewski's work there is no separation of the syntax and semantics of the object-language in the fashion we know it after Tarski: at each stage, a Leśniewskian system is a syntactico-semantic unity.[35]

The second difference between Leśniewski's metatheory and Tarski's metamathematics is that Leśniewski's metatheory is geared *exclusively* towards the proper construction of Leśniewski's *own systems*. Logic, for Leśniewski, is True and One: he did not strive towards a formal metatheory of utmost generality in order to capture all possible deductive systems, including non-classical ones. By contrast, if there is anything left at all to be True and One in Tarski, this is at most general metamathematics itself, not the sciences ruled by it. In 1928 Tarski showed that almost all basic concepts of metamathematics as he conceived of it can be defined in terms of *sentence* and *consequence*, and he gave axioms for the consequence relation *itself* (1930b, 22–4; 1956, 30–2). In fact, Tarski's definition of a deductive system is broad enough to include metamathematics, so the latter turns into a science *S* itself:

> The analysis and critical evaluation of *methods applied in practice in the construction of deductive sciences* ceased to be the exclusive or even the main task of methodology. The methodology of the deductive sciences became a *general theory of deductive sciences in an analogous sense as arithmetic is the science of numbers and geometry is the science of geometrical configurations*. In contemporary methodology we investigate deductive systems *as wholes* as well as the sentences which constitute them.
>
> (Tarski 1941, 138, my emphasis)

Forcing things a little, the words in italics in the passage just quoted can be seen as summing up the differences with Leśniewski's metatheory. First, Leśniewski saw system-building as being the main or sole task of methodology, and, secondly, he did not conceive of metamathematics as a deductive science itself. These differences are related and presuppose, in fact, a difference in the conception of what a deductive system is. Tarskian metamathematics as a science depends on giving a formalization of the requirements (1) and (4) of the Classical Ideal in model-theoretical terms, and on giving a semantical treatment in model-theoretical terms of the notion of 'following logically from', related to requirement (3b) of the Classical Ideal. This, in turn, is possible only if deductive systems are objects of investigation suitable to this end, that is, if they are set-theoretical objects. So the difference lies in the conception of deductive systems *as* set-theoretical *wholes* in this sense. Leśniewski's nominalistic systems are not such wholes: they are collections of inscriptions actually jotted down by someone, that is, of spatiotemporal tokens, they *grow* constantly and so does, in principle, their vocabulary. For this reason they were not apt for Tarski's metamathematical investigations in *The Concept of Truth*, for they were not even deductive systems in his sense.

---

[35] No syntax–semantics confusion ensues, though. Leśniewski had a perfectly clear and careful idea of the distinction between syntactic and semantic, and still operated consistently with notions having both aspects, like that of semantic category. Cf. Luschei 1962, 90 *and ff.*

THE AJDUKIEWICZ CONNECTION

Leśniewski had a clear idea of what the task of a logician was: building axiomatic systems according to the Classical Ideal of Science. On the basis of what we saw in the previous sections we can safely say that Tarski was, by training, utterly conversant with this way of doing logic and with the Classical Ideal itself. Still, Leśniewski, as we have seen, had no role in motivating Tarski's clarification of semantical notions in the axiomatic context. But, then, who or what pushed Tarski in this direction? The question, which has been raised in a recent chapter by Solomon Feferman (this volume), has to my mind a brief answer: Ajdukiewicz. An especially important role was played by Ajdukiewicz's *habilitation* dissertation, *From the Methodology of Deductive Sciences* (1921), in which the analysis of satisfaction, truth, definability, and logical consequence in precise terms was presented as important to the needs of axiomatics. As we shall see, shedding light on Ajdukiewicz's contribution will help us shed light on the content of that which I have called 'the New.'

The connection between Ajdukiewicz and Tarski has been already mentioned by Tadeusz Batóg (1995), and, with particular reference to axiomatics, by Paolo Mancosu:

> It can safely be asserted that the clarification of semantic notions was not seen as a goal for mathematical axiomatics. In 1918, Weyl gestures towards an attempt at clarifying the meaning of 'true judgement' but he does so by delegating the problem to philosophy (Fichte, Husserl). An exception here is Ajdukiewicz (1921), who however was only accessible to those who read Polish. Ajdukiewicz stressed the issues related to a correct interpretation of the notions of satisfaction and truth in the axiomatic context. This was to leave a mark on Tarski, who was thoroughly familiar with this text.
>
> (Mancosu, Zach, and Badesa 2008, 134)

A number of circumstances support the hypothesis that it was Ajdukiewicz who inspired Tarski. Tarski quotes Ajdukiewicz's dissertation in his truth monograph (1933, 87 note 78; 1935, 359 note 78; 1956, 237 note 1). On April 4, 1921 Tarski gave a lecture (probably his first public lecture) on Ajdukiewicz's book, entitled "On the notion of demonstration (in response to the dissertation of K. Ajdukiewicz)" (Jadacki 2003, 115; trans. 143). Tarski wrote that the deduction theorem, published in 1930, was in fact formulated in the 1921 lecture in connection with ideas found in Ajdukiewicz's book (Tarski 1930*b*, note to page 24; 1956, note to page 32). In addition to these matters of citation, aspects of Tarski's thought can be explained by Ajdukiewicz's influence. In particular, and most importantly, Tarski's notion of consequence, is, as known, similar to the notion of *Ableitbarkeit* given by Bolzano (1837). Indeed, the similarity is so striking that the question of what precisely the historical connection between Bolzano and Tarski was has been a favorite Bolzanological theme.[36] The question is answered by turning to Ajdukiewicz 1923, a popular

---

[36] For a comparison, see Siebel 2002, 590 ff.

textbook in which he gives a semantical formulation of consequence similar both to Bolzano's and to Tarski's (Batóg 1995, 55–6):

> Formal implication is a relation between propositional functions. One may define it in the following way: $\phi(x)$ is formally implied by $f(x)$ if, for every possible substitution of some value for a variable (or variables) x, either $f(x)$ is false or $\phi(x)$ is true. This formal implication is—as it seems—the source of the *common notion of consequence*. One may try to define it in such a way: b follows from a if there are propositional functions $f(x)$ and $\phi(x)$ (they may contain more than one variable) such that $\phi(x)$ is formally implied by $f(x)$ and after the substitution of some value for a variable (or variables) $f(x)$ becomes the statement expressing the proposition a and $\phi(x)$ becomes the statement expressing the proposition b. (as quoted by Batóg 1995: 56, my emphasis)

The central aim of Ajdukiewicz (1921) is the analysis of "the meaning of the expression 'exists' in the deductive sciences." The third and last part of the book presents Ajdukiewicz's proposal, prepared for by the first two parts ('I. The concept of proof in the logical sense (methodological draft)', 'II. On consistency proofs of axioms'). A number of points in this work are relevant for the connection with Tarski. A thorough analysis of all of them would require a separate chapter: in the few pages left I shall deal only with three, of which the third is the most relevant for my overall purpose (which says nothing on the intrinsic importance of the other two, and of the points I can barely mention or cannot mention at all).

First, as is well known, there has been a lively debate on what Tarski meant by the 'common concept' of consequence. Recently, Ignacio Jané has pointed out that the concept of consequence Tarski wanted to capture was the one common *in axiomatics* (Jané 2006).[37] The Ajdukiewicz connection I bring in here is the historical confirmation that this is indeed the case. Particularly important in this connection is, of course, Ajdukiewicz 1923's semantical formulation of the "common concept of consequence" that I just quoted above, but as far as I know it is Ajdukiewicz 1921 that provides a link with the axiomatic context. The concept of existence that Ajdukiewicz wants to clarify in his 1921 was the concept in use *in axiomatics*, and the same holds for the other concepts he discusses, including truth and logical consequence: neither real existence in the sense in which we say that lighthouses, thoughts or planets exist, nor finding out connections between this notion and that of axiomatics is what interests him. For, Ajdukiewicz claims, it is doubtful whether 'exists' as it is used in deductive theories has anything in common with its meaning in everyday language (Ajdukiewicz 1921, 46; trans. 33). The same holds for the other concepts he discusses.

This brings us to the second point. Ajdukiewicz's analysis of the notion of existence leads him to attempt the clarification of a number of other notions relevant to "current" axiomatics, including logical consequence and truth on the basis of satisfaction (truth and satisfaction "two concepts, usually identified" (1921, 56; trans. 40)). His interest in the notion of existence is motivated by consistency proofs for axiomatic systems, and it revolves, in fact, around what we would now call a model of a theory.

---

[37] As to which sort of axiomatics is meant here, see the third point below. Jané 2006 does not take into account the specific context of Polish axiomatics, and he makes no mention of Ajdukiewicz.

In the second part of the book, a proof of consistency of a theory is reduced to finding an *example* which does not verify at the same time two mutually contradictory propositional functions (or, which is the same, no such pair is *satisfied* by it). Ajdukiewicz sums up the second part of his book as follows:

Considering that, on the one hand, in the existence of the given object [the example, *ab*] we see a warrant for its not satisfying contradictory sentences, while—on the other hand—in consistency proofs this warrant is seen in the fact that the example satisfies the axioms of another system [as in Hilbert's proofs, *ab*], we arrive at the conjecture that existence (in the sense in which this term is used in deductive theories) consists in the object's being an element of the domain of some deductive theories whose consistency is assumed. (1921, 42; trans. 32, with changes)

Ajdukiewicz claims that attributing existence to objects amounts to giving the conditions of a *good definition*, that is, to find out to which definitions there correspond existing objects (1921, 55; trans. 40). He uses the notion of definition and that of *logical following* to define the notion of *satisfaction*, on the basis of which, in turn, *truth* and *domain of a theory* are defined. The link with Tarski seems apparent, but note that Ajdukiewicz gives all of these notions a *syntactic* analysis. For instance:

[*Truth*] "The proposition $f(P)$ is true means—there exists an object P satisfying $f(x)$." (1921, 56; trans. 40)

[*Satisfaction*] "Object P satisfies the propositional function $f(x)$ means—$f(P)$ follows from the definition of P." (1921, 56; trans. 40)

[*Logical consequence* ("following in purely logical sense")] "b is a logical consequence of a iff '$a \supset b$' is a logical theorem (or an axiom)." (1921, 19; trans. 19)[38]

The reason why Ajdukiewicz avoids giving a semantic analysis to these notions, in particular why he avoids employing the concept of truth in their definition, is that he thinks that the concept of truth is "not so clear" with reference to the sphere of objects he investigates, "however clear it is when applied to statements referring to the real world" (1921, 47, trans 34). In particular, Ajdukiewicz thinks it is inapplicable to the view of axiomatics he holds.

This connects to the last point I shall mention, which is the most relevant for our understanding of what I called 'the New.' As Hilbert and Bernays (1934, 1–2) put it, axiomatics comes in two kinds: *contentual* and *formal* (see also Jané 2006, 17). Let us take contentual axiomatics to be captured by the Classical Ideal of Science mentioned in the previous section: in this view—which was, as we saw, Leśniewski's—axioms express true propositions involving primitive terms (or: true propositions about the entities which primitive terms are about). Formal axiomatics is a view of axiomatics in which, instead, the (specific) terms of a (non-logical) theory are mere placeholders, that is, in fact, variables, so that axioms are *not propositions, but propositional functions*. This view was adopted at that time, among others, by Hilbert, with whom Ajdukiewicz had studied in Göttingen in 1913–14, Mario Pieri (quoted by Ajdukiewicz), and the so-called American Postulate Theorists

---

[38] On this definition and the discovery of the deduction theorem, see Batóg 1995, 57–8.

(Mancosu 2006, 240–4; Jané 2006, 19 ff). As Mancosu and Jané have explained, this was also Tarski's view (Mancosu 2006, 243; Jané 2006, 30–5).

The view taken by Ajdukiewicz is an extremely formalistic one. "To the symbols" occurring in axiomatics, says Ajdukiewicz, "we do not ascribe *any meaning*" (1921, 11; trans. 13).[39]

> Symbols of deductive theories are [...] symbols not by 'meaning' or 'denoting' anything, but by playing a definite 'role', by occurring in strictly defined relations.[40]
>
> It is customary to say that the axioms of the formalized, deductive sciences are judgements or propositions, propositional functions etc. Our own view does not allow to say so. We associate meaning with the word 'proposition'; a proposition must assert or deny something; [...] A symbol is a proposition if among its components there is an element which has intuitive sense and which expresses assertion or denial. Since among the components of formalized axioms no such element with intuitive meaning occurs, no axiom may be regarded as a proposition in the intuitive sense [...] of the word [...] [axioms] are but certain combinations of signs so pronounced so that *they sound like propositions* [...] to axioms we cannot ascribe truth or falsity unless in some metaphorical sense. [3.] since there is no place for the concept of truth in formalized deductive theories, there will be no place for the concept of evidence either, the elimination of which is welcome since every evidence is subjective and relative.
>
> (Ajdukiewicz 1921: 12–13; Eng. trans. 14, reproduced here with changes; emphasis in the original)

The view just sketched reflects what Ajdukiewicz calls (absolutely) *abstract* deductive sciences. He distinguishes between abstract and *applied* deductive sciences on the basis of whether the logical primitives contained in the axioms are considered to be meaningful symbols or not. Since the axioms of every deductive science contain logical symbols, whenever those symbols are endowed with an intuitive meaning (that is, *applied logic* is at issue), the deductive science at issue is applied. In applied sciences, however, axioms are still not propositions, but propositional functions. As such, though the axioms are neither true nor false, they can become true or false depending on the various *interpretations* of the variables appearing in them (1921, 20; trans. 20).

Ajdukiewicz's applied theories correspond to Hilbert and Bernays' formal axiomatics as characterized above. As an example of theories of this kind, Ajdukiewicz mentions the system of geometry for which Hilbert proves consistency:

> Let $A(X)$ denote the logical product of the axioms of geometry whose consistency is to be shown. These axioms are not definite propositions but are susceptible to various 'interpretations' i.e. they are propositional functions defined for a system of variables such as 'point', 'straight' etc. This whole system of variables is represented by the letter X in the symbol $A(X)$. The totality of objects signified by it forms the 'domain' of geometry. The domain

---

[39] Reported here with changes; emphasis in the original.

[40] Ajdukiewicz points out that despite their lack of meaning, their not being meaningful like expressions in everyday language, such inscriptions are still symbols and not ornaments, because, like pieces in a chess game, they play a definite role (1921, 11; trans. 13). Here is, thus, another source for Tarski's pointing out to Neurath that the characterization of pictures of sentences as 'ornaments' was not an original Viennese formulation, see Tarski 1992, 26. The reference given explicitly by Tarski is a lecture by Łukasiewicz from December 8, 1924 (Łukasiewicz 1925).

of geometry is thus a set of variables whose values are again sets, relations etc. The axioms are, therefore, neither true nor false but turn into true or false if values are substituted for all variables.

(1921, 29; trans 23–4, with changes)

The reason why Ajdukiewicz defines truth and satisfaction syntactically as we saw above is that he wants these concepts to be applicable in abstract theories. He says:

> We cannot disregard the difficulty involved in the common (*potoczny*) definition of truth of a proposition, respectively satisfaction, which sounds: a proposition is true to which something in reality corresponds, namely that which is asserted by the proposition. The nature of this correspondence alone is not easy to grasp, though this definition may be appropriate to define the truth of empirical propositions.
>
> If, however, we want to apply this definition to theorems of deductive theories, we shall meet great difficulties looking for the reality in which the correlates of aprioristic sentences are to be found. Great difficulties arise already for deductive theories that are, so to speak, semi-applied, such as e.g. a geometry in which a straight line is defined as any object satisfying such and such axioms. There are no such objects [...] in the normal real world; even if we grant being to Euclidean straight lines in the real world, we would have to refuse it to Riemannian straight lines. Perhaps this difficulty may be side-tracked by assuming that it is only relations which are asserted by the theorems of aprioristic sciences and that relations may exist even if no objects exist between which they hold. Nevertheless, there are insurmountable difficulties in interpreting the definition of truth mentioned above with reference to the theorems of absolutely pure deductive theories. [...] it is only in a metaphorical i.e. improper sense that we may speak of truth in deductive theories.

(1921, 55–66, trans. 40, with changes)

After this passage, Ajdukiewicz's definitions follow.

Now, a possible reaction prompted by the passage just quoted is this: what if one focused on applied theories (since "for [abstract] sciences the problem [...] has no relevance, it is not even meaningful" (Tarski 1933, 17; 1935, 281; 1956, 166)), found a way to water down the concept of 'reality', and gave a mathematical treatment of the result? Then truth and satisfaction in semantic terms would not be problematic anymore, and the "common definition" of truth as correspondence could be given new life. And this is, indeed, what Tarski did.[41]

The conclusion I want to draw from all this is that the New to which Tarski adhered, influenced in this by Ajdukiewicz among others, amounts to formal, as opposed to contentual, axiomatics. The interesting thing now would be to see whether we can say more about the difference between the Old and the New in terms of the Classical Ideal of Science in a way relevant for our comparison of Leśniewski and Tarski. A thorough account would exceed the scope of this chapter; thus here I shall give just a sketch of how I think such an account should be developed.

---

[41] Recall, again, that in 1923 Ajdukiewicz gives a *semantic* definition of logical consequence. In the light of what I say here, the reason why in 1923 Ajdukiewicz gives a semantic definition instead of the syntactic one he gives in 1921 seems to me to be not that in popular textbooks one is less afraid of paradoxes, but rather that Ajdukiewicz 1923 focuses on applied theories (against Batóg 1995, 56).

In formal axiomatics conditions (2) and (3) of the Classical Ideal of Science still hold. On suitable construal, (5) and (1)—necessity and homogeneity—can hold as well, in particular if (5) is taken to express a minimal take on aprioricity as following from most general laws, and (1) is just taken to mean that we must be able to indicate the subject-matter of a science. The difference between formal and contentual axiomatics would consist, then, in formal axiomatics' eschewing the semantic requirement (4), together with the epistemological ones, (6) and (7). Taking axioms to be propositional functions goes, clearly, against (4). It goes against both (6) and (7) as well, since it does not seem possible to say, in this case, that one *knows*, *grasps* or *has epistemic access to* all terms or concepts of the science in question or that one knows that its propositions are true. One might assume the axioms to be true, but to assume that an axiom is true and to know that is true are two very different things (however one can come to know that that axiom is true). For in what sense would the primitive terms or concepts be known or grasped, since in formal axiomatics we take the axioms simply to be propositional functions and the 'terms' they contain simply to be mere placeholders? We might think of construing this in such a bromidic way that it simply restates (1), that is, we mean that the domain or the field of the theory is known. But (7) doesn't just mean that one knows in this most general sense alone what a theory is supposed to be about. It means, in the case of the terms of science, that these terms must be meaningful at the outset, and this meaning graspable. Nothing of the sort remains from the perspective of formal axiomatics.

Now an appealing suggestion here might be that a Tarskian truth-definition restores content to the "terms" of a system of formal axiomatics. For one might think that the demand that terms be meaningful at the outset as found in contentual axiomatics can be satisfied by the notion of *giving an interpretation* of a formal axiomatic system. But this notion of interpretation is entirely foreign to contentual axiomatics as it presupposes, in a Tarskian framework, a quite specific view of the relation between a term and its meaning, that is, the model-theoretical one. Under this construal, systems in formal axiomatics remain, from the point of view of contentual axiomatics, empty shells. An interpreted formal system in this sense is not a system of meaningful propositions of the sort demanded in contentual axiomatics. Note that it makes no difference here whether the terms (i) *are* interpreted, or (ii) there is an *intended interpretation* for them, or (iii) they are *uninterpreted, but they are expected to be interpreted*. If 'meaning' is given by set-theoretical mappings, and the mapping can in principle be changed *ad libitum* without any question of primacy between the 'intended' or 'original' interpretation (Tarski 1937, 331–2), then we are still in formal axiomatics. All that matters to formal axiomatics is the shell game itself, not ensuring that axiomatic structures encode knowledge adequately.

One declared aim of the proponents of formal axiomatics had been, indeed, the elimination from the Classical Ideal of those epistemological concerns that had been associated, in the course of history, with faculties and epistemic processes like *imagination* and *intuition*. For these epistemological *desiderata* the meaningfulness and truth of the propositions involved was a *conditio sine qua non*. The proponents of formal axiomatics held, however, that such intuitive or imaginative elements disturbed inference processes (Jané 2006, 18–19; Mancosu, Zach, and Badesa 2008, 5). These inferentially extraneous elements were therefore to be eliminated by voiding axioms of content. For this reason, Tarski's insistence on deductive systems being systems of "meaningful sentences" should not be taken as an expression of favor for contentual axiomatics,[42] for his aim—in keeping with the *Zeitgeist*—was reducing *linguae* to *calculi*.

CONCLUSION

Leśniewski's and Tarski's general attitudes to logic, methodology, and semantics show fundamental differences. I proposed to account for this by using as a framework what I introduced as the Classical Ideal of Science, and related to it the difference between contentual and formal axiomatics. It turned out that, though Tarski seems to follow Leśniewski in adhering to the Classical Ideal, in Tarski's hands this Ideal changed dramatically in shape and intent. In particular, I observed that the establishment of Tarski's semantics dovetailed with the (then growing) tendency to expunge epistemological aspects from axiomatics, thereby setting aside the epistemological aspects of the Classical Ideal.

In his metatheory Leśniewski limited the formalization of the Classical Ideal to the parts I likened to the notion of *calculus ratiocinator*, and he concentrated exclusively on his systems. In his metamathematics Tarski extended, instead, the formalization of the same Ideal to various notions related to what I see as the semantic side, or its *lingua characteristica* parts; and, moreover, he broadened the applicability of this formalization to various deductive systems. Tarski carried out his project with a metamathematical approach in Hilbert's sense, that is, his analyses were carried out in *mathematical* terms. This had a strong impact not only on the Classical Ideal and its status, leading to the abandonment of its epistemic aspects, but also on the relationship between the Classical Ideal and the sciences obeying it, and on the relationship between sciences, in particular logic and mathematics. A major motivation for Tarski to embark on this transformation came, I claimed, from Ajdukiewicz's methodological work on formal axiomatics.

My account, I argued, is preferable to van Heijenoort–Hintikka-like accounts because it does justice both to Leśniewski's metatheoretical work and to Tarski's foundationalist leanings. Formal metatheory was fundamental to Leśniewski, for it is thanks to the care with which metatheory is formalized that the systems are paradox-free without need for 'unintuitive' or *ad hoc* axiomatic restrictions. And since metatheory and semantics are not necessarily Tarski's *model-theoretical* metatheory and semantics, Leśniewski could do both without the use of model-theoretical tools.

---

[42] In the light of what we saw in this section and earlier the following passage seems revealing: "Instead of 'meaningful sentence' we could say 'well-formed sentence'. I use the word 'meaningful' to express my agreement with the doctrine of intuitionistic formalism mentioned above" (Tarski 1930*a*, 363 fn 2; Tarski 1956, 62 fn 3).

References

Ajdukiewicz, Kazimierz (1921) *Z metodologii nauk dedukcyjnych*. Lwów, Nakładem Polskiego Towarzystwa Filozoficznego. Eng. trans. by J. Giedymin, *From the Methodology of the Deductive Sciences*, Studia Logica, 19, 1966: 9–46.

—— (1923) *Główne kierunki filozofii w wyjątkach z dzieł ich klasycznych przedstawicieli. Teoria poznania, logika, metafyzika*. Lwów, Jakubowski.

Badesa, Calixto (2004) *The Birth of Model Theory—Löwenheim's Theorem in the Frame of the Theory of Relatives*. Princeton, Princeton University Press.

Banach, Stefan and Alfred Tarski (1924) "Sur la décomposition des ensembles de points en parties respectivement congruentes." *Fundamenta Mathematicae* 6: 244–77.

Batóg, Tadeusz (1995) "Ajdukiewicz and the Development of Formal Logic." In *The Heritage of Kazimierz Ajdukiewicz*. V. Sinisi and J. Woleński (eds.). Amsterdam, Rodopi: 53–67.

Bellotti, Luca (2003) "Tarski On Logical Notions." *Synthese* 135: 401–13.

Betti, Arianna (2004) "Leśniewski's Early Liar, Tarski and Natural Language." *Annals of Pure and Applied Logic* 147: 267–87.

—— (2008a) "Leśniewski's *Characteristica Universalis*." *Synthese*, forthcoming.

—— (2008b) "Leśniewski's Systems and the Classical Model of Science." Forthcoming in *The Golden Age of Polish Philosophy—Kazimierz Twardowski's philosophical legacy*. S. Lapointe, M. Marion, W. Miśkiewicz and J. Woleński (eds.). Berlin, Springer.

Blok, Willem J. and Don Pigozzi (1988) "Alfred Tarski's Work on General Metamathematics." *The Journal of Symbolic Logic* 53: 36–50.

Bolzano, Bernard (1837) *Wissenschaftslehre*. Sulzbach, J. E. v. Seidel. Partial English translation: Rolf George trans., *Theory of Science*, Berkeley and Los Angeles, University of California Press, 1972.

De Jong, Willem, R. (1986) "Hobbes's Logic: Language and Scientific Method." *History and Philosophy of Logic* 17: 123–42.

—— and Arianna Betti (2008) "The Classical Model of Science—A Millennia-Old Model of Scientific Rationality." *Synthese*, forthcoming.

Duda, Roman (2004) "On the Warsaw Interactions of Logic and Mathematics in the Years 1919–1939." *Annals of Pure and Applied Logic* 127: 289–301.

Feferman, Anita Burdman and Feferman, Solomon (2004) *Alfred Tarski—Life and Logic*. Cambridge, Cambridge University Press.

Feferman, Solomon (2004) "Tarski's Conceptual Analysis of Semantical Notions." In *Sémantique et épistémologie*. A. Benmakhlouf (ed.). Casablanca, Editions Le Fennec [distrib. J. Vrin, Paris]: 79–108. Reprinted this volume.

Givant, Steven (1991) "A Portrait of Alfred Tarski." *The Mathematical Intelligencer* 13: 16–32.

Hilbert, David and Bernays, Paul (1934) *Grundlagen der Mathematik*. Berlin: Springer.

Hintikka, Jaakko (1996) *Lingua Universalis vs. Calculus Ratiocinator: An Ultimate Presupposition of Twentieth-Century Logic*. Dordrecht: Kluwer Academic Publishers.

Jadacki, Jacek Juliusz (2003) "Alfred Tarski w Warszawie." In *Alfred Tarski: dedukcja i semantyka (déduction et sémantique)*. J. J. Jadacki (ed.). Warszawa, Semper: 112–32 (French trans. "Alfred Tarski à Varsovie" *ibid*.: 139–62).

Jané, Ignacio (2006) "What is Tarski's *Common* Concept of Consequence?" *The Bulletin of Symbolic Logic* 12: 1–41.

Korte, Tapio (2008) "*Begriffsschrift* as a *Lingua Characteristia*" *Synthese*, forthcoming.

Kuratowski, Kazimierz (1973) *A Half Century of Polish Mathematics—Remembrances and Reflections*. Oxford/New York/Toronto/Sydney/Paris/Frankfurt, Pergamon Press and Warsaw, PWN Polish Scientific Publishers.

Leśniewski, Stanisław (1927–31) "O Podstawach Matematyki." *Przegląd Filozoficzny* 30: 164–206, 1927 (Chs. I–III); 31, 1928: 261–91 (Ch. IV); 32, 1929: 60–101 (Chs. V–IX); 33, 1930: 77–105 (Chs. VI–IX); 34, 1931: 142–70 (Chs. X–XI); Eng. trans. in Leśniewski (1992): 181–382.

—— (1929) "Grundzüge eines neuen Systems der Grundlagen der Mathematik." *Fundamenta Mathematicae* 14: 1–81. Eng. trans. in Leśniewski (1992): 410–92.

—— (1930) "Über die Grundlagen der Ontologie." *Sprawozdania z posiedzeń Towarzystwa Naukowego Warszawskiego—Comptes rendus des séances de la société des sciences et des lettres de Varsovie, Wydział III—Nauk Matematyczno-Fizycznych* 23, 1–3: 111–32. Eng. trans. in Leśniewski (1992): 606–28.

—— (1938) "Einleitende Bemerkungen zur Fortsetzung meiner Mitteilung u. d. T. 'Grundzüge eines neuen Systems der Grundlagen der Mathematik'." *Collectanea Logica* 1 (1939): 1–60. Eng. trans. in Leśniewski (1992): 649–710. The offprints of the article are dated 1938.

—— (1939) "Grundzüge eines neuen Systems der Grundlagen der Mathematik, §12." *Collectanea Logica* 1: 61–144. Eng. trans. in Leśniewski (1992): 492–605.

—— (1992) *Collected Works*, 2 vols., S. J. Surma, J. T. Szrednicki, D. I. Barnett, and V. F. Rickey (eds.), Dordrecht, Kluwer.

Lindenbaum, Adolf and Alfred Tarski (1926) "Communication sur les recherches de la Théorie des Ensembles." *Sprawozdania z Posiedzeń Towarzystwa Naukowego Warszawskiego/Comptes rendus des séances de la société des sciences et des lettres de Varsovie, Wydział III—Nauk Matematyczno-Fizycznych/Classe III—Sciences Mathématiques et Physiques* 19: 299–330.

Łukasiewicz, Jan (1925) "O pewnym sposobie pojmowania teorii dedukcji." *Przegląd Filozoficzny* 38: 134–38.

——, Franciszek Smolka, and Stanisław Leśniewski (1938) "U źródeł logiki trójwartościowej." *Filozofia Nauki* II 1994: 227–40.

Luschei, Eugene C. (1962) *The Logical Systems of Leśniewski*. Amsterdam, North-Holland.

Mancosu, Paolo (2006) "Tarski on Models and Logical Consequence." In *The Architecture of Modern Mathematics*. J. Gray and J. Ferreiros (eds.). Oxford, Oxford University Press: 237–70.

——, Zach, Richard and Badesa, Calixto (2008) "The Development of Mathematical Logic from Russell to Tarski: 1900–1935." Forthcoming in *The History of Modern Logic*. L. Haaparanta (ed.). New York and Oxford, Oxford University Press.

Moore, Gregory H. (1982) *Zermelo's Axiom of Choice: Its Origins, Development, and Influence*. New York, Springer.

Scholz, Heinrich (1936–7) "Die Wissenschaftslehre Bolzanos. Eine Jahrhundert-Betrachtung" *Semester-Berichte* (Münster i. W.) 9. Semester, Winter: 1–53. Repr. in *Mathesis Universalis* (1961), Schwabe, Basel: 219–67.

Siebel, Mark (2002) "Bolzano's Concept of Consequence." *The Monist* 85: 580–99.

Simons, Peter (1987) "Bolzano, Tarski, and the Limits of Logic." *Philosophia Naturalis* 24: 378–405. Now in *Logic and Philosophy in Central Europe from Bolzano to Tarski*, Dordrecht/Boston/London, Kluwer, 1992.

Sinaceur, Hourya (2001) "Alfred Tarski: Semantic Shift, Heuristic Shift in Metamathematics." *Synthese* 126: 49–65.

Sinisi, Vito F. (1969) "Leśniewski and Frege on Collective Classes." *Notre Dame Journal of Formal Logic* 10: 239–46.

Sobociński, Bolesław (1956) "On Well-Constructed Axiom Systems." *Yearbook of the Polish Society of Arts and Sciences Abroad/Rocznik Towarzystwa Polskiego na Obczyźnie* 6: 54–65.

Sundholm, Göran (2003) "Tarski and Leśniewski on Languages with Meaning versus Languages without Use." In *In Search of the Polish Tradition—Essays in Honour of Jan Woleński on the Occasion of his 60th Birthday*. J. Hintikka, T. Czarnecki, K. Kijania-Placek, T. Placek, and A. Rojszczak † (eds.). Dordrecht, Kluwer Academic Publishers.

Tarski, Alfred (1923) "O wyrazie pierwotnym logistyki." *Przegląd Filozoficzny* 26: 68–89. Eng. trans. in Tarski 1956: 1–23.

—— (1924a) "Sur les ensembles finis." *Fundamenta Mathematicae* 6: 45–95.

—— (1924b) "Sur quelques théorèmes qui équivalent à l'axiome du choix" *Fundamenta Mathematicae* 5: 147–54.

—— (1929) "Les fondements de la géométrie des corps (résumé)." In *Księga Pamiątkowa Pierwszego Polskiego Zjazdu Matematycznego, Lwów, 7–10.IX.1927*, Krakow: 29–33.

—— (1930a) "Fundamentale Begriffe der Methodologie der deduktiven Wissenschaften. I." *Monatshefte für Mathematik und Physik* 37: 361–404.

—— (1930b) "Über einige fundamentalen Begriffe der Metamathematik." *Sprawozdania z Posiedzeń Towarzystwa Naukowego Warszawskiego, Widział III Comptes Vendus des Séances de la Soziété des Sciences et des Lettres de Varsovie Nauk Matematyczno-fizycznych* 23: 22–9.

—— (1930/31) "O pojęciu prawdy w odniesieniu do sformalizowanych nauk dedukcyjnych." *Ruch Filozoficzny* 12: 210–11.

—— (1931) "Sur les ensembles définissables des nombres reels I." *Fundamenta Mathematicae* 17: 210–39.

—— (1932) "Der Wahrheitsbegriff in den Sprachen der deduktiven Disziplinen." *Akademischer Anzeiger der Akademie der Wissenschaften in Wien* 2, Sitzung der mathematisch-naturwissenschaftlichen Klasse vom 21. Jänner 1932.

—— (1933) *Pojęcie prawdy w językach nauk dedukcyjnych*. Warsaw, Nakładem Towarzystwa Naukowego Warszawskiego.

—— (1935) "Der Wahrheitsbegriff in den formalisierten Sprachen." *Studia Philosophica* 1: 261–405. Enlarged translation of Tarski (1933).

—— (1936) "O pojęciu wynikania logicznego." *Przegląd Filozoficzny* 39: 58–68. Page numbers refer to the reprint in A. Tarski—*Pisma logiczno-filozoficzne*—Tom I—Prawda, Wydawnictwo Naukowe PWN, Warszawa, 1995: 186–202 (ed. by J. Zygmunt).

—— (1937) "Sur la méthode déductive." In *Travaux du IXe Congrès International de Philosophie*. Paris, Hermann et Cie. 6: 95–103 (page numbers refer to the reprint in Tarski (1986b)).

—— (1941) *Introduction to Logic and to the Methodology of Deductive Sciences*. New York, Oxford University Press. Page numbers refer to the second edition (1946).

—— (1944) "The Semantic Conception of Truth and the Foundations of Semantics." *Philosophy and Phenomenological Research* 4: 341–76. Page numbers refer to the reprint in Tarski (1986b): 661–99.

—— (1956) *Logic, Semantics, Metamathematics—Papers from 1923 to 1938*. Oxford, Clarendon Press.

—— (1986a) "What are Logical Notions?" *History and Philosophy of Logic* 7: 143–54.

—— (1986b) *Collected Papers*, S. Givant and R. Mackenzie (eds.), 4 vols. Birkhäuser, Basel.

—— (1992) "Drei Briefe an Otto Neurath." *Grazer Philosophische Studien* 43: 1–32.

—— (2002) "On the Concept of Following Logically." *History and Philosophy of Logic* 23: 155–96.

van Heijenoort, Jean (1967) "Logic as Language and Logic as Calculus." *Synthese* 17: 324–30.

Vaught, Robert L. (1974) "Model Theory before 1945". In L. Henkin, J. Addison, C. C. Chang, W. Craig, D. Scott and R. Vaught (eds.), *Proceedings of the Tarski Symposium*, American Mathematical Society (Providence), 153–72.

—— (1986) "Tarski's Work in Model Theory." *Journal of Symbolic Logic* 51: 869–82.

Woleński, Jan (1995) "Mathematical Logic in Poland 1900–1939: People, Circles, Institutions, Ideas" *Modern Logic* 5: 363–405. Page numbers refer to the reprint in *Essays in the History of Logic and Logical Philosophy*, Cracow, Dialogikon, 1999.

—— (2003) "Lvov-Warsaw School." In *The Stanford Encyclopedia of Philosophy*, Edward N. Zalta (ed.), Summer 2003 Edition <http://plato.stanford.edu/entries/lvov-warsaw/>.

Zygmunt, Jan (1995) "Alfred Tarski—Szkic biograficzny." In *Alfred Tarski—Pisma logiczno-filozoficzne, Tom I, Prawda*. Warsaw, PWN: pp. vii–xx.

4

Tarski's Conceptual Analysis of Semantical Notions

Solomon Feferman

Dedicated to the memory of Robert L. Vaught (1926–2002)–fellow student, dear friend, colleague

THE PUZZLES OF "WHY" AND "HOW"

The two most famous and—in the view of many—most important examples of conceptual analysis in twentieth-century logic were Alfred Tarski's definition of *truth* and Alan Turing's definition of *computability*. In both cases a prior, extensively used, informal or intuitive concept was replaced by one defined in precise mathematical terms. It is of historical, mathematical, and philosophical interest in each such case of conceptual analysis to find out *why* and *how* that analysis was undertaken. That is, to what need did it respond, and in what terms was the analysis given? In the case of Turing, the "how" part was convincingly provided in terms of the general notion of a computing machine, and one can give a one-line answer to the "why" part of the question. Namely, a precise notion of computability was needed to show that certain problems (and specifically the *Entscheidungsproblem* in logic) are *uncomputable*; prior to that, the informal concept of computability sufficed for all positive applications. I shall argue that there was no similarly compelling *logical* reason for Tarski's work on the concept of truth, and will suggest instead a combination of *psychological* and *programmatic* reasons. On the other hand, the "how" part in Tarski's case at one level receives a simple one-line answer: his definition of truth is given in general set-theoretical terms. That also characterizes his analyses of the semantical concepts of *definability*, *logical consequence*, and *logical operation*. In fact, all of Tarski's

Expanded text of a lecture for the Colloque, "Sémantique et épistémologie," Casablanca, April 24–6, 2002, sponsored by the Fondation du Roi Abdul-Aziz al Saoud pour les Etudes Islamiques et les Sciences Humaines. I wish to thank the Fondation for its generosity in making that stimulating Colloque possible, as well as its organizers and, in particular, Dr. Hourya Benis Sinaceur for the invitation to take part in it.

work in logic and mathematics is distinguished by its resolute employment of set-theoretical concepts. However, the form in which these were employed shifted over time and the relations between the different accounts are in some respects rather puzzling. It is my aim here to educe from the available evidence the nature and reasons for these shifts and thereby to throw greater light on both the "why" and "how" questions concerning Tarski's conceptual analyses of semantical notions, especially that of truth. The main puzzle to be dealt with has to do with the relations between the notions of truth in a structure and absolute truth.[1]

THE INFLUENCE OF THE SET-THEORETICAL TOPOLOGISTS

A year ago, I gave a lecture entitled "Tarski's conception of logic" for the Tarski Centenary Conference held in Warsaw at the end of May 2001 (Feferman 2004). In that I emphasized several points relevant to the questions I'm addressing here, and will take the liberty in this section of repeating the following one almost verbatim. Namely, I traced Tarski's set-theoretic approach to conceptual analysis back to his mathematical studies at the University of Warsaw during the years 1919–24, alongside his logical studies with Stanislaw Leśniewski and Jan Łukasiewicz. Tarski's choice of concentration on mathematics and logic in this period was fortuitous due to the phenomenal intellectual explosion in these subjects in Poland following its independence in 1918. On the side of logic this has been richly detailed by Jan Woleński in his indispensable book about the Lvov-Warsaw school (1989). A valuable account on the mathematical side is given in the little volume of "remembrances and reflections" by Kazimierz Kuratowski, *A Half Century of Polish Mathematics* (1980). The grounds for the post-war explosion in Polish mathematics were laid by a young professor, Zygmunt Janiszewski. He had obtained a doctor's degree in the then newly developing subject of topology in Paris in 1912, and was appointed, along with the topologist, Stefan Mazurkiewicz, to the faculty of mathematics at the University of Warsaw in 1915. It was Janiszewski's brilliant idea to establish a distinctive Polish school of mathematics and to make an impact on the international scene by founding a new journal called *Fundamenta Mathematicae* devoted entirely to a few subjects undergoing active development.[2] Namely, it was to concentrate on the modern directions of set theory, topology, mathematical logic, and the foundations of mathematics that had begun to flourish in Western Europe early in the twentieth century.

Tarski's teachers in mathematics at the University of Warsaw were the young and vital Wacław Sierpiński, Stefan Mazurkiewicz, and Kazimierz Kuratowski; Sierpiński and Mazurkiewicz were professors, and Kuratowski was a docent. In 1919, the year that Tarski began his studies, the old man of the group was Sierpiński, aged

---

[1] Increasing attention has been given in recent years to the shifts and puzzles in Tarski's work on conceptual analysis. Most useful to me here have been the articles of Hodges (1985/86) and (2004), Gómez-Torrente (1996), de Rouilhan (1998), Sinaceur (2001), and Sundholm (2003), and the book of Ferreiros (1999), especially pp. 350–6. Givant (1999) provides a very clear account of the progression of Tarski's work over his entire career.

[2] Sadly, Janiszewski died in the flu epidemic of 1919–20 and did not live to see its first issue.

thirty-seven; Mazurkiewicz was thirty-one, while Kuratowski at twenty-three was the "baby". The senior member in the Warsaw mathematics department, Wacław Sierpiński, was especially noted for his work in set theory, a subject that Tarski took up with a vengeance directly following his doctoral work on Leśniewski's system of protothetic. Though Cantorian set theory was still greeted in some quarters with much suspicion and hostility, it was due to such people as Sierpiński in Poland and Hausdorff in Germany that it was transformed into a systematic field that could be pursued with as much confidence as more traditional parts of mathematics.

The main thing relevant to the present subject that I emphasized in my Warsaw lecture concerning this background in Tarski's studies is that in the 1920s, the period of his intellectual maturation in mathematics, topology was dominated by the set-theoretical approach, and its great progress lay as much in conceptual analysis as in new results. We take the definitions of the concepts of limit point, closed set, open set, connected set, compact set, continuous function, and homeomorphism—to name only some of the most basic ones—so much for granted that it takes some effort to put ourselves back in the frame of mind of that fast-evolving era in which such definitions were formulated and came to be accepted. Of course, some of the ideas of general topology go back to Cantor and Weierstrass, but it was not until the 1910s that it emerged as a subject in its own right. Tarski couldn't have missed being impressed by the evident success of that work in its use of general set theory in turning vague informal concepts into precise definitions, in terms of which definite and often remarkable theorems could be proved.

THE PARADIGMATIC CASE OF DIMENSION

In particular, a very interesting case of conceptual analysis in topology took place during Tarski's student days. This concerned the idea of the dimension of various geometrical objects and began with a puzzle over how to define in precise terms what it means to be a curve as a one-dimensional set. Informally, the idea of a curve that had been used up until the 1800s was that of a figure traced out by a moving point. As part of the progressive rigorization of analysis in the nineteenth century, Camille Jordan had proposed to define a curve (for example, in the plane or in space) as the continuous image of a line segment. When Giuseppe Peano showed, quite surprisingly, that the continuous image of a line segment could fill up a square in the plane, a new definition was urgently called for. There were a number of candidates for that, but the one relevant to our story and one that succeeded where Jordan's definition failed is that provided by Karl Menger in Vienna in 1921.[3] The details of his definition, which explains in quite general topological terms which sets in a topological space can be assigned a natural number as dimension, are not important for the present story. What is important is that Menger was soon in communication with the Warsaw topologists, including Kuratowski and Bronisław Knaster (who was a close

---

[3] See Menger (1994), pp. 38ff An essentially equivalent definition was given by Paul Urysohn in Moscow, independently of Menger's and around the same time.

friend of Tarski), and that his conceptual analysis of the notion of dimension had a direct impact on their work. In Tarski's 1931 chapter on definable sets of real numbers, the notion of dimension (among other intuitive geometrical notions) is specifically referred to as a successful example of conceptual analysis.[4] My conclusion is that for Tarski, topology was paradigmatic in its use of set theory for conceptual analysis.

In my Warsaw lecture I went in some detail into the *form* of Tarski's use of set theory in his analyses of the concepts of truth, logical consequence, and of what is a logical notion, and how it was that in this last, Tarski assimilated logic to higher set theory to (what I regard as) an unjustified extent.[5] But that only answers the "how" part of our basic question at one level, and for a fuller answer, one must probe deeper in each case. That will only be done here for the concept of truth; for the case of logical consequence, see the rewarding discussion in Gómez-Torrente (1996).

WHEN DID TARSKI DEFINE TRUTH IN A STRUCTURE?

In the case of truth, the first puzzle has to do with the *prima facie* discrepancy between what logicians nowadays usually say Tarski did, and what one finds in the *Wahrheitsbegriff*. This was brought out in the very perspicacious chapter by Wilfrid Hodges, "Truth in a structure" (1985/86), which begins with an informal explanation of the current conception. Hodges then goes on to report (p. 137) that:

[a] few years ago I had a disconcerting experience. I read Tarski's famous monograph "The concept of truth in formalized languages" (1935) to see what he says himself about the notion of truth in a structure. The notion was simply not there. This seemed curious, so I looked in other papers of Tarski. As far as I could discover, the notion first appears in Tarski's address (1952) to the 1950 International Congress of Mathematicians, and his chapter "Contributions to the theory of models I" (1954). But even in those papers he doesn't define it. In the first chapter he mentions the notion only in order to explain that he won't be needing it for the purpose in hand. In the second chapter he simply says "We assume it to be clear under what conditions a sentence ... is *satisfied* in a system ..."

Hodges continues, "I believe that the first time Tarski explicitly presented the mathematical definition of truth in a structure was his joint chapter (1957) with Robert Vaught." In fact, the general notion of structure for a first-order language L is already described in Tarski's 1950 ICM address, essentially as follows: a structure A is a sequence consisting of a non-empty domain A of objects together with an assignment to each basic relation, operation, and constant symbol of L of a corresponding relation between elements of A, operation on elements of A, or member of A, resp. Moreover, while Hodges is correct in saying that the notion of truth in a structure is not defined

---

[4] See p. 112 of the English translation of (1931) in Tarski (1983). Interestingly, he remarks in this work that, in contrast to the geometrical examples—in which there are competing conceptual analyses in mathematical terms because the informal notions are a confused mix—the "arbitrariness" in that of definability and related logical notions is "reduced almost to zero" because the intuitions to which they respond "are more clear and conscious."
[5] For my critique of Tarski's analysis of what is a logical notion, see Feferman (1999).

there, Tarski does talk of the antecedent notion of satisfaction in a structure as if it is well understood, since he refers to the association with each formula φ of the set of all sequences from A which satisfy φ in A.

At any rate, what *is* of interest, as Hodges makes clear, is that these notions of structure, and of satisfaction and truth in a structure, do not seem to appear explicitly prior to the 1950s in Tarski's work. This is doubly puzzling, since, as a common informal notion, the idea of a structure being a model of a system of axioms well precedes that, and surely goes back to the nineteenth century. Most famously in that period, one had the stunning revelation of various models for non-Euclidean geometry. Then came Dedekind's characterization of the natural numbers and the real numbers, in both cases structurally as models, unique up to isomorphism, of suitable (second-order) axioms. That was followed by Hilbert's model-theoretic considerations concerning his axioms for geometry. The structural view of mathematics took hold in the early part of the twentieth century in algebra, analysis, and topology with the formulation of axioms for groups, rings, fields, metric spaces, normed spaces, topological spaces, and so on, and with the systematic exploration of their various models. Within logic, the informal concept of model has been traced back by Scanlan (1991) to the American "postulate theorists," launched by work of Huntington in 1902 and Veblen in 1904; this involved, among others the concept of consistency in the sense of satisfiability and that of categoricity.[6] Finally, it was central to the famous theorems of Löwenheim of 1915 and Skolem in 1920 on existence of countable models, and of course to Gödel's completeness theorem published in 1930.

WAS TARSKI A MODEL-THEORIST?

For Tarski, on the face of it, there were two loci of interest in the relation between axiom systems and their models.[7] The first was geometry, which began to figure in his work almost as soon as he started publishing research papers in 1924. In the year 1926–7 he lectured at Warsaw University on an elegant new axiom system for Euclidean geometry which, in distinction to Hilbert's famous system of 1899, was formulated without the use of set-theoretical notions, i.e. in first-order logic, and he considered various models of subsystems of his system in order to establish some independence results. But this work was not published until the 1960s, when Tarski had clearly shifted to the current model-theoretic way of thinking (see Tarski and Givant 1999).

The second locus was the method of elimination of quantifiers to arrive at decision procedures for all first-order statements true in certain models or classes of models. That method had been developed initially by Skolem, who applied it to the monadic

---

[6] Incidentally, Scanlan points out that Tarski was aware of that work in his abstract (1924), where a second-order system of axioms is given for the order relation on ordinals which, when restricted to the accessible ordinals, is said to be categorical "au sens de Veblen-Huntington."

[7] In addition to the two here, Hodges (personal communication) has suggested a third: propositional logics and their matrix models.

theory of identity, and Langford, who applied it to the theory of dense order. The method was pursued intensively in Tarski's seminar during the years 1926–8, beginning with his extension of Langford's work to the class of discrete orders. The most famous results of that seminar were Presburger's decision procedure for the structure of the integers with the operation of addition and the order relation, and Tarski's procedure for the real numbers with the operations of addition and multiplication, obtained by 1931.[8] The first intended exposition of the latter was the monograph (1967) whose scheduled publication in 1940 was postponed indefinitely because of the war. Rather than containing a model-theoretic statement of the results, this is devoted to establishing the *completeness of axiom systems* for elementary algebra and geometry, and by its methods as providing a *decision procedure for provability* in those systems. Chronologically, it is not until the full exposition of the elimination of quantifiers method in the report (1948), prepared with the assistance of J. C. C. McKinsey, that it is presented frankly as a decision procedure for truth in the structure of real numbers. There is a corresponding marked difference in the titles between the two publications.

TARSKI'S ACCEPTANCE OF TYPE THEORY AS A GENERAL FRAMEWORK

Hourya Sinaceur (2000, pp. 8–9) has emphasized this difference between Tarski's point of view prior to 1940 and his shift to our current way of thinking, perhaps around that time. Indeed, in the primary relevant pre-war publications, specific mathematical theories are always regarded by Tarski within an axiomatic framework, often expanded to the simple theory of types, and he refrained from speaking of structures as if they were independently existing entities. As is documented fully in Ferreirós (1999), pp. 350–6, this was the accepted way of formulating things for a number of logicians and philosophers in the 1930s, under the powerful residual influence of *Principia Mathematica*.[9] We have only to look at the title of Gödel's incompleteness chapter (1931) for the most famous example, where the system actually referred to is a form of the simple theory of types based on the natural numbers as the individuals. Tarski carried so far the identification of mathematical concepts with those

---

[8] See Vaught (1986) for the historical development, though presented in current model-theoretic terms.

[9] The shift from the ramified theory of types (RTT), as a basis for "all" of mathematics in the *Principia*, to the simple theory of types (STT) was given impetus in publications by Chwistek in 1920 and Ramsey in 1926, but was also spread informally by Carnap among others. (Reck (2004) traces the ideas for STT back to Frege's *Begriffsschrift* and to lectures that Frege gave in Jena in the early 1910s, lectures that Carnap attended.) In the well-known logic text by Hilbert and Ackermann published in 1928, analysis is formulated within RTT, but in the second edition ten years later, it is formulated in STT over the rational numbers as the individuals. Others who adopted some form of type theory to some extent or other as a general logical framework in the 1930s were Carnap, Church, Gödel, Quine, and Tarski; initially, Tarski was also strongly influenced by Leśniewski's theory of semantical categories in this respect. For a full account of type theory at its zenith, cf. Ferreirós (1999) 350–6.

that can be developed in the simple theory of types, that he wrote the following in his introductory textbook on logic, after sketching how the natural numbers can be treated as classes of classes and thus based "on the laws of logic alone":

> the... fact that it has been possible to develop the whole of arithmetic, including the disciplines erected upon it—algebra, analysis, and so on—as a part of pure logic, constitutes one of the grandest achievements of recent logical investigations.
>
> (1941, p. 81)[10]

Insofar as this statement regards the simple theory of types, necessarily with the axiom of infinity, as a part of pure logic, Tarski here blithely subscribes to the logicist program, thereby ignoring the fact that the infinity axiom is not a logical principle and that the platonist ontology normally seen to be required to justify the impredicative comprehension axioms of the theory put their logical status into question.[11] At the same time, the formulation of mathematical notions in axiomatic terms as deductive theories on their own or within a wider "logical" framework, such as the theory of types, may be related to Tarski's (later) professed nominalistic, anti-platonistic, tendencies (cf. Feferman 1999b, p. 61), but that in turn is clearly in tension with his thorough-going use of set theory in practice, and his acceptance from the beginning of Zermelo's axioms as a framework for his extensive purely set-theoretical work.[12] It should be remarked that over a long period Tarski tended to regard the simple theory of types with the axiom of infinity and Zermelo's axioms as merely alternative ways of formalizing the general theory of sets; see, for example his chapter with Lindenbaum on the theory of sets (1926, p. 299) and his posthumous chapter on what are logical notions (1986b, p. 151).

YES, TARSKI *WAS* A MODEL-THEORIST

So far, we have been revolving around the side question as to why Tarski, in his primary publications prior to 1948, did not take a straightforward informal model-theoretic way of presenting various of his notions and results. But, in fact there is secondary published evidence in that period that Tarski *did* think in just those terms, including the following:

(i) in the abstract (1924) he speaks of the categoricity of a (second-order) system of axioms for the ordinals up to the first inaccessible (cf. n. 8 above); this work is presented in more detail in Section 4 of Lindenbaum and Tarski (1926);

---

[10] The passage, translated directly from the 1937 German edition, is left unchanged through the most recent fourth English edition.

[11] One must confront this form of logicism with the equivocal statements Tarski made in his 1966 lecture "What are logical notions?", published posthumously as (1986b); see pp. 151–3, and Feferman (1999), pp. 48–9.

[12] It also seems to me to be in disaccord with his approach to metamathematics in ordinary set-theoretical terms, in contrast to Hilbert's finitist program for metamathematics, for which formulation of mathematics in axiomatic terms was the essential point of departure.

(ii) in editorial remarks (to a 1934 chapter of Skolem's in *Fundamenta Mathematicae*) concerning results obtained in the seminar led by Tarski at the University of Warsaw in 1927–8 he characterizes the models of the first-order truths of $\langle \omega, < \rangle$ (cf. (1986), vol. 4, p. 568);

(iii) in a further editorial remark (ibid.), Tarski says that he obtained an upward form of the Löwenheim–Skolem theorem in the period of the aforementioned seminar;

(iv) in the appendix to (1935/36) Tarski (informally) defines for each order type α the set $T(\alpha)$ of all elementary properties true of any pair consisting of a set X and binary relation R that orders X in order type α; he then defines two order types α and β to be *elementarily equivalent* when $T(\alpha) = T(\beta)$ and gives various examples for which this holds and for which it doesn't hold; in a footnote he says that these notions can be applied to arbitrary relations, not just ordering relations; at the end of this Appendix it is stated that these ideas emerged in the Warsaw seminar of 1926–8, but that he was able to state them in "a correct and precise form" only with the help of the methods later used to define the notion of truth;

(v) in the same Appendix, using the fact that $T(\omega) = T(\omega + \omega^* + \omega)$ Tarski concludes that the property of being a well-ordering is not expressible in first-order terms;

(vi) in an abstract with Mostowski published in the *Bulletin of the American Mathematical Society* for 1949, reporting on results obtained in 1941, a decision procedure for truths in all well-ordered systems $\langle A, < \rangle$ is said to have been established, and the relation of elementary equivalence between such systems is characterized in terms of order-types (cf. (1986) vol. 4, p. 583); by (v) there is no prior axiom system for these systems with which one is dealing.

My own conclusion from this part of the evidence is that Tarski, just like the early model-theorists who preceded him, worked comfortably with the informal notion of model for first-order and second-order languages at least since 1924. Perhaps it was only the use of type theory as the logical standard of the times that caused him to refrain from a frankly model-theoretic way of presenting his results in that area. Moreover, there were no uncertainties or anomalies in informal model-theoretic work that would have created any urgent need for conceptual analysis of semantical notions to set matters right. In Robert Vaught's valuable survey (1974) of model theory before 1945, he points out that Skolem, for example, worked comfortably with the notion of truth in a model, though—by contrast—uncertainly with the notion of proof, and the latter is a principal reason that he missed establishing the completeness theorem for first-order logic. More generally, according to Vaught (1974, p. 161), since the notion of truth of a first-order sentence σ in a structure A "is highly intuitive (and perfectly clear for any definite σ), it had been possible to go even as far as the completeness theorem by treating truth (consciously or unconsciously) essentially as an undefined notion—one with many obvious properties." Even Tarski, as quoted in n. 5 above, from (1931), agreed with that.

TO WHAT NEEDS WAS TARSKI RESPONDING?

So *why*, then, to return to our basic question, did Tarski feel it necessary to provide an explicit definition of this notion? Vaught's own answer (1974, pp. 160–1), is that:

Tarski appears to have been unhappy about various results obtained during the seminar [of 1926–28] because he felt that he did not have a precise way of stating them (see the last page of Tarski (1935/1936)). ... Tarski had become dissatisfied with the notion of truth as it was being used [informally].[13]

If anything is clear about Tarski both from all his publications and of the experience of those who worked with him, it is that he was unhappy about anything that could not be explained in precise terms, and that he took great pains in each case to develop every topic in a very systematic way from the ground up.[14] More than conceptual analysis was at issue: in his several papers of the 1930s on the methodology of deductive sciences (or calculus of systems), his aim was to organize metamathematics in quite general terms within which the familiar concepts (concerning axiomatic systems) of consistency, completeness, independence, and finite axiomatizability would be explained and have their widest applicability. I was referring in part to Tarski's drives to do things in this way when I said that he was responding to *psychological* and *programmatic* needs rather than a *logical* need in the case of his conceptual analyses. But there are further psychological components involved. Namely, despite his training in logic by philosophers, Tarski was first and foremost a mathematician who specialized principally in logic, and he was first and foremost very concerned to interest mathematicians outside of logic in its concepts and results, most specifically those obtained in model theory. However, he may have thought that he could not make clear to mathematicians that these results were part of mathematics until he showed how all the logical notions involved could be defined in precise mathematical (i.e. general set-theoretical) terms.[15] On the other hand, there is no evidence that mathematicians of the time who might have been interested in the relevant model-theoretic results turned away from them as long as such definitions were lacking or worried about them for other reasons.

DEFINABILITY (AND TRUTH?) IN A STRUCTURE—FOR MATHEMATICIANS

What had worried logicians, and mathematicians more generally, in the early twentieth century was the appearance of paradoxes in the foundations of mathematics. The

---

[13] Givant (1999) p. 52 offers related reasons for Tarski's aims in this respect.

[14] As one example, not long ago I had occasion to look back at the notes I took from Tarski's lectures on metamathematics at University of California, Berkeley in the year 1949–50. Fully a month was taken up in those lectures with developing concatenation theory from scratch before one arrived at the syntax of first-order languages.

[15] Vaught (1974) p. 161, makes a similar remark, adding, "[i]t seems clear that this whole state of affairs [of the use prior to Tarski's work of semantical notions as undefined concepts] was bound to cause a lack of sure-footedness in meta-logic." But see n. 5 above.

famous ones, such as those of Cantor, Burali-Forti, or Russell, were set-theoretical. Around the same time attention was drawn to semantical paradoxes, such as those of the Liar, or of Richard and Berry, concerning truth and definability, respectively. While these did not seem to have anything to do with questions of truth and definability in the algebraic and geometrical structures of the sort with which Tarski had been dealing, they affected him in the following way, that I consider to be a further psychological aspect of the "why" problem. Namely, from early on he seemed to think that it was the metamathematical (i.e. syntactic) form in which those concepts were defined that was a principal obstacle to mathematicians' appreciation of the subject, if not outside of the purview of mathematics altogether. Thus, for example, at the outset of his 1931 chapter "On definable sets of real numbers," he writes:

Mathematicians, in general, do not like to deal with the notion of definability; their attitude toward this notion is one of distrust and reserve. The reasons for this aversion are quite clear and understandable. To begin with, the meaning of the term 'definable' is not unambiguous: whether a given notion is definable depends on the deductive system in which it is studied... It is thus possible to use the notion of definability only in a relative sense. This fact has often been neglected in mathematical considerations and has been the source of numerous contradictions, of which the classical example is furnished by the well-known antinomy of Richard. The distrust of mathematicians towards the notion in question is reinforced by the current opinion that this notion is outside the proper limits of mathematics altogether. The problems of making its meaning more precise, of removing the confusions and misunderstandings connected with it, and of establishing its fundamental properties belong to another branch of science—metamathematics.[16]

Tarski goes on to say that "without doubt the notion of definability as usually conceived is of a metamathematical origin" and that he has "found a general method which allows us to construct a rigorous metamathematical definition of this notion." But then he says that

by analyzing the definition thus obtained it proves to be possible... to replace it by [one] formulated exclusively in mathematical terms. Under this new definition the notion of definability does not differ from other mathematical notions and need not arouse either fears or doubts; it can be discussed entirely within the domain of normal mathematical reasoning.

Technically, what Tarski is concerned with in this 1931 chapter was to explain *first* in *metamathematical* terms and then in what he called *mathematical* terms the notion of definable sets and relations (or sets of finite sequences) in the specific case of the real numbers. More precisely, the structure in question is taken to be the real numbers with the order relation, the operation of addition, and the unit element, treated axiomatically within a form of simple type theory over that structure. The metamathematical explanation of definability is given in terms of the notion of satisfaction, whose definition is only indicated there. Under the mathematical definition, on the other hand, the definable sets and relations (of order 1 in the type structure) are simply those generated from certain primitive sets of finite sequences corresponding to the atomic formulas, by means of Boolean operations and the operations of

---

[16] Quotations are from the English translation in Tarski (1983), pp. 110–11.

projection and its dual.¹⁷ Later in the chapter, it is indicated how to generalize this to definability over an arbitrary structure, as introduced by the following passage:

> In order to deprive the notion of elementary definability (of order 1) of its accidental character, it is necessary to relativize it to an arbitrary system of primitive concepts or—more precisely—to an arbitrary family of primitive sets of [finite] sequences. In this relativization we no longer have in mind the primitive concepts of a certain special science, e.g. of the arithmetic of real numbers. The set Rl is now replaced by an arbitrary set V (the so-called universe of discourse or universal set) and the symbol Sf is assumed to denote the set of all finite sequences s [of elements of] V; the primitive sets of sequences are certain subsets of Sf.¹⁸

Since Tarski's *metamathematical* explication of the concept of *definability in a structure* makes use of satisfaction, I take it that the notion of truth in a structure is present implicitly in that 1931 chapter. Indeed, in a footnote to the introduction he says of the metamathematical definition that "an analogous method can be successfully applied to define other concepts in the field of metamathematics, e.g., that of *true sentence* or of a *universally valid sentential function*." Universal validity can only mean valid in every interpretation, and for that the notion of *satisfaction in a structure* is necessary. It would have been entirely natural for Tarski to spell that out for his intended mathematical audience at that stage, if he had simply regarded metamathematics as part of mathematics by presenting the syntax and semantics of first-order languages as a chapter of set theory.

TRUTH *SIMPLICITER*—FOR PHILOSOPHERS

Why, then, did he offer instead the puzzlingly different definition of truth that we came to know in the *Wahrheitsbegriff* (1935)? Actually, as Tarski makes plain in a bibliographical note to the English translation (1983, p. 152), its plan dates to 1929. According to that note, he made presentations of the leading ideas to the Logic Section of the Philosophical Society in Warsaw in October 1930 and to the Polish Philosophical Society in Lwów in December 1930. Thus it is entirely contemporaneous with the work just described on definability in a structure (and implicitly of truth in a structure), but now directed primarily to a philosophical rather than a mathematical audience.¹⁹

---

¹⁷ In the follow-up chapter with Kuratowski, these are shown to be imbedded in the hierarchy of projective sets in Euclidean space.

¹⁸ From the translation of Tarski (1931) in (1983), p. 135. Incidentally, in my work on Hermann Weyl, I have drawn attention to his publication (1910) where a general notion of definability in a structure is proposed. Weyl's purpose was to replace Zermelo's vague notion of definite property in his axiomatization of set theory. (See van Heijenoort (1967) p. 285.) For some strange reason this chapter never came to the attention of logicians such as Tarski concerned with semantical notions.

¹⁹ This must be amplified slightly: though the *Wahrheitsbegriff* was, as well, directed to a philosophical audience via its publication in *Studia Philosophica*, the first announcement of its ideas and results was published in 1932 in the mathematical and physical sciences section of the

Later presentations would also be similarly oriented, including his 1936 lecture to the Congrès Internationale de Philosophie Scientifique and his 1944 chapter "The semantic conception of truth and the foundations of semantics" published in *Philosophy and Phenomenological Research*.

Clearly, Tarski thought that as a side result of his work on definability and truth in a structure, he had something important to tell the philosophers that would straighten them out about the troublesome semantic paradoxes such as the Liar, by locating for them the source of those problems. Namely, on Tarski's view, everyday language is inherently inconsistent via ordinary reasoning about truth as if it were applicable to all sentences of the language. The notion of truth can be applied without contradiction only to restricted formalized languages of a certain kind, and the definition of truth for such languages requires means not expressible in the languages themselves. This, of course, makes truth prima-facie into a relative notion, namely relative to a language, as definability was emphasized above to be a notion relative to a structure. Nevertheless, as presented in the *Wahrheitsbegriff*, it should in my view be considered to be an absolute notion, albeit a fragment of such. How can that be? The difference is that we are not talking about truth in a structure but about truth *simpliciter*, as would be appropriate for a philosophical discussion, at least of the traditional kind. This is borne out by a number of passages, of which those that follow are only a sample (quoted from the English translation in 1983).

After explaining the need to restrict to formalized languages of a special kind to avoid the paradoxes, Tarski writes:

> ... we are not interested here in "formal" languages and sciences in one special sense of the word "formal", namely sciences to the signs and expressions of which no meaning is attached. For such sciences the problem here discussed [of defining truth] has no relevance, it is not even meaningful. We shall always ascribe quite concrete and, for us, intelligible meanings to the signs which occur in the languages we shall consider.
>
> (1983, pp. 166–7)

The definition of truth is illustrated in Section 3 of the *Wahrheitsbegriff* for the language of the calculus of classes of the domain of individuals within the simple theory of types. Nothing is said about the nature of that domain except that it must be assumed to be infinite (cf. p. 174, n. 2 and p. 185); we may presume it to contain all concrete individuals. The variables of the language of the calculus of classes are then interpreted to range over arbitrary subclasses of the domain of individuals. Now, by way of contrast with the notion that he is after, Tarski takes a bow in the direction of a relative notion of truth in the following passage:

> In the investigations which are in progress at the present day in the methodology of the deductive sciences (in particular in the work of the Göttingen school grouped around

Viennese Academy of Sciences, and the 1933 Polish monograph (of which the *Wahrheitsbegriff* was a translation), was published by the corresponding section of the Warsaw Academy of Sciences; see the items [32] (p. 917) and [33m] (p. 932) of the bibliography (Givant 1986) of Tarski's works.

Hilbert) another concept of a relative character plays a much geater part than the absolute concept of truth and includes it as a special case. This is the concept of *correct* or *true sentence in an individual domain a*. By this is meant... every sentence which would be true in the usual sense if we restricted the extension of the individuals considered to a given class *a*, or—somewhat more precisely—if we agreed to interpret the terms 'individual', 'class of individuals', etc. as 'element of the class *a*', 'subclass of the class *a*', etc., respectively.[20]

(1983, p. 199)

And, further on in this connection, we have:

... the general concept of correct sentence in a given domain plays a great part in present day methodological researches. But it must be added that this only concerns researches whose object is mathematical logic and its parts.... The concept of correct sentence in every individual domain... deserves special consideration. In its extension it stands midway between the concept of provable sentence and that of true sentence...

(1983, pp. 239–40)

It's clear from these quotations that what Tarski is after in the *Wahrheitsbegriff* is an absolute concept of truth, relativized only in the sense that it is considered for various specific formalized languages of a restricted kind. There is of course then the question of how the meanings of the basic notions of such languages are supposed to be determined. It is also part of Tarski's project to "not make use of any semantical concept if I am not able previously to reduce it to other concepts" (1983, p. 153). Thus, meanings can't be given by assignments of some sort or other to the external world. Tarski's solution to this problem is to specify meanings by translations into an informally specified metalanguage associated with the given language, within which meanings are supposed to be already understood[21] (1983, pp. 170–1).

Incidentally, and this is a separate issue worthy of discussion but not pursued here, these languages are not only taken to be interpreted, but are also supposed to carry a deductive structure specified by axioms and rules of inference. Tarski requires that "[t]he sentences which are distinguished as axioms [should] seem to us to be materially true, and in choosing rules of inference we are always guided by the principle that when such rules are applied to true sentences the sentences obtained by their use should also be true"[22] (1983, p. 167). His purpose in including the deductive structure is to show what light the notion of truth throws on that of provability, but of course that is unnecessary to the task of defining truth for a given language given solely by the assumed meaning of its basic notions and the syntactic structure of its sentences.

---

[20] In a footnote, Tarski warns off the philosophical part of his audience from the relativized definition, saying that "it is not necessary for the understanding of the main theme of this work and can be omitted by those readers who are not interested in special studies in the domain of the methodology of the deductive sciences..."

[21] Hartry Field (1972) has emphasized this as Tarski's way of solving the problem of supplying meaning, and argued that it is inadequate for a physicalist theory of truth. It is a separate, and debatable issue, whether Tarski's program to establish semantics on a scientific basis, as described in his (1936), would require him to meet Field's demands for such a theory.

[22] In the German of (1935), the first part of this reads: "Aussagen, die als Axiome ausgezeichnet wurden, scheinen uns inhaltlich wahr zu sein..."

WAS TARSKI A LOGICAL UNIVERSALIST?

In a first draft of this article I argued that the distinction between treating truth in an absolute rather than relative sense has to do with that first elicited by Jean van Heijenoort in his short but innovative article, "Logic as calculus and logic as language" (1967*b*, 1985).[23] In brief, according to van Heijenoort, for Frege and Russell logic is a *universal language*, while the idea of logic as calculus is the approach taken in the pre-Fregean work of Boole, De Morgan, and Schröder, later taken up again in the post-Russellian work beginning with Löwenheim and Skolem. To quote van Heijenoort,

> Boole has his universal class, and De Morgan his universe of discourse, denoted by '1'. But these have hardly any ontological import. They can be changed at will. The universe of discourse comprehends only what we agree to consider at a certain time, in a certain context. For Frege it cannot be a question of changing universes. One could not even say that he restricts himself to *one* universe. His universe is the universe. Not necessarily the physical universe, of course, because for Frege some objects are not physical. Frege's universe consists of all that there is, and it is fixed.
>
> (1985, pp. 12–13)

Russell's adaptation of this was in the ramified theory of types. The Frege–Russell viewpoint is certainly understandable if their systems are regarded as embodying purely logical notions. The work of the *Wahrheitsbegriff*, I argued, is presented in that universalist tradition, though the framework is modified to that of the pure simple theory of types (STT) rather than the ramified one, and is used informally rather than formally; in addition, as we have already remarked, the axiom of infinity is assumed in order to make use of the natural numbers within the theory. All that is in apparent conflict with the change of perspective represented by the Postscript to the *Wahrheitsbegriff*, as detailed in the very persuasive article on Tarski and the universalism of logic by Philippe de Rouilhan (1998), which was brought to my attention in the meantime.[24] On reconsideration, I have to agree that my claim needs to be qualified, though not necessarily radically; as this, too, is a side issue, but one that I want to address here, I will be as brief as the matter allows.

In the body of the *Wahrheitsbegriff* (cf. especially pp. 215 ff of (1983)), Tarski subscribed to Leśniewski's theory of semantical categories (credited to Husserl in its origins). Considered formally, this contains STT, called by Tarski in some places the theory of sets (e.g. p. 210, n. 2) and elsewhere the general theory of classes (e.g. pp. 241–2). Part of the significance of the theory of semantical categories is supposed to be its universality:

> The language of a complete system of logic should contain—actually or potentially—all possible semantical categories which occur in the languages of the deductive sciences. Just this fact

---

[23] See also van Heijenoort (1976) and Hintikka (1996).
[24] See also Sundholm (2003), which is of interest as well for its account of the relations between Tarski and Leśniewski.

gives to the language mentioned a certain "universal" character, and it is one of the factors to which logic owes its fundamental importance for the whole of deductive knowledge.

(1983, p. 220)

Every semantical category can be assigned a natural number as order, the order of expressions of that category: the order of individual terms is lowest and the order of a relational expression is the supremum of the orders of its arguments plus one. The order of a language consonant with the doctrine of semantical categories is the supremum of the orders of the expressions in that language, thus either a natural number or the first infinite ordinal ω. A metalanguage in which truth is to be defined for a given language must be of higher order than the order of the language. In Section 4 of the *Wahrheitsbegriff*, Tarski sketched how to define truth for languages of finite order, while in Section 5 he argued that there is no way to do that for languages of infinite order, since there is no place for a metalanguage to go, if it is to be part of the universal language.

But in the Postscript to the *Wahrheitsbegriff*, Tarski abandoned the theory of semantical categories, so as to allow for languages of transfinite order in some sense or other. The nature of such is only sketched there and the details are problematic; the difficulties are well explained by de Rouilhan in the article mentioned above. In any case, the idea of a universal language is clearly abandoned in the Postscript, and in that sense, Tarski is not a universalist. But if one reformulates van Heijenoort's basic distinction in the way that Sundholm (2003) does, as being one between *languages [used] with meaning vs. languages without use [i.e. as objects of metamathematical study]*, in my view we still find Tarski positioned on the former side of the 'versus', *even* in the Postscript. How then is it that Sundholm places him on the latter side? The choice of position depends on whether one emphasizes, as I do here, that *qua* philosopher, in the *Wahrheitsbegriff* Tarski is after the concept of absolute truth, or as Sundholm does, that *qua* (meta-) mathematician, he is after a concept of truth relative to a language. And that has to do with the tension between the two sides of Tarski's efforts with respect to the semantical notions, the one represented in the *Wahrheitsbegriff* for philosophers, and the original one for mathematicians, described above.

Next we see how this opposition affects the single most famous feature of the *Wahrheitsbegriff*, the truth scheme.

WHAT THE TRUTH SCHEME DOES AND DOESN'T DO

It is for *truth simpliciter* treated within the framework of STT that Tarski can formulate the conditions required of a "materially adequate" definition of truth for a language L of finite order in its metalanguage (inside STT). That takes the form of the scheme (or "convention", in his terminology)

(T) x is a true sentence if and only if p,

each instance of which is given by substituting for 'x' the name of a sentence in L, and for 'p' the sentence itself as it is given in its metalanguage. In particular, in the case that L is the language of a specific mathematical structure *A* whose underlying domain, relations, operations, and distinguished elements are taken to be given in the metalanguage, we can replace 'true sentence' in the left hand side of (T) by 'sentence true in *A*'. That is how it is done for the structure underlying the calculus of classes used to illustrate the definition of truth in the *Wahrheitsbegriff*, whose domain is the class of all subclasses of the class of individuals, and whose only relation is that of inclusion. But nothing like (T) is suggested by Tarski for the notion of truth in arbitrary structures. In particular, no analogue to the scheme (T) is suggested for the notion of truth in the calculus of classes relative to an arbitrary domain *a*. An obvious modification would take the form,

(T_{rel}) for all *a*, x is a true sentence in the domain *a* if and only if p,

each instance of which is given by substituting for 'x' the name of a sentence in the fragment L, and for 'p' the *relativization of that sentence* to the variable *a*. It is apparent from this example that if the definition of truth in a structure more generally were to be presented in the framework of STT, the formulation of a corresponding truth scheme would be all the more cumbersome, and would not have the striking obviousness of (T) as a criterion for the definition of truth. In this case, the tension between Tarski the philosopher and Tarski the (meta)-mathematician is resolved, out of mere simplicity, in favor of the former.

THE PHILOSOPHICAL IMPACT

The actual reception by philosophers of the form of the definition of truth given to us in the *Wahrheitsbegriff* was initially mixed, and remains so to this day. Some, like Karl Popper, took to it fairly quickly; they had first met in Prague in 1934 at a conference organized by the Vienna Circle, with whose tenets regarding the nature of science Popper was in dispute. When they met again in 1935 during an extended visit that Tarski made to Vienna, Popper asked Tarski to explain his theory of truth to him:

> and he did so in a lecture of perhaps twenty minutes on a bench (an unforgotten bench) in the *Volksgarten* in Vienna. He also allowed me to see the sequence of proof sheets of the German translation of his great chapter on the concept of truth, which were then just being sent to him.... No words can describe how much I learned from all this, and no words can express my gratitude for it. Although Tarski was only a little older than I, and although we were, in those days, on terms of considerable intimacy, I looked upon him as the one man whom I could truly regard as my teacher in philosophy. I have never learned so much from anybody else.
>
> (Popper 1974, p. 399)

Rudolf Carnap, a central figure in the Vienna Circle, was another philosopher who took reasonably quickly to Tarski's theory of truth. He had been favorable to Tarski's general approach to the methodology of deductive sciences since their first meeting in 1930. Later, during Tarski's visit to Vienna in 1935, Carnap became a convert to the theory of truth and urged Tarski to present it at the forthcoming first Unity of Science conference to be held in Paris. Relating the circumstances that led him to accept and promote the theory published in Tarski's "great treatise on the concept of truth," Carnap wrote in his intellectual autobiography:

When Tarski told me for the first time that he had constructed a definition of truth, I assumed that he had in mind a syntactical definition of logical truth or provability. I was surprised when he said that he meant truth in the customary sense, including contingent factual truth.... I recognized that [Tarski's approach] provided for the first time the means for precisely explicating many concepts used in our philosophical discussions.

(Carnap 1963, p. 60)

Carnap urged Tarski to report on the concept of truth at the forthcoming congress in Paris. "I told him that all those interested in scientific philosophy and the analysis of language would welcome this new instrument with enthusiasm, and would be eager to apply it in their own philosophical work." Tarski was very skeptical. "He thought that most philosophers, even those working in modern logic, would be not only indifferent, but hostile to the explication of his semantical theory." Carnap convinced him to present it nevertheless, saying that he would emphasize the importance of semantics in his own chapter, but Tarski was right to be hesitant. As Carnap reports:

At the Congress it became clear from the reactions to the papers delivered by Tarski and myself that Tarski's skeptical predictions had been right. To my surprise, there was vehement opposition even on the side of our philosophical friends. Therefore we arranged an additional session for the discussion of this controversy outside the offical program of the Congress. There we had long and heated debates between Tarski, Mrs. Lutman-Kokoszyńska, and myself on one side, and our opponents [Otto] Neurath, Arne Naess, and others on the other.

(Carnap 1963, p. 61)

The bone of contention was whether the semantical concepts could be reconciled with the strictly empiricist and anti-metaphysical point of view of the Vienna Circle. In the write-up (1936) of his talk "The establishment of scientific semantics" for the conference, Tarski tried to make the views compatible, but he still found it necessary to respond to critics as late as 1944 in the expository article, "The semantic conception of truth and the foundations of semantics."

I think it is fair to say that since then, at least Tarski's scheme (T) has been central to many philosophical discussions of the nature of truth,[25] though the philosophical significance of his definition of truth—or whether absolute truth is even definable—continues to be a matter of considerable dispute.

SEMANTICS WITHOUT SYNTAX: ONE MORE TRY AT A "NORMAL" MATHEMATICAL DEFINITION

In the address (1952) that he made to the 1950 International Congress of Mathematicians held in Cambridge, Massachusetts, Tarski tried once more to interest mathematicians in model theory by developing its notions in "normal" mathematical

---

[25] Cf., for example, the collections of articles in Blackburn and Simmons (1999) and in Lynch (2001).

terms, i.e. without reference to the syntax of first-order languages. These notions are applied to any given similarity class of structures A, called there algebraic systems. In ordinary metamathematical terms, a subclass S of the given similarity class is said to be an *arithmetical class* (or elementary class) if for some sentence σ of the corresponding first-order language L, S consists of all structures A such that σ is true in A. As Tarski describes what he is after,

> [t]he notion of arithmetical class is of a metamathematical origin; whether or not a set [sic!] of algebraic systems is an arithmetical class depends upon the form in which its definition can be expressed. However, it has proved to be possible to characterize this notion in purely mathematical terms and to discuss it by means of normal mathematical methods. The theory of arithmetical classes has thus become a mathematical theory in the usual sense of this term, and in fact it can be regarded as a chapter of universal algebra.
>
> ((1952), p. 705, reprinted in (1986) p. 461)

The means by which this is accomplished is by a kind of uniform extension across the given similarity class of the basic relations and operations on them that Tarski had used to explain the notion of definable relation in (1931). In the terminology of (1952) these are given by *arithmetical functions* F whose domain is the given similarity class and which for each A in that class has for its value a subset of the set A^ω of infinite sequences of elements of the domain A of A. Tarski explains frankly (op. cit., pp. 706–7) that these functions are obtained by imitation of the recursive metamathematical definition of satisfaction, in terms of which each such F is determined by a formula φ of L with F(A) equal to the set of all sequences which satisfy φ in A. The collection of all arithmetical functions is denoted **AF**. Certain F in **AF**, called *simple functions*, can be distinguished as corresponding to sentences, and for these, F(A) is either empty or A^ω. Finally, an arithmetical class S is defined to be one such that for some simple function F, A is in S if and only if $F(A) = A^\omega$.

As Quine said of Russell's Axiom of Reducibility, this entire procedure is "indeed oddly devious,"[26] and at a crucial point in the development, the effort at "normal" mathematization even breaks down. Namely, one of the main results of (1952) is Theorem 13, the *compactness theorem for arithmetical functions*, which takes the form that if **K** is any subclass of **AF** whose intersection is the function Z that assigns the empty set to each structure, then there is a finite subset **L** of **K** whose intersection is Z. The compactness theorem for **AC** is a corollary. Of this, Tarski says, "[a] mathematical proof of Theorem 13 is rather involved. On the other hand, this theorem easily reduces to a metamathematical result which is familiar from the literature, in fact to Gödel's completeness theorem for elementary logic." No indication is given as to what "mathematical proof" of this theorem Tarski had in mind; the first published candidate that might be considered to qualify for such would be the one using ultraproducts (a "mathematical" notion) given ten years later by Frayne, Morel, and Scott (1962). But even that depends on the fundamental property of ultraproducts relating truth in such a product to truth in its factors, the formulation and proof of which makes essential use of syntax.

---

[26] In van Heijenoort (1967a), p. 151.

Though some of the language and notation such as AC and AC_δ introduced in Tarski (1952) has survived in the model-theoretic literature, that of the vehicle of arithmetical functions has not, and—as far as I can tell—the impact of this approach on mathematicians outside of logic was nil.[27] Tarski himself abandoned it soon enough in favor of normal metamathematical explanations connecting semantics to syntax, finally fully spelled out in Tarski and Vaught (1957). Of course, all of that is unproblematically a part of ordinary set-theoretical mathematics, in accord with Tarski's basic vision of the subject.

CODA

In bringing his work on truth *simpliciter* to the attention of philosophers via the simply stated truth scheme (T), Tarski *did* have the kind of impact he would have liked to have had with the mathematicians in trying to interest them in semantical notions for structures via their purported mathematization. On the other hand, the enormously successful theory of models that began to take off in the 1950s, propelled by the basic work of Tarski and Abraham Robinson among others, and that has now reached applications to algebra, number theory, and analysis of genuine mathematical interest, makes common use, not of its "mathematized" version of arithmetical functions and classes, but of its basic "metamathematical" version of satisfaction and truth in a structure, and has been accepted by mathematicians without qualms about those notions. So Tarski's continual concerns in that respect were, in my view, quite misplaced.

To conclude, I must return to the question implicitly raised at the beginning of this chapter by the statement that—in the view of many—Tarski's definition of truth is one of the most important cases of conceptual analysis in twentieth-century logic.[28] Namely, how important is it? I have been told by more than one colleague (no names, please) over the years that Tarski was merely belaboring the obvious. I have to agree that there is some justice to this criticism, at least if we're thinking about the notions of satisfaction and truth in a structure—after all, the definitions are practically forced on us. But even if that's granted, Tarski's explication of these concepts, at least in the way that it was presented in the 1950s, has proved to be important as a paradigm for all the work in recent years on the semantics of a great variety of logical and computational languages as well as parts of natural language. And it has raised interesting questions about possible other approaches to informal semantics when that

---

[27] And within logic it had a specific unfortunate result: at Tarski's behest, Wanda Szmielew reformulated her important elimination of quantifiers procedure for Abelian groups in terms of the language of arithmetical functions (Szmielew 1955), turning something already rather complicated syntactically into an unreadable piece that, perversely, served even further to hide the underlying mathematical facts. Eklof and Fisher (1972) subsequently re-established her results by means of understandable standard model-theoretic techniques in a way that also brought the needed facts into relief.

[28] Tarski himself thought that his two most important contributions to logic were the decision procedure for the elementary theory of real numbers and his definition of truth.

paradigm doesn't seem to apply in any direct way (cf. Hodges 2004). Finally, as I have detailed in my (1999b), the definitions of satisfaction and truth have had some essential technical applications within standard metamathematics, including, besides nondefinability results à la Tarski, their use in Gödel's original (and nowadays preferred) definition of constructible set and in the use of partial truth definitions for non-finite and non-bounded axiomatizability results for various theories. Though there may not have been a compelling reason for the definitions in early model theory, they now constitute a *sine qua non* of our subject.[29]

References

Blackburn, S. and Simmons, K. (eds.) (1999) *Truth*, Oxford University Press (Oxford).
Carnap, R. (1963) Intellectual autobiography, in *The Philosophy of Rudolf Carnap* (P. A. Schilpp, ed.), *The Library of Living Philosophers*, vol.11, Open Court (La Salle).
de Rouilhan, P. (1998) Tarski et l'universalité de la logique, in *Le Formalisme en Question. Le tournant des années trente* (F. Nef and D. Vernant, eds.), Vrin (Paris), 85–102.
Eklof, P. C. and Fisher, E. R. (1972) The elementary theory of Abelian groups, *Annals of Mathematical Logic* 4, 115–71.
Feferman, S. (1999a) Logic, logics and logicism, *Notre Dame Journal of Formal Logic* 40, 31–54.
—— (1999b) Tarski and Gödel between the lines, in *Alfred Tarski and the Vienna Circle* (J. Woleński and E. Köhler, eds.), Kluwer Academic Publishers (Dordrecht), 53–63.
—— (2004) Tarski's conception of logic, *Annals of Pure and Applied Logic* 126, 5–13.
Ferreirós, J. (1999) *Labyrinth of Thought. A History of Set Theory and its Role in Modern Mathematics*, Birkhäuser Verlag (Basel).
Field, H. (1972) Tarski's theory of truth, *Journal of Philosophy* 69, no. 13, 347–75.
Frayne, T., Morel, A., and Scott, D. (1962) Reduced direct products, *Fundamenta Mathematicae* 51, 195–228.
Givant, S. (1986) Bibliography of Alfred Tarski, *Journal of Symbolic Logic* 51, 913–41.
—— (1999) Unifying threads in Alfred Tarski's work, *The Mathematical Intelligencer* 21, 47–58.
Gödel, K. (1931) Über formal unentscheidbare Sätze der Principia Mathematica und verwandter Systeme I, *Monatshefte für Mathematik und Physik* 38, 173–98.
Gómez-Torrente, M. (1996) Tarski on logical consequence, *Notre Dame Journal of Formal Logic* 37, 125–51.
Henkin, L., Addison, J., Chang, C. C., Craig, W., Scott, D., Vaught, R. (1974) *Proceedings of the Tarski Symposium*, American Mathematical Society (Providence).
Hintikka, J. (1996) *Lingua Universalis vs. Calculus Ratiocinator: An ultimate presupposition of twentieth-century philosophy*, Kluwer (Dordrecht).
Hodges, W. (1985/86) Truth in a structure, *Proceedings of the Aristotelian Society*, new series 86, 131–51.
—— (2004) What languages have Tarski truth definitions?, *Annals of Pure and Applied Logic* 126, 93–113.
Kuratowski, K. (1980) *A Half Century of Polish Mathematics*, Pergamon Press (Oxford).

---

[29] I wish to thank Geoffrey Hellman, Wilfrid Hodges, Paolo Mancosu, Göran Sundholm, and Johan van Benthem for their useful comments on a draft of this chapter.

Lindenbaum, A. and Tarski, A. (1926) Communication sur les recherches de la théorie des ensembles, *Comptes Rendus des Séances de la Soc. des Sciences et Lettres de Varsovie, Class III*, 19, 299–330.

Lynch, M. P. (ed.) (2001) *The Nature of Truth*, MIT Press (Cambridge, MA).

Menger, K. (1994) *Reminiscences of the Vienna Circle and the Mathematical Colloquium*, Kluwer Academic Publishers (Dordrecht).

Popper, K. (1974) Comments on Tarski's theory of truth, in Henkin *et al.* (1974), 397–409.

Reck, E. (2004) From Frege and Russell to Carnap: Logic and logicism in the 1920s, in *Carnap Brought Home: The View from Jena* (S. Awodey and C. Klein, eds.), Open Court (Chicago), 151–80.

Scanlan, M. J. (1991) Who were the American postulate theorists?, *Journal of Symbolic Logic* 56, 981–1002.

Sinaceur, H. (2000) Address at the Princeton University Bicentennial Conference on problems of mathematics (December 17–19, 1946), by Alfred Tarski, *Bulletin of Symbolic Logic* 6, 1–44.

—— (2001) Alfred Tarski: Semantic shift, heuristic shift in metamathematics, *Synthese* 126, 49–65.

Sundholm, G. (2003) "Tarski and Leśniewski on languages with meaning versus languages without use," in *In Search of the Polish Tradition—Essays in Honour of Jan Woleński on the Occasion of his 60th Birthday*. (J. Hintikka, T. Czarnecki, K. Kijania-Placek, T. Placek, and A. Rojszczak, eds.). Kluwer Academic Publishers (Dordrecht).

Szmielew, W. (1955) Elementary properties of Abelian groups, *Fundamenta Mathematicae* 41, 203–71.

Tarski, A. (1924) Sur les principes de l'arithmétique des nombres ordinaux (transfinis), *Annales de la Société Polonaise de Mathématique*s 3, 148–9. [Reprinted in Tarski (1986), vol. 4, 533–4.]

—— (1931), Sur les ensembles définissables de nombres réels. I, *Fundamenta Mathematicae* 17, 210–39; revised English translation in Tarski (1983), 110–42.

—— (1935) Der Wahrheitsbegriff in den formalisierten Sprachen, *Studia Philosophica* 1, 261–405; revised English translation in Tarski (1983), 152–278.

—— (1935/36) Grundzüge des Systemenkalküls. Erster Teil, *Fundamenta Mathematicae* 25, 503–26/ Zweiter Teil, *F.M.* 26 283–301; revised English translation as a single article in Tarski (1983), 342–83.

—— (1936) Grundlagen der Wissenschaftlichen Semantik, *Actes du Congrès Internationale de Philosophie Scientifiques*, vol. 3, Hermann and Cie (Paris), 1–8; revised English translation in Tarski (1983), 401–8.

—— (1941) *Introduction to Logic and to the Methodology of Deductive Sciences*, Oxford University Press (New York).

—— (1944) The semantic conception of truth and the foundations of semantics, *Philosophy and Phenomenological Research* 4, 341–75.

—— (1948) A decision method for elementary algebra and geometry (prepared with the assistance of J. C. C. McKinsey), RAND Corp. (Santa Monica); 2nd rev. edn, 1951, University of California Press (Berkeley).

—— (1952) Some notions and methods on the borderline of algebra and metamathematics, *Proceedings of the International Congress of Mathematicians, Cambridge, Mass. 1950*, vol. 1, Amer. Math. Soc. (Providence), 705–20.

—— (1954) Contributions to the theory of models I, *Indagationes Mathematicae* 16, 572–81.

—— (1967) The Completeness of Elementary Algebra and Geometry, Inst. Blaise Pascal (Paris). (Reprint from page proofs of a work originally scheduled for publication in 1940 in the series, *Actualités Scientifiques et Industrielles*, Hermann and Cie, Paris, but which did not appear due to the wartime conditions.)

—— (1983) *Logic, Semantics, Metamathematics*, 2nd edn (John Corcoran, ed., J. H. Woodger, trans.) Hackett Publishing Company (Indianapolis).

—— (1986*a*) *Collected Works*, vols. 1–4, (S. R. Givant and R. N. McKenzie, eds.), Birkhäuser (Basel).

—— (1986*b*) What are logical notions?, *History and Philosophy of Logic* 7, 143–54.

Tarski, A. and Givant, S. (1999) Tarski's system of geometry, *Bull. Symbolic Logic* 5, 175–214.

Tarski, A. and Vaught, R. L. (1957) Arithmetical extensions of relational systems, *Compositio Mathematica* 13, 81–102.

van Heijenoort, J. (ed.) (1967*a*) *From Frege to Gödel. A Source Book in Mathematical Logic 1879–1931*, Harvard University Press (Cambridge). (Third printing, 1976.)

—— (1967*b*) Logic as calculus and logic as language, *Boston Studies in the Philosophy of Science* 3, Reidel Pub. Co., 440–6; reprinted in J. van Heijenoort (1985), 11–16.

—— (1976) Set-theoretic semantics, in *Logic Colloquium '76* (R. O. Gandy and M. Hyland, eds.), North-Holland (Amsterdam), 183–90; reprinted in van Heijenoort (1985), 43–53.

—— (1985) *Selected Essays*, Bibliopolis (Naples).

Vaught, R. L. (1974) Model theory before 1945, in Henkin *et al.* (1974), 153–72.

—— (1986) Tarski's work in model theory, *Journal of Symbolic Logic* 51, 869–82.

Weyl, H. (1910) Über die Definitionen der mathematischen Grundbegriffe, *Mathematisch-naturwissenschaftliche Blätter* 7, 93–95 and 109–13; reprinted in H. Weyl (1968), Gesammelte Abhandlungen, vol. I, 298–304.

Woleński, J. (1989) *Logic and Philosophy in the Lwow-Warsaw School*, Kluwer Academic Publishers (Dordrecht).

5

Tarski's Theory of Definition

Wilfrid Hodges

This chapter reviews what Alfred Tarski said about the theory of definitions during the years 1926–38. It is not the chapter I was expecting to write. I had believed that Tarski had his own well-formed views on definitions, and that I would be able to collect them together from his papers. Not so: his statements about central questions in the theory of definitions are often indirect and sometimes frankly careless. By contrast he was extremely careful about any questions to do with the relationship between object theory and metatheory. So his true interests reveal themselves.

For the theory of definitions, the effect is a little like playing the violin with gloves on—if you can really play well with them on, you must be terrific with them off. And so the work of Tarski that revolves around definitions, whatever its motives, did have a fundamental effect on our understanding of definitions. One measure of this is that these papers of Tarski are prominent in Robert Vaught's masterly summary [73] of Tarski's contributions to model theory—a part of mathematical logic with definitions close to its heart. Another discipline linked with the theory of definitions is formal semantics; when eventually the history of this discipline is written, Tarski should be named as one of its founders.

PRELIMINARIES

The texts

All mathematicians write definitions. Alfred Tarski is remarkable, even among mathematical logicians, for the number of his papers that discuss the notion of definition itself.

Sometimes I had to chase up Polish originals. For help on this front I am hugely indebted to Sasha Ivanov, Zofia Adamowicz, Jan Zygmunt, and Jan Woleński. Without help from Barbara Bogacka in the translations I would have been lost. I also thank Johan van Benthem, John Corcoran, Heinz Dieter Ebbinghaus, Solomon Feferman, Ivor Grattan-Guinness, David Hitchcock, Patrick Suppes, and David Wiggins for various pieces of information. But blame me for any errors.

(a) First there are several papers from the 1930s that discuss what is meant by 'definition' or 'definable' in the context of deductive theories. These include [70] (1926), [52] (1934) and the book [55] (1936). They also include what amounts to a two-part chapter [50], [26] (1931) on definability in the real numbers.

(b) The move from deductive theories to model theory in the mid century made it necessary to reformulate ideas from (a). The papers [62] (1952) and [63] (1954) and the book [71] (1953) contain some of these reformulations.

(c) During the 1930s Tarski wrote a number of papers giving formal definitions of semantic notions from scientific methodology. These papers include [51] (1933) on truth, [56] (1936) on logical consequence, and [57] (1936) on semantic notions in general. This group of papers is relevant to us for two reasons. First, they introduce two very general criteria that the definitions must meet. And second, the machinery driving these papers also plays a role in some of the papers under (a).

Besides these papers there are several outliers. Two papers [47] and [48], both from 1930, systematize some definitions in the theory of deduction. A chapter [72] (1957) adapts the definition of truth to 'truth in a structure' for purposes of model theory. A late contribution to group (c) is the definition of a logical notion ([67], which is Corcoran's edited transcription of a talk given by Tarski in 1966). A chapter [61] (1948) discusses definability in arithmetic. Apart from some brief remarks in [67], none of these outliers discuss the notion of definition itself. There was also a manuscript of a projected book by Richard Montague, Dana Scott, and Tarski on set theory, which I saw in the Reichenbach Library at UCLA in 1967. My recollection is that it had a fairly full treatment of definitions by induction on the ordinals and other well-founded structures; but so far my enquiries have failed to get any further information about it.

At the centre of all this work stands the monograph [51] on the definition of truth. I will refer to it as *the big Truth chapter*. Tarski later wrote two expository papers on philosophical aspects of this work [60], [65], but they add nothing new on the notion of definition itself.

All the papers of group (a) above were primarily about deductive theories. So we turn to these.

Deductive theories

During Tarski's formative years, most leading workers in logic and foundations of mathematics accepted a certain view of the nature of logic and its place in the exact sciences. According to this view, a branch of knowledge is properly formalized by being put into the form of a *deductive theory*. A deductive theory consists of the following items:

- A set of symbols; each symbol is either a logical symbol or punctuation, or a *primitive*; the primitive symbols have meanings which we believe we understand, and these meanings are relevant to the subject matter of the branch of knowledge.

- A syntax which defines the set of sentences that can be formed from the symbols.
- A set of *axioms*, consisting of some sentences that we believe we know to be true; some of these are distinguished as *logical axioms*.
- A set of *inference rules*.

The *theorems* of the deductive theory are the sentences that are deducible from the axioms by the rules of inference. Ideally all the known truths of the branch of knowledge should be theorems of the system.

There are descriptions of this paradigm in Hilbert [12], Weyl [75], Łukasiewicz [32] i. 2, Tarski [55] chapter 6, and Church [6] §07. Frege and Peano both subscribed to versions of the paradigm. From Frege onwards there was a good deal of experimentation with different logical axioms and rules. Tarski normally required that the logic in his deductive theories would be a logic of finite types with an axiom of extensionality.

The terminology was a little vague, and Tarski tried to impose some precision by distinguishing levels. At the bottom level of a deductive theory is the set of theorems; he called this the *deductive system*. One level up, we attach the syntax and the consequence relation; the result is the deductive theory itself. (Footnote to [54] on p. 343 of [66].) Tarski also refers to a deductive theory as a deductive 'science' (*nauka*) or 'discipline' (*badanie*).

Above both these levels lies the study of the deductive theory, for example the informal justification of the axioms and the rules of inference, and any general statements about what is or is not a theorem of the deductive theory. Łukasiewicz used the phrase 'methodology of the system' to cover any treatment of the deductive theory as an object of investigation (for example [32] iii. 6); he seems to have assumed that this treatment would be informal.

Tarski took over the term 'methodology' and used it vigorously throughout his career. But he also wanted a term for *formalized* methodology, and for this he coined the name 'metatheory', which is clearly based on Hilbert's 'metamathematics'. He claimed to be the first person to axiomatize his metatheory ([66] p. 173 footnote 3). He also distinguished between 'mathematical' concepts, which are those in the deductive theory itself, and 'metamathematical' concepts which belong in the metatheory and are used to reason about the deductive theory. The levels are particularly important to distinguish in [50], and this is one of the two papers to which Tarski made 'more serious changes' in Woodger's translations in 1954 ([64] p. viii).

Tarski distinguished informal methodology from formal by the fact that informal methodology is done in Polish. In other words, its language is 'everyday' (*potoczny*) or 'intuitive'. (At pp. 24, 70 of [51] we find 'everyday intuition', *intuicją potoczną*.) Tarski sometimes described technical terms of logic as 'everyday' and 'intuitive' when they were expressed in Polish. For example in [50] he comments on the phrase 'A finite sequence of objects satisfies a given sentential function':

... the intuitive meaning (*sens intuitif*) of the above phrase seems clear and unambiguous.

([66] p. 116f)

One can find similar uses of 'intuitive' in Łukasiewicz. The everyday language and the intuitions that Tarski has in mind are those of researchers in the foundations of

Tarski's Theory of Definition

mathematics, not those of the general public, and not even those of philosophers. (This is argued in more detail by Jané [20] and Hodges [18].)

So now we have four levels:

- intuitive methodology,
- formal metatheory,
- deductive theory,
- set of theorems.

Below I will sometimes call the deductive theory the 'object theory' to contrast it with the metatheory, though this is not Tarski's own usage.

What is the point of deductive theories? Tarski never discusses the question. Even in his elementary textbook he falls back on the authority of history ([55] §32, [59] §36):

(1) By way of a compromise between [the] unattainable ideal and the realizable possibilities, certain principles regarding the construction of mathematical [sciences] have emerged, and they may be described as follows...

So we will have to read between the lines. There are useful clues in some phrases that read oddly today.

For example Tarski refers several times to 'practising' or 'performing' (*uprawiać*) a deductive theory; sometimes he adds 'on the basis of its language'. ([51] pp. 32, 37, 61, 132, 139, 158; unfortunately the force of *uprawiać* has gone missing in the journey from Polish to English through German.) The implication is that a deductive theory is a *kind of activity*. (The Polish internet carries nearly 20,000 pages containing the phrase *uprawiać bezpieczny seks*, 'practise safe sex'.) There is really only one activity that he can have in mind. This is the activity of sitting in front of a piece of chapter and writing a sequence of formulas in the language of the theory, where each formula is either an axiom of the theory, or a definition, or something derived from previous rows by rules of the theory.

There are not many published examples of this activity. Frege's *Grundgesetze* [9] has some, and so does *Principia Mathematica* [76], but in both cases the authors spoil the effect by interpolating a good deal of German or English. For virtually pure examples one should look at papers of Leśniewski. For example his chapter [29] pp. 399–409 on abelian groups is eleven pages long, and nearly eight pages consist of a formal derivation using nothing outside the formal language and symbols to index the applications of the rules.

Tarski's use of the word 'mathematics' for deductive theories is not just a turn of phrase. He says in his elementary text ([55] §32, [59] §36):

(2) ... *not only is every mathematical [science] a deductive [science], but also, conversely, every deductive [science] is a mathematical [science]* [Tarski's italics] ... We shall not enter here into a discussion of the reasons in favor of this view...

Today this seems just silly. Some of the greatest achievements of mathematics have involved *designing* a new formalism, and designing a formalism seems not to be a thing you can do within the confines of a deductive theory. I think we will be able to understand Tarski's position here much better when we bring Leśniewski into the picture later in this chapter.

We should note one other curious phrase that Tarski uses. He several times speaks of establishing a thing 'empirically' (*na drodze empirycznej, la voie empirique*), for example at [50] p. 229 or at [51] pp. 21, 26, 31, 71, 134 (the pages in [66] are 129, 158, 162, 165, 196, 248). The examples don't all fit any obvious pattern, but in several of them Tarski seems to be describing an informal deduction. At [50] he refers to a conclusion that we would most naturally reach by an induction in the metatheory; he says we can give ourselves 'subjective certainty' of the conclusion by sampling instances instead of doing the induction. ([51] p. 196 is similar but with metatheory and metametatheory in place of deductive theory and metatheory.) I find this hard to swallow; we can't *choose* which arguments to use in order to give ourselves subjective certainty. (On the other hand one can see a kind of argument that exactly fits Tarski's description when one looks back to the days before logicians had proof by induction. See [17] §9ff on Walter Burley's use of the method I call Sample.)

It does seem that in all his discussion of deductive theories Tarski has in mind a highly idealized mathematician. This mathematician is able to do mathematics by 'practising' deductive theories, and can switch his or her cognitive faculties on or off at will. This picture will fill out as we see what Leśniewski fed in, and how Tarski himself used the ideal mathematician.

One reason why Tarski didn't defend his view of deductive theories was that he believed all other logicians accepted it ([51] p. 32):

(3) ... *języki sformalizowane konstruuje się, jak dotąd, wyłącznie po to, by na ich gruncie uprawiać sformalizowane nauki dedukcyjne* (... hitherto formalized languages have been constructed, without exception, in order to practise *formalized deductive sciences* [Tarski's emphasis] on the basis of them)

Note the *uprawiać*. In 1956 ([64] p. 166) Woodger translated the word not as 'practise' but as 'study'. This is at the wrong level—as for example practising safe sex is very different from studying it. Already in the 1950s fundamental parts of Tarski's vision were fading from view.

Here is Carnap confirming the general drift of Tarski's interests in 1930 ([5] p. 30):

Alfred Tarski came to Vienna in February 1930, and gave several lectures, chiefly on metamathematics. We also discussed privately many problems in which we were both interested. Of special interest to me was his emphasis that certain concepts used in logical investigations, e.g., the consistency of axioms, the provability of a theorem in a deductive theory, and the like, are to be expressed not in the language of the axioms (later called the object language), but in the metamathematical language (later called the metalanguage).

Carnap has probably foreshortened things a little. In 1930 Tarski was certainly claiming that certain things need to be done in the metatheory, but he was also pointing to the usefulness of pushing things down from the metatheory into the object theory where possible. ([50] p. 211 '... *il me semble que cette méthode permet d'aboutir à certains résultats que l'on ne réussirait pas d'obtenir, si l'on opérait avec la conception métamathématique des notions étudiées.*')

BEFORE TARSKI

Tadeusz Kotarbiński

Tadeusz Kotarbiński (1886–1981) was a well-respected philosopher who kept in close touch with the Polish logicians of his time, although he later admitted 'To be quite frank, formal logic never appealed to me strongly' ([24] p. 11). In 1929 Kotarbiński published a textbook of philosophical logic under the title *Elementy Teorji Poznania, Logiki Formalnej i Metodologji Nauk* [23]. Tarski lists this book in the bibliography of the original Polish edition of his textbook ([55] p. 9), commenting that it covers a broad range of topics. He invites the reader to look at the sections on reasoning, deductive theories, and the deductive method.

About fifty pages of the book are devoted to definitions, and they serve as a good record of the general state of the theory of definitions in Poland at the time. Kotarbiński's main discussion of definitions ([23] pp. 29–58) is in the context of natural languages. He explains that a definition is

(4) an answer to the question what a given linguistic expression means.

([23] p. 29)

The linguistic expression is called the *definiendum*, and the text explaining its meaning is the *definiens*. (This is standard terminology, and I use it henceforth.) He classifies definitions along two dimensions. The first dimension is in terms of their linguistic form, and the second is in terms of their purpose.

He begins with the classification by form. The definiendum must be named, not used. (Apparently he saw this as a corollary of (4).) Sometimes the definiens is named too, as in

(5) 'Travailler' means the same as 'to work'.

but sometimes the definiens is used and not named, for example

(6) 'John is crazy' means that John behaves in outlandish ways.

In both (5) and (6) the definiendum is in quotation marks, but only (5) has quotation marks on the definiens. Kotarbiński calls a definition 'lexicographic' (*słownikowa*) if the definiens is named, and 'semantic' (*semantyczna*) if it is used but not named ([23] pp. 29f). He remarks that in a broader sense, both kinds of definition are 'semantic' because they both give the meaning of the definiendum.

There remains the case where the definiendum is used but not named. Definitions of this kind, Kotarbiński says, are like the Berlitz method of language learning, where we pick up the meaning of the word 'pecks' by meeting it in sentences 'The hen pecks', 'The sparrow pecks', 'The pigeon pecks', and so on. His main reason for including it at all is the claim sometimes made, that axioms of a deductive theory help to clarify the meanings of the primitive symbols of the system. For example, he says, one might write 'If one of three points on a straight line is between the second and the third, then it is between the third and the second' as a step towards explaining the meaning of the word 'between'. Because of this kind of application, Kotarbiński refers to this form of definition as 'axiomatic' (*aksjomatyczna*).

Since a definition should name its definiendum, axiomatic definitions are strictly not definitions at all. So, Kotarbiński comments, it would be more precise to call them 'axiomatic pseudodefinitions'. They do occur in mathematics, but generally under severe restrictions. He discusses the forms that they take in mathematical writing.

Curiously Tarski (for example in the big Truth chapter, [51] p. 103, [66] p. 223) uses the expression 'pseudodefinition' in a sense incompatible with Kotarbiński's. For Tarski a typical example of a pseudodefinition is a comprehension axiom

(7) $\exists R \forall x \forall y \, (Rxy \leftrightarrow \phi(x,y))$

where ϕ is a formula with at most x, y free and no occurrences of R. With the help of a suitable axiom of extensionality, this axiom implies the existence of a unique relation R consisting of those pairs (a, b) such that $\phi(a, b)$ holds. But there is no definiendum here; the symbol R is a bound variable, not a defined term. Tarski says that this usage of 'pseudodefinition' comes from Leśniewski. The term 'pseudodefinition' doesn't appear in the index to Leśniewski's Collected Works [29], so I have to leave it there.

After his three-way classification of the forms of definitions, Kotarbiński introduces 'classical definitions' (p. 43). He mentions Plato and Aristotle; a footnote cites Trendelenburg, Überweg, and others. He calls attention to the 'eliminability property' (*charakter rugujący*) of classical definitions, which he expresses as follows (p. 50):

(8) The definiens doesn't contain the expression being defined.

He quotes with approval a remark of Padoa [39] that one only knows how to use an expression when one knows how to do without it.

Next he turns to the classification of definitions by purpose ([23] pp. 51ff). Here he distinguishes between two kinds of definition, 'analytic' and 'synthetic', thus:

(9) ... analytic definitions ... answer the question what a given term means in some existing usage, while synthetic definitions ... answer the question what a given term means in some proposed system of expression.

([23] p. 52)

He may have taken this distinction from the Jäsche Logic of Kant [21] p. 631. He shows no awareness of its use in late writings of Frege.

A later section of Kotarbiński's book ([23] pp. 292–5) discusses definitions in deductive theories. His discussion is clearly based on a debate between Łukasiewicz and Leśniewski, which we turn to next.

Jan Łukasiewicz

Jan Łukasiewicz (1878–1956) was one of Tarski's teachers, and a close colleague during the 1920s. In 1929 he wrote:

(10) In discussions with my colleagues, especially Professor S. Leśniewski and Dr A. Tarski, and often in discussions with their and my own students, I have made clear to myself many a concept, I have assimilated many a way of formulating ideas, and I have learned about many a new result, about which I am today not in a position to say to whom the credit of authorship goes.

([32] preface p. xi)

He was a strong advocate of deductive theories. He even had his students analysing ancient logic in the light of them, to the despair of historians of logic.

In particular Łukasiewicz discussed how we can add definitions to a deductive theory, and what we gain by it. In his textbook of 1929 he observed that added definitions 'serve to abbreviate certain expressions belonging to a given theory', and 'may contribute some new intuitions to the theory and thus add to the terms belonging to the theory in question terms which have a meaning outside that theory' ([32] p. 32). I guess that the following is an example of what Łukasiewicz means by adding 'terms which have a meaning outside that theory'. It purports to be Tarski's first effort at conceptual analysis. Łukasiewicz explains that he was looking for a definition of the concept of 'possibility in general'.

(11) The definition in question was discovered by Tarski in 1921 when he attended my seminars as a student at the University of Warsaw. Tarski's definition is as follows:

$Mp = CNpp$.

Expressed verbally this says: 'it is possible that p' means 'if not-p then p'.

([33], p. 55 in [35])

What Tarski had done was to find, for any formula ϕ of Łukasiewicz's three-valued calculus, another formula which is true if and only if ϕ is not false. For most readers it must seem that the conceptual advance here is in the setting up of the three-valued logic, not in Tarski's entirely technical observation.

In 1951 Łukasiewicz ([34] p. 164) recalled a problem raised by Leśniewski in 1921 in connection with the addition of Sheffer strokes to *Principia Mathematica* by Sheffer and Nicod. The problem goes back to the format that *Principia Mathematica* uses for introducing new expressions by definition. Suppose for example that they had wanted to define A (disjunction in Łukasiewicz notation) in terms of C and N ('if...then' and negation). They would have written

(12) $CNpq = Apq$ Df

where (in Łukasiewicz's words)

(13) The symbol '.=.Df' is associated with a special rule of inference allowing the replacement of the *definiens* by the *definiendum* and vice versa.

Now the problem is that by this route the authors would not have succeeded in defining A in terms of C and N; they would have defined it in terms of C, N and '.=.Df'.

Leśniewski gave his own account of his question in [29] p. 418. The question is ingenious: it forces us to think what we mean by 'definition' in the context of deductive theories. In fact there were two broad views in circulation, which should have been recognized as incompatible but were not. The first view (that of *Principia*) is that the addition of a definition to a deductive theory does *nothing* to the deductive theory; it simply sets up a convention that we can rewrite the expressions of the deductive theory in a shorter way and perhaps more intuitive way. People who took this view were divided about whether the definition is the convention or a piece of text (like (13)) that expresses the convention. The second view is that adding a definition to a deductive theory creates a new deductive theory which contains a new symbol. In this case too one could think of the definition as the process of adding the new symbol and any attached formulas, or one could single out some particular formula in the new system as the 'definition' of the new symbol. Leśniewski distinguished formulas from permissions etc. by calling the latter 'directives' (*dyrektywy*). Tarski seems to avoid this word; for him a definition is always a piece of text. (At [66] p. 166 footnote 1 Tarski summarizes the difference between the first and second views.)

At two meetings of the Polish Philosophical Society [30], [31] in 1928, Łukasiewicz spoke on definitions. These talks gave Leśniewski and Tarski an opportunity to put their views. As they are reported, Tarski's contributions were purely technical and were limited to propositional logic.

Stanisław Leśniewski

Stanisław Leśniewski (1886–1939) is difficult to get the measure of. His contemporaries agree about his intelligence, his insight, and his commitment. But he was apt to take up unorthodox positions and pursue them with unorthodox zeal. He was happier developing his own systems than engaging with anybody else's.

In his major chapter [27] (1929) Leśniewski expresses some views on deductive theories. He refers several times to private conversations with Tarski, and to contributions made by Tarski. Tarski's quoted contributions are all mathematical: for example he finds a set of axioms, or he simplifies an argument. In no case does Tarski contribute any thoughts on deductive theories themselves.

Leśniewski says ([27], p. 487 of [29]):

(14) Having endeavoured to express some of my thoughts on various particular topics by representing them as a series of propositions meaningful in various deductive theories, and to derive one proposition from others in a way that would harmonize with the way I finally considered intuitively binding, I know no method more effective for acquainting the reader with my logical intuitions than the method of formalizing any deductive theory to be set forth.

Here Leśniewski mentions two activities: the first is deriving propositions within a deductive theory, and the second is creating a formal deductive theory. He regards both of them as ways of conveying his 'intuitions'. The second, if done properly, ensures that the formal deductions 'harmonize' with his intuitive deductions. Before and after this passage, he emphasizes that in the process of formalization the axioms, definitions, and final directives 'have for me an irresistible intuitive validity', and the theses have 'a certain specific and completely determined sense' and are 'genuinely meaningful'.

If we cut out the autobiography and set these remarks beside Tarski's assumptions and practice, a broadly consistent picture emerges. The working mathematician proves things informally at first. But proofs need to be checked and communicated to other people. For these purposes mathematicians have agreed on certain ways of formalizing their intuitive thoughts, namely as deductive theories. They have agreed, for example, that we should take as primitive terms of our deductive theories expressions that 'seem to us to be immediately understandable', and we should take as axioms statements 'whose truth appears to us evident' (in Tarski's words, [55] §32 = [59] §36). The ideal mathematician sets up deductive theories that meet these requirements, and then 'practises' these deductive theories.

I believe that the previous paragraph summarizes what Leśniewski ([29] p. 487) meant when he called himself an intuitionistic formalist, and what Tarski meant when he said in 1930 ([48], p. 62 of [66]) that this was his position too. From the written record it seems that Tarski accepted Leśniewski's account wholesale. Building on it, he concentrated on the middle ground between the intuitions and the formal theory. There are some natural questions here: Is there a need for a middle formalized level between the intuitions and the formalized object theory? (This is the question that Carnap said Tarski was excited about in 1930.) How much scope is there for material to be moved from one level to another?

There is a suggestive analogy with Turing's work in the mid 1930s. Turing studied the computing mathematician, not the proving mathematician. But Turing's problem, like Leśniewski's, was to describe a formalism that exactly matched the content of the mathematician's mind.

Later we will see how Leśniewski's requirement that his definitions have an irresistible intuitive validity reappears as Tarski's requirement of *trafność* ('material adequacy'). It's impossible to measure how far Leśniewski's other views on definitions influenced Tarski without establishing what those other views were, and this is difficult.

In a footnote to the 1941 English edition [59] of his elementary textbook [55] of 1936, Tarski wrote (§40):

(15) A very high level in the process of formalization was achieved in the works of the late Polish logician S. LEŚNIEWSKI (1886–1939); one of his achievements is an exact and exhaustive formulation of the rules of definition.

Tarski added the final clause 'one of . . . rules of definition' in the 1941 edition, presumably as a mark of respect for a teacher who had died just two years earlier.

At first sight one might think Tarski is telling us that Leśniewski made an exact and exhaustive study of the rules of definition. But if he did, none of it survives except some extremely detailed and precise rules of definition for Leśniewski's own systems ([27]), and a different set of rules for Łukasiewicz's theory of deduction ([28]). The only concession that Leśniewski makes to any kind of general perspective is to remark that

(16) The directives I give here for a system based on [negation and implication] can very easily be transposed to a system based on others...

([28], p. 631 in [29])

Even this much generalization might be difficult. Leśniewski's rules are very closely tied to fine details of Łukasiewicz's axioms; for example there is a reference to the '11th word of Axiom (*L*)'. He offers no motivation whatever for any of the rules. Either he is phenomenally bad at explaining himself, or else he just doesn't have a general theory of definition at all and he can only handle it system by system. My own feeling is that the correct methodology is to assume the latter until someone proves otherwise.

This doesn't in any way contradict the text of Tarski's footnote. Tarski says not that Leśniewski 'formalized' the rules of definition (which would imply that he had informal rules to formalize, and this I doubt), but that he gave a 'formulation' of the rules of definition. In other words Leśniewski wrote out—exactly and exhaustively, as Tarski says—the rules of definition for the particular systems that he studied. On that reading, Tarski's footnote is certainly true. Mostowski [36] p. 251 says something similar:

(17) Leśniewski strongly emphasized that a rule of definion needs to be formalized rigorously. He gave a precise formulation of these rules with reference to the systems of logic that he himself created...

In short, Leśniewski probably had no general theory of definition.

But maybe we give up too quickly. In the beautiful chapter on Definition in his logic textbook [46] (1957), Patrick Suppes offers the following information (p. 153):

(18) Two criteria which make more specific these intuitive ideas about the character of definitions are that (i) a defined symbol should always be eliminable from any formula of the theory, and (ii) a new definition does not permit the proof of relationships among the old symbols which were previously unprovable; that is, it does not function as a creative axiom.*

[Footnote:] *These two criteria were first formulated by the Polish logician S. Leśniewski (1886–1939); he was also the first person to give rules of definition satisfying the criteria.

This footnote of Suppes is widely quoted but completely wrong. The rule (i) about eliminability is central in Pascal's treatment of definition in [41] in the mid 17th century. Pushing back still further, Porphyry [42] in the third century AD already says that defined terms should be eliminable, and he has a name (*antistréphein*) for the processes of introducing them into and eliminating them from a sentence ([42] p. 63 l. 20ff).

The situation with rule (ii) about creativity is worse still. Again the notion is older than Leśniewski, though by a narrower margin. Frege denounced creative definitions (he called them *schöpferischen Definitionen*) at some length in §143 of his *Grundgesetze* II [9], which Leśniewski knew well. But Leśniewski disagreed profoundly with Frege about creative definitions. On Leśniewski's view,

(19) if one introduces definitions at all, then they should be as creative as possible.
(*skoro się już definicye wprowadza, to winny one być jaknajbardziej twórcze.*)

([31])

He would certainly not have endorsed Suppes' rule (ii).

In 1996 I met Suppes at the Congress on Logic, Methodology and Philosophy of Science in Florence and asked if he could point me to the source of his footnote; he said no, and recently he confirmed this. I don't know of any place earlier than Suppes' book where one finds these two conditions (non-creativity and eliminability) paired as the conditions for a sound definition, so I think it's conceivable that Suppes invented this criterion of soundness himself. My memory is fairly clear that in the late 1960s I learned that in reasonable logics a definition is non-creative and eliminable in a theory T if and only if it's equivalent in T to an explicit definition (see [15] p. 103f for the proof), but I can't trace where or how I learned it.

The same pair of conditions appears in footnote 3 on page 307 of [66] in 1983, replacing a different pair in Woodger's 1956 translation [64]. I was unable to find out why this change was made. (Zygmunt's edition of [52] seems to imply that eliminability and non-creativity are in the Woodger translation, and I regret that I said the same in [15]; I should have checked.)

DEFINITIONS IN DEDUCTIVE THEORIES

Padoa's method

In 1900 Alessandro Padoa proposed a method for showing that a primitive of a deductive theory doesn't have a definition in terms of the other primitives. In his own words,

(20) to prove that the system of undefined symbols is irreducible with respect to the unproved propositions it is necessary and sufficient to find, for each undefined symbol, an interpretation of the system of undefined symbols that verifies the system of unproved propositions and that continues to do so if we suitably change the meaning of only the symbol considered.

([38], p. 122 of van Heijenoort)

Here 'irreducible' means 'we cannot deduce from the system a relation of the form $x = a$, where x is one of the undefined symbols and a is a sequence of other such symbols (and logical symbols)'. The 'unproved propositions' are the axioms.

The most straightforward reading of Padoa's claim is that a certain semantic condition is equivalent to a certain syntactic one. Call the primitive 'S'. Then the two conditions are:

(21) (Semantic) The deductive theory has two models which agree on all symbols except S, and disagree on S.

('Model of T' here means what Padoa calls an 'interpretation of the language of T which verifies T', at least when he is being careful. Also a more careful formulation would make it clear that the two models have the same domain.)

(22) (Syntactic) No definition of S in terms of the other primitives is a theorem of the deductive theory.

We should smell a rat here. In 1900 no significant equivalences between semantic conditions and syntactic ones were known. In particular the condition (22) might fail because the theory has no models at all, without this showing up as a syntactic inconsistency.

The direction

(23) Semantic $\Rightarrow$ Syntactic

is safe for most deductive theories; to use a later terminology, (24) follows from the soundness of the proof calculus. This direction is what is usually known as *Padoa's method*. The other direction is unsafe; today we know many logical calculi for which it is false. In [38] Padoa gives a brief proof of the safe direction which probably says as much as one can say without going into details of the proof rules. Then he says 'Conversely' and repeats the proof of the safe direction in slightly different words.

A second weakness of Padoa's account is his notion of a definition. Different logical systems allow different kinds of definition. Padoa's text, read literally, suggests that for him a definition is always an equation between the definiens and the definiendum; mathematicians sometimes call a definition in this form a *closed-form definition*. Logical calculi with an iota operator can usually bring definitions to this form, but most other systems can't.

The main divide is between logics that have variables and equality in the type of the symbol being defined, and logics that don't. With variables and equality there are several ways of saying 'S is the unique thing such that . . .'. For example we have:

(24) $\forall X(X = S \leftrightarrow \phi(X))$.

(25) $\phi(S) \wedge \forall X(\phi(X) \rightarrow X = S)$.

In many calculi (26) is a consequence of a sentence θ if and only if the following formula is provable from θ (but note that the formula is not a sentence and is not provably equivalent to (26)):

(26) $\phi(S) \wedge (\phi(X) \rightarrow X = S))$.

There are also forms that depend on the type of the symbol S. For example if S is a binary relation symbol we have

(27) $\forall x \forall y(S(x,y) \leftrightarrow \psi(x,y))$.

In logics with variables and equality in the type of S, the sentences (25) and (26) can both be converted to an equivalent sentence of the form (28), and vice versa. But without variables of the type of S, (28) is the only possible one of these three forms.

Two of Tarski's papers [70], [52] discuss Padoa's method. Suppose the issue is whether a symbol S is definable in terms of the other symbols in a theory T.

(α) In [70] (cf. p. 305 in [66]) Tarski introduces a new theory T' which is the same as T except that S is replaced by a new symbol S' throughout. He considers the condition

(28) $T \cup T' \vdash \phi(S) \wedge S = S'$.

(Here and below, $T \vdash \phi$ means that ϕ is derivable from T by the proof rules of the system.) He notes that (29) is necessary and sufficient for a definition of the form (25) to be deducible from T. He gives no proof, but this is an elementary exercise. Necessity is clear. For sufficiency, if (29) holds then so does the same with S! replaced throughout by a variable X, and then—taking into ϕ the parts of T that are used in the derivation—(27) is logically derivable.

(β) In both [70] and [52] he points out that if T is finite and yields a definition of S in the form (26), then without loss we can take $\phi(X)$ to be the conjunction of T with X in place of S throughout, and this definition is then a logical theorem. (This follows from the argument for (α), by taking the whole of T into ϕ.)

(γ) In [52] but not in [70] he points out an adjustment of (β) that quantifies out any symbols that are not needed in the definition of S.

Now where in all this is Padoa? Remember that Padoa's condition (22) was semantic, in terms of finding two different interpretations for the symbol S. Tarski's discussion ignores this completely and devotes itself to equivalences between purely syntactic conditions.

It might be said that it's easier to infer the falsehood of (29) from Padoa's condition than it is to infer the nonexistence of a definition (25) for S. Let's test this. Assume we have verified Padoa's condition by giving models I and J for the theory T, which differ on S but agree on all other symbols.

To infer the nonexistence of a definition (25) *derivable from T*, we note that if there were such a definition, we could use it to infer the interpretation of S in I from the interpretations of the other symbols in I, and likewise in J. But then S would have the same interpretation in I and in J, contradiction.

To infer the falsehood of (29), we note that since T' has the same structure as T, we can convert J to a model J' of T'. Since I and J' agree on the interpretations of the common symbols of T and T', $I \cup J'$ forms a model of $T \cup T'$. But then (29) must be false, because S and S' have different interpretations in this model.

The first argument is essentially Padoa's. I hope the reader will agree with me that the second argument is less direct and carries a heavier load of abstract apparatus. So if it's a matter of using Padoa's condition to infer a syntactic condition by a purely intuitive argument—and neither Padoa nor Tarski offers us anything more than an intuitive argument for this step—then Padoa's formulation is better than Tarski's. In short when Tarski claims, in both [70] and [52], that he has given a theoretical basis for Padoa's argument, his claim is false. Tarski has done nothing at all to strengthen or justify Padoa's inference.

Tarski claims in [70] that he has shown the 'generality' of Padoa's method. I presume he means that he has proved necessity and sufficiency. But the condition that he has shown necessary and sufficient for S to be undefinable from other terms is not Padoa's condition. Indeed it couldn't be, given that the system that Tarski uses has no completeness theorem.

Tarski claims in [52] that he has given Padoa's method in an 'extended' form, since he considers when S is definable from a particular subset of the other symbols in T. The same comment applies: he generalizes one of his syntactic characterizations to this case, but he makes no attempt to work out how Padoa's condition could be adjusted to meet this case.

In one important respect Tarski's method is much less general than Padoa's. Tarski restricts himself explicitly to systems of type theory which have variables and equality in the type of the defined symbol. Some such restriction is needed for Tarski's equivalences between syntactic conditions, but it is certainly not needed for Padoa's method. In fact Padoa's method is usually applied to first-order theories which have no higher-order variables.

It seems to me that this work of Tarski is never cited except for historical purposes, or by people who have misunderstood its contents. (I recently saw a chapter on internet ontologies that attributes Beth's Theorem to [52].) The mathematical interest is almost trivial; certainly an ingenious and ambitious mathematician like Tarski would never have published it as a contribution to mathematics. It adds nothing to the intuitive justification or the practical application of Padoa's method. So why did Tarski publish it?

The following reconstruction works, and is the only one I know that does. (There are further details in [19].) When Tarski says he is providing a 'theoretical base' ([70] p. 112) or a 'theoretical justification' ([66] p. 300) for Padoa's method, he means a *justification in terms of the practice of a deductive theory*. Now imagine Padoa with a theory T, maybe on chapter already, and two interpretations of T (say I and J) that agree on all the symbols except S. What happens next? Probably Padoa can find some way of writing down what I and J are. But how does he write that they are *interpretations of* T? Part of what this means is that Padoa attaches two different senses to the symbol S. But first, the methodology of deductive theories forbids us to read a symbol in two different senses. And second, the fact that Padoa interprets S in a certain way is not itself a part of the deductive theory, it's a mental preliminary to using the theory.

So Padoa's two models don't figure in the deductive theory at all. The task that Tarski sets himself is to analyse what *mathematics* takes place when Padoa uses his method, and to express that mathematics in a suitable deductive theory. Tarski observes that Padoa can choose I to interpret S and J to interpret a copy S' of S. Then Padoa can reason in the deductive theory $T \cup T'$. Since Padoa claims to know that I is a model of T, Padoa must have grounds for believing this, and those grounds can be written as added axioms for the deductive theory. (Tarski glosses over this step; I assume he reckoned he could take it as read.) Likewise with J. Since Padoa claims to know that I differs from J, with the help of adequate axioms he can prove $S \neq S'$ in the theory $T \cup T'$. But if T defines S, so that $T \cup T'$ gives the same definition for both S and S', then Padoa can practise $T \cup T'$ to prove $S = S'$; contradiction. Paragraphs (β) and (γ) above are variants of the same general picture.

In short, Tarski is not claiming to make Padoa's original proposal any more plausible. He is claiming to transfer as much as possible of Padoa's method into the form of calculations within a deductive theory. The effect of Tarski's analysis of Padoa's method is to *eliminate the model theory*.

Definable sets in a deductive theory

In 1931 Tarski and Kazimierz Kuratowski published three papers on definable sets, in French and consecutively in the same journal [50], [26], [25]. (The papers were written in 1930.) The first, by Tarski, sets out four different ways of expressing 'definable set' in connection with a deductive theory for the reals. Using Tarski's own terms we can describe the four ways as Intuitive, Metamathematical, Mathematical, and Accidental Mathematical. (Tarski distinguishes a second version of the Mathematical, which we can call the Geometric—more on this below.) The second chapter, by both authors, concentrates on the connection between the Metamathematical and the Mathematical definitions; it gives a fuller proof of their equivalence and a number of examples. The third, by Kuratowski, uses the connection proved in the second chapter as a tool for placing various sets of real numbers in the projective hierarchy. The procedures that Kuratowski describes are known today as the *Tarski–Kuratowski algorithm* (Kechris [22] p. 355).

Since the third chapter is by Kuratowski and the second is mainly an elaboration of a result in the first chapter, we can concentrate on the first chapter.

The Intuitive way of expressing 'definable set of reals' runs as follows. Write $T_\mathbb{R}$ for the deductive theory of the reals as ordered abelian group. Suppose for example that $\phi(x)$ is a formula of the language of $T_\mathbb{R}$ with one free individual variable x. Then $\phi(x)$ *defines* the set of all real numbers that satisfy $\phi(x)$. Note that ϕ is not a definition in any of the senses discussed by Kotarbiński, since it defines a set rather than a symbol.

We reach the Metamathematical version by formalizing the Intuitive one. More precisely the task is to formalize the notion of satisfaction, so Tarski is referring here to his truth definition (which hadn't yet been published). Tarski says that the formalization presents some unforeseen difficulties and it is still not clear whether it can be done ([50] p. 217). A page of small print ([50] pp. 218f) explains that the problem lies in the logic of higher types. If one restricts $T_\mathbb{R}$ to some bounded order, for example taking it to be first-order, then these problems vanish. The English version in 1956 ([64] VI) removes any suggestion that there is still work to be done on the formalization of the Intuitive version.

Next comes the Mathematical version. The Metamathematical version is available for definable relations, not just definable sets. But for this generalization Tarski has to explain what a relation is 'mathematically' (i.e. in $T_\mathbb{R}$), and for this he needs the notion of a sequence. He assumes that $T_\mathbb{R}$ is equipped to talk about the natural numbers, so that he can take a sequence to be a set of ordered pairs which expresses a function whose domain is a set of natural numbers. He takes a relation to be a set of sequences which all have the same domain. So far this is just book-keeping. The main problem is to define 'definable relation' using only symbols of $T_\mathbb{R}$ and symbols definable in $T_\mathbb{R}$.

Tarski first defines an infinite family of symbols corresponding to atomic formulas. For example $M_{3,5}$ names the set of all functions $f: \{3, 5\} \to \mathbb{R}$ with $f(3) \leqslant f(5)$. Then he defines some operations on relations. For example the binary operation $\bar{o}$ takes $M_{3,5}$ and $M_{5,8}$ to the set of all functions $f: \{3, 5, 8\} \to \mathbb{R}$ such that $f(3) \leqslant f(5) \leqslant f(8)$. (Note that $\bar{o}$ is the intersection of $M'_{3,5}$ and $M'_{5,8}$ where $M'_{3,5}$ is the set of all functions $f: \{3, 5, 8\} \to \mathbb{R}$ with $f \upharpoonright \{3, 5\} \in M_{3,5}$, and likewise with $M'_{5,8}$.)

Using these defined symbols, Tarski introduces a symbol Df whose intended meaning is 'first-order definable relation on the reals'; the definition of Df is that it is the smallest set that contains the sets $M_{i,j}$ etc. and is closed under the operations $\bar{o}$ etc. Then the 'first-order definable sets of reals' are defined to be the images of the members of Df whose domain is a singleton. Tarski remarks that we can extend these definitions straightforwardly to 'second-order definable relation' and so on; the papers [26] and [25] take up the second-order case. He adds that the Richard paradox shows that we can't hope to give a Mathematical definition of 'definable relation' covering all orders simultaneously.

The Geometric definition is a recasting of the Mathematical, with slightly different primitive notions that are explained in geometric language (for example 'projection' corresponding to existential quantification, and in [69] an operation that adds a dimension). Tarski says ([50] p. 238) that Kuratowski called his attention to the geometric interpretation.

Finally we reach the Accidental Mathematical version. This is limited to first-order definability, and it is based on the 'method of quantifier elimination'. Tarski took over this technique from Skolem [45] and applied it systematically to various theories in his Warsaw seminar from 1926 to 1928. In quantifier elimination we find, for each first-order formula ϕ, not the set of finite sequences that satisfy it, but another formula ϕ^* which is equivalent to ϕ in the deductive theory. The formula ϕ^* is a Boolean combination of formulas of some simple form—for the case of real numbers, it's a Boolean combination of atomic formulas. So the definable sets are exactly those defined by formulas ϕ^*; the mathematical content of the formulas ϕ^* then gives us the Accidental Mathematical description. For example if the deductive theory is $T_\mathbb{R}$, then the definable sets of reals are precisely the finite unions of intervals whose endpoints are either rational numbers or $\pm\infty$.

Why 'accidental'? If we change the primitives of $T_\mathbb{R}$, then in general we get different definable sets; for example if we add multiplication as a new primitive, we can define any intervals with algebraic endpoints. Of course this is true whether or not we use quantifier elimination. But with only quantifier elimination to guide us, we don't know what sorts of definable sets to expect when the primitives change. Tarski shows that his Mathematical description doesn't have this glitch. In fact it generalizes to a uniform Mathematical definition of the relation 'x is a set of finite sequences which is definable from $y_1, \ldots, y_n$', where $y_1, \ldots, y_n$ range over objects of appropriate type over the real numbers. This gladdens the heart of a modern model theorist, but strictly it isn't model theory yet. Tarski still assumes that there is a fixed universe (the real numbers), and the $y_1, \ldots, y_n$ are variables rather than non-logical constants.

So now we have our four kinds of description of 'definable set of reals'. In the last three cases Tarski puts his description in the form of a definition adding a new symbol to the theory. Now a definition is in effect an axiom, and we learned earlier in this chapter that axioms should be statements 'whose truth appears to us evident'. So we still have to establish that these three definitions appear to us to be evidently true.

In this chapter Tarski concentrates on establishing the 'subjective certainty' of the Mathematical definition, commenting that this is a question of 'capital importance' ([50] p. 229). Taking the Metamathematical definition as evidently true, we can deduce the same for the Mathematical definition by a metamathematical argument. (To show that every formula defines a relation in Df, we go by induction on the complexity of the formula. In the other direction we use induction on the construction of elements of Df. In spite of its 'capital importance', Tarski leaves details to the reader—as he does also for the Artificial Mathematical definition.)

Of course this metamathematical argument has no force at all unless we are already convinced that the Metamathematical definition is true. But the truth of the Metamathematical definition is the same thing as the truth of Tarski's truth definition. Clearly [50] is not the right place to investigate this, so Tarski simply says (p. 229):

(29)   ... admettons que la justesse matérielle de la définition métamathématique des ensembles définissables ... soit hors de doute.

Woodger translates

(30)   ... let us suppose that the material adequacy of the metamathematical definition of definable set ... is beyond doubt.

([64] 129)

That can't be right; it means that the truth of the Mathematical definition is proved only under an assumption which is never justified or discharged. Perhaps 'grant' for 'suppose' would rescue the sense.

After mentioning the metatheoretic justification of the Mathematical definition, Tarski asks a curious question: Suppose we wanted to assure ourselves of the correctness of the Mathematical definition, but using only Mathematical arguments, what could we do? To make sense of this question we have to be clear what is supposed to happen on the page and what belongs in the mind of the mathematician. An argument that takes place purely in the deductive theory is going to use only notions expressible in that theory, and this doesn't include the notion of a formula. So part of what has to be proved here *isn't even expressible in the theory*. Tarski's question means: What calculations in the deductive theory could we *use* to convince ourselves of the correctness of the Mathematical definition? Tarski's answer refers to formulas, but he must mean formulas that we *use* in these calculations, and not formulas that we *talk about*. Here is one of his examples, written as a calculation in $T_\mathbb{R}$ with some of his definitions unpacked:

(31) $0 \leqslant v_0 \quad \leftrightarrow \exists v_1(v_1 = v_1 + v_1 \wedge (v_1 = v_0 \vee v_1 < v_0))$
   $\leftrightarrow v_0 \in \{a \in \mathbb{R} : \exists b \in \mathbb{R}((b, b, b) \in \{(s, t, u) : s = t +_\mathbb{R} u\}$
   $\wedge \quad (a, b) \in (\{(s, s) : s \in \mathbb{R}\} \cup \{(s, t) : s <_\mathbb{R} t\}))\}$

The idea is that by doing enough 'purely automatic' calculations of this sort in $T_\mathbb{R}$, we can eventually convince ourselves that the definition of Df is accurate, even if we aren't up to carrying out an induction in the metatheory. This is one of the arguments that Tarski describes as 'empirical' ([50] p. 229ff).

CONDITIONS ON DEFINITIONS

Definitions of basic notions

As we noted earlier, mathematicians have a habit of giving definitions. Most mathematical definitions are local, in the sense that they are meant to serve some purpose in a particular argument and might never be used again. Contrast these with a global definition, which is meant for everyone to use from then onwards.

For example one might give a definition that extends a known technique to a broader range of applications. Lebesgue's definition of the Lebesgue integral is an obvious example. Arguably we can put Kuratowski's definition of topological spaces in this category too; it generalizes concepts originally introduced for spaces over the reals or the complex numbers.

Or one might want to tidy up a notion that had given rise to mistakes. Peano gave a number of definitions with this aim in mind—for example a definition of surface area that was meant to correct a mistake in an earlier definition by Serret. Frege also justified his definitions as a protection against errors; but whereas Peano could point to mathematical mistakes in the literature, Frege's complaints were about mistaken philosophical comments on foundational matters.

Or thirdly, one might want to give a definition as a conceptual analysis. Łukasiewicz's definition of causality might be an example; Frege's definition of number certainly was.

Or fourthly, one might want to give what Carnap [4] p. 3 calls an 'explication', namely 'the transformation of an inexact, prescientific concept, the *explicandum*, into a new concept, the *explicatum*'.

In the late nineteenth and early twentieth centuries, many of the mathematicians and logicians who gave global definitions did so within the framework of deductive theories. Frege and Łukasiewicz are examples. Lebesgue didn't, and in practice some of Peano's definitions hardly pay more than lip service to the formalism that he required in theory. Where writers did use a deductive theory, the deductive theory wasn't the goal of the exercise; rather it was a tool to help them achieve whatever aim they had in giving their definitions.

Tarski seems to be the exception here. He published a number of papers giving definitions of metamathematical notions. But unlike the other writers just mentioned, his sole aim seems to have been to formalize these notions in the way required by intuitionistic formalism. Feferman [7] rightly emphasizes that conceptual definitions of various kinds must have been an important part of Tarski's education. But what strikes me is how little direct effect they seem to have had on him.

(i) Tarski shows no interest in generalizing the notions that he formalizes. There is an interesting test case at the beginning of [54], p. 343 footnote 1 of [66], where he defines 'deductive theories' as models of a certain set of axioms. This certainly generalizes the usual notion of a deductive theory; but Tarski immediately pulls back and says

(32) In order not to depart too much from the usual meaning of the term 'deductive theory', I have in mind only those models of the axiom system which are constituted by certain sets of expressions and operations on expressions.

This is not lack of interest in the arbitrary models. As an intuitionistic formalist, Tarski *has to make this restriction*. Otherwise there is no way that he can convince himself (note the 'in mind', *im Sinne*) that the axioms are true statements about the consequence relation.

(ii) Tarski hardly ever identifies errors made by earlier writers, except for the general point that traditionally philosophers have ignored the need to relativize to a particular language—in effect, to a particular deductive theory ([66] pp. 110, 164f, 402). At [66] p. 416 he does criticize a definition given by Carnap; but there is no suggestion that he wrote the chapter to correct Carnap's error.

(iii) Nor do his definitions contain any conceptual analysis, except to the extent that he gives his definitions in terms of stated primitive notions. Definition in terms of primitives is not a form of conceptual analysis unless you take your primitives to be conceptually prior to the defined notions, and this is something that Tarski never does. In the introduction to the big Truth chapter ([66] p. 152f) the only requirement he puts on the primitives used to define truth is the Leśniewskian requirement that their senses 'must admit of no doubt'. This requirement includes the requirement that the primitives should not be semantic, because of doubts about the robustness of semantic concepts.

In fact the language of concepts and mental contents is conspicuously missing from Tarski's discussions, except where one would expect it from a faithful intuitionistic formalist: where he checks that his primitive notions are clear and that we have adequate grounds for believing his axioms and definitions.

(iv) Carnap himself ([4] p. 5) gives Tarski's truth definition as an example of an explication in Carnap's sense. But this is certainly wrong. Tarski restricts his truth definition to deductive theories, precisely to avoid ambiguities in the notion of 'true'. Of course it may happen that some sentence of your favourite deductive theory turns out not to have a truth value in spite of your best efforts; but then in general the Tarski truth definition won't give it a truth value either, because the subject-specific assumptions in the metatheory are the same as those in the object theory. ([66] pp. 165, 211.)

This restriction to the metatheory of deductive theories, in order not to have to deal with the 'usually ambiguous and inexact terms of colloquial language', is already in place in Tarski's earlier work on the definition of syntactic notions in the metatheory, [66] p. 60. Tarski has more to say about this. Near the beginning of [50] he contrasts his aim with that of earlier geometers who defined 'movement', 'line', 'surface', or 'dimension' ([66] p. 112):

(33) In geometry it was a question of making precise the spatial intuitions acquired empirically in everyday life, intuitions which are vague and confused by their

very nature. Here we have to deal with intuitions more clear and conscious, those of a logical nature relating to another domain of science, metamathematics. To the geometers the necessity presented itself of choosing one of several incompatible meanings, but here arbitrariness in establishing the content of the term in question [viz. 'definable'] is reduced almost to zero.

The same seems to apply to all Tarski's semantic definitions in the 1930s. In the case of logical consequence he admits there are various notions, but he builds the ambiguity into his own definition as a parameter to be fixed ([66] p. 420).

In Tarski's early treatment of the consequence relation, he notes that for setting up the metatheory one of his tasks (though not the only one) is to make notions precise ([66] p. 60). But this is a requirement of intuitionistic formalism.

Trafność

A definition of the set A in the form

(34) $\forall x(x \in A \leftrightarrow \phi(x))$

is true if and only if the members of A are exactly the objects that satisfy $\phi(x)$. This is the condition often known as *extensional correctness* of the definition. We saw that in [52] this was not Tarski's preferred form of definition; there he uses the form

(35) $\forall X(X = A \leftrightarrow \forall x(x \in X \leftrightarrow \phi(x)))$.

In the presence of the axiom of extensionality, (35) is true if and only if (36) is true. We recall that in [52] Tarski explicitly requires the system to contain the axiom of extensionality. So in this case too, Leśniewski's requirement amounts to this: that we should be sure that the definition is extensionally correct.

Seeing the importance of this condition, Tarski coined a name for it: *trafność*. Allow me a short linguistic interlude on this word and its translation.

The Polish noun *trafność* is the noun from the adjective *trafny*, whose literal meaning is 'on target' or 'accurate'. The word derives from the same Germanic root as the modern German *treffen*, which has much the same meaning. In fact Leopold Blaustein in his German translation of the big Truth chapter rendered *trafny* as *zutreffend* (except when he didn't; see p. 117 below). The word *trafny* is entirely appropriate for a definiens that captures the notion it was intended to capture. Kotarbiński does use it a few times in more or less this sense, but not as a technical term. (See for example [23] pp. 27, 36, 55.)

Some four years after he arrived in the United States in 1939, Tarski wrote in English an expository chapter on his truth definition [60]. He must have consulted about how to express *trafny*, and somebody told him to translate it as 'adequate'. (Woodger's translation in [64] follows Tarski by translating Blaustein's German *zutreffend* as 'adequate'.) In modern English this is an unfortunate choice; 'adequate' is far too vague to suggest extensional correctness. But Tarski's unknown advisor had his reasons. Extensional correctness was a condition that Aristotle required for definitions, and Aristotelian philosophers often expressed it by saying that the definiens must be 'equal' to the definiendum. In the twelfth century Abelard (e.g. [1] p. 591) used the term *adaequatus* for this condition; in Latin it means 'made equal'. The term had a revival in the post-Renaissance period. Unfortunately the endemic confusions and vaguenesses in the logic of this period often make it impossible to tell exactly what the authors mean by 'adequate'. Consider two examples:

(36) • [The definitions are] adequate...; in other words...they in fact grasp the current meaning of the notion as it is known intuitively.
 • A definition must be...adequate; that is, it must agree to all the particular species or individuals that are included under the same idea.

One of these quotations is from the eighteenth century logician and poet Isaac Watts ([74] I.6.v), and the other is from Tarski [50]. Maybe I make a point best by leaving the reader to guess which is which. But note that in the first quotation we are given no clue whether the 'meaning' is the sense or the extension, and in the second we need to be told whether the 'species' are subsets of the extension or components of the concept. I conclude that the presumed advice given to Tarski was historically accurate but not helpful for modern readers; and that Tarski's customary high standards of clarity don't always reach as far as his explanations of definitions.

The judgement that a particular definition in a deductive theory is *trafny* is a judgement in the intuitive metatheory of the deductive theory. So there is no question of proving it formally unless we formalize the metatheory, and in Tarski's view we can't do this without first thoroughly formalizing the deductive theory itself. Tarski discusses this point several times in the big Truth chapter. The deductive theory in which we give the truth definition is actually a metatheory, so a theory in which we show that the truth definition is *trafny* will be a meta-metatheory.

Recall that Leśniewski's requirement, as applied to definitions, is that we should be sure that the definitions are extensionally correct. Sometimes I read that Tarski must have meant his truth definition to be intensionally correct in some sense; otherwise why would he have required (in his Convention T) that we can prove its *trafność* from the axioms of the metatheory? The argument has some force, but not much. The reason Tarski wants the instances of Convention T to be provable in the metatheory is that he thinks this will convince us that the truth definition is extensionally correct. Now it may or may not be the case that we can only be convinced of the extensional correctness of a definition if the definition is also intensionally correct. It may or may not be true that Tarski thought this was the case. Perhaps somebody remembers discussing the question with him over coffee. It's clear that he didn't think the question was worth discussing in [51].

Formal correctness

Tarski certainly wasn't the first person to propose extensional correctness as a requirement on definitions. But he added to this a view that does seem to be original with him. Namely, the *only* other condition he requires on definitions of semantic terms is that they should be 'formally correct' (*formalnie poprawny*), i.e. they should satisfy the formal requirements on any definition added to a deductive theory. This condition appears nearly thirty times in the big Truth chapter. (It appears a dozen times more in the English version because of a glitch in the translation; see p. 117 below.)

But what are the requirements for adding a definition to a deductive theory? We have seen already that Łukasiewicz and Leśniewski didn't agree about the answer: Łukasiewicz thought a definition needed to be non-creative, but Leśniewski welcomed creative definitions. So when Tarski referred to formal correctness in [51], and added no further remarks about what he meant, he was passing the buck. There was no consensus about what this condition amounted to.

For example, can a recursive definition be formally correct? The answer has to be yes, because Tarski's definition of satisfaction is recursive. He does comment ([66] pp. 175–7 footnotes) that recursive and inductive definitions are open to 'methodological misgivings', but he suggests that these misgivings can be overcome by converting the definitions to 'normal' definitions by a device found in *Principia Mathematica*. So perhaps a definition is 'formally correct' if and only if it is equivalent in the deductive theory to a 'normal' definition. But then Tarski doesn't tell us what a normal definition is either.

In [52] he did give a normal form, but it makes sense only in logics of sufficiently high order. Also he didn't call it normal; his name for it was 'possible definition' (*eventualna definicja*). He used the same name for the forms that he gave much later in [71] p. 20 for first-order logic, as in (41) below; these latter are the forms usually called 'explicit definitions' today. Tarski does use the phrase 'explicit definition' in 1938, in French and again without any explanation ([2] p. 55). I haven't discovered any Polish phrase that he uses for 'explicit' definition, unless it be *normalny*.

There is one more clue. Tarski added a footnote to the German translation [53] of [52] in 1934, and it said ([53] p. 99 footnote 12):

(37) ... the two conditions for a correct definition—consistency and eliminability ...

... die beiden Bedingungen einer korrekten Definition—Widerspruchsfreiheit und Rücku- bersetzbarkeit ...

Here *korrekt* is the word Tarski chooses elsewhere to represent *poprawny*. We know about eliminability. The problem is that eliminability and consistency together are *not* equivalent to being equivalent to an explicit definition. Explicit definitions of relation symbols are never creative, but one easily constructs eliminable and consistent relation definitions that are creative. For example let T be any theory whose language contains a sentence ϕ such that neither ϕ nor $\neg\phi$ is deducible from T. Define a new relation symbol R by

(38) $\forall x \forall y (R(x,y) \leftrightarrow x = y) \wedge \phi$.

Now if ψ is any sentence containing occurrences of R, let ψ' be the result of replacing every formula $R(s,t)$ in ψ by the equation $s = t$. It's a consequence of (39) that ψ is equivalent to ψ', so (39) has the eliminability property. Also $T \cup \{\phi\}$ is syntactically consistent, and then we can consistently add the first conjunct of (39) because this conjunct is an explicit definition.

The condition 'consistency' survived into Woodger's translation in 1956 ([64] p. 307 footnote 3). But Woodger and Tarski had only limited contacts and we can't infer that 'consistency' was still Tarski's considered opinion. (See p. 105 above.)

It could be that Tarski's mind was elsewhere when he wrote that footnote to the German translation. But there is another possible explanation. An old doctrine going back at least to Christian Wolff held that 'inconsistent' definitions like '(planar) triangle with two right angles' don't define concepts, because our minds are incapable of forming contradictory concepts ([77] §43). Now of course one should distinguish between (i) definitions that are inconsistent and (ii) definitions that introduce a term that is necessarily not true of anything. But the sad fact is that people were still confusing these two things as late as the early twentieth century. In 1906 Frege wrote a devastating review of an article of Schoenflies that revolved around exactly this muddle, but for some reason the review was never published ([11]). Some version of the same confusion may lie behind the doctrine of 'implicit definition' in the philosophy of science, which taught that a set of axioms 'partially' or 'approximately' define a concept as long as they are consistent. Tarski's friend Ernest Nagel wrote an exposition of implicit definition in 1939 [37].

We know that in 1938 Tarski rejected this doctrine of implicit definition, but his reason was a methodological one that might not apply to a definition by a formula in a deductive theory. (Cf. (53) below.) So it's possible that even in 1934 Tarski did subscribe to this very ill-judged old doctrine about definitions. I hope he didn't, but I can't throw any further light.

In [15] I commented on some discrepancies between the Polish original and Woodger's English translation of the big Truth chapter. More recently I checked against Blaustein's German translation, and the reason for the discrepancies became clear. Blaustein translates *poprawny* as *korrekt*, while he translates *trafny* sometimes as *zutreffend* and sometimes as *richtig*. Woodger realizes that there are only two concepts involved, but he mistakenly thinks that *richtig* is a variant of *korrekt* and not of *zutreffend*. So he translates both *richtig* and *korrekt* as 'correct'; in one place where Blaustein has *korrekte und richtige Definition*, Woodger writes just 'correct definition' ([64] p. 224 l. 1).

Besides this example, the places where Woodger's 'correct' represents Tarski's *trafny* in [64] are:

(39) (page, line): 154, 11; 188, 18; 195, 5; 214, 22; 230, 4; 235, 33; 236, 3; 246, 18 and 24; 248, 9; 254, 2; 263 footnote line 7; 266, 2.

Unfortunately all these confusions have survived into the Corcoran edition [66], with the same page and line references. The fact that Tarski did nothing to correct them is one indicator of how little he was interested in definitions for their own sake. On p. 16 l. 14 of [51] Tarski himself writes *poprawnie* when he must surely mean *trafnie*.

PULLING THE THREADS TOGETHER

Model-theoretic definability

After the 1930s deductive theories largely disappeared from the scene. Most people who took a view on the overall structure of mathematics came to adopt a different paradigm. Put briefly, they came to recognize only one deductive theory for the whole of mathematics. This theory was usually first-order set theory, either Zermelo–Fraenkel set theory or a related system with classes added. The theory was a back-stop: nobody actually carried out formal deductions in it, but if one couldn't see how to formalize an argument in set theory then the argument was suspect.

This was a global change in mathematics, but one can track it within Tarski's papers on definition. In [52], reflecting work from the 1920s, Tarski uses deductive theories whose logic is simple type theory. This is still his framework in [50], although now he is handling set-theoretic operations on sets of finite sequences. In §2 of this chapter he needs the natural numbers to label terms of sequences, and he suggests that we might prefer to add to the deductive theory 'a system of arithmetic axiomatically constructed upon it' ([66] p. 120). In the big Truth chapter he needs a metatheory that can handle infinite sequences as well, so that the set-theoretic content goes up once again; instead of type theory, the logic of the metatheory is 'any sufficiently developed system of mathematical logic' ([66] p. 170). He also uses what he calls the 'set-theoretic' device of diagonalization, but only at the meta-meta-level ([66] p. 248). The Postscript in the German translation of 1935 explicitly puts into the metatheory the notion of transfinite ordinals 'taken from the theory of sets' ([66] p. 269). By this stage, the metatheory might as well be set theory. But so far Tarski uses the set theory only in order to handle sets, sequences, etc. of individuals, and not (for example) as a possible source of new models. Also it's clear that Tarski's only reason for using set theory at all is that he needs some specific set-theoretic devices that are unavailable or inconvenient in most logical calculi.

Under the new set-theoretic paradigm, mathematicians came to regard the more specialist axiom systems, for example the axioms of fields, as definitions within set theory. More precisely, the symbols for the field operations are meaningless. But we can choose a set K and assign to the constant symbols 0, 1 two elements of K, and to the function symbols $+, \cdot$ two binary functions on K. If this assignment makes the field axioms true, we say that K together with the assignments is a *model* of the field axioms. Then we can define a *field* to be a model of the field axioms. So the axioms become a definition of the class of fields.

Tarski's first published response to this new situation, his address [62] to the International Congress of Mathematicians in 1950, was something of a holding operation. Just as in the 1930s he saw the main task as being to find out what could be done in the mathematics and what needed metamathematics; the title of the chapter, 'Some notions and methods on the borderline of algebra and metamathematics', announces this clearly. The chapter reads oddly today. The really interesting model-theoretic results—including some of the earliest applications of the compactness theorem—are tacked on at the end, just as the quantifier elimination for the reals makes only a brief appearance towards the end of [50].

Tarski brings up to date the Mathematical version of definability from [50], taking into account that the deductive theory (the 'mathematics') is now set theory. Already in [50] he had introduced the set $\mathcal{D}(\mathcal{F})$ of relations definable from a set $\mathcal{F}$ of primitive relations on a fixed universe. Now he allows the universe to vary too, so that $\mathcal{F}$ is now a class of structures; but the remaining details are largely unchanged. Tarski shows how with these notions he can define, purely set-theoretically and with no reference to models or to satisfaction of formulas, what we now call a 'first-order definable class of structures'. Hence he can give a purely set-theoretic definition of elementary equivalence: two structures are elementarily equivalent if and only if they are in exactly the same first-order definable classes. He can also state the compactness theorem without any reference to first-order languages, and this language-free version is the one he uses for the model-theoretic applications at the end of the chapter. He shows how quantifier elimination can be used to prove facts about certain first-order definable classes, and he emphasizes that the method of quantifier elimination itself is not 'mathematical' but 'metamathematical' ([62] pp. 716, 719). The Mathematical version of definability in [50] rested on a prior Metamathematical notion, and clearly the same has to be true for the notion of definability in [62]. But Tarski leaves this unexplained. He makes no attempt to bring the notion 'M is a model of ϕ' into mathematics, and he even places the completeness theorem for first-order logic in metamathematics rather than mathematics ([62] p. 710). Just as with his earlier treatment of Padoa's method [52], he seems to be aiming to remove as much as possible of the model-theoretic content from model-theoretic arguments.

The missing metamathematical definition of 'M is a model of ϕ' arrives a few years later, sketched in [63] and then fully explained in [72]. In [63] Tarski is still emphasizing from his first line onwards that model theory lies in metamathematics, and that the model-theoretically defined classes of structure can also be given a 'purely mathematical definition' ([63]). But by this date (1954) all the energy has gone out of the distinction. Tarski doesn't address the question how one could reason in a metatheory outside set theory; in fact all the metamathematics in [63] could perfectly well be done within set theory. In [72] metamathematics isn't even mentioned, and phrases like 'purely algebraic' ([72] p. 82) and 'purely mathematical' ([72] p. 96) read like descriptions of flavour rather than foundational status.

By the 1950s first-order logic had become the logic of preference. In [70], [52], and [26] Tarski had used forms of definition that were good only for higher order logics, but they had the advantage that the same form worked for all types of primitive symbol. First-order logic needed one form of definition for individual constants, one for relation symbols, and one for function symbols, thus:

(40) $\forall x(x = c \leftrightarrow \phi(x))$
$\forall y_1 \ldots \forall y_n(R(y_1, \ldots, y_n) \leftrightarrow \phi(y_1, \ldots, y_n))$
$\forall x \forall y_1 \ldots \forall y_n(x = F(y_1, \ldots, y_n) \leftrightarrow \phi(x, y_1, \ldots, y_n))$
where c, R, F doesn't occur in ϕ.

It happened to be Tarski who wrote down these forms and the conditions under which they can be added to a theory without being creative, in 1953. ([71] p. 20. In fact he gave only the second and third forms, but the first is a special case of the third.) These forms are usually known today as 'explicit definitions'. Tarski (for reasons not clear to me) calls them 'possible definitions'. Also he doesn't use the phrase 'creative definition'; in fact I don't know any place where Tarski does use this phrase before 1983 ([66] p. 307 footnote 3).

When set theory became accepted as the deductive theory for all mathematics, the choice of axioms was handed over to the consensus of the mathematical community. If an individual mathematician didn't feel that the axioms had an 'irresistible intuitive validity' (in Leśniewski's phrase), that was a personal matter, nothing to do with mathematics. So Leśniewski's intuitionistic formalism no longer had any work to do, except perhaps for a few people interested in the foundations of set theory. Tarski duly announced in 1956 ([64] p. 62) that he no longer subscribed to it.

The birth of the truth definition

In this section I step back and try to reconstruct the sequence of events that produced Tarski's truth definition and its corollaries. Of course this can only be on the basis of the evidence that we have today, and tomorrow we may know more.

Until early 1929 Tarski's interest in definitions had been confined to two projects. The first was the project of investigating how definitions fit into the hierarchy of deductive theory and metatheory. A debate between Leśniewski and Łukasiewicz had set this project moving, and some of Tarski's early contributions appear in two reports [30] and [31] of talks by Łukasiewicz. The early work [70] giving a 'theoretical justification' for Padoa's method belongs here too.

The second project was to devise a general formal metatheory for deductive theories. Like the construction of any deductive theory, this project involved four things: to choose the primitives, to write axioms in terms of the primitives, to define other notions in terms of the primitives, and to derive further truths from the axioms by logical deduction. Most pages of Tarski's two papers [47], [48] on this project are devoted to deriving theorems. But one of the best tests of a theory is its ability to handle the required concepts, so the choice of primitives and definitions was a crucial part; in fact the word 'concepts' appears in the titles of both papers. (Both papers were published in 1930, but Tarski first reported on this project to the Polish Mathematical Society in 1928, [66] footnote p. 30.) The concepts considered in these papers are all syntactic—the only primitives are 'sentence' and 'consequence' (by rules of inference). There is no hint that one might extend the project to take in the concept of truth.

It may be that by early 1929 Tarski had already wondered about the possibility of giving metatheoretic definitions of semantic notions. But there is no positive evidence of this, and at that date there was no obvious route to take towards such definitions. The first place I know of where Tarski refers to a project of finding such definitions, or more precisely the

(41) task of laying the foundations of a scientific semantics, i.e. of characterizing precisely the semantical concepts and of setting up a logically unobjectionable and materially adequate way of using these concepts

is in 1935 ([57], p. 402 of [66]).

Step 1

In 1929 Kotarbiński's book [23] appeared, and Tarski read it. (Perhaps he took it on his honeymoon for light reading.) After reading and digesting the material on definition, he came to Kotarbiński's chapter 3 on truth, and here he saw the formulation

(42) Jan thinks the truth if and only if Jan thinks that such-and-such is the case and such-and-such is in fact the case.

([23] p. 127, referred to at [51] p. 155 of [66])

He recognized 'thinks that . . .' as a kind of quotation-mark context, and saw that a formalization of (43) in a metatheory would have on the left side a sentence of the language of the deductive theory in quotation marks, and on the right an equivalent expression of the metalanguage. An instance looks like this:

(43) 'It is snowing' is a true sentence if and only if it is snowing.

Tarski will have been sensitized to the quotation marks by Leśniewski's emphasis on their role in semantic paradoxes ([51] footnote on pages 154f of [66]).

I suggest that this feature struck Tarski as so important that he quickly adopted some terminology for it. He took over Kotarbiński's term for a definition where an expression appears in quotation marks on the left and is explained without them on the right: *semantic definition*. True, a genuine semantic definition in Kotarbiński's sense would have the form

(44) 'x is a true sentence' means that p.

with the quotation marks in a different place from Tarski's formulation. But Tarski was impressed enough by the similarity that he applied Kotarbiński's term to a definition along the lines of his (44). As Tarski himself explains it, he regards (44) and related formulations as semantical definitions because they

(45) establish a direct correlation between the sentences of the language and the names of these sentences

(among other things; see the detailed footnote 2 on pp. 237f of [66]). This seems to be the reason why he talks of his 'semantic conception of truth', for example in the title of [60]. The name has puzzled people ever since.

Step 2

Next, I suggest that Tarski saw very quickly that the method of quantifier elimination delivered sentences of exactly the required form. (He must have had quantifier elimination coming out of his ears at the time, thanks to his Warsaw seminar of 1926–8 and his own continuing work in the area.) More precisely, quantifier elimination yields results of the form

(46) ϕ is equivalent to ϕ^*.

For example if the deductive theory is that of the ordered additive group of real numbers, then

(47) '$\exists x(3x > 5 \wedge 6x = y)$' is equivalent to '$y > 10$'.

All that is needed is to name the formula ϕ on the left, and on the right to use a formula of the metalanguage which expresses the same as ϕ^*. Now the formula '$y > 10$' has a free variable. So to express the equivalence of the formulas, we need to quantify this variable in the metatheory. Thus we need to say

(48) For all reals ρ, ρ satisfies '$\exists x(3x > 5 \wedge 6x = y)$' if and only if $\rho > 10$.

Note how the word 'satisfies' pushes its way in here, to connect the metalanguage variable ρ with the metalanguage name of the object language formula '$\exists x(3x > 5 \wedge 6x = y)$'.

Step 3

To get a definition of 'true', rather than an infinite number of special cases like (49) (which Tarski calls 'partial definitions', [66] p. 155ff), we need a single sentence that captures all the special cases. It would be tempting to suggest at this point that Tarski took over from quantifier elimination the idea of using a definition by recursion on the complexity of the formula on the left side. Namely, in quantifier elimination the formula ϕ^* is defined recursively from ϕ. But the truth must be more complicated than that, because in 1929 Tarski didn't define ϕ^* from ϕ by recursion on complexity.

We know how Tarski handled quantifier elimination in his Warsaw seminar in 1926–8, because Presburger [44] sets out the machinery and ascribes it to Tarski (in his footnote on p. 97, which Tarski himself endorses at footnote 21 on p. 42 of [58]). As Presburger has it, there is no definition of ϕ^* at all. The main theorem of his quantifier elimination states that for every ϕ there is a ϕ^* such that etc. etc. This is proved by a descent argument on the number of quantifiers in the quantifier prefix of ϕ, which is assumed to be in prenex form. Presburger does add the comments that the proof is effective, and that one can easily extract from it a decision procedure (*Verfahren*). If we suppose that these comments came from Tarski, and that Tarski had in mind a deterministic algorithm, and that Tarski had thought about how to describe such an algorithm formally, then we could probably infer that it had crossed Tarski's mind to define a suitable ϕ^* from ϕ by recursion on complexity; but there are too many ifs here. In [50] p. 233, discussing the quantifier elimination for the ordered additive group of reals, Tarski merely says that every first-order sentence in the language can be proved or refuted. (His added remark about a 'mechanical method' at [64] p. 134 dates from 1956.)

The first explicit mention of proof by induction in connection with quantifier elimination may well be on page 25 of [58] in 1939; this is induction on the number of

Tarski's Theory of Definition 123

quantifiers in the formula, but it might as well be induction on the complexity of the formula. However, it's reasonable to suppose that already by 1929 Tarski had thought about how to formalize quantifier elimination within the metatheory, and for this he would have had to use proof by induction. On page 17 of [58] Tarski says he will simplify the proof 'since an exact presentation of the proof would be very troublesome and complicated', and he adds that he will give the reasoning 'a mathematical and not a metamathematical form'; we have seen places where Tarski uses similar language to mean that an induction will be left out.

Though Tarski couldn't have taken over definition by recursion from quantifier elimination to the truth definition, I suggest that something close to this did happen. To lift from a truth definition covering finitely many cases to one covering infinitely many, some kind of induction or recursion would be the natural tool, as Tarski himself says at [66] pp. 188f. I suggest he looked at the quantifier elimination that he already understood, and checked how an inductive argument might go. Given ϕ^* and ψ^*, the quantifier elimination takes $\phi^* \wedge \psi^*$ to serve as $(\phi \wedge \psi)^*$; and similarly with the other Boolean expressions. Translating as in Step 2, this means we can infer from

(49) For all α, α satisfies '$\phi(x)$' if and only if $\phi^*(\alpha)$;

For all α, α satisfies '$\psi(x)$' if and only if $\psi^*(\alpha)$

to

(50) For all α, α satisfies '$(\phi \wedge \psi)(x)$' if and only if $(\phi^* \wedge \psi^*)(\alpha)$.

In quantifier elimination something quite different happens when we add a quantifier to the formula on the left. I suggest it occurred to Tarski that there was no point in trying to copy quantifier elimination at this step, because this is exactly where quantifier elimination does different things for different theories. Instead he could copy the passage from (50) to (51), and add a quantifier on the right to match the one added on the left.

There is some shuffling around to be done here, but it is all mathematical bookkeeping of a kind that Tarski excelled at. From (50) to (51) is one step of an induction on the complexity of the formula on the left. This has to be an induction in the metametatheory. (That's clear from the footnote on page 188 of [66], once we change the last word 'metatheory' to 'meta-metatheory' as in the Polish and the German.) Trying to set up this induction, one finds that one needs to define the formulas on the righthand sides of (50) etc. recursively in terms of the formulas on the left. But now there is no problem about doing that, because the formula on the right is calculated from that on the left in a particularly simple way.

By this route Tarski would have reached his truth definition for first-order formulas. The extension to higher order languages is in principle straightforward, though of course there are technical details to work out.

There are two reasons for suggesting Steps 2 and 3. The first is that they provide a clear route from what Tarski was doing in early 1929 to the truth definition that he announced in 1930. The second is that traces of this route are in his papers. The quantifier elimination for the ordered additive group of real numbers is in the first of his published papers to mention the truth definition, namely [50]. A version of Skolem's quantifier elimination for the theory of sets with inclusion is part of Tarski's opening example in the big Truth chapter, [66] pp. 202–8. In both cases Tarski stresses that the quantifier elimination is an 'accidental' feature of the theories in question and couldn't lead to a general definition (of truth or of definability); the 'accidental' part of quantifier elimination is exactly what Step 3 eliminates.

Step 4

Having both a metatheory for provability and a metatheory for truth, Tarski could combine them and show by induction on the complexity of proofs that every provable statement of a deductive theory is true ([66] p. 236). (What? Even if the theory has false axioms? Yes, even then; the axioms of the metatheory include those of the object theory.) As Tarski himself notes, there was a template for this argument already in Hilbert and Ackermann [13] pp. 65–8.

At this point we can start to put dates. Step 3 was already in place before the work reported in [26], and Kuratowski announced results from this chapter to the Polish Mathematical Society in June 1930 ([50] footnote 2 on page 211); so Steps 1 to 3 occupied Tarski from sometime in 1929 to sometime before mid 1930. Tarski's Historical Notes on p. 277f of [66] say that in 1929 he had already 'arrived at the final formulation of the definition of truth along with most of the remaining results presented in this work . . .'. It's probably not safe to include under 'final formulation' anything that Tarski doesn't explicitly mention in these Notes. But he claims here that by the end of 1929 he had 'the definition of the concept of truth for the case where the means available in the metalanguage are sufficiently rich'. That seems to cover Step 3, with the reservation that in [52], which is probably from mid 1930, he still had doubts about some technical details connected with type theory. He definitely claims in the Notes that he had Step 4 in 1929. The timing is tight, but the adaptation of quantifier elimination is so straightforward that it's entirely believable.

Step 5

I suggest that Tarski's next move was to follow his earlier concerns, particularly his treatment of Padoa's method: I suggest that he tried to push as much of the truth definition as he could into the object theory. His style was to start with quite simple syntactic manipulations. I think he noted that he could eliminate the metalanguage variable ρ in (49) by rewriting the formula as a correlation between the metatheory name of the formula and a set of reals expressible in the object theory:

(51) '$\exists x(3x > 5 \land 6x = y)$' correlates with $\{r \in \mathbb{R} : r > 10\}$.

The recursion doesn't need many changes from the previous version, but technically (52) uses a higher type of object than (48) did. So Tarski never offered it as a form of the truth definition, but he did notice that the class of correlated sets was straightforwardly definable in the object theory. This is exactly the reduction from the Metamathematical notion of definability to the Mathematical notion in [50]. The correlation must have appeared in Tarski's proof that the Mathematical notion is correct given the Metamathematical one. Incidentally this correlation seems to have been the first explicitly compositional semantics (in the sense of Partee et al. [40]) for a formal language.

As we noted earlier, the methodology of deductive theories required Tarski to show that his definitions of intuitive notions were extensionally correct. For the Mathematical definition of definability he could show this by a straightforward deduction from the Metamathematical; but then there was the problem of justifying the Metamathematical. This was the problem of proving the *trafność* of the definition of satisfaction. That problem first appears in this context in [52], and I suggest that Tarski came to the problem by this route. It was natural for him to mention formal correctness as another requirement on a definition, though at this point he saw no need to be specific about what formal correctness was. In [49], which is Tarski's abstract of a talk that he gave to the Warsaw Philosophical Society in October 1930 on his truth definition, Tarski used the phrase 'accurate from an intuitive point of view' (*trafny z intuicznego punktu widzenia*); at this date he wasn't yet using *trafny* on its own as a technical term.

Probably sometime early in 1930 Tarski mentioned the results of Step 5 to Kuratowski, and the result was the Tarski–Kuratowski algorithm [26], [25]; these papers were published in 1931, but very likely written in late 1930, since Kuratowski reported on them to the Polish Mathematical Society in October 1930 ([25] footnote 1 on page 249). Tarski must have been overjoyed to have a concrete mathematical application of the truth definition so soon. But nothing else along these lines materialized during the 1930s, and for a while Tarski's further thinking about the truth definition was all at a foundational level. For example in early 1931 Gödel's incompleteness result led him to prove his theorem on the undefinability of truth, and incidentally vindicated the care he had put into calculating how the order of the formulas affected the order of their truth definition.

The next few years seem to have been fallow for the truth definition. One can see things that he might have done but didn't. First, in the 1935 German translation of his write-up on Padoa's method [52] he saw the need to say something about formal correctness, but what he said showed that he hadn't given much thought to the matter. Second and more striking, the write-up itself in 1934 was just as anti-model-theoretic as the 1926 abstract. At this late date Tarski still had no interest in using the truth definition to formalize Padoa's notion of 'interpretation of a theory'.

Step 6

Something in 1934 or early 1935 made Tarski realize that he could use the truth definition to define 'model of a theory' and then use this definition in turn to express 'Every model of ϕ is a model of ψ'. The definition of 'model' appears in [56] (a congress talk delivered in 1935) and the textbook [55] of 1936. By intuitionistic formalism the primitive symbols already have a meaning, and this meaning may be completely unrelated to the 'model' we are interested in. So Tarski has to do something along the lines of re-interpreting the primitives. In fact he avoids doing this literally, and instead he replaces the primitives by variables; this is a move we saw anticipated in his treatment of Padoa's method in 1925 [70].

In 1934 Tarski had given a talk to the Polish Mathematical Society on the definition of elementary equivalence of relations. One would like to know whether the 1934/35 notion of 'model' played any role in this definition. Unfortunately he never wrote up even an abstract of this talk, and the summary at the end of [54] is cryptic. The definition of elementary equivalence that he gave in 1950 [62] doesn't use the 1934/35 notion of 'model'.

Tarski saw the 1935 congress as an opportunity to advertise the truth definition. He thought that a good framework for the definition would be as part of a project to give definitions of semantic notions. The chapter [57] does present this view quite convincingly, until one notices that without exception all the definitions mentioned are corollaries of the truth definition. So there is no real evidence of any project here, except to exploit the truth definition within the framework of deductive theories and their metatheories.

Step 7

In 1950 Tarski defined (by implication only, as we noted in the previous section) the notion of the class of structures defined by a theory. He may have come to this definition before 1950, but not much before. In 1938 he expressly disowned the notion that a theory can define a class of structures ([2] p. 55):

(52) [The view that axioms can be seen as implicit definitions] ne pourrait être justifiée que si la notion de définition, étudiée par la méthodologie contemporaine des sciences déductives, avait subi au préalable une extension essentielle. Or, entre les définitions proprement dites (c'est-à-dire définitions 'explicites') et les systèmes d'axiomes il y a des différences si essentielles—surtout au point de vue méthodologique—qu'il ne paraît pas avantageux d'embrasser les uns et les autres du même nom de définition.

In [14] I suggested that these remarks pointed to Tarski's caution. Cautious he was. But today I would rather say that the remarks show how deeply Tarski's early metalogical work was controlled by the deductive theory-metatheory framework that he learned from Łukasiewicz and Leśniewski, right down to the end of the 1930s.

Conclusions

All in all, it seems that during the 1920s and 1930s Tarski had no particular interest in definitions for their own sake. They forced themselves on him because of his intense work on metatheory. They came in two guises: as parts of a deductive theory or a metatheory, and as a problem about where to fit them into the theory/metatheory scheme. But apart from what he had to say about them in these two connections, he could be quite careless with them. One example is his apparently thoughtless treatment of formal connectness of definitions. Another is his vague explanation of *trafność* (37).

Tarski's detachment from traditional views on definitions was probably a benefit. From Aristotle onwards there had been a tendency to explain definitions in terms of a list of mistakes to avoid. Tarski has none of that. (It reappears in Carnap's list of 'requirements for an explicatum', [4] p. 7.) Instead Tarski has the novel and clinical pair of requirements, formal correctness and (where it applies) *trafność*. True, Tarski himself is vague about some details of them. But we can fill in most of the details by referring to the methodology of deductive theories. I think myself that there are hidden depths in these two requirements; in [16] I compared the requirement of formal correctness with its counterparts in lexicography.

The formal requirements on a definition include two things that were ignored quite disgracefully by most logicians before Tarski. One is to match the variables in the definiendum with those in the definiens. Frege, Peano, and Leśniewski were careful about this, certainly. But for example in 1929 Carnap wasn't; his definition of 'definition' on page 7 of his *Abriss* [3] says nothing about variables. Today the requirements are part of the culture, and Tarski's formulations are generally the ones used. The other requirement often ignored in the early days was that a definition shouldn't be added to a theory unless the theory already entails the existence and uniqueness assumptions implied by the definition. This is part of non-creativity, which Tarski admittedly missed stating; but again his explicit rules ([71] p. 20) have it right, and those are the rules that we follow today.

Let me mention two other aspects of definition where Tarski made important contributions. These were things he did under the pressure of his work on metatheory and perhaps without much awareness of the broader context.

The first has to do with mentally identifying things, and it ties in with the origins of model theory. By the 1920s there were three known procedures that involved varying the meanings of the primitive symbols of an axiom system. (1) To prove the consistency of a theory T, we think of an interpretation for the primitives that makes T true. Variants of this appear in proofs of non-deducibility, and in Padoa's method for proving undefinability. (2) We use an axiom system to define a class of structures, namely all those structures that make the axioms true when the primitives are interpreted as names of parts of the structures. (3) We define an isomorphism as a relation between structures that matches up interpretations of the same primitives. Tarski accepted all these procedures, as indeed did all mainstream logicians. But what is very noticeable about Tarski in his early career is his unwillingness to count (1) or (2) as mathematical, or even to include them in a formalized metamathematics. For him they were purely intuitive procedures. During the period from 1935 to 1950 one sees him being forced to move larger and larger portions of them into mathematics, but only where the formalization meets his own methodological standards. One might say that his very unwillingness to accept model-theoretic methods as mathematics made him build them on firm foundations—and of course we are all the beneficiaries.

A matter that I hope somebody will investigate is the striking parallel between Tarski's methodological misgivings over (1) and (2) and those of Frege in his controversy with Hilbert over the foundations of geometry [10]. Besides the similar general context, there are parallels of detail. Like Tarski, Frege complained that Hilbert's procedures lay outside 'mathematics' and would require a new deductive discipline ([10]

1906 p. 426). Like Tarski, Frege avoided the notion of 'two different interpretations' of a symbol S by introducing a copy of S to carry the second meaning ([10] 1906 p. 426f.). Like Tarski, Frege objected that a set of axioms doesn't define a class of structures in any adequately explained sense of 'define' ([10] 1903 p. 321). But as far as I know, Tarski was unaware of the existence of this work of Frege.

The second has to do with the interpretation of compound phrases. Traditional logicians had distinguished between simple and compound expressions. It was common to say that compound expressions can't be defined, but rather they get their meanings from the meanings of the simple expressions that are composed to form them. With a very few exceptions, no traditional logicians gave more than a trivial account of how the composition of words conveys a composition of meanings. A breakthrough came with the idea that the syntax of a language is autonomous. Given the syntax, one can go on to describe the meanings of compound phrases in terms of their syntactic structure and the meanings of the phrases that compose them. This view goes by the name of *compositionality* (for example in Partee *et al.* [40]). The view only became possible with the appearance of formal grammars. These were used in logic some decades before they became standard in linguistics, but it was probably only in the 1920s that logicians gained enough expertise with formal grammar and induction on complexity to be able to handle compositionality. (Above I suggested that quantifier elimination provided Tarski with a set of exercises in the use of induction on complexity. In 1921 Emil Post made the use of complexity of formulas a selling point for his doctoral dissertation [43].) The first fully compositional semantics was in Tarski's truth definition. It reached Barbara Partee through Tarski's student Richard Montague.

Finally, consider two personal references that Tarski makes. The first is in his introductory remarks to [50] where he explains that 'mathematicians in general' don't like to operate with the notion of definability. He adds that he hopes his chapter will convince the reader that there is no need to have reservations about this notion. Who does he have in mind? What mathematicians had avoided using the notion of definability because they mistrusted it? (Heinz Dieter Ebbinghaus kindly tells me that Zermelo, after his visit to Warsaw in 1929, reported having discussions with Knaster, Leśniewski, and Tarski, and that there is some evidence that Zermelo's notion of *definit*—'well-defined'—was one of the things they discussed. Did one of them mistrust it? Presumably not Zermelo himself, since he based his set theory on it.)

My guess is that Tarski meant himself in the first instance. As we saw with Padoa's method, his notion of giving a theoretical foundation to a notion was to incorporate it into a suitable deductive theory. He mistrusted definability in the same way as he mistrusted model-theoretic methods: it was fine to use informally, but it wasn't a solid mathematical notion until Tarski showed how to make it one.

Second, Anita and Solomon Feferman [8] p. 41 discuss why Tarski dedicated the 1954 volume of English translations of his papers 'To his teacher Tadeusz Kotarbiński' rather than to his more significant teachers Łukasiewicz and Leśniewski. They attribute it to the breakdown in personal relations between Tarski and those two during the 1930s. But why Kotarbiński?

If the reconstruction in this chapter is correct, then Kotarbiński did do for Tarski something that neither Łukasiewicz nor Leśniewski achieved. Kotarbiński's chapter 3 was the catalyst that enabled Tarski to grow from a clever mathematical problem-solver into a giant in the world of ideas.

References

[1] Peter Abelard, *Dialectica*, ed. L. M. De Rijk, Van Gorcum, Assen 1970.
[2] Marcel Barzin, 'Langage et réalité' (with comments by Tarski and others), in *Les Conceptions Modernes de la Raison*, ed. Raymond Bayer, Hermann, Paris 1939, pp. 20ff.
[3] Rudolf Carnap, *Abriss der Logistik mit Besonderer Berücksichtigung der Relationstheorie und ihrer Anwendungen*, Springer, Vienna 1929.
[4] —— *Logical Foundations of Probability*, University of Chicago Press, Chicago 1950.
[5] —— 'Intellectual autobiography', in *The Philosophy of Rudolf Carnap*, ed. P. A. Schilpp, Open Court, La Salle, Illinois 1963, pp. 3–81.
[6] Alonzo Church, *Introduction to Mathematical Logic vol. I*, Princeton University Press, Princeton 1956.
[7] Solomon Feferman, 'Tarski's conceptual analysis of semantical notions', in *Sémantique et Épistémologie*, ed. Ali Benmakhlouf Éditions Le Fennec, Casablanca 2004, pp. 79–108. Reprinted this volume.
[8] Anita Burdman Feferman and Solomon Feferman, *Alfred Tarski: Life and Logic*, Cambridge University Press, Cambridge 2004.
[9] Gottlob Frege, *Grundgesetze der Arithmetik: I* 1893, Hermann Pohle, Jena *II* 1903.
[10] —— 'Grundlagen der Geometrie', *Jahresbericht der deutschen Mathematikervereinigung* 12 (1903) 319–24, 368–75; 15 (1906) 293–309, 377–403, 423–30.
[11] —— Review of Schoenflies, 'Die logischen Paradoxen der Mengenlehre', transl. in Gottlob Frege, *Posthumous Writings*, ed. Hans Hermes *et al.*, Blackwell, Oxford 1979, pp. 176–83.
[12] David Hilbert, 'Axiomatisches Denken', *Mathematische Annalen* 78 (1918) 405–15.
[13] David Hilbert and Wilhelm Ackermann, *Grundzüge der Theoretischen Logik*, Springer, Berlin 1928.
[14] Wilfrid Hodges, 'Truth in a structure', *Proceedings of Aristotelian Society* 86 (1985/86) 135–51.
[15] —— 'What languages have a Tarski truth definition?', *Annals of Pure and Applied Logic* 126 (2004) 93–113.
[16] —— 'Formally correct definitions', in *The Logica Yearbook 2003*, ed. Libor Běhounek, Filosofia, Prague 2004, pp. 35–43.
[17] —— 'Detecting the logical content: Burley's "Purity of Logic"', in *We Will Show Them, Essays in Honour of Dov Gabbay*, vol. II, ed. Sergei Artemov *et al.*, College Publications, Kings College London 2006, pp. 69–115.
[18] —— 'The scope and limits of logic', *Handbook of the Philosophy of Science: Philosophy of Logic*, ed. Dale Jacquette, Elsevier, Amsterdam 2007, pp. 41–63.
[19] —— 'Tarski on Padoa's method: a test case for understanding logicians of other traditions', in *International Conference on Logic, Navya-Nyāyā and Applications: A Homage to Bimal Krishna Matilal, Kolkata 2007, Volume I*, ed. Mihir K. Chakraborty *et al.*, College Publications, Kings College London (forthcoming).
[20] Ignacio Jané, 'What is Tarski's *common* concept of consequence?', *Bulletin of Symbolic Logic* 12 (2006) 1–42.

[21] Immanuel Kant, *Lectures on Logic*, ed. J. Michael Young, Cambridge University Press, Cambridge 1992.
[22] Alexander Kechris, *Classical Descriptive Set Theory*, Springer, New York 1995.
[23] Tadeusz Kotarbiński, *Elementy Teorji Poznania, Logiki Formalnej i Metodologji Nauk*, Wydawnictwo Zakładu Narodowego Imienia Ossolińskich, Lwów 1929.
[24] —— 'Notes on the development of formal logic in Poland in the years 1900–39', in McCall [35] pp. 1–14.
[25] Kazimierz Kuratowski, 'Evaluation de la classe borélienne ou projective d'un ensemble de points à l'aide des symboles logiques', *Fundamenta Mathematicae* 17 (1931) 249–72.
[26] K. Kuratowski and A. Tarski, 'Les opérations logiques et les ensembles projectifs', *Fundamenta Mathematicae* 17 (1931) 240–8.
[27] Stanisław Leśniewski, 'Grundzüge eines neuen Systems der Grundlagen der Mathematik', *Fundamenta Mathematicae* 14 (1929) 1–81.
[28] —— 'Über Definitionen in der sogenannten Theorie der Deduktion', *Comptes rendus des séances de la Société des Sciences et des Lettres de Varsovie*, cl. iii, 24 (1931) 289–309; trans. as 'On definitions in the so-called theory of deduction', [29] ii, pp. 629–648.
[29] Stanisław Leśniewski, *Collected Works*, ed. S. J. Surma *et al.*, Kluwer, Dordrecht 1992.
[30] Jan Łukasiewicz, 'Rola definicyj w systemach dedukcyjnych', *Ruch Filozoficzny* 11 (1928/9) 164.
[31] —— 'O definicyach w teoryi dedukcyi', *Ruch Filozoficzny* 11 (1928/9) 177–8.
[32] —— *Elementy Logiki Matematycznej*, PWN, Warsaw 1929; trans. as *Elements of Mathematical Logic*, Pergamon Press, Oxford 1963.
[33] —— 'Philosophische Bemerkungen zu mehrwertigen Systemen des Aussagenkalküls', *Comtes rendus des séances de la Société des Sciences et des Lettres de Varsovie*, cl. iii, 23 (1930) 51–77; trans. 'Philosophical remarks on many-valued systems of propositional logic' in [35] pp. 40–65.
[34] —— *Aristotle's Syllogistic from the Standpoint of Modern Formal Logic*, Clarendon Press, Oxford 1951.
[35] Storrs McCall (ed.), *Polish Logic 1920–1939*, Clarendon Press, Oxford 1967.
[36] Andrzej Mostowski, *Logika Matematyczna*, Monografie Matematyczne, Warsaw 1948.
[37] Ernest Nagel, 'The formation of modern conceptions of formal logic in the development of geometry', *Osiris* 7 (1939) 142–224.
[38] Alessandro Padoa, 'Essai d'une théorie algébrique des nombres entiers, précédé d'une introduction logique à une théorie déductive quelconque', *Bibliothèque du Congrès international de philosophie, Paris, 1900*, vol. 3, Armand Colin, Paris 1902, pp. 309–65; translated in part as 'Logical introduction to any deductive theory' in *From Frege to Gödel*, ed. Jean van Heijenoort, Harvard University Press, Cambridge, Mass. 1967, pp. 118–23.
[39] —— 'La logique déductive dans sa dernière phase de développement', *Revue de Métaphysique et de Morale* 19 (1911) 828–32; 20 (1912) 48–67, 207–31.
[40] B. H. Partee, A. Ter Meulen, and R. E. Wall, *Mathematical Methods in Linguistics*, Kluwer, Dordrecht 1990.
[41] Blaise Pascal, *De l'Esprit Géométrique et de l'Art de Persuader*, in *Oeuvres Complètes*, Seuil, Paris 1963, pp. 348–59.
[42] Porphyry, *Porphyrii Isagoge et in Aristotelis Categorias Commentarium*, ed. Adolfus Busse, Reimer, Berlin 1887.

[43] Emil Post, 'Introduction to a general theory of elementary propositions', *American Journal of Mathematics* 43 (1921) 163–85; reprinted in *From Frege to Gödel*, ed. Jean van Heijenoort, Harvard University Press, Cambridge, Mass. 1967, pp. 264–83.

[44] M. Presburger, 'Über die Vollständigkeit eines gewissen Systems der Arithmetik ganzer Zahlen, in welchem die Addition als einzige Operation hervortritt', *Comptes Rendus du Premier Congrès des Mathématiciens des Pays Slaves, Warszawa 1929*, Warsaw (1930) 92–101; supplementary note ibid. 395.

[45] Thoralf Skolem, 'Untersuchungen über die Axiome des Klassenkalküls und über Produktations- und Summationsprobleme, welche gewisse Klassen von Aussagen betreffen', in *Skrift Videnskapsselskapet, I. Mat.-nat. klasse 1919* no. 3 (1919).

[46] Patrick Suppes, *Introduction to Logic*, Van Nostrand, Princeton, NJ 1957.

[47] Alfred Tarski, 'Über einige fundamentale Begriffe der Metamathematik', *Comptes Rendus des séances de la Société des Sciences et des Lettres de Varsovie* 23 (1930) iii, 22–9.

[48] —— 'Fundamentale Begriffe der Methodologie der deduktiven Wissenschaften I', *Monatshefte für Mathematik und Physik* 37 (1930) 361–404.

[49] —— 'O pojęciu prawdy w odniesieniu do sformalizowanych nauk dedukcyjnych', *Ruch Filozoficzny* 12 (1930/1) 210–11, reprinted in [68] pp. 3–8.

[50] —— 'Sur les ensembles définissables de nombres réels. I', *Fundamenta Mathematicae* 17 (1931) 210–239; translated as 'On definable sets of real numbers' in [66] pp. 110–42.

[51] —— *Pojęcie prawdy w językach nauk dedukcyjnych*, Prace Towarzyctwa Naukowego Warszawskiego, Wydzial III Nauk Matematyczno-Fizycznych 34, Warsaw 1933; page references are to the reprint in [68] pp. 13–172.

[52] —— 'Z badań metodologicznych nad definiowalnością terminów', *Przegląd Filozoficzny* 37 (1934) 438–60.

[53] —— 'Einige methodologische Untersuchungen über die Definierbarkeit der Begriffe', *Erkenntnis* 5 (1935) pp. 80–100; translated as 'Some methodological investigations on the definability of concepts' in [66] pp. 296–319.

[54] —— 'Grundzüge des Systemenkalküls, Zweiter Teil', *Fundamenta Mathematicae* 26 (1936) 283–301; translated in 'Foundations of the calculus of systems' in [66] pp. 342–83.

[55] —— *O logice matematycznej i metodzie dedukcyjne*j, Biblioteczka Mat. 3–5, Książnica-Atlas, Lwów and Warsaw 1936.

[56] —— 'O pojęciu wynikania logicznego', *Przegląd Filozoficzny* 39 (1936) 58–68; reprinted in [68] pp. 186–202; translated as 'On the concept of logical consequence' in [66] pp. 409–20.

[57] —— 'O ugruntowaniu naukowej semantyki', *Przegląd Filozoficzny* 39 (1936) 50–7; reprinted in [68] pp. 173–85; translated as 'The establishment of scientific semantics' in [66] pp. 401–8.

[58] —— *The Completeness of Elementary Algebra and Geometry*, Hermann, Paris 1940 (unpublished because of the war, but reprinted in limited edition by CNRS, Paris 1967).

[59] —— *Introduction to Logic and to the Methodology of the Deductive Sciences*, translation and revision of [55], Oxford University Press, New York 1941.

[60] —— 'The semantic conception of truth', *Philosophy and Phenomenological Research* 4 (1944) 13–47.

[61] —— 'A problem concerning the notion of definability', *Journal of Symbolic Logic* 13 (1948) 107–11.

[62] —— 'Some notions and methods on the borderline of algebra and metamathematics', *Proceedings of International Congress of Mathematicians*, vol. 1, American Mathematical Society, Providence, RI 1952, pp. 705–20.

[63] —— 'Contributions to the theory of models I', *Indagationes Mathematicae* 16 (1954) 572–81.

[64] —— *Logic, Semantics, Metamathematics: Papers from 1923 to 1938*, trans. and ed. John Woodger, Clarendon Press, Oxford 1956.

[65] —— 'Truth and proof', *Scientific American* 220 (6) (1969) 63–77.

[66] —— *Logic, Semantics, Metamathematics: Papers from 1923 to 1938*, ed. John Corcoran, Hackett Publishing Company, Indianapolis, Indiana 1983.

[67] —— 'What are logical notions?', *History and Philosophy of Logic* 7 (1986) 143–54.

[68] —— *Pisma Logiczno-Filozoficzne, Tom 1 'Prawda'*, ed. Jan Zygmunt, Wydawnictwo Naukowe PWN, Warsaw 1995.

[69] Alfred Tarski and Kazimierz Kuratowski, 'Les opérations logiques et les ensembles projectifs', *Fundamenta Mathematicae* 17 (1931) 240–8.

[70] Alfred Tarski and Adolf Lindenbaum, 'Sur l'indépendance des notions primitives dans les systèmes mathématiques', *Annales de la Société Polonaise de Mathématique* 1926, pp. 111–13.

[71] Alfred Tarski, Andrzej Mostowski, and Raphael Robinson, *Undecidable Theories*, North-Holland, Amsterdam 1953.

[72] Alfred Tarski and R. L. Vaught, 'Arithmetical extensions of relational systems', *Compositio Mathematica* 13 (1957) 81–102.

[73] Robert Vaught, 'Alfred Tarski's work in model theory', *Journal of Symbolic Logic* 51 (1986) 869–82.

[74] Isaac Watts, *Logic; or, The Right Use of Reason*, Bumpus, London 1724.

[75] Hermann Weyl, *Philosophie der Mathematik und Naturwissenschaft*, Leibniz Verlag, Munich 1928.

[76] Alfred North Whitehead and Bertrand Russell, *Principia Mathematica* to *56, Cambridge University Press, Cambridge 1962 (original 1913).

[77] Christian Wolff, *Erste Philosophie oder Ontologie* §§1–78, Felix Meiner, Hamburg 2005.

6

Tarski's Convention T and the Concept of Truth

Marian David

In this chapter, I discuss in some detail the original version of Tarski's condition of adequacy for a definition of truth, his *Convention T*. I suggest that Tarski designed Convention T to serve two functions at once. I then distinguish two interpretations of Tarski's work on truth: the standard interpretation and a non-standard, alternative interpretation. On the former, but not on the latter, the very title of Tarski's famous article about *the* concept of truth harbors a lie.

CONVENTION T

Convention T, the original version of Tarski's condition of adequacy for a definition of truth, can be found in his article "The Concept of Truth in Formalized Languages" (Tarski 1983), henceforth CTF. This 1983 English translation revises a 1956 English translation which is based on the German version of the article (Tarski 1935)—compared to the German, the revised English translation adds and/or changes a substantial number of footnotes. The German version, in turn, is an expansion of the Polish original (Tarski 1933), adding an important Postscript.[1]

I quote Convention T as it appears in §3 of CTF, together with a large bit of the paragraph preceding it which helpfully brings up most of the notions and ideas that play an important role in Tarski's thinking leading up to the convention, including ones that did not quite make it into the convention itself (CTF: 187–8):

Let us try to approach the problem from a quite different angle, by returning to the idea of a semantical definition as in §1. As we know from §2, to every sentence of the calculus of classes there corresponds in the metalanguage not only a name of this sentence of the structural descriptive kind, but also a sentence having the same meaning... In order to make clear

---

[1] In the Postscript, the central theorem of §5 is withdrawn and replaced by the theorem now familiar to logicians as *Tarski's Theorem*, saying roughly that truth is definable for an object-language if, but only if, the metalanguage in which truth is to be defined is "essentially richer" than the object-language; cf. CTF, p. 273; and Tarski's 1944, sec. 10.

the content of the concept of truth in connexion with some one concrete sentence of the language with which we are dealing we... take the scheme ['*x is a true sentence if and only if p*'] and replace the symbol '*x*' in it by the name of the given sentence, and '*p*' by its translation into the metalanguage. All sentences obtained in this way... naturally belong to the metalanguage and explain in a precise way, in accordance with linguistic usage, the meaning of phrases of the form '*x* is a true sentence' which occur in them. Not much more in principle is to be demanded of a general definition of true sentence than that it should satisfy the usual conditions of methodological correctness and include all partial definitions of this type as special cases; that it should be, so to speak, their logical product. At most we can also require that only sentences are to belong to the extension of the defined concept...

Using the symbol '*Tr*' to denote the class of all true sentences, the above postulate can be expressed in the following convention:

CONVENTION T. *A formally correct definition of the symbol 'Tr', formulated in the metalanguage, will be called an* adequate definition of truth *if it has the following consequences:*
(α) all sentences which are obtained from the expression '$x \in Tr$ if and only if p' by substituting for the symbol 'x' a structural-descriptive name of any sentence of the language in question and for the symbol 'p' the expression which forms the translation of this sentence into the metalanguage;
(β) the sentence 'for any x, if $x \in Tr$ then $x \in S$' (in other words '$Tr \subseteq S$').

The formulation of Tarski's adequacy condition which appears as Convention T at the end of this passage differs in various respects from the formulations he gives in his more popular writings on truth (cf. Tarski 1936, 1944, 1969). It also differs from the formulations that can be found in most discussions of his writings by others. Let us take a closer look at the whole passage.

The language of the calculus of classes to which Tarski refers at the beginning of our passage is completely specified in §2 of CTF. It is a formalized language containing an in-principle inexhaustible stock of variables, 'x_I', 'x_{II}', 'x_{III}', ..., called the first, second, third,... variable, and only four constants: the universal quantifier, 'Π', the negation sign, '*N*', the sign of alteration (disjunction), '*A*', and the inclusion sign, '*I*'. The variables are interpreted to range over classes of individuals so that, e.g. the sentence 'Π$x_I Ix_I x_I$' says that, for all classes *a*, *a* is included in *a*. You may note that this is a rather meek language, lacking the means to express that an individual or a class is an element of some class. Tarski refers to this language as "the language of the calculus of classes"—as if there were only one. He does not give it a proper name. I will give it a proper name for definiteness: I will call it *Calish*. Distinguish Calish from the calculus of classes. Calish is a formalized language expressing the calculus of classes in Polish notation. The calculus of classes can be expressed in various other languages, using alternative, and maybe more familiar, notations. Calish is the *object-language* for which Tarski will construct a definition of truth in §3 of CTF—but Tarski does not yet use the term 'object-language' in CTF (though he will use it later, in Tarski 1944).

A *structural-descriptive name* of an expression names an expression by spelling it out, without using quotation marks. The following structural-descriptive name—'the expression which consists of five successive expressions, namely the universal quantifier, the first variable, the inclusion sign, the first variable, and the first variable'—names the sentence 'Π$x_I Ix_I x_I$'. The sentence belongs to Tarski's object-language, Calish. Its structural-descriptive name belongs to the *metalanguage* to which Tarski alludes at the beginning of our passage. Tarski usually refers to this language simply as "the metalanguage". He does not give it a proper name. I will call it *Meta-Calish*. This is the language *in* which Tarski will construct his definition of truth *for* Calish. Meta-Calish is also characterized in §2 of CTF—in some (but not in complete) detail. It contains structural-descriptive names of all expressions of Calish. It also contains expressions that allow one to translate all the sentences of Calish into different sentences of Meta-Calish with the same meaning. The sentence 'for all classes a, a is included in a' is such a sentence: it is a translation into Meta-Calish of the Calish sentence '$\Pi x_i I x_i x_i$'. Meta-Calish also contains various other expressions, among them expressions that occur in the schema quoted in part (α) of Convention T and in the sentence quoted in part (β).

The following sentence,

(1) The expression which consists of five successive expressions, namely the universal quantifier, the first variable, the inclusion sign, the first variable, and the first variable, is a true sentence if and only if, for all classes a, a is included in a,

is an instance of the schema '*x is a true sentence if and only if p*'. The instance is constructed in accordance with the instructions Tarski gives in the paragraph preceding the convention. As he puts it in that paragraph, it is supposed "to make clear the content of the concept of truth in connexion with some one concrete sentence of the language with which we are dealing," namely in connection with the sentence '$\Pi x_i I x_i x_i$,' of Calish. So (1) is one of those sentences he refers to in the paragraph as *partial definitions* which "explain in a precise way, in accordance with linguistic usage, the meaning of the phrase '*x is a true sentence*' which occurs in them." Later, in Tarski (1944: sec. 4), the schema will be labeled 'T', and partial definitions like (1) will be called 'equivalences of the form T'—nowadays, they are often referred to as *T-sentences* or as *T-biconditionals*.[2]

Structural-descriptive names become excruciatingly cumbersome with increasing length.[3] In practice, quotation-mark names are much easier to decode than structural-descriptive names, which is why they are used much more frequently. So, one might alternatively think of (1) in terms of the more perspicuous:

(1*) '$\Pi x_i I x_i x_i$,' is a true sentence if and only if, for all classes a, a is included in a.

---

[2] The label 'biconditional' is much better than 'equivalence'. The latter misleadingly suggests a *relational* claim, i.e. a claim of the form '[NAME] is equivalent with [NAME]'. But being a biconditional, i.e. taking the form '[SENTENCE] if and only if [SENTENCE]', (1) does not make a relational claim. Tarski is aware of the potential for confusion on this score (cf. Tarski 1946: chap. 2) and complains about what appears to be an instance of such a confusion in his 1944, section 15. He nevertheless keeps referring to the T-biconditionals as "equivalences" in his 1944 and 1969.

[3] Actually, the official structural-descriptive name of '$\Pi x_i I x_i x_i$,' is still more cumbersome than the one I have given. According to CTF, p. 172, it has to be constructed by repeated application, with embeddings, from 'the expression which consists of two successive expressions x and y'. (If you try to work this out, you will find it difficult to get a grammatical result.) To avoid such complexities, Tarski introduces various symbolic devices for abbreviating structural-descriptive names; e.g. the abbreviated name of '$\Pi x_i I x_i x_i$,' looks like this: 'un⌢(v_1⌢(in⌢(v_1⌢ v_1)))'.

This is the form in which T-biconditionals are more usually given.[4] Note, however, that part (α) of Convention T refers to structural-descriptive names rather than quotation-mark names. This is because the official metalanguage Tarski has specified in §2 of CTF, Meta-Calish, employs only structural-descriptive names to talk about the expressions of Calish. As far as CTF is concerned, T-biconditional (1) is an official partial definition, whereas T-biconditional (1*) is merely a helpful device for fixing ideas.

Tarski uses the symbols 'Tr' and 'S' in his formulation of Convention T. Taking the second first, note that the expression '$x \in S$' appears in part (β) of the convention without introduction. We are supposed to remember from §2 of CTF that Tarski treats this expression as a notational variant (a symbolic abbreviation) of 'x is a sentence'—he does this even though '$x \in S$' is short for 'x is an element of the class of all x such that x is a sentence'. Both, 'x is a sentence' and its abbreviation, '$x \in S$', belong to Meta-Calish; they have been defined together to pick out the sentences (closed well-formed formulas) of Calish—at the point where Tarski gives this definition, in §2 of CTF, he has already told us that the expression 'is an element of', together with its symbolic abbreviation, '$\in$', together with various other expressions from general set theory, such as 'the class of all x such that', belong to Meta-Calish.[5]

As Tarski indicates in the sentence introducing Convention T, the expression '$x \in Tr$' is short for 'x is an element of the class of all x such that x is a true sentence.' He nevertheless treats it as a notational variant of 'x is a true sentence', and in part (α) of Convention T, he uses it, rather than the more familiar 'x is a true sentence', to formulate the schema for constructing T-biconditionals such as (1).

When Tarski characterizes his metalanguage in §2 of CTF, he does not list 'true sentence', or 'Tr', among the expressions belonging to Meta-Calish—not yet. For, as he sees it, the issue of whether truth is definable for Calish within Meta-Calish is the

---

[4] It is still more usual to give them for cases in which one's object-language is contained in one's metalanguage; e.g. the biconditional

'$\Pi x, Ix, x,$' is a true sentence if and only if $\Pi x, Ix, x,$

would be a Meta-Calish T-biconditional, if Meta-Calish did contain Calish, as it does not, and if it did contain quotation-mark names, as it does not.

[5] Tarski defines 'sentence' and 'S' in one breath in Definition 12 (CTF: 178), saying "*x is a sentence (or a meaningful sentence)—in symbols $x \in S$—if and only if* . . ." He appears quite unconcerned by what looks to be a difference in ontological commitment between 'x is a sentence' and 'x is an element of the class of all sentences'. There is a similarly unconcerned passage in his *Introduction to Logic*, where he says that any sentential function with one free variable can be transformed into an equivalent function of the form '$x \in K$', where in place of 'K' we have a constant denoting a class, so that one may "consider the latter formula as the most general form of a sentential function with one free variable" (Tarski 1946: 70–1). By the way, there is nothing in the definiens of Definition 12, or in the definiens of the definiens, corresponding to the parenthetical 'meaningful': the definition of 'sentence' is given "by means of purely structural [i.e. syntactic] properties" (CTF: 166). The parenthesis is merely a reminder that the expressions to which the syntactically defined term 'sentence' applies, i.e. the well-formed formulas of the formalized language Calish, are assumed to have their ordinary meanings: "We shall always ascribe quite concrete and, for us, intelligible meanings to the signs which occur in the languages we shall consider" (CTF: 167).

question of whether a logically simple expression, like 'true sentence' or '*Tr*', can be properly introduced into Meta-Calish. Or, to put this somewhat differently, the question is whether Meta-Calish, as characterized in §2, can express the concept of truth for Calish with some combination of expressions already available in it—this is what Convention T is about. If it can, then a simple expression like 'true sentence', or '*Tr*', can be properly introduced into Meta-Calish. So, in a sense, these expressions don't matter: they merely serve as outward signs that the relevant concept is already expressible in Meta-Calish without them. In another sense, however, these expressions do matter, for they serve to remind us of the concept whose definability in Meta-Calish is being investigated.

Tarski's use of '$x \in Tr$' as a variant for 'x is a true sentence', in combination with part (β) of Convention T, also serves to remind us that he thinks of 'x is a true sentence' as a logically simple, fused, predicate, along the lines of 'x is a truesentence', rather than the logically complex predicate 'x is true and x is a sentence'. If he did think of the predicate in the second way, part (β) of Convention T would be entirely superfluous.

Convention T is formulated as a sufficient condition: it says that a definition of '*Tr*' will be called an adequate definition of truth, *if* it has the sentences described in (α) and the sentence mentioned in (β) as consequences; it does *not* say *only if*. This may come as a surprise, because Tarski's adequacy condition is almost always presented as a sufficient *and necessary* condition. Moreover, at other places Tarski himself puts it as a sufficient and necessary condition; e.g. the first time he formulates it in Tarski 1944 (sec. 4).[6] What are we to make of this? There are two clear indications in CTF that Tarski intends Convention T as a sufficient *and* necessary condition. First, in the text preceding the convention, he says, commenting on the relevant instances of the schema '*x is a true sentence if and only if p*', that "not much more in principle is to be demanded of a general definition of true sentence than that it should...include all partial definitions of this type as special cases..." If this much *is to be demanded*, then having the sentences described in (α) as consequences is intended as a necessary condition. He goes on to say: "At most we can also require that only sentences are to belong to the extension of the defined concept..." So, having the sentence mentioned in (β) as consequence is also intended as a necessary condition. Second, in a footnote appended to Convention T, Tarski says that "after unimportant modifications" of its formulation, the convention would "become a normal definition" (CTF: 188)—and a normal definition requires an 'if and only if'. Tarski's later book, *Introduction to Logic*, provides additional, circumstantial evidence. There he points out that mathematicians prefer the word 'if' when laying down definitions: "what we have here," he says, "is a tacit convention to the effect that 'if' or 'in case that', when used to join definiendum and definiens, are to mean the same as the phrase 'if, and only if' ordinarily does" (Tarski 1946: 36). As far as Convention T as stated in CTF is concerned, there

---

[6] However, the second and more official formulation in section 4 of Tarski's 1944 states it only as a sufficient condition. But then again, when he briefly restates it in section 9, he specifically reminds us of the necessity of the condition.

seems to be sufficient evidence that, despite first appearances to the contrary, it is intended as a necessary as well as a sufficient condition.[7]

Convention T is focused on the "adequacy," or "material adequacy" (CTF: 186), of a definition of truth. It takes for granted that the definition whose material adequacy is under consideration is *formally correct*. Tarski tells us surprisingly little about this presupposed condition of formal correctness. In the paragraph preceding Convention T, he alludes briefly to "the usual conditions of methodological correctness," but he does not indicate what they are. He does say, early on in CTF, that a question about the definability of a concept "is correctly formulated" only if one gives a list of the terms one intends to use in constructing the definition, that these terms "must admit of no doubt," and that he will not make use of any semantic concept in his definition if he is "not able previously to reduce it to other concepts" (CTF: 152–3). Later, in §2 of CTF, he lists the expressions belonging to his metalanguage, Meta-Calish under two general headings: expressions of "a general logical character" and expressions of "a structural-descriptive character" (CTF: 169–73). Though it would surely have been pertinent at this point, he does not remind us of the condition of formal correctness: he does not bother to emphasize that the vocabulary of Meta-Calish, which he has just specified, does not contain any undefined or unreduced semantic expressions. Quite a bit later, when commenting on (the original version of) his negative theorem in §5, he describes his article as an attempt to "reduce" semantic concepts to structural-descriptive, i.e. syntactic, concepts (CTF: 252). He points out that the attempt fails in the end. Though the reduction succeeds with respect to certain "poor" object-languages, such as the language of the calculus of classes, i.e. Calish, it does not go through with respect to "rich" object-languages, such as the language of the general theory of classes, i.e. the language of general set theory (CTF: 253–4). The reduction fails in cases of the latter sort because, as the theorem tells us, it is impossible in such cases to give an adequate definition of truth on the basis of the metatheory, *if* the metatheory is consistent (CTF: 247). In other words, the envisaged definition itself or, more generally, the metatheory to which the definition would belong, would be inconsistent. As far as I can see, this failure seems to concern the condition of formal correctness rather than material adequacy, or if it concerns material adequacy then only because material adequacy presupposes formal correctness according to Convention T. Tarski does not comment on this—though it is noteworthy that he tends to talk in terms of 'correctness' rather than 'adequacy' when reflecting on the import of his negative theorem.

In §2 of CTF, Tarski distinguishes between the meta*language*, on the one hand, and the meta*theory*, on the other (CTF: 167). The latter contains a system of axioms and definitions formulated in the metalanguage. For example, the definitions of sentencehood and truth for Calish belong to the metatheory for Calish formulated in Meta-Calish. What it the metatheory a theory of? At bottom, it is an axiomatic theory of the syntactic structure of the object-language: "What we call the metatheory is, fundamentally, the *morphology of language*—a science of the form of expressions"

---

[7] For more discussion of this topic see Patterson (2006) who, looking also at later versions of Tarski's condition of adequacy, arrives at a somewhat different conclusion.

(CTF: 251). Its axioms do not contain any semantic notions. Such notions appear only in the definitions and then only if they are ultimately defined in terms of (reduced to) non-semantic notions. The *axioms* of the metatheory play a rather important though strangely unacknowledged role in Convention T. Before we get to this point, let us take a closer look at these axioms.

There are two groups of axioms in the metatheory: one group Tarski calls "the general logical axioms"; the other one he calls, somewhat awkwardly, "the specific axioms of the metalanguage" (CTF: 173). Tarski does not list any examples from the first group, the one he *calls* "the general logical axioms." He merely says that they are well-known (referring us to Whitehead and Russell's *Principia Mathematica*) and that they "suffice for a sufficiently comprehensive system of mathematical logic" (CTF: 173). It is clear, however, that in addition to typical logical axioms this group is also supposed to contain the axioms of general set theory. Since general set theory is *not* regarded as belonging to logic nowadays, this makes Tarski's label for this group of axioms rather problematic from our present point of view. But there is no deliberate misdirection involved here. At the time when Tarski composed the original version of CTF, it was fairly widely held that general set theory does belong to logic; and Tarski evidently held the view too—one may note that the inventory of his metalanguage lists expressions such as 'is included in', 'is an element of', 'class', 'infinite class', 'ordered pair', 'sequence', and 'natural number' under the heading "expressions of a general logical character" (cf. CTF: 170–1). One may note also that about a decade later Tarski was already rather more skeptical about the logical nature of set theory.[8]

Tarski does provide a list of the axioms of the second group, the one he calls "the specific axioms of the metalanguage." They describe syntactic properties and relations of his object-language, Calish. The first four are mostly concerned with identity conditions of simple and complex expressions of Calish. Axiom 1, for example, says that the negation sign, the alternation sign, the sign for the universal quantifier, and the inclusion sign "are expressions, no two of which are identical" (CTF: 173). The upshot is that claims about the non-identity of intuitively different expressions of Calish become enshrined as axioms in the Meta-Calish metatheory. Note that non-identity claims, syntactic or otherwise, are not logical truths and that Tarski does not regard these syntactic non-identity claims as logical truths, even though he lays them down as axioms: he clearly separates the axioms from this second group from the ones he refers to, problematically, as the general logical axioms. Note also that Tarski's labels for the two groups of axioms suggest that he had the following picture in mind. The axioms of the first group, including the axioms of set theory, belong in one form or another to any metatheory, independently of the specifics of the object-language and metalanguage at hand. The axioms of the second group, on the other hand, will depend entirely on the specifics of the metalanguage at hand,

---

[8] In his 1944, endnote 12, Tarski first points out that he is using 'logical' in "a broad sense" in which it comprehends "the whole theory of classes and relations (i.e., the mathematical theory of sets)". But he then remarks that he is "personally inclined" to use the term in "a much narrower sense, so as to apply it only to what is sometimes called 'elementary logic', i.e., to the sentential calculus and the (restricted) predicate calculus."

that is, ultimately on the specifics of the object-language to which the metalanguage belongs, because these axioms specify the syntax of the object-language in terms of its metalanguage.

Let us return now to Convention T. In the paragraph preceding the convention, Tarski says that an adequate definition of truth should *include* as special cases all partial definitions, that is, all the relevant T-biconditionals such as (1). In Convention T itself, Tarski does not use the term 'include', he uses the term 'consequence'. Specifically, he says that a definition of '*Tr*' will be called an adequate definition of truth if it has as *consequences* the sentences described in part (α) of the convention, i.e. the relevant T-biconditionals, as well as the sentence mentioned in part (β). Tarski does not mention the axioms of the metatheory here and his wording creates the impression that he means all these sentences to be consequences of the definition taken just by itself.

This is curious. For when one looks at the definition Tarski constructs in §3 of CTF, which he says is an adequate definition of truth, it turns out that it does not by itself have the T-biconditionals as consequences. At least, it does not have them as formal consequences—they are not *derivable* from the definition of truth alone: the axioms (and more definitions) of the metatheory are needed as additional premises. Moreover, at times Tarski *does* mention the metatheory or its axioms in this connection. When he describes the problem of constructing a definition that satisfies Convention T, later in CTF, he says that "it is a question of *whether on the basis of the metatheory of the language we are considering the construction of a correct definition of truth in the sense of convention T is in principle possible*" (CTF: 246); and when he states a version of his adequacy condition in the short chapter "The Establishment of Scientific Semantics," he talks of the "provability" of the T-biconditionals "on the basis of the axioms and rules of inference of the metalanguage" (Tarski 1936: 404). Still, in Convention T itself, Tarski suppresses any reference to the axioms of the metatheory. Why he does this is unclear, and it seems fair to say that the wording of Convention T is rather misleading on this point.[9]

One might suggest that Tarski may have had in mind some relation other than derivability when using the term 'consequence' in Convention T. This seems unlikely: he would have said so. Moreover, note that in §2 of CTF he does define 'consequence' for formulas of Calish so that it refers to a formal/syntactic derivability relation (CTF: 182). Admittedly, the term 'consequence' thus defined in §2 applies to a relation between items of Tarski's object-language, Calish, whereas the term 'consequence' that appears in Convention T applies to a relation between items of Tarski's metalanguage, Meta-Calish. Still, it would be very strange if Tarski had reused the same term in Convention T while having something fundamentally different in mind.

---

[9] In the various versions of his adequacy condition which Tarski gives in his later papers, he makes again no reference to the axioms of the metatheory. He says that an adequate definition "implies" the T-biconditionals, that the T-biconditionals "follow from" an adequate definition (Tarski 1944: sec. 4), that an adequate definition "enables us to ascertain" the T-biconditionals (Tarski 1969: 106), and that such a definition will "imply" the T-biconditionals "as consequences" (Tarski 1969: 114).

One might suggest that Tarski in Convention T thought it alright to suppress references to the additional premises that are needed for deriving the T-biconditionals from an adequate definition of truth because these additional premises consist of *axioms*. This may be so; he does not say. But consider that one would normally regard this as acceptable only if the axioms involved express logical truths. As I have pointed out, Tarski at the time subsumed set theory under logic. This may make it understandable why Convention T does not mention any need for set-theoretic axioms when talking about the consequences of an adequate definition of truth. But Tarski did not subsume the principles that describe the syntax of an object-language under logic. This makes it difficult to understand why Convention T does not mention the need for the syntactic axioms.[10]

The issue is of some interest because it bears on how one should conceive of an adequate definition of truth in relation to the T-biconditionals. In the paragraph preceding Convention T, and at various other places in CTF, Tarski describes a definition of truth that is adequate in the sense of Convention T as the "logical product" of the T-biconditionals and refers to the latter accordingly as "partial definitions."[11] But, as we have seen, among the axioms needed for deriving the T-biconditionals there are axioms that are not logical in nature, not even by Tarski's own lights at the time he composed CTF. It is then hard to see how Tarski can avoid the censure that it is both misleading and wrong to describe a definition of truth that is adequate in the sense of Convention T as the *logical* product of the T-biconditionals.[12]

THE DOUBLE-LIFE OF CONVENTION T

I think that Tarski wants Convention T to play a double-role, to serve two important functions at once. On the one hand, and taken strictly, it is supposed to provide

---

[10] Set-theoretic axioms in the metatheory are needed for deriving the relevant T-biconditionals from the definition of truth only if the object-language under consideration is similar in complexity to Calish; they are not needed for object-languages with finitely many sentences. Syntactic non-identity axioms, however, are needed even for finite object-languages (as long as they contain more than one sentence). Take an object-language with only two sentences, say 's_1' and 's_2', and assume for simplicity's sake that the metalanguage contains the object-language and contains quotation-mark names, "s_1" and "s_2", of the two sentences of the object-language. Tarski tells us that the following will be an adequate definition of truth for this language (cf. CTF: 188):

$x \in Tr$ if and only if ($x =$ 's_1' and s_1) or ($x =$ 's_2' and s_2).

Deriving the T-biconditional with respect to sentence 's_1', i.e.,

's_1' $\in Tr$ if and only if s_1,

requires the premises "$s_1 =$ 's_1'" and "'s_1' $\neq$ 's_2'". The former is an instance of a logical truth; it will thus be covered by some non-problematic member of the group of "general logical axioms". The second is not an instance of a logical truth; it would have to be laid down as a (the sole) syntactic non-identity axiom for this particular object-language.

[11] See, e.g., CTF, pp. 155, 157, 163, 165, 187, 236, 238, 253.

[12] I should point out though that Tarski tends to qualify his talk of the definition as a logical product with a "so to speak", cf. CTF, pp. 187 and 238.

a necessary and sufficient condition of adequacy for a definition of truth for Tarski's object-language, Calish, formulated in Tarski's metalanguage, Meta-Calish. In this role the condition is radically non-general. On the other hand, and taken more loosely, the condition is also supposed to function in the role of an exemplar, to intimate or convey something rather more general than it strictly speaking expresses.

Imagine you first encountered Convention T all by itself. (You may want to reread the convention at this point but without the preceding paragraph.) You would then still understand it in a rough way, but you would also notice that the crucial terms 'the metalanguage' and 'the language in question' have strangely floating or indefinite or indeterminate reference. One of the first questions coming to your mind ought to be: "Which metalanguage?" and "What is meant by *the* language in question?" These questions receive no answer from within Convention T. When, on the other hand, Convention T is seen in the context of CTF, the references of these terms becomes clear. The term 'the language in question' refers back to Tarski's object-language, Calish, the language of the calculus of classes, which he mentions explicitly early in the preceding paragraph and which he has specified in §2 of CTF. The term 'the metalanguage' picks up on his allusion to *the* metalanguage, also early in the preceding paragraph, which allusion in turn refers back to Meta-Calish, his metalanguage, which he has characterized in §2 of CTF.

Remember in addition the following three points: the symbol 'S', which appears without introduction in part (β) of Convention T, was defined earlier, in §2 of CTF, to pick out the sentences of Calish; the sentence quoted in part (β) belongs to Meta-Calish; and the schema for the T-biconditionals, quoted in part (α), also contains material belonging to Meta-Calish.[13]

In sum, Convention T, taken strictly, is maximally specific. Moreover, it is maximally specific along two different dimension. The first one is more frequently acknowledged: strictly speaking, Convention T talks only about the adequacy of a truth definition for the one object-language Calish. Recognizing this, one might still think that Convention T is at least general along another dimension: one might still think that it gives a condition of adequacy for truth definitions for Calish in arbitrary metalanguages. But no, it doesn't, for it refers back to, and contains quoted material from, the one metalanguage Meta-Calish.[14]

This second dimension of specificity can be illustrated by comparing Convention T with Konvention W from the German version of Tarski's article (Tarski 1935: 305–6). Tarski's object-language in the German version is the same as in CTF, Calish: it is specified as containing the same four constants of Polish notation and the same in-principle inexhaustible stock of variables. But Tarski's metalanguage in the German version is not the same as the one in CTF. Take the Calish sentence '$\Pi x_, I x_, x_,$'. The metalanguage specified in §2 of the German version, call it *Meta-Kalisch*, does not contain the structural-descriptive name 'the expression which

---

[13] Though, as far as I can tell, the 'p' in the schema does not belong to Meta-Calish. Tarski lists a 'p' as belonging to his metalanguage on p. 173 of CTF, but since it is said to represent (a sequence of) natural numbers, this can't be the 'p' that appears in the schema.

[14] Gupta and Belnap mention this point briefly in their 1993, p. 2, footnote 4.

consists of...' (etc.), nor does it contain the sentence 'for all classes *a*, *a* is included in *a*'. Instead, it names the sentence 'Π*x,Ix,x,*' with the structural-descriptive name 'der Ausdruck, der...' (etc.) and translates the sentence as 'für jede beliebige Klasse *a*, *a* ist in *a* enthalten'. Moreover, in place of Meta-Calish expressions such as 'if and only if', 'is an element of' (or ' ∈ '), 'class of all *x* such that', and many others, Meta-Kalisch contains 'dann und nur dann, wenn', 'ist ein Element von' (or 'ε'), 'Klasse aller solchen *x*, dass', and many others. The vocabularies of the two metalanguages are almost completely different.

This has the following consequences. Convention T and Konvention W state different conditions. That's because Convention T quotes expressions belonging to Meta-Calish but not to Meta-Kalisch and Konvention W quotes expressions belonging to Meta-Kalisch but not to Meta-Calish—part (α) of Konvention W mentions the schema '*x* ε *Wr* dann und nur dann, wenn *p*' and part (β) mentions the sentence '*für ein beliebiges x, wenn x* ε *Wr, so x* ε *As*'. And this means, furthermore, that a definition of truth for Calish that is adequate by the lights of Convention T will not be judged adequate by the lights of Konvention W, and vice versa. Take the definition of truth for sentences of Calish Tarski constructs in §3 of CTF. It is formulated in Meta-Calish. Assume that it is an adequate definition according to Convention T, as Tarski maintains. It then has among its consequences all the Meta-Calish sentences constructed from the schema '*x* ∈ *Tr if and only if p*' by following the instructions given in part (α) of Convention T, as well as the one Meta-Calish sentence quoted in part (β) of the convention. However, since Meta-Calish does not contain the relevant expressions of Meta-Kalisch, the Meta-Calish definition does *not* have as consequences the sentences of Meta-Kalisch referred to in parts (α) and (β) of Konvention W: these sentences cannot even be formulated in Meta-Calish. This goes the other way round too. In §3 of the German version of CTF, Tarski constructs a definition of truth for Calish in Meta-Kalisch. Assume that it is an adequate definition according to Konvention W, as Tarski maintains. It then has among its consequences all the Meta-Kalisch sentences constructed from the schema '*x* ε *Wr dann und nur dann, wenn p*' by following the instructions given in part (α) of Konvention W, as well as the one Meta-Kalisch sentence quoted in part (β) of Konvention W. But since Meta-Kalisch does not even contain the relevant expressions of Meta-Calish, the definition given in the German version of CTF does *not* have as consequences the sentences of Meta-Calish referred to in parts (α) and (β) of Convention T. Note that there is no conflict here: Convention T merely implies that the German version's definition of truth for Calish, which is constructed in Meta-Kalisch, is not an adequate definition of truth for Calish in Meta-Calish (and vice versa), which is obvious enough. But the situation illustrates how radically specific Convention T really is.

If Convention T is thus radically specific, why didn't Tarski make this more explicit? His condition of adequacy is surely one of the centerpieces of his article, Why did he use the indeterminate terms 'the metalanguage' and 'the language in question' whose precise reference is determined only through the larger context in which the condition occurs? This has to do with the other, more elusive role Convention T is supposed to play.

In the second sentence of CTF Tarski signals that he is going to give a definition of truth only for one individual language: when announcing that it is the task of his article to construct a materially adequate and formally correct definition of the term 'true sentence', he inserts the rather important qualification "with reference to a given language" (CTF: 152). But later, in §2—after he has abandoned the attempt to give a definition of truth for ordinary languages and has announced that he is going to restrict his attention to formalized languages and has drawn the object-language/metalanguage distinction, but just before he begins to lay down his particular object-language and his particular metalanguage—he remarks that it is possible to give a method for defining truth "for an extensive group of formalized languages" (CTF: 167). He then says that giving a general description of this method and of the languages for which the method works "would be troublesome and not at all perspicuous" (CTF: 168), and that he therefore prefers to introduce us to this method in another way:

> I shall construct a definition of truth of this kind in connexion with a particular concrete language and show some of its most important consequences. The indications which I shall then give in §4 of this article will, I hope, be sufficient to show how the method illustrated by this example can be applied to other languages of similar logical construction.
>
> (CTF: 168)[15]

It is this strategy—using a particular exemplar to convey something more general—which is, I believe, the reason that leads Tarski to formulate Convention T as he does, with the referentially indeterminate terms 'the metalanguage' and 'the language in question'. On the one hand, the actual context in which the convention occurs *does* provide the proper determinate references for Tarski's use of these terms in the convention, namely Meta-Calish and Calish respectively, while on the other hand, the indeterminate terms also impart a suggestion of generality to the convention. (Remember that one can uses phrases like 'the dog...' to refer to a particular, contextually salient dog but that one can also use them to express generalizations about all, or all typical dogs.) Strictly speaking, this suggestion of generality is of course wrong: Convention T does not really talk about truth definitions for arbitrary object-languages, not even about truth definitions for some range of object-languages that are of a logical construction similar to Calish, nor does it talk about truth definitions for Calish in arbitrary metalanguages or in a range of metalanguages similar to Meta-Calish. Nevertheless, the suggestion of generality carried by the indeterminate terms manages to impart the desired message, namely that, by taking Convention T as our model, we could relatively easily formulate "analogous" adequacy conditions for other object-language/metalanguage pairs. The message comes through even

---

[15] Compare also the Introduction of CTF, where he says that "there is a uniform method of construction of the required definition" for each of the "poorer" languages for which truth is definable and announces that he will "carry out this construction for a concrete language in full and in this way facilitate the general description of the above method which is sketched in §4" (CTF: 153–4). How "uniform" the method of construction actually is is a difficult question.

though we—and I believe Tarski too—would be very hard pressed to spell out what 'analogous' actually amounts to here.[16]

The use of referentially indeterminate terms in this double-role is in fact a pervasive feature of CTF: it can be found in Tarski's use of the most basic vocabulary of his metalanguage. When he first introduces Calish, on p. 168 of CTF, he says it contains the universal quantifier, 'Π', (together with some other signs). Six pages later, when he introduces the structural-descriptive terms of Meta-Calish, he lists (among many others) the term 'the sign of the universal quantifier'. This is a referentially indeterminate term: it could refer to other signs, e.g. to '∀', or to Russell's '( )'. But Tarski takes for granted that the reference of his uses of 'the sign of the universal quantifier' throughout CTF has been fixed by the initial context on p. 168 where he set up Calish to contain 'Π', referring to it as *the* universal quantifier. Note how this allows this indeterminate term to play its second, generality-intimating role. If you wanted to construct a definition of truth, not for Calish but for an object-language that expresses the calculus of classes in a different way, say by using '∀', you would, as it were, only have to go to p. 168 of CTF, where Tarski introduced his Calish, and replace 'Π' with '∀', thereby referring to '∀' as *the* universal quantifier. You could then introduce and use the term 'the sign of the universal quantifier', which would now belong to your new metalanguage, in the same way in which Tarski used it: following his method, setting up your metalanguage much like he did his, the indeterminate term would now refer back to '∀', rather than 'Π', when used by you in your context.

Consider also Tarski's use of the term 'sentence'. In §2 of CTF, he defines the Meta-Calish term '*x* is a sentence', or '*x* ∈ *S*', as the limiting case of a sentential function, one without free variables. The definition, Definition 12 (CTF: 178), invokes the previously defined term, '*x* is a sentential function', whose definition uses (abbreviations of) the referentially indeterminate terms 'the sign of the universal quantifier', 'the negation sign', 'the alteration sign', and 'the inclusion sign', which in Tarski's context refer to the basic vocabulary of Calish, i.e. to 'Π', '*N*', '*A*', and '*I*', respectively (cf. Def. 10, p. 177 of CTF). So Definition 12, even though its definiendum looks like this, '*x* is a sentence' (or '*x* ∈ *S*'), defines this term so as to pick out only the sentences (the closed well-formed formulas) of Calish. Nevertheless, when he gives a brief preview of what he is about to do, Tarski announces that he will "obtain the concept of *sentence*" as a special case of the notion of a sentential function (CTF: 176). A natural reaction to this remark would be: "Well, not really, for you are really going to obtain a much more restricted concept, something like *sentence of Calish*." This reaction may well be justified. But note that Tarski does not, and does not seem to want to, talk in these terms. He does not phrase the definiendum of Definition 12 as '*x* is a sentence of the language of the calculus of classes'. He wants the definition to play the generality-intimating role—he himself will go on to use the term 'sentence', without

---

[16] What would it take to formulate a free-standing, context independent, general condition of adequacy for arbitrary object- and metalanguages? At the very least it would require finding a formulation that does not invoke expressions of Meta-Calish or of any other particular metalanguage. Readers may want to try their hands on this.

qualifications, to refer to the sentences of Calish (primarily in §3) and later to refer to the sentences of various other languages (in §§4 and 5). The wording of Tarski's definition does not require any modifications, even when employed by others working with other object-languages: in their contexts, the definition will end up picking out the sentences of their object-languages.[17]

Consider also Tarski's use of the terms 'true sentence' and 'truth'. As I said above, Tarski signals right at the beginning of CTF that he is only going to define 'true sentence' with reference to a given language. He reminds us of this at the beginning of §3. But then, in the remainder of §3, he talks as if this qualifying restriction simply were not there, mentioning it again only once more, at the very end of §3 (cf. CTF: 208). Remember the sentence that immediately precedes Convention T. Tarski says there that he is going to use the symbol 'Tr', the symbol he is about to mention in the formulation of the convention, "to denote the class of all true sentences." *All* true sentences? Not really: only the true sentences of Calish. Tarski must be using 'all' or 'true sentence' (or both) in a seriously restricted way.[18] Note also the use of the term 'truth' in Convention T: as Tarski formulates it, the convention looks like it specified the condition under which a formally correct definition of 'Tr' is an adequate definition of *truth*, period. Tarski does not say "an adequate definition of truth for Calish" or something like that: he just says "an adequate definition of truth."

The definition Tarski finally states in §3 of CTF, Definition 23, defines truth only for his specific object-language, Calish—this for reasons analogous to the reasons why his definition of 'sentence' is restricted to the sentences of Calish. Definition 23 defines "x is a true sentence—in symbols $x \in Tr$—if and only if $x \in S$ and every infinite sequence of classes satisfies x" (CTF: 195). The definiens presupposes the term 'sequence f satisfies the sentential function x', and the definition of this term, Definition 22, appeals to the basic vocabulary of Calish (CTF: 193–5). Hence, this term, and consequently the term 'true sentence', or 'Tr', is defined only for Calish. Nevertheless, when Tarski talks about his definition of 'true sentence', no restriction to Calish is mentioned. In the paragraph before Definition 23, where he gives a brief explanation of what he is about to do, he simply says that "the concept of truth" will be reached on the basis of Definition 22; and right after he has stated his definition, he claims that it is an "adequate definition of truth in the sense of convention T" (CTF: 194, 195). No restriction to his object-language, Calish, is mentioned at all.

---

[17] This also works, albeit in a rather more limited fashion, when it comes to Definition 10, which defines 'x is a sentential function'. Since this definition relies on the referentially indeterminate terms 'the inclusion sign', 'the negation sign', etc., the wording of the definition can remain, even if one has earlier introduced an object-language containing, say, '$\subseteq$' and '$\sim$' instead of Calish which contains 'I' and 'N'. But changes in the wording of Definition 10 will become necessary, if one's object-language has a different grammar than Calish, or if it is not a language talking about classes at all so that the term 'the inclusion sign' becomes inappropriate.

[18] In the paragraph preceding the convention he considers what is to be demanded "of a general definition of true sentence [sic]." He means a "general" definition in the sense that it concerns all sentences of Calish, as opposed to the "partial definitions" he has just been talking about, each one of which concerns only one sentence of Calish.

Again, it seems he proceeds in this manner because he wants both Convention T and his definition of 'true sentence', or 'Tr', to play the double-role: on the one hand, and taken strictly, they are maximally specific, playing their role in the project of defining 'true sentence' for Calish within Meta-Calish, while on the other hand, they are supposed to serve as exemplars and are thus formulated without explicit references to Calish and Meta-Calish so as to convey or suggest a more general message that would be rather difficult to state explicitly.

TRUE IN L?

Standard discussions of Tarski's work on truth present Tarski as having defined a term of the form 'true in L', or a concept *truth in L*, for one specific object-language, Calish, *and* as having shown how to define terms of the form 'true in L', or concepts of *truth in L*, for a range of (well-behaved) object-languages.[19] Convention T is then presented accordingly to state a condition under which a definition, formulated in a metalanguage, of a term of the form 'true in L' is an adequate definition of *truth in L*. Note that you must *not* think here of the 'L' as a genuine variable so that 'x is true in L' would express a relation holding between sentences and languages — Tarski did not define anything properly expressible by 'x is true in y', with variable 'y' ranging over languages. Instead, you must think of 'L' as a dummy letter, so that 'x is true in L' is a schematic way of hinting at various one-place predicates, 'x is true in _____', where a name naming some object-language goes into the gap, e.g. 'x is true in Calish'.[20]

Presenting Tarski in this way implies that he has not defined the concept *truth*, or *true sentence*, but a different and much more restricted concept, which we might want to call '*truth in Calish*', but that he has also given us guidelines for defining additional such concepts — concepts we can name only after we have named the object-languages we are interested in. For example, we could specify certain (well-behaved) fragments of English and German, name them 'E_0', 'E_1', ..., 'G_0', 'G_1', ..., and follow Tarski's guidelines to define concepts which we might call '*truth in E_0*', '*truth in G_0*', and so on.

A troubling question arises: What do all these concepts have in common that justifies our using the word 'truth' or 'true' when naming or expressing them? One's first inclination is to respond that these concepts have the following feature in common: the definitions of the terms expressing them, the definitions of 'x is true in Calish' 'x is true in E_0', 'x is true in G_0', etc., all satisfy Convention T. But that can't be quite right, for Convention T is about a definition of 'x is a true sentence', or '$x \in Tr$', for

---

[19] Though not, of course, for arbitrary object-languages: that, according to Tarski's Theorem, cannot be done consistently, since it would allow defining a term 'true in L' for the language indicated by 'L' itself, which would lead into paradox.

[20] Note that the 'in' in 'true in L' is very different from the 'in' in 'is defined in a metalanguage'. Note also that the standard way of presenting Tarski in terms of 'true in L' swallows the word 'sentence' that appeared in Tarski's 'true sentence' — remember Tarski's 'Tr'.

Calish, formulated within Meta-Calish: it says nothing about 'x is true in E_0' or 'x is true in G_0'. Taking Convention T as a model and using Tarski's general guidelines (given in §4 of CTF) for specifying metalanguages for object-languages of certain types, we may be able to formulate an "analogous" convention covering 'x is true in E_0', and an "analogous" convention covering 'x is true in G_0', and so on. But these conventions will all be different and different from Convention T, which makes it difficult to spell out how precisely they help answering the troubling question.[21]

I will not pursue this troubling question. Instead, I want to raise an interpretive question, though I won't be able to resolve it: Did Tarski himself hold the view which the standard way of presenting him has him advocate as a matter of course? Interestingly, that's not at all easy to tell, because Tarski himself does not actually use truth-terms with built-in language parameters, terms of the form 'true in L'.

To get a better idea of what is involved here, consider a formulation such as

d is an adequate definition of truth for Calish (within the metalanguage Meta-Calish)

This can be parsed in two ways: (*a*) as saying that d adequately defines *truth-for/in-Calish* within Meta-Calish; or (*b*) as saying that d adequately defines *truth* and does so for Calish and within Meta-Calish.

The standard interpretation of CTF opts for (*a*) on the following grounds. Tarski has not defined 'x is true in y' with variable 'y' ranging over different object-languages. Instead, he has given a recipe for defining a one-place predicate, 'x is a true sentence', or '$x \in Tr$', for a range of different object-languages. His recipe involves constructing different definitions which appeal to the expressions of their respective object-languages and assign different extensions to their respective occurrences of the definiendum 'x is a true sentence'—after all, different object-languages contain different sentences. But this means, so the standard interpretation, that *different concepts* are being defined which, to avoid ambiguity, are best expressed by different definienda, different so-called "truth-predicates" with different built-in language parameters, such as 'x is true in Calish' and 'x is true in E_0'.

The alternative, non-standard, interpretation I want to consider here opts for (*b*) in light of the way Tarski expresses himself throughout CTF. He uses 'true sentence', rather than some term or terms of the form 'true in L', and he confidently talks about *the* concept of truth, implying that there is only one: Tarski's way of talking gives few indications that he held a view according to which there are somehow different "truth-concepts" for different languages.

Tarski does, of course, indicate in a number of places that his definition is restricted somehow to his individual object-language, Calish, but not in a way suggestive of the

---

[21] The basic point was well-raised by Quine (1951: 32–7), albeit with respect to Carnap's attempt at explicating *analyticity*. Adapted to Tarski-style definitions, Quine's objection would go like this: Tarski shows us how to define 'x is true in Calish', 'x is true in E_0', 'x is true in G_0', and so on, but not the general notion of truth, not 'x is true in y', with variable 'x' and 'y'. The newly defined term 'x is true in Calish', or rather 'x-is-true-in-Calish', "might better be written untendentiously as 'K' so as not to seem to throw light on the interesting word ['true']" (Quine 1951: 33). Compare my 1996, where I ask why Quine did not raise this objection against the idea that Tarski-style definitions throw any light on truth.

standard 'true in L' interpretation. At the very beginning of the Introduction of CTF, he says his task is "to construct—with reference to a given language—*a materially adequate and formally correct definition of 'true sentence'*" (CTF: 152). At the beginning of §3, he says that he is about to turn to "the construction of the definition of true sentence, the language of the calculus of classes still being the object of investigation" (CTF: 186). At the very end of §3, he says that he has succeeded "*in doing for the language of the calculus of classes what we tried in vain to do for colloquial language: namely to construct a formally correct and materially adequate semantical definition of the expression 'true sentence'*" (CTF: 208). Note the absence of truth terms with built-in language parameters. With hindsight, one might even think that Tarski goes to some lengths to avoid formulations of the form 'true in L'—and this holds not only for CTF but also for his later writings on the subject (cf. Tarski 1944, 1969).[22]

The standard interpretation relies on the principle: *if different extensions, then different concepts*—applied to our case: if different definitions constructed in accordance with Tarski's recipe assign different extensions to different occurrences of 'true sentence', then they define different concepts; that is, they assign different concepts to their respective occurrences of 'true sentence', so that their respective definienda are less ambiguously expressed by different terms such as 'true in Calish', 'true in E_0', etc. The standard interpretation, one might say, assumes that option (*b*) boils down to option (*a*): *defining* truth *for* Calish or *for* E_0 amounts to defining *truth in Calish* or *truth in* E_0.[23]

The standard interpretation has to explain why Tarski's way of expressing himself throughout CTF does not make it at all apparent that he is telling us how to define a range of different "truth-concepts"; it has to explain why he systematically refrains from using terms of the form 'true in L'. To explain this, one may cite Tarski's strategy of proceeding by exemplar: Tarski's Convention T is really about *truth in Calish*, and Tarski really defines the concept *truth in Calish*, within Meta-Calish, but

---

[22] Tarski uses an 'in L' formulation in "Truth and Proof"—but not to talk of 'true in L'. Instead, he uses it to restrict the domain of the quantifier in a definition of 'true': "For every sentence x (in the language L), x is true if and only if..." (Tarski 1969: 106–7). Searching for a source of the contemporary custom of talking in terms of 'true in L', I find Carnap talking of the definition of 'true in S' in his *Introduction to Semantics*. However, Carnap's 'S' is supposed to indicate a "semantical system"—a system of rules, formulated in a metalanguage, containing the syntactic "rules of formation" of a language as well as "rules of designation" and "rules of truth" (Carnap 1942: 22–5). Early in the book, Carnap distinguishes between *languages*, which at first appear to be syntactically individuated, and *semantic systems* of languages: it seems that it should be possible for there to be different semantic systems of the same language, so that one would naturally expect Carnap to talk in terms of 'true in S of L'. But as one reads on, it turns out that Carnap tends to individuate languages in terms of semantic systems, which is why he has only 'true in S'. Note that this is doubly different from Tarski: first, Tarski does not use terms with built-in parameters anyway; second, if he did, it would not be Carnap's 'true in S'—after all, a semantic system contains metalinguistic rules and is defined in semantic terms.

[23] Carnap gives a version of Convention T in the form: "A predicate pr_i in [a metalanguage] M is an *adequate* predicate (and its definition an adequate definition) for the concept of *truth* with respect to an object language $S =_{Df} \ldots$" (Carnap 1942: 27). Three pages earlier, Carnap has said that the rules of truth of a system define 'true in S'. Note how Carnap seems to assume that defining *truth* with respect to S amounts to defining 'true in S'. (Note also the uncertain status of 'S': Does it indicate a language or a semantic system of a language?)

he suppresses explicit references to Calish and Meta-Calish so as to convey a more general recipe for defining a range of different concepts—convey it by means of formulations that appear general even though in the context in which he uses them they really refer to specifics.

On the alternative interpretation, the different definitions of 'true sentence' that can be constructed in accordance with Tarski's recipe are concerned with one concept, *truth*, or better *true sentence*, but they define this one concept *for* different object-languages. The alternative interpretation, one might say, assumes that option (*b*) can stand on its own and does not boil down to option (*a*). This seems feasible only if the transition 'different extensions → different concepts' is not generally reliable—as indeed it isn't: where *context sensitivity* comes into play the transition appears to fail. A context-sensitive term such as 'today' may plausibly be said to express one and the same concept even though different occurrences of the term have different extensions. So, on the alternative interpretation, Tarski keeps using the definite article ('*the* concept of truth') and the term 'true sentence', rather than a term of the form 'true in L', because he in effect treats 'true sentence' (as well as other terms, such as 'sentence' and 'the sign of the universal quantifier', etc.) as context sensitive: the term 'true sentence' expresses one concept, the concept *true sentence*, whose extension varies depending on which language is the salient one in a given context: in the context of much of CTF its extension is the set of true sentences of Calish; in other contexts, its extension might be the set of true sentences of E_0, or of G_0.[24]

I claimed earlier, by way of motivating the alternative interpretation, that Tarski does not use terms of the form 'true in L'. There is one possible exception to this claim, namely the very title of §3 of CTF, "The Concept of True Sentence in the Language of the Calculus of Classes" (CTF: 186), which can be parsed in the 'true in L' way, i.e. as talking of the concept *truth in Calish*. However, it can also be understood along the lines of the alternative interpretation, i.e. as talking of the concept *true sentence* defined for the language Calish. In favor of the second reading one can point out that the title is constructed in analogy to the title of the whole article, "The Concept of Truth in Formalized Languages," which is not to be understood in the 'true in L' way, for according to the standard interpretation there is no such concept as the concept *truth in formalized languages*.

Let us look at a few passages from Tarski that seem to bear on the two competing interpretations. In the Introduction of CTF, Tarski talks of *the* meaning of the term 'true' and of *the* concept of truth; he also says this:

The extension of the concept to be defined depends in an essential way on the particular language under consideration. The same expression can, in one language, be a true statement, in

---

[24] The view that 'true' is context sensitive is suggested, albeit somewhat indirectly, by Parsons (1974); it is explicitly advanced and worked out in more detail by Burge (1979). But both authors focus specifically on the behavior of 'true' in liar reasoning. The non-standard interpretation of Tarski under consideration here would maintain that a form of contextualism about 'true' and *truth* is suggested throughout CTF by Tarski's persistent use of 'true sentence', without built-in parameter, and of 'the concept of truth', without parameter but with the definite article.

another a false one or a meaningless expression. There will be no question at all here of giving a single general definition of the term. The problem which interests us will be split into a series of separate problems each relating to a single language.

(CTF: 153)[25]

On the standard interpretation, Tarski is speaking rather misleadingly at the beginning of this passage: there is no such thing as *the* concept to be defined; instead, there are different concepts with different extensions—when Tarski is talking about "separate problems", he is talking about constructing different definitions of different concepts. On the alternative interpretation, he is not talking misleadingly. There is such a thing as *the* concept to be defined, the concept *true sentence*—when he is talking about "separate problems", he is talking about constructing different definitions of the same concept but *for* different object-languages.[26]

Late in CTF, towards the end of §5, Tarski adds some remarks about cases where whole classes of object-languages, instead of one single object-language, are under consideration. He says:

As I have already emphasized in the Introduction, the concept of truth essentially depends, as regards to both extension and content, upon the language to which it is applied. We can only meaningfully say of an expression that it is true or not if we treat this expression as a part of a concrete language. As soon as the discussion concerns more than one language the expression 'true sentence' ceases to be unambiguous. If we are to avoid this ambiguity we must replace it by the relative term 'a true sentence with respect to the given language'.

(CTF: 263)

Note again Tarski's use of the definite article: he talks of *the* concept of truth as depending on the language to which it is applied. He also says the language dependence in question pertains to both the extension and *the content* of the concept of truth. Taken literally, this implies a distinction between the content of the concept and the concept itself.[27] If content can be equated with *intension* (cf. Tarski 1944: sec. 3), the passage can be taken to indicate that Tarski is committed to the transition 'different extension → different intension' but not to the transition 'different

---

[25] The occurrence of the word 'statement' in this passage is a bit disconcerting, for Tarski's 'true' is supposed to apply to sentences. However, my Polish informant tells me that 'statement' is a contribution by the translator: the Polish original (Tarski 1933) has 'zdanie', which corresponds exactly to English 'sentence' and is the term Tarski always uses in connection with 'true'. (The choice of 'Aussage' for 'zdanie' in the German translation (Tarski 1935) is quite unfortunate.) Thanks to Dr. Arkadiusz Chrudzimski, University of Salzburg.

[26] A similar passage can be found in one of Tarski's later writings. Having announced that he will apply the term 'true' to sentences, he says: "Consequently, we must always relate the notion of truth, like that of a sentence, to a specific language; for it is obvious that the same expression which is a true sentence in one language can be false or meaningless in another" (Tarski 1944: sec. 2). Note how, on the standard interpretation, the reference to *the* notion of truth is misleading; not so on the alternative interpretation.

[27] The actual phrase "the content of the concept of truth" shows up only in one place in CTF, namely in the paragraph preceding Convention T. In Tarski's 1944, sec. 3, we find the phrase "the meaning (or the intension) of the concept of truth."

intension → different concept', hence not to the transition 'different extension → different concept'. These aspects of the passage suggest the alternative rather than the standard interpretation.

On the other hand, Tarski also says in the passage that the term 'true sentence' becomes *ambiguous* when more than one object-language is under consideration, and he refers to the disambiguated term as a *relative* term. These remarks might suggest the standard 'true in L' interpretation. However, things are not very clear-cut here. As to the remark that the term 'true sentence' becomes *ambiguous* when more than one object-language is under consideration, it depends on what Tarski means by 'ambiguous'. If he means that different occurrences of the term expresses different concepts, then the remark points towards the standard interpretation. If he merely means that different occurrences of the term have different extensions (or intensions), then the remark is compatible with the non-standard interpretation: if 'true sentence' is contextual, then different occurrences of the term can have different extensions (and even different intensions) while expressing one and the same concept.

As to the remark about the disambiguated term being *relative*, the continuation of the passage shows that Tarski is thinking there of constructing a single metalanguage common to the object-languages under consideration. He seems to be saying that, with such a metalanguage in hand and provided the object-languages are well-behaved ones, we should be able to define a genuine relational term, 'x is true in language y', albeit one whose range of application, i.e. the range of the variable 'y', will be restricted to the object-languages under consideration (cf. CTF: 263–4). This does not really fit well with the 'true in L' interpretation on which 'true in L' is not a relational term at all but merely a stand-in for various one-place predicates. One might also note that Tarski's remark about the disambiguated relative term is programmatic. He goes on to point out that "quite new complications might arise" when attempting to construct a definition of such a relative term, and he mentions specifically complications connected with "the necessity of defining the word 'language'" (CTF: 263–4). Again, this does not fit smoothly with the 'true in L' interpretation on which one would expect Tarski to mention difficulties connected with the notion of a language much earlier in his article and in a more prominent place.[28]

In sum, it seems to me quite difficult to tell whether Tarski's own intentions are better represented along the lines of the standard interpretation or along the lines of the alternative interpretation. Judging from how Tarski typically expresses himself, there is quite a bit to be said for the latter—though the evidence doesn't seem to

---

[28] In the second half of §3 of CTF, after he has constructed his definition of "true sentence" for Calish and has proved various theorems involving this term, Tarski refers to the defined concept as "the absolute concept of truth" and proceeds to define and discuss "another concept of a relative character," namely "the concept of *correct or true sentence in an individual domain a*," which applies, roughly speaking, to sentences that would be true in the ordinary sense if the quantifiers of the object-language under consideration were restricted to range over individuals of domain a (CTF: 199). Note that the relativization involved in 'x is true in domain a', where the variable 'a' ranges over sets of individuals to be associated with the quantifiers of Calish, is along a quite different dimension than the relativization involved in 'x is true in language y', where the variable 'y' ranges over an array of object-languages.

be conclusive. The issue concerns the question of *concept individuation*. Tarski provides us with guidelines for constructing different definitions: Does he think these definitions will define one concept, *truth* or better *true sentence*, but define it *for* different object-languages, or does he think the different definitions will define different concepts of the form '*truth in L*'? As I said, it is difficult to tell. Tarski's later chapter, "The Semantic Conception of Truth," contains an endnote suggesting that this question may, in the end, have no answer:

> The words "notion" or "concept" are used in this chapter with all of the vagueness and ambiguity with which they occur in philosophical literature. Thus, sometimes they refer simply to a term, sometimes to what is meant by a term, and in other cases to what is denoted by a term. Sometimes it is irrelevant which of these interpretations is meant; and in certain cases perhaps none of them applies adequately. While on principle I share the tendency to avoid these words in any exact discussion, I did not consider it necessary to do so in this informal presentation.
>
> (Tarski 1944; endnote 4)

With respect to the last remark, we may observe that, while Tarski's CTF surely aims to be an "exact discussion," it does not exhibit much of a tendency to avoid the vague and ambiguous word 'concept'.

A CONVENTION?

Tarski's Convention T is commonly described and treated as a condition of adequacy for a definition of truth. But Tarski labels it a *convention* and phrases it accordingly, using the words "will be called an adequate definition of truth." So, taken literally and seriously, Convention T does not state a condition under which something actually *is* an adequate definition of truth, it merely states a condition under which something *will be called* an adequate condition of truth.

Somewhat curiously, Tarski's does not comment at all in CTF on why he gives his condition the form of a convention. He does not use conventionalist language elsewhere in CTF (except for the sentence that introduces the convention, where he refers to it as a "postulate"); and his practice seems to belie to some extent his labeling and wording of Convention T: a reader of CTF will come away with the overall impression that Tarski intended to do rather more than merely recommend a convention. Since, moreover, he does not use the label 'Convention T' again in his later writings and calls the version he gives in "The Semantic Conception of Truth" a *criterion* for material adequacy (cf. Tarski 1944: sec. 4), one might even think that he was not serious when presenting it as a convention in CTF. But this would be rash. Although he drops the label, he keeps the conventionalist wording, using phrases such as "we shall say", "we will consider . . . as adequate", and "we stipulate", when presenting versions of his condition after CTF.[29]

---

[29] Compare Tarski's 1936, p. 404; 1944, secs. 4 and 9; 1969, pp. 106 and 114. When Tarski raises an issue of material adequacy in one of his earlier papers, "On Definable Sets of Real Numbers," he treats it as a factual, not as a conventional issue: "Now the question arises whether

Why does Tarski present his condition as a mere convention? One could try to understand this element of conventionalism as a reflection of his view about the indefinability of our concept of truth. The idea would be roughly this. According to Tarski, there cannot be a definition of a term '*Tr*' that actually *is* as a formally correct and materially adequate definition of *the* concept *true sentence*, that is, of *our* concept *true sentence*—that concept is not definable. Since the best we can hope for is various definitions of various *Ersatz*-concepts, it would be pointless to give a condition under which a definition actually *is* an adequate definition of truth. What Tarski does instead is to present a condition for *calling* a definition that assigns some other concept to '*Tr*' an "adequate definition of truth" (and to intimate further conditions for calling further definitions that assign still other concepts to '*Tr*' "adequate definitions of truth"). In effect, Convention T constitutes the recommendation to refer to some concept with our familiar term 'truth', even though the concept is not the/our concept *true sentence*, on the grounds that the concept is sufficiently similar to our concept to function as an Ersatz-concept.

This sort of account of the conventionalist aspect of Convention T would fit the standard interpretation mentioned earlier, on which Tarski shows us how to construct definitions of various Ersatz-concepts of the form '*truth in L*', none of which is the/our concept *true sentence*. The account does not fit the alternative interpretation, on which Tarski *does* show us how to construct various definitions of the/our concept *true sentence*, albeit definitions that define that concept *for* certain object-languages—the concept being indefinable for various other object-languages, e.g. languages as rich as our ordinary languages. This might be counted as a point in favor of the standard interpretation. However, Tarski's own remarks from his later writings do not indicate any connection between the conventionalist aspect of Convention T and his views on the indefinability of truth.

In Tarski's late chapter "Truth and Proof," we find some remarks concerning the goal and the logical status of an explanation of the meaning of a term. Tarski observes there that at times such an explanation "may be intended as an account of the actual use of the term involved," while at other times such an explanation "may be of a normative nature, that is, it may be offered as a suggestion that the term be used in a definite way" (Tarski 1969: 102). He then says that the explanation of the meaning of 'true' he wants to give "is, to an extent, of mixed character"; and he continues: "What will be offered can be treated in principle as a suggestion for a definite way of using the term 'true', but the offering will be accompanied by the belief that it is in agreement with the prevailing usage of the term in everyday language" (Tarski 1969: 102).

So far, one could maybe still see these remarks as being motivated along the lines of the account sketched above. But, as one reads on, it turns out that something else

---

the definitions just constructed . . . are also adequate materially; in other words *do they in fact grasp the current meaning of the notion as it is known intuitively?*" (Tarski 1931: 128–9). But note that the definitions he is concerned with there are not definitions of truth.

is on Tarski's mind. He reminds us that there are different *conceptions* of truth—the classical Aristotelian conception, the pragmatist conception, and the coherentist conception—and announces that he aims for the first one: "We shall attempt to obtain here a more precise explanation of the classical conception of truth" (Tarski 1969: 103).[30] As Tarski presents things, he conveys the impression that it is this *choice*—the choice to make precise the classical rather than some other conception of truth—that motivates him to put his condition into a conventionalist format. This is foreshadowed slightly at one point in CTF, where he says (albeit early on and long before he lays down Convention T) that he will be "concerned exclusively with grasping the intentions which are contained in the so-called *classical* conception of truth ('true—corresponding with reality') in contrast, for example, with the *utilitarian* conception (true—in a certain respect useful)" (CTF: 153). It comes out a bit more clearly—though it is not made explicit—in "The Semantic Conception of Truth," where he first mentions different conceptions of truth, then says he wants his definition "to do justice to the intuitions which adhere to the *classical Aristotelian conception of truth*" (Tarski 1944: sec. 3), and then proceeds to formulate a preliminary version of his condition/convention with reference to the classical conception: "... if we base ourselves on the classical conception of truth, we shall say..." (Tarski 1944: sec. 4). Moreover, in a later part of this chapter, he briefly discusses the question whether the conception of truth he focuses on, the *semantic* conception, which is supposed to be a modernized form of the classical conception, is the "right" one. He professes "not to understand what is at stake in such disputes," urges us "to reconcile ourselves with the fact that we are confronted, not with one concept, but with several different concepts which are denoted by one word," and maintains that "the only rational approach to such problems" is to "try to make these concepts as clear as possible" (Tarski 1944: sec. 14). Note the indication that, as far as his investigation is concerned, he has *chosen* to make precise the concept characterized by the classical conception of truth.

So, judging from indications present in Tarski's own works, the conventionalist aspect of Convention T seems intended to reflect that a choice has been made by Tarski, that he has chosen to make precise the classical conception of truth rather than some other conception. If this is indeed the case, then the conventionalist aspect of Convention T is motivated by considerations that appear to be quite neutral between the standard and the alternative interpretation. At least as far as I can see, this motivation does not seem to favor either one of the two interpretations. The difference between them can be rephrased: Is it Tarski's intention to show us how to define different concepts of the form '*truth in L*', each of which is an Ersatz for the concept intended by the classical/semantic conception of truth, or is it his intention to show us how to define, albeit *for* different languages, the one concept, *truth* or rather *true sentence*, intended by the classical/semantic conception?

---

[30] Tarski seems to distinguish implicitly between the *concept* of truth and a *conception* of truth. When he talks about a conception he has in mind something taking propositional form—a rough principle or definition that aims to tell us what truth is.

References

Burge, Tyler (1979) Semantical Paradox. *The Journal of Philosophy* 76. Reprinted in R. L. Martin, ed., *Recent Essays on Truth and the Liar Paradox*, Oxford: Clarendon Press 1984: 83–117.

Carnap, Rudolf (1942) *Introduction to Semantics*. Cambridge, Mass.: Harvard University Press.

David, Marian (1996) Analyticity, Carnap, Quine, and Truth. *Philosophical Perspectives, 10, Metaphysics*: 281–96.

Gupta, Anil and Belnap, Nuel (1993) *The Revision Theory of Truth*. Cambridge, Mass.: The MIT Press.

Parsons, Charles (1974) The Liar Paradox. *Journal of Philosophical Logic* 3. Reprinted in R. L. Martin, ed., *Recent Essays on Truth and the Liar Paradox*, Oxford: Clarendon Press 1984: 9–45.

Patterson, Douglas, E. (2006) Tarski on the Necessity Reading of Convention T. *Synthese* 151: 1–32.

Quine, W. Van (1951) Two Dogmas of Empiricism. Reprinted in *From a Logical Point of View*, 2nd edn, revised, Cambridge, Mass.: Harvard University Press 1980: 20–46.

Tarski, Alfred (1983) = CTF. The Concept of Truth in Formalized Languages. In *Logic, Semantics, Metamathematics*, translated by J. H. Woodger, 2nd edn, edited by J. Corcoran, Indianapolis: Hackett Publishing Company: 152–278 (1st edn by Oxford University Press, 1956). Revised translation of Tarski 1935.

—— (1969) Truth and Proof. *Scientific American* 220, June: 63–77. Reprinted in R. I. G. Hughes, ed., *A Philosophical Companion to First-Order Logic*, Indianapolis: Hackett (1993), 101–25. Page references are to the reprint.

—— (1946) *Introduction to Logic*. 2nd edn, New York: Oxford University Press (1st edn 1941).

—— (1944) The Semantic Conception of Truth. *Philosophy and Phenomenological Research* 4: 341–75.

—— (1936) The Establishment of Scientific Semantics. Reprinted in *Logic, Semantics, Metamathematics*, 2nd edn, Indianapolis: Hackett Publishing Company 1983: 401–8. (Polish original in *Przegląd Filozoficzny*, vol. 39: 50–7.)

—— (1935) Der Wahrheitsbegriff in den formalisierten Sprachen. *Studia Philosophica* I, Lemberg: 261–405. Expanded translation of Tarski 1933.

—— (1933) *Pojęcie prawdy w językach nauk dedukcyjnych* (On the Concept of Truth in Languages of Deductive Sciences): Warsaw.

—— (1931) On Definable Sets of Real Numbers. Reprinted in *Logic, Semantics, Metamathematics*, 2nd edn, Indianapolis: Hackett Publishing Company 1983: 110–42.

7

Tarski's Conception of Meaning

Douglas Patterson

Tarski's remarks on definition are a source for his views on meaning, something to which he otherwise gave little explicit attention, despite the fact that his views on truth and semantics became seminal for treatments of meaning in a number of disciplines. Here I will try to pull together Tarski's scattered remarks on meaning and definition and to set them in historical context, with the overall aim of coming to understand the implicit philosophy of language that underlies the work of the 1930s and early 1940s. I will first gather together Tarski's remarks on meanings, extensions, and concepts, taking his remarks on definition as central (§§1–4). The basic claim here will be that three conceptions of meaning are at work in Tarski's writing and that, while the three do form a coherent package, we need to be careful to keep their differences and their interaction in mind in attempting to understand any given passage. I'll then discuss Tarski's project of defining truth in particular (§5) and I will apply the accounts developed to resolve a number of familiar interpretive questions about Tarski's definition, for instance whether his definitional procedure wrongly makes what ought to be contingent truths about semantics into necessary or logical truths (§6). With those interpretive questions addressed, I will then supplement remarks I have made elsewhere (Patterson 2006a) on Tarski's presentation of the results on the indefinability of truth (§§7–8). With our study of his views on meaning and definition in hand we will be in a position fully to understand the significance, according to his views of the time, of the fact that no consistent definition of truth is possible under certain circumstances.

1. FORMAL DEFINITION: MEANING AS DETERMINED BY SENTENCES HELD TRUE

In many places in Tarski's work, claims about meaning are bound up with discussions of the formal definability of one expression in terms of others. The topic is most extensively treated in "Some Methodological Investigations on the Definability of Concepts." There formal definability comes to the derivability of an explicit definition of the *definiendum* from a theory:

Every sentence of the form:

$$(x) : x = a . \Leftrightarrow . \phi(x; b', b'', \ldots)$$

where '$\phi(x; b', b'', \ldots)$' stands for any sentential function which contains 'x' as the only real variable, and in which no extra-logical constants other than 'b'', 'b''', ... of the set B occur, will be called a *possible definition* or simply a *definition of the term* 'a' *by means of the set B*. We shall say that the term 'a' *is definable by means of the terms of the set B on the basis of the set X of sentences*, if 'a' and all terms of B occur in the sentences of the set X and if at the same time at least one possible definition of the term 'a' by means of the terms of B is derivable from the sentences of X.

(1983, 299)[1]

Note that formal definability is always relative to a theory, the set of sentences X. Case studies here include, for example, the number of primitives required to express geometry (1983, 306).[2] In application, many questions of interest when it comes to formal definition concern whether terms of one sort can be eliminated in favor of terms of another sort, where the sorts are of some independent interest, for instance the definability of terms from mechanics in terms of those from geometry (1983, 317–18). When we turn to Tarski's treatment of truth, the relevant elimination will be of semantic terms in favor of terms from "the morphology of language" (1983, 252).

The primary conception of meaning at work in Tarski's writing, I will argue, is that the meaning of a term within a language and theory is determined by the sentences involving a term that are held true—that is, that the meaning of a term is determined by a set of derivable sentences within that theory (including those treated as derivable from the empty set of sentences, the axioms). One characteristic passage comes in this familiar discussion of the role of the T-sentences in settling the meaning of "is true":

Let us try to approach the problem from quite a different angle, by returning to the idea of a semantical definition as in §1. As we know from §2, to every sentence of the language of the calculus of classes there corresponds in the metalanguage not only a name of this sentence of the structural-descriptive kind, but also a sentence having the same meaning... In order to make clear the content of the concept of truth in connexion with some one concrete sentence of the language with which we are dealing we can apply the same method as was used in §1 in formulating the sentences (3) and (4) (cf. p. 156). We take the scheme (2) {'x *is a true sentence if and only if p*'} and replace the symbol 'x' in it by the name of the given sentence and 'p' by its translation into the metalanguage. All sentences obtained in this way... naturally belong to the metalanguage and explain in a precise way, in accordance with linguistic usage, the meaning of the phrases of the form 'x is a true sentence' which occur in

---

[1] See Hodges (this volume) for a number of conditions Tarski understood to be equivalent to the one stated here.

[2] Here Tarski is most influenced by Veblen and the American Postulate Theorists such as Langford and Huntington, as well as by Hilbert. See Scanlan 2003 for the connection to the former which included, as Scanlan discusses, Tarski's intensive study of Langford's work in a seminar from 1927 to 1929. For the connection with Hilbert, see Sinaceur 2001 and forthcoming.

them. Not much more in principle is to be demanded of a general definition of true sentence than that it should satisfy the usual conditions of methodological correctness and include all partial definitions of this type as special cases; that it should be, so to speak, their logical product.

(1983, 187)

Here the idea is that 'true sentence' has the meaning that is determined for it by the T-sentences; a theory that includes them endows 'true sentence' with the correct meaning. This sort of meaning, the sort settled by the assertibility of certain sentences within a theory, is the sort of meaning that is supposed to be captured by formal definition; as he writes in 1944:

…we are able to put into a precise form the conditions under which we will consider the usage and the definition of the term *"true"* as adequate from the material point of view: we wish to use the term *"true"* in such a way that all the equivalences of the form (T) {*X is true if, and only if, p*} can be asserted, and *we shall call a definition of truth "adequate" if all these equivalences follow from it.*

(1944, 344)

A striking illustration of the importance of sentences held true for determining meaning comes from Tarski's discussion of the fact that eliminability is always relative to a theory. In a note, having said that "it is not difficult to see why the concept of definability, as well as all derived concepts, must be related to a set of sentences" he makes not the formal point that definability as treated in the article has to do with what is derivable from what and hence has to assume what set of sentences (possibly empty) is available for the provision of auxiliary premises, but rather says:

there is no sense in discussing whether a term can be defined by means of other terms before the meaning of those terms has been established, and on the basis of a deductive theory we can establish the meaning of a term which has not previously been defined only by describing the sentences in which the term occurs and which we accept as true.

(1983, 299)

A formal explicit definition thus codifies the meaning of a term as established by the sentences containing it which may be derived within a theory: the import of the formal definability of one term in terms of others within a theory is that the sentences that suffice to settle the meaning of these other terms also settle the meaning of the defined term when supplemented only by what Tarski calls a "possible" definition. Put also in terms that Tarski uses, the point in formal definition is to establish that a theory involving a certain term is equivalent to some theory involving strictly fewer primitive terms supplemented only by the definition (1983, 306). This way of thinking of definitions doesn't sit particularly well with a common view of definitions as somehow without content: a formally correct definition, though conservative over the theory to which it is added, and though it eliminates the defined term relative to the theory, makes all the difference between a sub-theory and a full theory and thus has whatever content the first lacks and the second has. The extendability of one theory to another through formal definition doesn't show *all by itself* that the second has some particular status assumed to hold of the first.³ This will be important when we consider the status of Tarski's definitions of semantic terms below.

In thinking of the sort of meaning that is passed from some terms to another via a formal definition as itself settled by these undefined terms figuring in sentences "which we accept as true" Tarski endorses the conception of the meaning of a term as implicitly defined by the axioms of the theory in which it is involved that is familiar from positivists such as Carnap, who writes, in 1934:

> Up to now, in constructing a language, the procedure has usually been, first to assign a meaning to the fundamental mathematico-logical symbols, and then to consider what sentences and inferences are seen to be logically correct in accordance with this meaning. Since the assignment of the meaning is expressed in words, and is, in consequence, inexact, no conclusion arrived at in this way can very well be otherwise than inexact and ambiguous. The connection will only become clear when approached from the opposite direction: let any postulates and any rules of inference be chosen arbitrarily; then this choice, whatever it may be, will determine what meaning is to be assigned to the fundamental logical symbols.
>
> (2002, xv)⁴

The view hails, in turn, from Hilbert and can be seen even in his famous remark that "it must be possible to replace in all geometric statements the words *points, lines, planes* by *tables, chairs, mugs*" (Ewald 1996, 1089) and in remarks like this one, from a 1903 letter to Frege:

> My opinion is that a concept can only be logically determined through its relations to other concepts. These relations, as formulated in determinate assertions, are what I call axioms. I thereby come to the conclusion that axioms... are definitions of concepts. I have not come to this opinion for the purposes of my own amusement; rather, I have found myself forced to accept it by the requirements of rigor in logical inference and in the logical structure of a theory.
>
> (Coffa 1986, 33–4)

This conception of meaning is thus familiar from thinkers whose influence on Tarski is well-documented and can be found expressed in passages from Tarski like the one presented above.⁵ The basic idea is that within a formal theory, some axioms are chosen and taken to be such that they "may be asserted" or "are considered as true"

---

³ Example: showing that a given system understood as "logic" can be extended to something that seems worth calling "arithmetic" doesn't show all by itself that "arithmetic is really just logic" in any sense that carries philosophical weight (e.g. avoidance of ontological commitment to abstract objects) rather than showing the reverse, namely, that logic is really implicitly arithmetic (with whatever problematic features "arithmetic" so understood has). To draw any such consequences from the possibility of formal definition, substantial theses both about the discipline to which the definitions are added and the character of the definitions are required. See §6.

⁴ I think this Carnapian attitude of "tolerance" is part of what is behind the remarks on rival conceptions of truth in the second part of Tarski 1944.

⁵ For Hilbert, see e.g. Tarski 1941, 120, 140, also Sinaceur 2001 and forthcoming, and for Carnap see the many favorable references to the *Abriß der Logistik* in Tarski 1983.

and the terms involved are taken to have meanings that are settled by their deductive role within the theory.[6] Compare, for instance, Carnap's own treatment of meaning in "Language I" in *Logical Syntax*:

By the logical **content** of {a sentence} or {set of sentences} (in I) we understand the class of non-analytic sentences (of I) which are consequences of {the sentence or sentences} respectively (in I). The "content" or "sense" of a sentence is often spoken of without determining exactly what is to be understood by the expression. The defined term 'content' seems to us to represent precisely what is meant by 'sense' or 'meaning'—so long as nothing psychological or extra-logical is intended by it.

(2002, 42)

This positivist, inferential conception of meaning is the first of the three we will find in Tarski's work. It ties meaning to a language only via the mediation of what is derivable from what there (including what is treated as derivable from no premises, that is, as axiomatic) and this, in turn, explains a fact often noted with some consternation by interpreters, that Tarski often speaks freely of languages as individuated by the theories consisting of their assertible sentences and as having properties, such as "inconsistency," that only sets of sentences can have (Tarski, 1983, 165, 1944, 349; see e.g. Burge 1984, 83–4 for the criticism).[7] Inconsistency of a language thus comes to nothing more than the derivability of contradictions from sentences that determine meanings.

This notion of meaning is often also related to notions of "adequate usage" as well (1944, 348, also 1983, 187 as quoted above)—and this even in "colloquial" language, where Tarski is often prepared to speak of certain sentences involving a term as determining its proper use and thereby setting the conditions in which a definition of the term must operate. Given Tarski's view that colloquial language is, in fact, inconsistent, a topic of central interest will be what to say in the case where the assertible sentences of a language are inconsistent. If they are, by the conception

---

[6] See Detlefsen 2004 and Coffa 1986 for two excellent discussions of the history here (both Detlefsen and Coffa trace the view back much further than Hilbert) as well as Friedman 1999 for related discussion. It's very easy to confuse this idea with the idea that the terms have meanings that *make the axioms true*. On at least some readings the latter was, in fact, Hilbert's view of mathematics. I don't think, by contrast, that it is what Carnap had in mind in *Logical Syntax*, but I can't settle interpretive issues concerning Carnap and Hilbert in this chapter; certainly, though, the quotations in the text don't suggest the stronger view. In any case, as I discuss below in §6, I believe that Tarski adhered only to the weaker view that an expression's deductive role in a theory determines its meaning. Tarski famously held that some languages are inconsistent, as will be discussed in a moment. If on Tarski's conception terms had meanings that made axioms *true* then inconsistent theories (and languages) would involve contradictions that were *true*. But if a contradiction is true, then by the T-sentences we can derive everything. Soames 1999 reads Tarski's remarks on inconsistent languages as involving commitment to true contradictions; see Ray 2003 and Patterson 2006 for responses.

[7] Of course, this alone doesn't justify, including full type theory, systems sufficient for arithmetic, and so on, under the heading of "language," but as is often noted, during the period under study here Tarski often seemed happy to include these things under the general heading of "logic" (see Feferman, this volume), and this assimilation would have made it more natural to attach the relevant deductive systems to "languages." Furthermore, to the extent that Tarski was attracted to formalist doctrines about implicit definition, the assimilation would also seem natural to him.

of meaning offered all sentences would seem to have degenerate meaning, since all sentences involving all terms will be "held true" as axioms or derivable from axioms.[8] We will thus need to sort out how Tarski can say both that "colloquial" language is like this and nevertheless maintain that "is true" in such language has a "meaning current in everyday life" (1983, 153).

2. SEMANTIC DEFINITION: MEANING AS EXTENSION

Semantic definition is the topic of "On Definable Sets of Real Numbers." As the title indicates, the concern here is not with the definition of terms by other terms, but rather of the objects and set-theoretic constructs therefrom which theories are intuitively *about*. The main sort of question about this sort of definition asks whether, given a language and an interpretation of its primitive vocabulary, various constructs out of this interpretation are in fact the extensions of expressions constructed out of the vocabulary via whatever formation rules are provided.[9]

In a way that strongly anticipates the fundamental negative results on the definability of truth, Tarski shows how, given an association between certain primitive sentential functions of a system sufficient for the arithmetic of the real numbers and the sets that satisfy them, the set of all arithmetically definable sets of reals can be defined recursively in non-semantic terms. As he notes, however, it follows on pain of a contradiction in the form of the Richard paradox that this set does not include all sets of reals (1983, 119). Tarski relates truth to satisfaction of a sentence by all sequences of objects in a way that parallels exactly the definition of truth in "The Concept of Truth in Formalized Languages" (1983, 117), so that in its essentials the celebrated definition of truth of the long article is already present in the shorter treatment of semantic definability. Since the more famous article goes into significantly more detail on the definition of truth, we can expect that merely introducing an expression that defines

---

[8] "Would seem to have" since strictly one could distinguish the deductive roles of various terms from one another even in a theory in which everything ultimately could be derived from the axioms. I won't make much of this possibility here since Tarski's remarks are too sparse to support such an interpretation and something else can be seen to underwrite his remarks on the meaningfulness of expressions in inconsistent languages (§3).

[9] The confusion on which "a" semantic definition is a special *kind* of formal definition is nevertheless quite common. A good example here is Coffa 1991, 294ff. Coffa misconstrues questions about semantic definition in terms of the introduction of "new" (294) expressions with specified extensions into a theory and thus, by page 296, becomes thoroughly ensnared in talk of "semantic definitions" ("M-definitions") as things one can *give* and of an individual such "definition" as involving a "definiens" (296). Nothing of the sort is involved: once syntax and lexical and compositional semantics are settled a language either contains an expression with a certain extension, or it doesn't—introducing a *new* expression with that extension is beside the point if it does, and constitutes a change to the semantics of the language (*not* a formally correct definition) if it doesn't. (These remarks aren't incompatible with Hodges' very helpful point (this volume) that Tarski chose the title "Semantic Conception of Truth" because Kotarbiński called formal definitions with quotes only in their definienda "semantic": Tarski's formal definitions of semantic concepts are "semantic" in this sense, but aren't to be confused with the semantic definition of a set by a predicate, as Hodges notes.)

the set of truths cannot have been Tarski's only aim there. What is added, as we'll see, is an extended discussion of the attempt to capture the "intuitive meaning" of semantic concepts in a certain sort of formal definition. We will return to this topic below.

Though Tarski often explicitly discusses meaning, and often explicitly discusses what he calls "semantics," claims to the effect that semantics *is* the study of meaning are missing in his work. Generally he characterizes semantics as the study of the relations between expressions and what they refer to (e.g. 1983, 193–4, 252, 401). Indeed, Tarski's relative silence on the connection, contrasted with the frequency of his comments on the relation of meaning to roughly "proof theoretic" notions such as derivability, assertibility, or being accepted as true, is sufficient to give the impression that, if anything, a *contrast* between meaning and semantics is intended. In view of the role that Tarski's inventions came to play in what most people now take to be the study of *meaning*, this of itself is an interesting historical note.

Nevertheless there are three important links between the conception of meaning as settled by sentences held true and the conception of meaning as extension in Tarski's work. The first is the applicability of Padoa's method for showing that terms are independent within a theory, one that Tarski further develops:

In order, by this method, to show that a term '*a*' cannot be defined by the terms of the set *B* on the basis of a set *X* of sentences, it suffices to give two interpretations of all extra-logical constants which occur in the sentences of *X*, such that (1) in both interpretations all sentences of the set *X* are satisfied and (2) in both interpretations all terms of the set *B* are given the same sense, but (3) the sense of the term '*a*' undergoes a change.

(1983, 300)

Since definitions explicitly express the meaning with which a term is endowed by a theory and their impossibility can be shown by the possibility of two interpretations of this sort, that a term isn't rendered fully meaningful by some sub-theory is in this way shown. Here defining a set or having an extension is treated as directly relevant to the meaningfulness of a term, in that if the meaning of a term is to be determined by the meanings of others, then the construct out of the domain of the interpretation of the language it semantically defines has to be determined by those they semantically define.[10]

The second link between the conceptions can be seen in an emphasis in the treatment of formal definition on the categoricity of a theory. Having by repeated

---

[10] Here, as with the next point, we need to understand the relation of these issues to something rightly stressed by Hodges (this volume): Tarski's aim in the article appears to be to give broadly "proof-theoretic" reconstructions of the semantic methods of establishing independence of terms set out by Padoa. That is, Tarski will recast the notion of *interpretation* he ascribes to Padoa in the passage above in terms of derivability of sentences of certain forms. This emphasizes the large extent to which for Tarski during this period formal definition takes precedence over semantic definition, something I will stress in what follows. Nevertheless, the authors to whom Tarski refers here and in the treatment of categoricity to be discussed in a moment do have a semantic take on the issues, and Tarski sees his own work as constrained to show, within the treatment of derivability in a theory, why those semantic methods are acceptable; witness here the remark that his results "provide a theoretical justification for the method of Padoa" (1983, 300).

applications of Tarski's version of Padoa's method established a set of mutually independent terms that are required for the expression of a full theory, we then have another question: does the theory fully determine the meanings of its primitive terms? This question Tarski sees as crucially tied to the categoricity of the theory:

> A non-categorical set of sentences (especially if it is used as an axiom system of a deductive theory) does not give the impression of a closed and organic unity and does not seem to determine precisely the meaning of the constants contained in it.
>
> (1983, 311)

Unique determination of a structure, elements of which its terms semantically define, is thus a desideratum in a theory and is one because only thus are the meanings of primitive terms fully determined. This emphasis on categoricity links the sort of meaning involved in the treatment of formal definition to semantic properties of axiom systems and hence makes clear that determination of semantic properties in admissible interpretations of an axiom system is an aspect of meaning broadly construed for Tarski.

Indeed, Tarski in this period regards it as desirable that a theory should not merely be categorical, but that it should be what he calls "complete with respect to its specific terms," which will be if it has no categorical extensions that contain terms that aren't eliminable in favor of the theory's original primitive terms (1983, 311). This is an interesting moment in Tarski's development, since the requirement is mathematically rather uninteresting: as he himself notes, theories only come to satisfy it when they are supplemented by axioms to the effect that nothing doesn't fall under certain basic terms, or more generally which determine how many things exist that don't do so (1983, 310–11 note 2). That is, in order to make a theory complete with respect to its specific terms, we will often have to add to it a statement as to exactly how many objects it doesn't cover—hardly a topic of intrinsic interest to the theory in question. Interestingly, this emphasis seems to have been picked up from Langford; see Scanlan 2003, 318, who notes that Langford himself made no use in further proofs of such a postulate in the 1926 chapter to which Tarski paid a great deal of attention in his 1927–1929 seminar.

The idea in the focus on "completeness with respect to specific terms" actually seems to be that the most "complete" theory would determine the structure of the whole universe up to isomorphism. There is some support for this conception of the importance of categoricity and "completeness with respect to specific terms" in the fact that in the passages in question Tarski attributes his appreciation for the significance of categoricity to the third, 1928 edition of Fraenkel's *Einleitung in die Mengenlehre*.[11] There Fraenkel sees categoricity as tending toward the realization of the "economy of thought which Mach considered to be the aim of all science" (1928, 350). Given Tarski's enthusiastic citation of Fraenkel, we can only assume that when, in the discussion of mechanics and its relation to geometry, categoricity is taken as a "criterion" for judging a formally presented scientific theory

---

[11] Fraenkel in turn, by the way, cites Tarski with approval in the relevant passages.

(1983, 318), Tarski is expressing endorsement of something like Fraenkel's report of Mach's sentiments—sentiments echoed as well in his famous remarks on the unity of science in "The Establishment of Scientific Semantics" (1983, 406).[12] (Note, though, that at 318 Tarski seems quite reasonably to doubt the value of a categorical formulation of mechanics as opposed to geometry.)

The third link can be seen when we notice something strange about Tarski's treatment of semantic definition itself, something that the subsequent acceptance of the semantic conception Tarski helped to introduce can make it difficult to spot. The basic construction Tarski gives in "On Definable Sets of Real Numbers" is very familiar to someone even moderately versed in basic formal methods for the study of semantics. One first sets out the syntax of the language under study and one assigns semantic values to the primitive expressions, e.g. sets to one-place predicates or "sentential functions," as Tarski says. A total assignment to open and closed sentences will then be determined as long as for every way of forming an open or closed sentence from simpler ones (e.g. conjunction, existential quantification) there is some set-theoretic operation on the values assigned to the component open sentences that determines the semantic value of the resulting open sentence:

> We notice that every sentential function determines the set of all finite sequences that satisfy it. Consequently, in the place of the metamathematical notion of a sentential function, we can make use of its mathematical analogue, the concept of a set of sequences. I shall therefore introduce first those sets of sequences which are determined by the primitive sentential functions. Then I shall define certain operations on sets of sequences which correspond to the five fundamental operations on expressions. Finally, in imitation of the definition of sentential function, I shall define the concept of definable set of sequences of order n.
>
> (1983, 120)

This sounds familiar: one interprets a language by giving a syntax, a lexical semantics, and a compositional semantics. In particular, there is nothing here about the semantic treatment being relative to a *theory* or a set of *axioms*. But in Tarski's mind, the whole construction *is* relative to a theory, that is, semantic definition is definition not merely in a language as individuated by a primitive vocabulary and a way of forming expressions, but in a language *and* a theory:

> The notion of definability should always be *relativized to the deductive system* in which the investigation is carried out. Now in our case it is quite immaterial which of the possible systems of the arithmetic of real numbers is chosen for discussion. It would be possible, for example, to regard arithmetic as a certain chapter in mathematical logic, without separate axioms and primitive terms. But it will be more advantageous here to treat arithmetic as an independent deductive science, forming as it were a superstructure of logic.
>
> The construction of this science may be thought of in more or less the following way: as a basis we admit some system of mathematical logic, without altering its rules of inference and

---

[12] The enthusiasm faded: in a note added for the first, 1956 edition of *Logic, Semantics, Metamathematics*, Tarski says that he no longer endorses the views on categoricity hinted at in the article. This is, no doubt, because in the meantime he has come to endorse first- as opposed to higher-order logic as the proper vehicle for the formalization of scientific discourse. See Mancosu 2005, Frost-Arnold, this volume, and my remarks below.

of definition; we then enlarge the system of primitive terms *and axioms* by the addition of those which are specific to arithmetic.

(1983, 113, emphasis mine)

Here the axioms are treated as somehow relevant, and the whole construction is interpreted as relative to a theory. But the construction Tarski actually gives in the body of the article makes no use of this; the semantic treatment is relative not to some *axioms*, but to the *interpretation* of the primitive vocabulary, just as the modern reader would expect.

It isn't far to seek what is going on here, given the treatment we have already seen of the conception of meaning that emerges in Tarski's treatment of formal definition. Tarski's *basic* conception of meaning is the one shared with Carnap: a term's meaning is determined by its role in a theory. The emerging *semantic* treatment, born in this and related articles, and which will ultimately supplant the earlier conception of meaning in the minds of those influenced by Tarski, is here still subservient to the more formalist conception in that Tarski is still thinking of that older conception when he considers how *primitive* terms get their meaning: thinking of the primitive terms as having a meaning only makes sense relative to a theory. This is the way I propose we make sense of the fact that Tarski in his informal remarks treats semantic definition as relative to a theory, but in his formal construction needs only to assume that it is relative to an assignment of semantic values to primitive terms: he holds, as in the treatment of formal definition, that primitive terms have the meaning they do because of their role in the axioms and deductive structure. In "On Definable Sets" Tarski is at an early stage in the development of the semantic treatment of the meaning of a term. Though we see in the article part of the birth of an articulate conception of meaning that need not be theory-relative, in Tarski's mind at the time the relativization to a theory is still necessary to make sense of the idea of primitive terms as *having* an interpretation. Later, mature model theoretic treatments that build on the results of the article will, by contrast, treat it as a matter of indifference how set-theoretic interpretants are assigned to primitive vocabulary, and will thus drop the idea that semantic definition is relative to a theory.[13]

3. MEANING AS CONCEPT EXPRESSED

A third strand in Tarski's conception of meaning runs through his use of words like "concept" and "notion," as well as his various remarks on "intuitive" meaning. The

---

[13] And what of the assumption of uniqueness of sets defined by the primitives built into the article, which ought to be problematic given that an axiom set may fail to be categorical, as we have seen Tarski understands? (The assumption comes when Tarski (1983, 121–3) assigns sets to primitive sentential functions using an abstraction operator that is supposed to form an expression the refers to "the" set of objects that satisfy the embedded sentential function.) My conjecture here is that at this early stage, given the embedding of the construction in higher-order logic, the possibility of a failure of categoricity isn't at the center of Tarski's attention; this will come only later with fuller understanding in the mathematical-logical community of the special features of first-order logic and Tarski's own shift to a preference for first-order languages.

easiest way to grasp the importance of this strand is to note how Tarski insists repeatedly that such things are clear at the outset of various inquiries, and how he continues to use these notions even in contexts where coherent formal and semantic definition are impossible, when languages are inconsistent. For instance, Tarski is happy to say at the outset of "The Concept of Truth in Formalized Languages" that

A thorough analysis of the meaning current in everyday life of the term 'true' is not intended here. Every reader possesses in greater or less degree an intuitive knowledge of the concept of truth and he can find detailed discussions on it in works on the theory of knowledge.

(1983, 153)

Likewise, in various other discussions, often before embarking on an involved project of formal definition, Tarski insists that the "intuitive meaning" or "concept expressed by" the definiendum is in some way perfectly clear. Before discussing semantic definition, which Tarski understands in terms of satisfaction, he says of the latter:

... we can try to define the sense of the following phrase: '*A finite sequence of objects satisfies a given sentential function.*' The successful accomplishment of this task raises difficulties which are greater than would appear at first sight. However, in whatever form and to whatever degree we do succeed in solving this problem, the intuitive meaning of the above phrase seems clear and unambiguous.

(1983, 117)

Likewise, "The Semantic Conception of Truth" states its aim with respect to the "notion" of truth in this familiar passage:

Our discussion will be centered around the notion of *truth*. The main problem is that of giving a *satisfactory definition* of this notion, i.e. a definition which is *materially adequate* and *formally correct* ... The desired definition does not aim to specify the meaning of a familiar word used to denote a novel notion; on the contrary, it aims to catch hold of the actual meaning of an old notion. We must then characterize this notion precisely enough to enable anyone to determine whether the definition actually fulfills its task.

(1944, 341)

This theme of rendering intuitions precise, related to use of the phrase "intuitive meaning," also receives extended discussion in the case of semantic definability:

The problem set in this article belongs in principle to the type of problems which frequently occur in the course of mathematical investigations. Our interest is directed towards a term of which we can give an account that is more or less precise in its intuitive content, but the significance of which has not at present been rigorously established, at least in mathematics. We then seek to construct a definition of this term which, while satisfying the requirements of methodological rigour, will also render adequately and precisely the actual meaning of the term. It was just such problems that the geometers solved when they established the meaning of the terms 'movement', 'line', 'surface', or 'dimension' for the first time. Here I present an analogous problem concerning the term 'definable set of real numbers'.

Strictly speaking this analogy should not be carried too far. In geometry it was a question of making precise the spatial intuitions acquired empirically in everyday life, intuitions which are vague and confused by their very nature. Here we have to deal with intuitions more clear and conscious, those of a logical nature relating to another domain of science, metamathematics. To the geometers the necessity presented itself of choosing one of several incompatible meanings, but here arbitrariness in establishing the content of the term in question is reduced almost to zero.

I shall begin by presenting to the reader the content of this term, especially as it is now understood in metamathematics. The remarks I am about to make are not at all necessary for the considerations that will follow—any more than empirical knowledge of lines and surfaces is necessary for a mathematical theory of geometry. These remarks will allow us to grasp more easily the constructions explained in the following section and, above all, to judge whether or not they convey the actual meaning of the term.

(1983, 112)

In all cases, a concept grasped in everyday life, or an intuitive notion or meaning, is to guide the construction of formal definitions. Grasp of the ordinary notion is, as emphasized in this last passage, not necessary for appreciation of the formal theory developed, but it is essential for judging the extent to which the "actual meaning of an old notion" is really "caught hold of."

Hence "intuitions," "intuitive meaning," ordinary "concepts" or "notions" are taken by Tarski to be largely perspicuous, especially in the cases in which we're interested here, and guide the setting out of formal definitions. It is in this respect that Tarski may still be influenced by Brentanian doctrines about the intuitive evidence of meaning, as Woleński and Simons (1989) assert.[14] In any case what is striking is that Tarski is perfectly willing to retain this talk even in cases where he can't make good on it in a formalization because formally correct definition is impossible. Although "Every reader possesses in greater or less degree intuitive knowledge of the concept of truth" nevertheless:

In §1 colloquial language is the object of our investigations. The results are entirely negative. With respect to this language not only does the definition of truth seem to be impossible, but even the consistent use of this concept in conformity with the laws of logic.

(1983, 153)

Note the clear implication here: there *is* a concept of truth that cannot be *used* in conformity with the laws of logic in its application to colloquial language. Thus there is more to the concept of truth than is captured in any particular formal definition for a language. The concept, a grasp of which is presupposed, guides the attempt to set out a formal definition but is not impugned by the impossibility of this definition for certain languages. Concepts, unlike the meanings of the terms that express them, are not relative to a language, whether formalized or not.

The main problem, as is well known, is that colloquial language allows construction of the antinomy of the liar: apparently legitimate substitutions into the T-schema "*s* is true if and only if p" where what is substituted for "p" translates *s* seem to produce logical falsehoods (1983, 157ff, 1944, 347ff). When, however, Tarski writes, in a familiar passage, that:

If these observations are correct, then *the very possibility of a consistent use of the expression 'true sentence' which is in harmony with the laws of logic and the spirit of everyday language seems to*

---

[14] I was too quick to dismiss this suggestion in 2006*b*.

> be very questionable, and consequently the same doubt attaches to the possibility of constructing a correct definition of this expression
>
> (1983, 165)

we are not to conclude that the concept of truth itself is somehow incoherent, but that nothing can properly express it in colloquial language. We must therefore be careful, when reading Tarski, not immediately to conclude, as Soames does in a discussion of Tarski, that "our ordinary notion of truth is defective precisely because its unrestrictedness gives rise to paradox" (1999, 99). Tarski appears to think that, somehow, the notion is just fine, and there is rather a problem with the language in which we try to express it.

Historically, this line of thought in Tarski's remarks on meaning seems to be the remnant of what he in one place endorses as Leśniewski's "intuitionistic formalism," which seems to have amounted to the view that one ought, despite formalization, regard the terms used in a formalism as having meanings one grasps *independently* of the formalization (1983, 62).[15] This is, of course, opposed to the formalist or positivist views built into Tarski's remarks on the sort of meaning a term gets *from* its role in a formalization, and in this strand of his thinking Tarski is therefore allied more with Frege and Russell as these two are contrasted with Hilbert in the very helpful Coffa 1986 (see especially 29ff): against Hilbert and the growing endorsement of the conception of axiomatizations of mathematical theories as implicit definitions of their primitive terms, one that came to full flower in the work of the logical positivists (see Friedman 1999 as well as the remarks above) Frege held out for the idea that terms have their own meanings that must be grasped independently of their role in the axioms.[16] In the idea that terms have meanings that are settled by their role in a theory, and which for the non-primitives can be expressed in formal definitions, Tarski has it Hilbert's way, while in the idea that a term nevertheless in some sense expresses the "concept" or "notion" or "intuitive meaning" that guides construction of its role in the formal system, Tarski also has it Frege's (and apparently Leśniewski's) way.

4. SUMMARY REMARKS

Tarski's conception of meaning thus involves one informal and two formal elements. The informal element is the grasp on what it is we intend to express in a language,

---

[15] Unfortunately there is nothing illuminating in the passage from Leśniewski that Tarski cites. See Hodges (this volume) for one interpretation. In a note added to the page for the 1956 edition Tarski says that he ceased to endorse the Leśniewskian conception.

[16] How does Frege expect us to reconcile this with the famous "context principle"? Probably the idea is still that terms have no *other* meaning than what they contribute to the thoughts expressed by complete sentences, but that it is a mistake to think that one can know what a given sentence says before knowing what the terms it employs mean, that is, what sort of contribution they make to sentential meaning in general. One must avoid the misunderstanding on which the context principle says that for no term and no sentence in which it figures does the term have any meaning that doesn't figure in *that* sentence. See Patterson 2005. Perhaps, though, Frege simply isn't consistent with his earlier views in his response to Hilbert; at any rate, I can't determine the proper interpretation of Frege here.

the "concept" or "notion" that is to be expressed by an expression of a certain language, while the formal are, first, a conception of meaning as determined by the sentences involving a term that are held true in a theory, and, second, a conception of meaning as determining extension. Thus, an expression can express a concept in a language, and in this sense have a meaning—the sort of meaning "current in everyday life," "intuitive" meaning—even when it doesn't formally express that meaning in a way that could be captured in a formal definition or that fully determines an extension. We've seen that Tarski is willing to say this when formal definition is impossible because the language is inconsistent, and we'll also see that he is willing to retain this talk when forced to fall back to theories that aren't categorical. Having a formally definable meaning and an extension is a desideratum, but it isn't essential to meaningfulness in the basic sense of expressing a concept. Call the three aspects of meaning, for convenience, *intuitive meaning*, *formal meaning*, and *semantic meaning*. The two formal aspects of meaning are intended to be harmonious, and to express the informal grasp of meaning which is always assumed. Under some conditions, as we will see, this harmonious expression eludes us.

A final note before we move on to truth in particular: the theories of both formal and semantic definition are metalinguistic with respect to the expressions of a language. When it comes to formal definition the basic issue is whether a certain sentence involving the defined term is provable in a theory, while when it comes to semantic definition the question is whether an expression has a certain extension. This means that when considering formal definition we need to make claims in a metalanguage ML about provability in a theory expressed in an object language L, while when considering semantic definability we need to make claims, in a metalanguage ML, about the extension in a domain D of an expression of L. This also means something that is surprisingly often forgotten: the *evaluation* of a definition as *good* or not, as "adequate" or "accurate," or the evaluation of a certain predicate with respect to its having the "right" or "intended" extension, takes place *not* in the language L in which the definition is stated, but in the metalanguage ML. For a simple example, consider that when we define a bachelor as an unmarried man, we want to say that this definition is *correct*. But this can only mean that it's really the case that all and only unmarried men are in fact *bachelors*. We can't justify the claim, of "bachelors are unmarried men" that it is a good definition of "bachelor" without ourselves saying "bachelor." There is nothing circular about this, and likewise, there will be nothing circular in saying, *of* a definition of truth, that it applies to all and only certain sentences that are *true* or that it expresses the concept of *truth*. Such a claim relates the intuitive meaning of "is true" to an expression in a formal language intended to have a formal meaning that is beholden to it. The evaluative claim is a claim about a definition made in a metalanguage; it is not somehow a circular element taken up in the definition itself. When the expressions we consider formally or semantically are *themselves* semantic predicates for relations between object language expressions and elements of D, we then have *three* languages plus the domain to keep straight: expressions of L that have extensions in D, semantic expressions in ML that have as their extensions relations between elements of L and D, and the further language MML in which we discuss all of this.

5. THE PROJECT OF DEFINING TRUTH

Turning now to his most famous definitional project, Tarski's stated aim is to provide a "formally correct" and "materially adequate" definition of the term "true sentence" (1983, 152). As he notes, and as accords with his discussion of formal definition, the project only makes sense conceived of as aiming to provide definitions relative to specified languages and theories. Relative to a language and theory, a formal definition will eliminate the term "is true" in extensional contexts,[17] and it will, as a formal definition, be evaluated for how well it captures the intuitive notion of truth. Understanding this project of definition in terms of the above treatment of Tarski's conception of meaning, then, we should expect the endeavor to have three aspects: an intuitively grasped concept of truth (intuitive meaning) is to be expressed within a formal language either by being implicitly defined by the axioms of a theory or explicitly defined in terms of the primitives of that theory (formal meaning), and the formalization, if possible, is to be categorical and is thus, in particular, to assign a structurally determinate extension to the expression so defined (semantic meaning). We should thus expect Tarski to have something to say on all *three* topics, and we should be wary of any interpretation that runs two or more of them together.

As for the "intuitive meaning" of "is true," Tarski describes his goal as being to find a "precise expression" (1944, 343) of the "intuitions which adhere to the *classical Aristotelian conception of truth*" (1944, 342). On this conception—which, note, Tarski associates not with his formal definitions, but with the concept which those definitions are intended to express—a good definition of truth:

We can express in the following words:

(1) *a true sentence is one which says that the state of affairs is so and so, and the state of affairs indeed is so and so.*

From the point of view of formal correctness, clarity, and freedom from ambiguity of the expressions occurring in it, the above formulation obviously leaves much to be desired. Nevertheless its intuitive meaning and general intention seem to be quite clear and intelligible. To make this intention more definite, and to give it a correct form, is precisely the task of a semantical definition.

As a starting-point, certain sentences of a special kind present themselves which could serve as partial definitions of the truth of a sentence or more correctly as explanations of various concrete turns of speech of the type '*x* is a true sentence'. The general scheme of this kind of sentence can be depicted in the following way:

(2) *x is a true sentence if and only if p.*

In order to obtain concrete definitions we substitute in the place of the symbol '*p*' in this scheme any sentence, and in the place of '*x*' any individual name of this sentence.

(1983, 156–7)

---

[17] The restriction to extensional contexts is an artifact of Tarski's focus on extensional theories. See below for the significance of this, and Belnap 1993 for more on the general theory of definitions as developed following Tarski.

This is Tarski's philosophical analysis of the notion of truth. Formal definitions are beholden to it in that they'll be evaluated as successful to the extent that they can be seen as expressing this conception. Thus the intuition to be rendered precise is that a sentence "p" is true if and only if p, that is, that the T-sentences settle the meaning of "is true," as noted in §1 (1983, 165, 187, also 1944, 348). This intuitive conception is a conception as to which sentences involving an expression "is T" are to be assertible in a language if "is T" is to be regarded intuitively as expressing the notion of truth. It's this that Tarski calls the "semantic conception" of truth. This conception is most importantly a view about under what conditions a definition of truth is good, and is not itself fully expressed in any particular such definition. Note also that given Tarski's views about what the expression of an intuitive concept within a language come to, namely, the assertibility within a theory expressed in that language of certain sentences involving the term, there is no such thing as a language containing an expression that expresses a certain concept without the corresponding meaning-determining sentences being theorems of the relevant theory. When it comes to truth, this means that there is, on Tarski's view of the period, no such thing as an expression that expresses the concept of truth except within a theory of which all of the T-sentences for a certain object language are theorems.

The details of Tarski's procedure when L and ML allow a definition with all desired features are so familiar, and so often rehearsed in the literature, that I'll simply give the briefest of summaries here.[18] One first defines the satisfaction of an open sentence by a sequence of objects, calling an open sentence satisfied by a sequence just when the relevant members of the sequence stand in the relation expressed by the open sentence. Saying this in any particular case involves *using* an expression that translates the predicative component of the open sentence. In Tarski's own example, there is only one lexical predicate, "$\subseteq$", so the relevant clause is, notational niceties aside, "$x \subseteq y$" is satisfied by a sequence including x and y iff x is a subset of y. Sentential connectives are handled in the familiar way (e.g. an open sentence with "or" as its main connective is satisfied just in case either one or the other disjoined open sentences is satisfied, etc.) and quantification is handled by looking at preservation of satisfaction across variations in sequences at the relevant positions (e.g. "there is an x such that $x \subseteq y$" is satisfied by a sequence iff "$x \subseteq y$" is satisfied by at least one of the sequences that differ from the sequence in question only with respect to what they assign to "x"). (Def. 22, 1983, 193). Suitable use of higher-order logic or set theory turns these recursive conditions into an explicit definition for membership in a set. A true sentence is then defined as a sentence satisfied by every sequence (Def. 23, 1983, 195). Since the definition of satisfaction makes liberal use of expressions that in fact *do* translate the corresponding expressions of the object language, the result is a definition that implies the T-sentences for L in ML (1983, 195–6).

I emphasized above (§4) that we need to keep in mind that the evaluation of a definition as *good*, that is, as endowing the defined expression with the *right* meaning, is always metalinguistic relative to the definition and hence, when semantic terminology is at issue, that it is always a matter to be taken up in MML. Attention to this

---

[18] See Soames 1999 for one of many accessible treatments.

point allows us to set aside one very familiar worry about Tarski's procedure, namely that the criterion of adequacy that expresses his philosophical conception of truth, namely, Convention T, *uses* the semantic notion of translation. The worry is that this is somehow incompatible with his goal of eliminating semantic terms (1983, 152–3).

Convention T states a sufficient condition for the "material adequacy" of a definition of "is an element of the set of true sentences." As such, it is a claim about the conditions under which a formal definition of this expression in *ML* will be a *good* one. But this goodness can in any case only be explicated *using* the notion of truth in *MML*—as we have seen, this is a feature of any evaluation of a definition. As long as a definition is formally correct, it correctly defines *something*; the question as to whether it is "materially adequate" is the question whether the intended meaning of the expression—its "intuitive meaning" or the "concept" or "notion" that is supposed to be expressed by the expression of *ML*—is actually present in the set of sentences held true in a theory, but the question can't be asked without saying what the meaning is, and this in turn can't be done without using an expression of *MML* that is assumed to have the intended meaning, just as we can't say that "bachelors are unmarried men" is a good definition unless we're willing to say that bachelors are in fact unmarried men. In Tarski's case this appeal to intuitive meaning comes in the assumption that when one sentence translates another it also states its truth condition, the assumption without which Convention T makes no sense. We can't say of Def. 23 that it gets it *right* unless, apprised of the definition of satisfaction we're willing to assent to the claim that sentences satisfied by all sequences are in fact *true*. But the simple equation of truth with "satisfaction by all sequences" is hardly going to effect this; it is the link to the T-sentences that renders the definition intuitively satisfying (relative, that is to the Aristotelian conception of truth assumed throughout). If, with Tarski, we accept this conception, then we can recognize, of an expression defined so as to imply the T-sentences, that it expresses the relevant concept, since we are ourselves willing to accept that, where what is substituted for "p" translates *s*, the sentence *s* is true if and only if p. There is no circularity here; indeed, there would be no circularity in saying that it is a criterion of adequacy on a definition of truth that it imply those instances of "*s* is true in L if and only if p" that are in fact *true*.[19] This, by the way, is also what is missing from the otherwise parallel definition of truth in "On Definable Sets of Real Numbers": in "The Concept of Truth in Formalized Languages" the same formal definition is explicitly evaluated for its capturing the intuitive notion of truth via the T-sentences. What is added to a definition of truth already present but not treated as particularly interesting in the earlier article is explicit argument to the effect that the definitions given capture not just *extension* but *intuitive meaning*.[20]

---

[19] Putnam betrays the common confusion that this would somehow be circular at 1994, 319–20.

[20] It follows that material adequacy and extensional adequacy are not the same, despite common assumptions to the contrary. 1983, 129 is perfectly clear that "material adequacy" is a matter of intuitive meaning as opposed to merely correct determination of extension, for instance. Some of the remarks in Patterson 2006*b* now strike me as confused on the difference. It is correct, as I claim there, that implication of the T-sentences for an object language *L* by a formal definition in a metalanguage *ML* isn't necessary for *ML* to have a term that has as its extension the set of all and only true

Now everything here goes well as long as *L* and *ML* cooperate in allowing a formal definition with the desired features. These are *material adequacy, formal correctness*, and *extensional adequacy*, the three criteria of adequacy that correspond to the three aspects of meaning. Material adequacy requires that the "actual meaning of an old notion" be "caught hold of," in the words of Tarski 1944: intuitive meaning must be preserved. Formal correctness requires that the term defined actually be endowed with a coherent role in the theory in which it figures. Extensional adequacy, in turn, is the requirement that the term have the extension it should, given the intuitive meaning it expresses. When *L* and *ML* cooperate, all three aspects of Tarski's conception of meaning are brought together harmoniously. When they don't, choices must be made. In §§7–8 below we will examine what choices Tarski made. First, though, I will apply the above reading of Tarski's views on meaning and definition and on the definition of truth in particular to the consideration of some familiar criticisms of Tarski.

6. SOME STANDARD CRITICISMS

Attention to the details of Tarski's conception of meaning corrects a number of stock criticisms of Tarski. To focus discussion, let us consider these familiar claims from Putnam:

Since (2) {"'snow is white' is true-in-L iff snow is white"} is a *theorem of logic* in meta-L (if we accept the definition, given by Tarski, of "true-in-L") since no axioms are needed in the proof of (2) except axioms of logic and axioms about spelling, (2) holds in all possible worlds. In particular, since no assumptions about the *use* of the expressions of *L* are used in the proof of (2), (2) holds true in worlds in which the sentence "Snow is white" does not mean that snow is white.

...all a logician wants of a truth definition is that it should capture the *extension* (denotation) of "true" as applied to L, not that it should capture the *sense*—the intuitive notion of truth (as restricted to L). But the concern of philosophy is precisely to discover what the intuitive notion of truth is. As a philosophical account of truth, Tarski's theory fails as badly as it is possible for an account to fail. A property that the sentence "Snow is white" would have (as long as snow is white) no matter how we might use or understand that sentence isn't even doubtfully or dubiously "close" to the property of truth. It just isn't truth at all.

(1994, 333)

sentences of *L*. It is also right, as I claim there, that though it is necessary that the truth predicate of *ML* be defined in such a way that the T-sentences for *L* are implied if the definition is to express the "classical Aristotelian" conception of truth, Tarski held that there could be alternatives to the classical Aristotelian conception. However, a number of passages (e.g. p. 8) clearly display a conflation of material adequacy and extensional adequacy, often abetted by undifferentiated use of the term "adequacy." I've tried to correct that oversight here. I thus stand by the claim that Tarski should not be read as holding that material adequacy according to Convention T is necessary for extensional adequacy. I likewise stand by the claim that only on the assumption that the Aristotelian conception of truth is correct is implication of the biconditionals necessary for material adequacy. As I wrote then, implication of the T-sentences was for the Tarski of this period a *philosophical* rather than a logical or mathematical desideratum.

One idea here is that since Tarski makes "'snow is white' is true-in-L iff snow is white" a *definitional* truth, it is thereby according to his definition a *necessary* truth, whereas it is obviously a contingent fact that the sentence has that truth-condition, since the truth condition of a sentence depends on the meanings and ultimately the uses of its expressions and expressions have their meanings contingently because these uses are themselves contingent. Tarski is therefore guilty of turning obviously contingent truths into necessary truths and his procedure is thereby deeply flawed.

In order to evaluate the criticism, we must appreciate first, though Tarski does not emphasize the fact in the writings of the period, that a formal definition is relative not only to a language and a theory, but to a set of contexts as well (Belnap 1993, 121), and that in Tarski's case the set of contexts is always the extensional contexts. Definability of a term relative to a language and theory in extensional contexts need not imply that it is definable in intensional contexts, should the language and theory include these.[21] The point is easy to miss since Tarski is always interested in theories formulated in extensional languages, but it nevertheless crucially affects the interpretation of his definitions. What Putnam's complaint ignores is that Tarski's definition eliminates the truth-predicate *in extensional contexts only* and is thus neutral as to necessity or contingency, since Tarski's background theories themselves have no resources to distinguish necessary from contingent truth.

It is thus an unwarranted imposition to infer from the fact that the T-sentence follows from axioms for formal syntax *plus* Tarski's definition that it, in the language in which it is given, expresses something that "holds in all possible worlds." It's an even more unwarranted imposition to claim, as Putnam and others do, that according to Tarski Putnam's (2) "is a theorem of logic"—granting, temporarily, the use of the term "logic" to cover Tarski's use of higher-order type theory plus a theory of formal syntax.[22] It isn't, and Tarski never said it was: it is a theorem of *"logic" so construed plus the truth-definition*. The truth definition itself is not a truth of "logic," that is, it is not a truth of higher-order type theory plus formal syntax, any more than "bachelors are unmarried men" is a truth of logic. (A dictionary is not a list of *logical* truths.) It's clear from our discussion in §1 that though Tarski holds that definitions must be eliminable and conservative, he does *not* hold that they're somehow without content so that anything that follows from a theory with them added is somehow without further assumption of the same status as the theorems of the unsupplemented theory. On the contrary: showing a term to be eliminable via a definition relative to a sub-theory of a theory is showing that the sub-theory can be conservatively enriched to the whole theory by the definition alone; it *isn't* showing that somehow there is no difference at all between the sub-theory and the theory. In particular, even if Tarski did think that the logical and mathematical truths of the background theory of *ML* were

---

[21] Notice that in Putnam this is buried under the bizarre assumption that "logicians" are only interested in extensions.

[22] I'd thus take issue with the remarks at Heck 1997, 537, and Etchemendy 1988, 57, for instance, though I do not have space here for a full discussion of their forms of the views we find in Putnam.

necessary truths—though he never says anything of the sort, in accord with the general Polish aversion to intensional contexts, and though, as emphasized above, nothing in the theories themselves forces this understanding of them—it doesn't follow that he thought his definitions were necessarily true, and hence it doesn't follow that he thought the T-sentences were necessarily true. Since T-sentences are so clearly *not* necessary truths, we shouldn't take Tarski to be guilty of the error of proceeding on the assumption that they are without compelling reason.[23]

Now it might appear that this doesn't square well with the conception of formal meaning discussed in §1, for it might be thought that if a term has whatever meaning is required to make the axioms and theorems of some theory *true*, then there is no such thing as contingent truth: no such thing as a term preserving its meaning while a sentence the truth of which is supposed to determine its meaning goes from true to false. But this would be a confusion based on a misunderstanding of the conception of formal meaning: what determines meaning on this conception is not the *truth* of certain sentences, but their being *held true*, that is, their being treated as axioms and theorems. Carnap, for instance, doesn't say that the *truth* of "postulates" settles the meaning of terms; it's their being "chosen" to function as postulates that does so. A term's formal meaning is a matter of derivability relations in a formal theory, and Tarski's view is that we can't change *those* without changing the meaning of the term. This isn't unreasonable, amounting as it does merely to the view that a change in inferential role is a change in meaning.[24] Definitions could hardly be expected to work any other way.

A sentence's semantic meaning, that is, its truth condition, likewise needs to be settled by its formal meaning, since formal meaning is supposed to determine semantic meaning as discussed in §2. But a sentence's *truth value* is not likewise supposed on any view of Tarski's to be settled by our holding it true. Hence Tarski could allow that a theory, stated in *ML*, that has the T-sentences as theorems is such that the meaning of "is true in L" is settled within the theory *by that theory's having the*

---

[23] Often interpreters (e.g. Raatikainen, this volume) try to save Tarski from Putnam's criticism by arguing that on his view languages are individuated by the meanings of their terms. Given this, defining truth in a given language in a way tied essentially to the (actual) meaning of a certain list of terms in that language looks acceptable, even given the assumption—the one I have argued here is mistaken—that definitions express necessary truths. The view, however, can't make sense of the fact that Tarski clearly, and quite reasonably, allows that "as far as natural language is concerned . . . this language is not something finished, closed, or bounded by clear limits. It is not laid down what words can be added to this language and thus in a certain sense already belong to it potentially" (1983, 164). If Tarski allows that terms can be *added* to a language, then he can't be thinking that it's a necessary truth about a language that it have exactly the lexicon it actually has. Of course Tarski holds that colloquial language can't be treated rigorously for reasons such as unclarity as to what its terms are; the point is that we shouldn't saddle him with the view that languages necessarily have exactly a certain set of terms with exactly certain meanings, as the standard defense does. Rather, I've suggested, we should recognize that definitions in extensional theories need only be true, not necessarily or logically true. (If it grates to hear definitions called "true," then just read: definitions need only allow only the derivation of truths as opposed to necessary or logical truths.)

[24] Note that this unobjectionable, weak view leaves it open whether meaning determines inferential role or *vice versa*.

axioms and deductive structure it has. But this doesn't commit him to saying that the theory could not go from true to false (if *L* were to change) or that it could not have been false (had *L* been different), or for that matter that it could not *be* false (if, in fact, it has as theorems some instances of "*s* is true in L if and only if p" where what is substituted for "p" *doesn't* translate *s*).[25] These are claims, made in *MML*, about the *truth values* sentences of the theory expressed in *ML* might have or have had, and allowing that they are contingent is perfectly compatible with the idea that expressions of the theory mean what they mean (have the formal meanings they do) because of their deductive role within the theory. Now, of course, if we were to recognize that the meanings of the sentences of *L* had changed, so that the *ML* truth-definition that implied them had gone false, we'd have to change the definition in ML, and *that*, given the conception of formal meaning, would be a change in the formal meaning of "is true in L" in ML, since *different* sentences of the form "*s* is true in *L* iff p" would become the ones that needed to be implied by the truth-definition. So Tarski is committed to the claim that we can't change the definiens in the definition of truth without changing the meaning of the definiendum. This, again, however, is not unreasonable: if we substitute in a definition a definiens with a new meaning, then the definiendum may be given a different meaning by the definition.

If one more bit of evidence here is needed, one may turn to a note where Tarski explicitly discusses the relation between logical truth and the meaning of non-logical terms. At issue is whether an extension of a theory *X* to a new theory *Y* by addition of a single logical truth involving a single term not present in *X* makes *Y* "essentially richer"; of this he says:

> It is obvious that the new extra-logical constant cannot be defined by means of the terms of *X*. In fact the only sentence of *Y* in which this constant occurs is logically provable, and thus true independently of the meaning of the specific terms contained in it. It can thus be asserted that the meaning of the constant in question is not at all determined by the set *Y* of sentences.
>
> (1983, 309 n. 1)

On Tarski's primary conception of meaning, logical theorems do not endow non-logical expressions with *any* meaning. Hence if the definition of truth had the status that theorems of logical theories do, the truth-predicate, as a non-logical expression, would be assigned no determinate meaning by the definition. Since Tarski clearly doesn't think this, we must conclude that he doesn't think the definition, or the theorems that require it for their derivability within the theory that is enriched by it, are logical truths.

Turning next to the second paragraph of our quotation from Putnam, does Tarski's "philosophical" account of truth "fail as badly as it is possible to fail"? It is here that another widespread confusion, one which rests on failing to distinguish

---

[25] Though the last requires us not to take the idea of axioms as things that are "held true" too seriously, or the view is Moore-paradoxical; just read "held true" as "treated as derivable within the theory."

Tarski's remarks on intuitive meaning from those on formal meaning, and more generally rests on failing to note the difference between what is in *ML* and what is in *MML*, is at work. Tarski's formal definitions of truth predicates are not, and were not conceived by him to be, his philosophical account of truth. His philosophical account of truth is the "Aristotelian" conception, and the central plank of this account is the claim that a definition of truth should somehow sum up what is stated in the T-sentences. Whatever its merits on its own terms, this is a "philosophical account," and it is clearly *not* exhaustively expressed or intended exhaustively to be expressed in any *particular* formal definition of truth. Hence of course focus on the formal definitions alone will find them wanting as accounts of truth; the mistake, though, is to think that the philosophical view is exhaustively expressed in a definition, as opposed to the account of the conditions under which a definition is good. The philosophy takes place in *MML* as, ultimately, the account of the conditions under which a definition in *ML*, relative to a set contexts in L and a theory expressed in L, is a good one. Davidson (1990, 282–95) seems to me to come close to getting this right, though he underplays the importance of the Aristotelian conception and hence understates the extent to which Tarski did in fact tell us more about truth than his definitions do.

It follows also that we can ignore the very familiar complaint that Tarski's definitions don't tell us what they have in common as definitions of truth for various languages, and, relatedly, that they don't themselves explain how they are to be adapted to extensions of a language or to new languages: they aren't supposed to.[26] Again, the Aristotelian conception in *MML* tells us under what conditions a definition is good—when it implies the T-sentences—and the interaction of this with facts about the syntax and semantics of L informs the construction of a definition for particular languages L in *ML*, whether they be entirely new, or related structurally or historically to other languages.

7. THEOREM I: TEXTUAL ISSUES

I hope the foregoing makes clear that more attention to Tarski's views on definition and meaning will improve our understanding of what he says about defining truth in particular. Tarski's interest in defining truth is an interest in seeing how well truth according to the "Aristotelian" conception can be expressed in a mathematically tractable way (1983, 252). As is well known, Tarski thinks this is impossible except under certain conditions, and "The Concept of Truth in Formalized Languages" contains what is often taken to be an early statement of the now familiar theorem that arithmetic truth is not arithmetically definable. In this section I will discuss an interpretive issue that was brought to my attention by Lionel Shapiro and by Gómez-Torrente 2004, the issue being that Tarski may not actually be establishing something as close to the usual result as is often assumed.

---

[26] See Davidson 1990 for a discussion with references.

The formal core of the indefinability result is set out at pages 247–51 of CTFL. The language at issue is "the language of the general theory of classes," a language like the previous ones but with no upper bound on the order of expressions occurring in it. Tarski introduces the simplifying assumption, not in force earlier in the article, that the object language is a fragment of the metalanguage.[27] Tarski states what is to be proved as follows:

THEOREM I. (α) *In whatever way the symbol 'Tr' is defined in the metatheory, it will be possible to derive from it the negation of one of the sentences which were described in the condition (α) of the convention* T;

(β) *assuming that the class of all provable sentences of the metatheory is consistent, it is impossible to construct an adequate definition of truth in the sense of convention* T *on the basis of the metatheory.*

(247)

Since the object language L—here, the language of the general theory of classes—suffices for the arithmetic of the natural numbers and the ML in which its metatheory is to be stated is assumed to have an a formally defined symbol "Tr", by the diagonalizing considerations Tarski attributes to Gödel, we will be able to prove in the deductive system of ML that, for some sentence s of L, s iff $\sim$F($\langle s \rangle$), where $\langle s \rangle$ is the Gödel code of s and "$\sim$F" is a purely arithmetic predicate such that $\sim$Tr(s) iff $\sim$F($\langle s \rangle$); hence we'll have s iff $\sim$Tr($\langle s \rangle$). But this sentence is the negation of the T-sentence for s. Hence if we assume that "Tr" is a an expression that defines the class of truths of L, the negation of a T-sentence will be a theorem of the deductive theory in ML.

The point here is, in the first instance, entirely syntactic, as Gómez-Torrente 2004 stresses in his reading of the theorem: arithmetization guarantees that for *every* predicate of ML a sentence of the relevant form will be provable. This is part (α) of the theorem as stated, and just so far it doesn't matter what we take "Tr" to mean. Continuing, however, if we *do* take "Tr" to express the notion of truth, then in accord with Tarski's conception of that, the T-sentences have to be theorems of ML as well. But then the deductive theory associated with ML will be inconsistent. In the form in which Tarski states the theorem and gives the proof at 247–51, part (β) makes this point in terms of setting out a definition that is adequate according to Convention T: if our definition of "Tr" implies the T-sentences, then the deductive system of ML is inconsistent. A more general point, however, is clearly in the offing, given Tarski's conception of what is required to express the intuitive meaning of "is true": arithmetization is incompatible with any defined predicate of ML having the intuitive meaning of "is true." (Recall here that having this intuitive meaning requires the T-sentences being theorems whether or not they follow from some definition.)

---

[27] I blush to admit that I overlooked this in my remark in footnote 12 to Patterson 2006*a*. My doing so allowed me mistakenly to take the sentence following (2) on 250 of CTFL as additional evidence that the semantic reading of the theorem (see below) was correct. As I note below, a fair amount of evidence for that reading remains.

At issue in this section will be two readings of the passage in which the theorem is stated and proved. The first will be one on which Tarski's aim is to state only a syntactic result of the above form. Call is the *syntactic* reading. (I associate this reading with Gómez-Torrente 2004, but also with an interpretation pushed by Lionel Shapiro in conversation, so the reader should not take "the syntactic reading" to be exactly Gómez-Torrente's view.) The second will be the *semantic* reading on which Tarski intends to establish something like the result usually attributed to him: if *ML* has an expression that semantically defines the truths of *L*, the metatheory is inconsistent. As Gómez-Torrente notes, the semantic reading is common ("the more general practice is to use the name 'Tarski's Theorem' for a result involving the defined, or even the intuitive notion of truth" (2004, 35)). I myself went in for the semantic reading in 2006*a* and 2006*b*, and Gómez-Torrente attributes it explicitly to Solomon Feferman and (I think, though the passage reads unclearly) to Raymond Smullyan (2004, 35).[28] On the syntactic reading, Tarski is merely making the point that arithmetization shows that the negation of a T-sentence will be provable for any predicate we wish to treat as a truth-predicate, and hence that we can't introduce a truth-predicate by a formal definition materially adequate according to Convention T. Gómez-Torrente puts it thus: "neither part (α) nor part (β)" of Theorem I "use any notion of truth, either primitive or defined" (2004, 33). He insists that Tarski wishes, in the passage from 247 onward, only to make the syntactic point about formal definition:

> It was a certain kind of fear of limitation that led Tarski to state his theorem in the way he did. What limitation? The limitation connected with his 1933 refusal to consider a mathematics for the metalanguage which did not go beyond finite type theory. In the case of languages like LGTC, this leads to the impossibility of formulating an indefinability theorem in terms of a defined truth predicate. Yet Tarski was clever enough (and this is no surprise, because he was quite clever) to find a formulation of the theorem in which no concept of truth, primitive or defined, appeared. Finding it may not have been very painstaking, certainly not as painstaking as Gödel's caution-induced work was painstaking. But the thing is he found it (and Gödel didn't).
>
> (2004, 36)

Tarski, the idea is, meant more or less only to insinuate the application of the result in the claim that the intuitive concept of truth could not be expressed in a language meeting the conditions of Theorem I: "Tarski found a fully cautious result which ... immediately suggests and receives its significance from implications for the intuitive concept of truth" (2004, 36). On the semantic reading, by contrast, the point is the more general one that no predicate supposed to express the concept of truth for *L* in *ML* can define the set of truths of *L* on pain of inconsistency in the deductive system of *ML*.

---

[28] My impression is that Gómez-Torrente is quite right that the passages of CTFL in question are usually taken to provide the usual semantic result that ML can't have a predicate that defines the set of truths of L when ML bears the relations to L at issue in the theorem. For what it is worth, a quick online survey on "Tarski's Theorem" turns up a great many explanations that attribute the semantic result to 1983, 247ff.

In order to begin sorting out the differences between these two readings, let us attend to one of the many noteworthy features of Theorem I and its proof as we have them: Theorem I as stated expressly concerns an expression "Tr" *introduced by a formal definition*: "in whatever way the symbol 'Tr' is defined in the metatheory, it will be possible to derive" the negation of a T-sentence (247). Indeed here, for reasons that might bear looking into, the Polish is even clearer than the German and the English, having "it will be possible to derive from this definition" and not merely "derive from it."[29] Now, a question ought to strike us: why do we need to assume, to prove what Tarski wants to prove here, that "Tr" is *defined* rather than *primitive*? After all, for the syntactic point about arithmetization alone it is a matter of complete indifference whether or not "Tr" is introduced by a definition: diagonalization still tells us that the negation of a T-sentence will be provable; this is why Tarski is able to say that the "way" in which "Tr" is defined can be "whatever" way we want it to be. But why does it need to be defined at all? If Tarski simply wanted to prove that, given arithmetization, any metatheory that had a *symbol* that was taken to be a truth predicate in the sense that all T-sentences for it were held true would be inconsistent, he should have chosen an arbitrary symbol of the metatheory and simply made the point that the diagonal lemma told him that for *every* predicate F of the metatheory there was a theorem of the form s iff $F(\langle s \rangle)$ and hence likewise for $\sim F$ one would have, for some s', s' iff $\sim F(\langle s' \rangle)$, and that, therefore, one could have all of the instances of s iff $F(\langle s \rangle)$ only at the price of inconsistency. There is no need here for the assumption that F has a formal definition, and in fact Gómez-Torrente's own "anachronistic reconstruction" of the proof makes no use of the claim that the expression concerned is defined rather than primitive.

But Tarski *does* assume that the symbol at issue has a formal definition. Why? The answer comes when we consider the contrast with what Tarski allows in Theorem III and keep in mind the reductive and philosophical aims of CTFL. In Theorem III and its discussion Tarski allows that we can introduce a primitive symbol "Tr" and take the T-sentences for it as new axioms. In this way, the *concept* of truth for *L* will be expressed in *ML*: Theorem III allows that *"the consistent and correct use of the concept of truth is rendered possible by including this concept in the system of primitive concepts of the metalanguage and determining its fundamental properties by means of the axiomatic method"* (266). But in this case, its being expressed does not suffice for the categoricity of the resultant theory, nor does it suffice (more obviously) for the elimination in favor of non-semantic expressions Tarski seeks in the article. The former point is the more important one for us at the present juncture. Tarski writes:

It seems natural to require that the axioms of the theory of truth, together with the original axioms of the metatheory, should constitute a categorical system. It can be shown that this postulate coincides in the present case with another postulate, according to which the axiom system of the theory of truth should unambiguously determine the extension of the symbol '*Tr*' which occurs in it, and in the following sense: if we introduce into the metatheory, alongside this symbol, another primitive sign, e.g. the symbol '*Tr'*' and set up analogous axioms for

---

[29] Lionel Shapiro brought this to my attention.

it, then the statement "$Tr = Tr'$" must be provable. But this postulate cannot be satisfied. For it is not difficult to prove that in the contrary case the concept of truth could be defined exclusively by means of terms belonging to the morphology of language, which would be in palpable contradiction to Theorem I.

(258)

At the end of this passage the connection with Theorem I is the syntactic one, to be sure: if we could prove the statement "$Tr = Tr'$" then we wouldn't (by (β) of Theorem I) be able formally to define the *concept* of truth in terms of the non-semantic vocabulary of the metatheory. But here this fact is linked to the semantic one that setting up the theory of truth axiomatically doesn't fully determine the *extension* of "Tr". That any ineliminable "Tr" that expresses the intuitive notion of truth by having its T-sentences as axioms lacks a determinate extension and hence is not formally definable within the deductive theory of *ML* (if that is consistent) is here taken to be a consequence of Theorem I, and this means that Tarski understands and grants that Theorem I has a semantic dimension. This isn't incompatible with Tarski having stated and proved the syntactic result as well; the point is, rather, that the more commonly attributed semantic result must be at issue in the passage.

The general context at both ends of the passage, then, suggests that Tarski has the usual semantic points (which we know from 1930's "On Definable Sets of Real Numbers," discussed earlier, he was perfectly able to appreciate) in mind: the statement of the Theorem requires a formal definition, and hence the eliminability of "Tr", which would be irrelevant if the syntactic point were the only one at issue, while the discussion of Theorem III, looking back at Theorem I, clearly assumes that it has an import that concerns the extension of "Tr" and not just the derivational structure of the metatheory. Let us now examine the proof itself, from 250, at more length:

Let us suppose that we have defined the class Tr of sentences in the metalanguage. There would then correspond to this class a class of natural numbers which is defined exclusively in terms of arithmetic. Consider the expression '$\cup_1^3 (\iota_n.\phi_n) \notin Tr$'. This is a sentential function of the metalanguage which contains 'n' as the only free variable. From the previous remarks it follows that with this function we can correlate another function which is equivalent to it for any value of 'n', but which is expressed completely in terms of arithmetic. We shall write this new function in the schematic form '$\psi(n)$'. Thus we have:

(1)   for any n, $\cup_1^3 (\iota_n.\phi_n) \notin Tr$ if and only if $\psi(n)$.

Since the language of the general theory of classes suffices for the foundation of the arithmetic of the natural numbers, we can assume that '$\psi(n)$' is one of the functions of this language. The function '$\psi(n)$' will thus be a term of the sequence ϕ, e.g. the term with the index k, '$\psi(n)$' $= \phi_k$. If we substitute 'k' for 'n' in the sentence (1) we obtain:

(2) $\cup_1^3 (\iota_k.\phi_k) \notin Tr$ if and only if $\psi(k)$.

The symbol '$\cup_1^3 (\iota_k.\phi_k)$' denotes, of course, a sentence of the language under consideration. By applying to this sentence condition (α) of convention T we obtain a sentence of the form '$x \in Tr$ if and only if p', where 'x' is to be replaced by a structural descriptive or any other individual name of the statement $\cup_1^3(\iota_k.\phi_k)$, but 'p' by this statement itself or by any statement which is equivalent to it. In particular we can substitute '$\cup_1^3(\iota_k.\phi_k)$' for 'x' and for 'p'—in view of the meaning of the symbol 'ι_k'—the statement 'there is an n such that $n = k$ and $\psi(n)$' or, simply '$\psi(k)$'. In this way we obtain the following formulation:

(3) $\qquad\qquad\qquad\qquad \cup_1^3(\iota_k.\phi_k) \in Tr$ if and only if $\psi(k)$.

Tarski continues by stating what he takes this to show (250–1):

> The sentences (2) and (3) stand in palpable contradiction to one another; the sentence (2) is in fact directly equivalent to the negation of (3). In this way we have proved the first part of the theorem. We have proved that among the consequences of the definition of the symbol 'Tr' the negation of the sentences mentioned in the condition (α) of the convention T must appear. From this the second part of the theorem immediately follows.

I'll mention again, that although Gómez-Torrente pays direct attention to the formulation of Theorem I at 247, he substitutes for the actual text of 250 an admittedly "anachronistic" (2004, 31) reconstruction.

Now the semantic reading of the passage depends on two assumptions. First, that when Tarski says "Let us suppose that we have defined the class Tr of sentences in the metalanguage" he means it: the hypothesis is that there is an expression of the metalanguage "Tr" that has as its extension the set of truths of the object language. Second, the semantic reading assumes that in stating that "we obtain" (3), a T-sentence, Tarski is saying that (3), like (1) and (2), will be a theorem of the metatheory under the hypotheses of the theorem. Hence, on the semantic reading of the proof Tarski assumes for reductio that "Tr" is an expression of ML that semantically defines the set of truths of L, and then proves that (2) and (3) will be theorems of the metatheory on this assumption.

The syntactic reading of the passage has to discount both of these appearances. On the first count, the syntactic reading of the proof has to take it as a slip when Tarski supposes that the *class* of Tr is defined; this could only be semantic definition. Indeed, if the syntactic reading is right, what Tarski ought be saying at the top of 250 is something like "Let us consider a symbol 'Tr', and let us assume nothing at all about its meaning." This is not, however, what he says; indeed, the only natural reading of the sentence is that it's an assumption of the proof to come that the expression "Tr" defines the set of truths, as it has been assumed to do for dozens of pages.[30] (Gómez-Torrente's "anachronistic reconstruction" simply supplies the use of an arbitrary predicate that the syntactic reading requires, contrary to what we have in the text itself.)

Second, if Tarski, in establishing (α) of the theorem wants only to establish that the *negation* of a T-sentence is derivable, then we have to take the discussion of (3) as being there to establish only that (2) *is* the negation of a T-sentence, and not that (3) itself will be a theorem if the class Tr is definable. This is a sustainable reading—if

---

[30] There is no difference here in the translations; the German and Polish have a class being defined, not a symbol.

for no other reason than because the text between (2) and (3) is awfully cryptic—but it ignores the fact that (1) and (2) are also "had" and "obtained"—there is no verbal difference in the treatment of (3) from that of (1) and (2)—and that in the note to 249 Tarski says:

> For the sake of simplicity we shall in many places express ourselves as though the demonstration that follows belonged to the metatheory and not to the meta-metatheory; in particular, instead of saying that a given sentence is provable in the metatheory, we shall simply assert the sentence itself.

This, I think, can only be taken to apply to the numbered theses on the next page, and to whatever extent Tarski "simply asserts" (1) and (2) he seems also "simply to assert" (3). Indeed, at 2004, 34 Gómez-Torrente explicitly notes that the general syntactic result in which he takes Tarski to be interested is "actually the next-to-last step in the proof" of (α). By this "next-to-last" step he has to mean (2) of 250. However, he says nothing about (3) or what it is for, nor does he say anything about why the result Tarski is supposedly proving is only the "next-to-last step" in the proof Tarski offers. The syntactic reading thus has significant difficulty with the Theorem's assumption that "Tr" is defined, as well as with several aspects of the statement of the proof at 250. The syntactic reading of Theorem I also leaves it a mystery why Tarski would discuss the relations between Theorem III and Theorem I in semantic terms of extension and categoricity.

That said, however, the overwhelming problem with the semantic reading of 250 is that it makes no sense either of the statement of the theorem at (247), which, in its separation into (α) and (β), clearly corresponds to the syntactic reading, or of Tarski's statement of what he has just shown at 250–1, which squares perfectly with the statement of the Theorem at 247. I'll have to break down here and admit that at the present juncture I simply don't know what to make of the passage. What we appear to have is a statement of a syntactic theorem (247) followed by a proof of a semantic result (250), followed by the assertion that proving the semantic result is proving the syntactic theorem (251). Neither reading, it seems to me, can make good sense of the whole text. The syntactic reading has the significant merit of making the structure of the proof at 250 match the statement of the theorem at 247 and the discussion of what has been proved at the top of 251, while the semantic reading is an unqualified failure in this regard. The syntactic reading cannot, however, explain why the theorem itself concerns a defined rather than a primitive predicate, why Tarski begins the main part of the proof on 250 by assuming that a *class* has been defined, or why Tarski appears to treat (3) as having the same status as (1) and (2), nor can it make good sense of the relationship between Theorem I and Theorem III as Tarski discusses the latter. I would like to be able to sort this out, but here I will have to leave the problematic text to the reader's consideration, pausing only to close with this note: even Gómez-Torrente agrees that the intuitive notion of truth is back in play by 254 (2004, 32), so the issue here concerns only whether in the very statement of Theorem I at 247 and its proof at 250–1 Tarski is trying to expunge the intuitive notion of truth in order to make a general syntactic point. My claim is merely that there is more evidence than the strict syntactic reading can allow that the intuitive notion of truth, and concerns about semantically defining the set of truths of the object language, are in play in these passages. The dispute between the syntactic and semantic readings of Theorem I and its proof, then, is relatively local.

8. THEOREM I: PHILOSOPHICAL ISSUES

Regardless of the details of the proof at 250, what is important for my purposes is what Tarski's remarks on Theorem I and Theorem III show about the relative priority for him, at the time of writing, of semantic, formal, and intuitive meaning. Recall that the occasion for the sketch of the proof of Theorem I is the recognition that when it comes to the language of the general theory of classes, the metalanguage has no expressions of higher type than the object language. The body of CTFL is set out in the simplified form of type theory Tarski preferred during the period (see Ferreirós 2001 for a general introduction to some of the themes and, e.g. 1983, 113 for one of many short discussions of the simplified type theory). Each expression of a language is set at a definite level of a hierarchy and takes arguments only from lower levels. No expression takes arguments of its own type, and thus no expression takes as arguments expressions of all types.

This stratification is not essential to the formal results, as is now well known, but it does crucially influence Tarski's evaluation of their significance at the time. For it is this that institutes the parallel with his remarks on colloquial language, as for instance here in his first brief sketch of the argument:

(1) a particular interpretation of the metalanguage is established in the object language itself and in this way with every sentence of the metalanguage there is correlated, in one-many fashion, a sentence of the language which is equivalent to it (with reference to the axiom system adopted in the metatheory); in this way the metalanguage contains as well as every particular sentence, an individual name, if not for that sentence at least for that sentence which is correlated with it and equivalent to it. (2) Should we succeed in constructing in the metalanguage a correct {*richtige*} definition of truth, then the metalanguage—with reference to the above interpretation—would acquire that universal character which was the primary source of the semantical antinomies in colloquial language (cf. p. 164). It would then be possible to reconstruct the antinomy of the liar in the metalanguage, by forming in the language itself a sentence x such that the sentence of the metalanguage which is correlated with x asserts that x is not a true sentence.

(1983, 248)[31]

---

[31] See Hodges, this volume, on the mistranslation of "richtige" as "correct" here. In fairness I'll emphasize that this seems to speak in favor of the syntactic reading of Theorem I, since in this passage, correctly translated, universality is blamed on the T-sentences following from the definition of truth, rather than from an expression's defining the set of truths.

Now the L at issue in Theorem I is a type-theoretic language of "infinite order," that is, there is no upper bound on the type to which expressions for classes can belong. Thus Tarski holds that there is no language with signs of higher order than L. Tarski is furthermore, at the time of writing, convinced that no sense can be made of the suggestion that somehow ML isn't constrained by the type-theoretic hierarchy. In discussing the implications of "the theory of semantical categories" for the definition of semantic notions, he writes:

> In order to make clear the nature of these difficulties, a concept must first be discussed which we have not hitherto had an opportunity of introducing, namely the concept of *semantical category*.
>
> This concept, which we owe to E. Husserl, was introduced into investigations on the foundations of the deductive sciences by Leśniewski. From the formal point of view this concept plays a part in the construction of a science which is analogous to that played by the notion of type in the system *Principia Mathematica* of Whitehead and Russell. But, so far as its origin and content are concerned, it corresponds (approximately) rather to the well-known concept of part of speech from the grammar of colloquial language. Whilst the theory of types was thought of chiefly as a kind of prophylactic to guard the deductive sciences against possible antinomies, the theory of semantical categories penetrates so deeply into our fundamental intuitions regarding the meaningfulness of expressions, that it is scarcely possible to imagine a scientific language in which the sentences have a clear intuitive meaning but the structure of which cannot be brought into harmony with the above theory.
>
> (1983, 215)

The importance of the theory of semantical categories to the indefinability result as Tarski conceives of it at the time is shown, among other things, by the fact that §5 contains a complete argument for the *formal* indefinability of truth for a language of infinite order that was written before the addition, inspired by Gödel, of the proof of Theorem I as we have it at 249–50, that simply notes that the methods used in the definitions given earlier in the article all assume the availability in ML of signs of higher type than those of L:

> When we try to define the concept of satisfaction in connexion with the present language we encounter difficulties which we cannot overcome.... In the language with which we are now dealing variables of arbitrarily high (finite) order occur: consequently in applying the method of unification it would be necessary to operate with expressions of 'infinite order'. Yet neither the metalanguage which forms the basis of our present investigations, nor any other of the existing languages, contains such expressions. It is not in fact at all clear what intuitive meaning could be given to such expressions.
>
> (1983, 243–4)

Thus, even before the addition of Theorem I and its proof, Tarski had convinced himself that no formal definition of semantic notions as applied to an L of infinite order was possible in a metalanguage of which one could make any sense. This emphasis persists with the addition of the results following Gödel (added after the work had gone to press, 247). Tarski thus, at this period, draws the conclusion that the set of truths of the general theory of classes *cannot* be defined.

Before we move on we should note that Tarski soon softened his attitude on this. In the Postscript he renounces allegiance to the sort of type theory assumed in the body of the work and allows the possibility of expressions of "infinite order," thereby significantly weakening the purport of the result. Historically, it is interesting that what Tarski really contemplates in the Postscript is adoption of a fully first-order take on the issues, but it takes him several years to see this (see Mancosu 2005 for Tarski's development here), speaking in the Postscript rather of symbols of transfinite order, and in a note of first-order systems as containing variables of "indefinite" order.

What I am interested in here is the insight that Tarski's take on the indefinability results as he understood them *before* the shift expressed in the Postscript and worked out later gives us into the conception of meaning which animated the project in the early 1930s. Whether Tarski is proving a syntactic result or a semantic one, one reaction he *doesn't* consider to the indefinability theorem as he understands it is to hold that the metalanguage might contain an expression that defined the set of truths of the object language without some of the T-sentences being theorems.[32] His stated response, as we have seen, in Theorem III involves giving up eliminability of "Tr" and extensional adequacy, the defining of a set by "Tr", in favor of retaining material adequacy in axiomatic form. Tarski doesn't consider relinquishing the assumption that all of the T-sentences are true, but rather considers giving up on the idea that "is true" is eliminable relative to his metatheory:

> An interpretation of Th. I which went beyond the limits given would not be justified. In particular it would be incorrect to infer the impossibility of operating consistently and in agreement with intuition with semantical concepts an especially with the concept of truth... The idea naturally suggests itself of setting up semantics as a special deductive science with a system of morphology as its logical substructure...
>
> (1983, 255)

Of course the approach has its drawbacks, in particular the lack of categoricity in the resulting axiomatic theory. In the terms of "Some Methodological Investigations on the Definability of Concepts," no semantic theory for a sufficiently rich L can be "complete with respect to its specific terms": we can express the intuitive concept by taking the T-sentences as axioms, but fully determinate meaning remains beyond our grasp.

Tarski finds this approach attractive, as noted above, because it allows one the possibility of "operating consistently and in agreement with intuition with semantical concepts and especially the concept of truth" (255). Faced with a choice between having a term that has the intuitive meaning of "is true" as expressed in the Aristotelian conception of truth and operating with one that defines a set and thereby achieves the twin goals of eliminability in favor of non-semantic terms and mathematical tractability, Tarski sticks with intuitive meaning. Tarski in this period chooses material adequacy over full extensional adequacy because, given his overall conception of the interaction of intuitive, formal and semantic meaning, it doesn't

---

[32] Gómez-Torrente (2004, 32) agrees here.

make any sense to say that a term has the *semantic* meaning that it's supposed to have if the term lacks a *formal* meaning that properly answers to the intuitive meaning it is supposed to express. "Is true" is supposed to apply to *truths* of course, but an expression not such that *all* putative T-sentences involving it are themselves true is not a language's version of "is true," according to the classical Aristotelian conception of truth that guides Tarski's project.

Tarski's view here is best understood by way of contrast with a reaction to which the contemporary reader will be tempted, which will run roughly as follows. Consider an object language L like Tarski's object language. The set of sentences of this language exists, and so do all of its subsets. Encouraged especially by a later and clearer grasp on the purely formal result, one not clouded by Tarski's continued emphasis, at the time, on formal definability, one can be struck by the following thought: surely *one* of those subsets is the set of truths of L. Trying to specify which set it is, with whatever complications may be necessary—recognizing a set of sentences such that neither they nor their negations are true, etc.—seems a worthy exercise. If Tarski's adherence to the "intuitive meaning" of the truth-predicate in ML renders it indefinable by forcing us to hold true all of the T-sentences for L, so much the worse for that intuitive meaning: what needs to go, rather, is commitment to the truth of all of the T-sentences for L in ML.[33]

Tarski's understanding of the result, however, in the early 1930s, is quite different: no term that has the intuitive meaning of "is true" as expressed in the T-sentences can determine the right extension if it determines any when L is of sufficient expressive power. On Tarski's view in this period, no expression not governed by the T-sentences expresses the concept of truth, and hence no set determined by such an expression could be the set of *truths*.[34] Whatever one would be doing in introducing such an expression, one wouldn't be defining truth. Of course, as he notes (1944, 355–6), there are alternatives to the Aristotelian conception of truth, but since Tarski is more committed to it than he is to the claim that it must be possible to define truth for a language of infinite order, he doesn't consider giving up the Aristotelian conception. Since Tarski wants that the concept of truth be expressed, even if it doesn't determine an extension, and even if it isn't eliminable in favor of less problematic

---

[33] I discuss this at more length elsewhere; it's a staple of current accounts that they buy the definability of the set of truths of the object language at the price of claiming that there are some sentences that cannot be used to state *their own* truth conditions in an extension of the language they are in, that is, that there are some sentences for which the T-sentences aren't true.

[34] Note here that we can, in terms of the foregoing, sort out an interpretive puzzle from earlier in the work. Tarski says, in his initial discussion of the liar, that it is the substitution of sentences containing "true sentence" into the T-schema, that is responsible for the paradox, but also says that "nevertheless no rational ground can be given why such substitutions should be forbidden in principle" (1983, 158). What's strange about this is that avoidance of contradiction seems as rational a ground for such prohibition as there could be, and that Tarski himself goes on to do just what he says here can be "given no rational ground" by restricting the definition of truth to situations where such substitutions aren't possible. The resolution is this: the substitutions can't rationally be forbidden because to do so conflicts with the capture of the intuitive meaning of "is true" via the T-sentences. A necessary condition of this is that ML be of higher order than L.

terms, he will reject in this period any expression that does determine a set but lacks the relevant formal properties in *ML*.

With Theorem I we are at the "limits of thought" as Tarski saw them in the early 30s: as *L* itself is already maximally expressive and the indefinability in *ML* is a result of this, *not even from MML* can we take ourselves to have a view of the set of truths of *L*, so that we could recognize some expression in *ML* as having it as its extension. All we have in *ML* and *MML* are predicates that either do or do not express the "intuitive" concept of truth. No perspective is available from which to say that an expression that doesn't have this intuitive meaning nevertheless picks out a *set* that we have reason to think is the set of truths. The content of the indefinability theorem in this period, then, is that no expression with the intuitive meaning of "is true" can adequately determine an extension except in the restricted case where the language to which it applies lacks certain expressive resources.

The view of meaning assumed in the works of the early 1930s is one on which the formal definition of a term's meaning, as codified by the set of sentences involving it that are held true, is primary. Determination of an extension relative to a theory is a mark of full meaningfulness, but that meaning be precise and tractable in this way is a desideratum in a theory intended to make rigorous sense, but is not necessary for Tarski to countenance a term as meaningful.

Attention to Tarski's remarks on definition and meaning thus facilitates understanding of his remarks on indefinability and how to react to it as well as on many other topics. In addition to making clear that his conception of meaning is more nuanced than it is usually taken to be, it has significant interpretive payoffs: standard criticisms can clearly be seen to be ineffective in light of it, and we can also see an understanding of the indefinability of truth for languages of sufficient expressive power in Tarski's work of the period which is significantly and interestingly different from the modern interpretation of such results.[35]

References

Awodey, Steve and Reck, Erich H. (2002) Completeness and Categoricity Part I: Nineteenth-Century Axiomatics to Twentieth-Century Metalogic. *History and Philosophy of Logic* 23, 1–30.
Belnap, Nuel (1993) On Rigorous Definitions. *Philosophical Studies* 72, 115–46.
Burge, Tyler (1984) Semantical Paradox. *Recent Essays on Truth and the Liar Paradox*, Martin ed. New York: Oxford University Press, 83–117.
Carnap, Rudolf (2002) *The Logical Syntax of Language*, Smeaton trans., Chicago: Open Court Press.
Coffa, Alberto (1986) From Geometry to Tolerance: Sources of Conventionalism in Nineteenth-Century Geometry. In Colodny ed., *From Quarks to Quasars*. Pittsburgh: University of Pittsburgh Press, 3–70.

---

[35] I thank Lionel Shaprio for a series of discussions, especially about the material of Section 7, and Andrew Arana for an enlightening series of discussions about Hilbert and formalism.

Coffa, Alberto (1991) *The Semantic Tradition from Kant to Carnap: To the Vienna Station.* Cambridge: Cambridge University Press.

Davidson, Donald (1990) The Structure and Content of Truth. *The Journal of Philosophy* 87, 279–328.

Detlefsen, Michael (2004) Formalism. *Oxford Handbook of Logic and the Philosophy of Mathematics*, Shapiro ed. New York: Oxford University Press, 236–317.

Etchemendy, John (1988) Tarski on Truth and Logical Consequence. *The Journal of Symbolic Logic* 53, 51–79.

––––– (1990) *The Concept of Logical Consequence.* Cambridge, MA: Harvard University Press.

Ewald, William (1996) *From Kant to Hilbert: A Source Book in the Foundations of Mathematics.* New York: Oxford University Press.

Ferreirós, José (2001) The Road to Modern Logic: An Interpretation. *Bulletin of Symbolic Logic* 7, 441–84.

Fraenkel, Abraham A. (1928) *Einleitung in die Mengenlehre*, 3rd edn repr. 1946. New York: Dover Publications.

Frege, Gottlob (1980) *The Foundations of Arithmetic*, Austin trans. 2nd rev. edn. Evanston: Northwestern University Press.

Friedman, Michael (1999) *Reconsidering Logical Positivism.* Cambridge: Cambridge University Press.

Frost-Arnold, Greg (2004) Was Tarski's Theory of Truth Motivated by Physicalism? *History and Philosophy of Logic* 25, 265–80.

Gómez-Torrente, Mario (2004) The Indefinability of Truth in the "Wahrheitsbegriff". *Annals of Pure and Applied Logic* 126: 27–37.

Heck, Richard G. Jr. (1997) Tarski, Truth and Semantics. *The Philosophical Review* 106, 533–54.

Langford, Cooper H. (1926) Some Theorems on Deducibility. *Annals of Mathematics* 28, 16–40.

Leśniewski, Stanisław (1929) Grundzüge eines neuen Systems der Grundlagen der Mathematik. *Fundamenta Mathematicae* 14, 1–81.

Mancosu, Paolo (2005) Harvard 1940–1941: Tarski, Carnap and Quine on a Finitistic Language of Mathematics for Science. *History and Philosophy of Logic* 26, 327–57.

Patterson, Douglas (2005) Learnability and Compositionality. *Mind and Language* 20, 326–52.

––––– (2006a) Tarski, the Liar, and Inconsistent Languages. *The Monist* 89, 1.

––––– (2006b) Tarski on the Necessity Reading of Convention T. *Synthese* 151, 1–24.

Putnam, Hilary (1994) A Comparison of Something with Something Else. In *Words and Life*, Conant ed. Cambridge, MA: Harvard University Press.

Ray, Greg (2003) Tarski and the Metalinguistic Liar, *Philosophical Studies* 115, 55–80.

Scanlan, Michael (2003) American Postulate Theorists and Alfred Tarski. *History and Philosophy of Logic* 24, 307–25.

Sinaceur, Hourya (2001) Alfred Tarski: Semantic Shift, Heuristic Shift in Metamathematics. *Synthese* 126, 49–65.

––––– (forthcoming) Tarski's Practice and Philosophy: Between Formalism and Pragmatism. In *Logicism, Intuitionism, and Formalism: What Has Become of Them?* Lindström, Palmgren, Segerberg and Stoltenberg-Hansen, eds. Springer Verlag.

Soames, Scott (1999) *Understanding Truth.* New York: Oxford University Press.

Szaniawski, K., ed. (1989) *The Vienna Circle and the Lvov-Warsaw School.* Dordrecht: Kluwer Academic Publishers.

Tarski, Alfred (1941) *Introduction to Logic and to the Methodology of Deductive Sciences*. 1995 repr. of 2nd edn of 1946, New York: Dover Publications.

—— (1944) The Semantic Conception of Truth and the Foundations of Semantics, *Philosophy and Phenomenological Research* 4.

—— (1983) *Logic, Semantics, Metamathematics*, 2nd edn Woodger trans., Corcoran ed. New York: Oxford University Press.

Woleński, J. and Simons, P. (1989) De Veritate: Austro-Polish Contributions to the Theory of Truth from Brentano to Tarski. In Szaniawski 1989, 391–442.

8

Tarski, Neurath, and Kokoszyńska on the Semantic Conception of Truth

Paolo Mancosu

Carnap's *Autobiography* reports that Tarski's presentation of the semantic conception of truth at the Paris Congress in 1935 gave rise to conflicting positions. While Carnap and others hailed Tarski's definition as a major success in conceptual analysis others, such as Neurath, expressed serious concerns about Tarski's project.[1] Tarski 1944 contains only indirect references to these debates and it avoids explicit mention of those whose objections were not formulated in print (such as Neurath).[2] My goal in this chapter is to review the debate that accompanied the international recognition of Tarskian semantics by using not only the published sources but also the extended correspondence between Neurath, Tarski, Lutman-Kokoszyńska, and Hempel.

It is well known that Tarski's theory of truth had a lasting impact on some members of the Vienna Circle, such as Carnap. It is less known, at least outside the community of historians of the logical positivist movement, that Tarski's work appeared in the midst of a debate on the nature of truth which involved several members of

---

I would like to thank Johannes Hafner, Greg Frost-Arnold, and Thomas Uebel for their useful comments on a previous version of this chapter. I would also like to thank Dr Brigitte Parakenings for her wonderful kindness during my stay in Konstanz (October 2004) while working on the Carnap and Neurath Nachlaß. I am also grateful to Solomon Feferman for helping me track down the Popper-Tarski correspondence. I would like to acknowledge the comments of the audiences of the Colloque International "Le rayonnement de la philosophie polonaise au XX siècle," (École Normale Supérieure, Paris, 2/7/05), of the Séminaire de Philosophie des Science of the IHPST, Paris (February 2005), of the Department of Philosophy at the Catholic University in Milan (April 2005), of the group 'Logic in the Humanities' at Stanford University (March 2006), and of the Escuela Latinoamericana de Logica Matematica in Oaxaca (August 2006), where I presented parts of this chapter. All passages quoted from the Carnap Nachlaß are quoted with the permission of the University of Pittsburgh (all rights reserved). All passages quoted from the Neurath Nachlaß are quoted with the permission of the Wiener Kreis Stichting (all rights reserved). Finally, I am very grateful to Sandra Lapointe for improving the style of the translations from German into English.

[1] See Carnap's *Autobiography* (1963, p.61) and *Introduction to Semantics* (1942, p. x). While my attention here will be exclusively on Neurath, I should point out that among the early objectors to the semantic theory of truth one finds Jørgensen, Juhos, Nagel, and Naess.

[2] Neurath 1936 is referred to in Tarski 1944 but only as reference for a survey of the discussions which took place in Paris in 1935. For a biography of Tarski see Feferman and Feferman 2004. For other aspects of Tarski's philosophical engagement see Mancosu 2005, 2008, Woleński 1993, 1995.

the Vienna Circle.[3] The most important interventions in this debate, prior to Tarski's work, were Schlick's article "On the foundations of knowledge" (1934), Neurath's reply "Radical physicalism and the 'real world'" (1934), and Hempel's "On the logical positivists' theory of truth" (1935). The influence of Tarski's work is evident in successive articles related to this debate such as Carnap's "Truth and confirmation" (1936) and Lutman-Kokoszyńska's "On the absolute concept of truth and some other semantical concepts" (1936b). Of course, as it will become clearer in what follows, there were substantial differences between the disputants as to the issues they addressed concerning truth and thus when I say that the debate concerned "the nature of truth" (as if there was a single notion being explicated), this should be taken with a grain of salt.

The key date here is the Paris meeting (the First International Congress of the Unity of Science) of 1935, where Tarski was invited to present his work on semantics and the theory of truth. Ayer says in his autobiography that "philosophically the highlight of the Congress was the presentation by Tarski of a chapter summarizing his theory of truth" (Ayer, 1977, p. 116).

Tarski's theory of truth seemed to many to give new life to the idea of truth as correspondence between language and reality. The discussion following Tarski's presentation was summarized by Neurath in his long overview of the Paris Congress published in *Erkenntnis*. Already at the Paris meeting, Neurath had suggested that

From the point of view of terminology he [Neurath] thinks that one should reserve the use of the term "true" for that Encyclopedia, among the many consistent ones which are controlled by protocol sentences, that has been chosen, so that each consequence of this Encyclopedia and each new sentence accepted into it would be called "true" and any one contradicting it would be called "false".

(Neurath 1936, 400)

It was this proposal for the use of "truth" (given in prior but similar formulations) that had led Schlick to attack, in 1934, Neurath's position as a "coherence" theory of truth. It is important to point out from the outset that from Schlick's point of view Neurath's proposal amounts to a philosophical position on the nature of truth (a coherence theory), and that some passages in Neurath support this reading. However, from Neurath's point of view the proposal is more radical and perhaps he would even reject the idea that he was defending a conception, or a theory, of truth. Indeed, as will become clearer below, Neurath was calling for a replacement of a methodology of science that thinks of itself as methodology of truth attainment by a scientific metatheory that systematically explores how warrant obtains and spreads both across systems and across communities of investigators. Whenever I use "conception of truth" in connection to Neurath, the reader should keep in mind the remarks just made.

Neurath's concerns about the Tarskian definition of truth were already obvious from the above mentioned report but their full articulation can only be

---

[3] See Hofmann-Grüneberg 1988, Grundmann 1996, Hempel 1982, Rutte 1991, and Uebel 1992. For an earlier account see Tugendhat 1960. For a general account of the protocol debate see Cirera 1994, Oberdan 1993, Uebel, 1992.

grasped from the correpondence he had with, among others, Tarski, Carnap, Lutman-Kokoszyńska, and Hempel on the subject.[4]

THE CORRESPONDENCE BETWEEN TARSKI AND NEURATH

The correspondence between Neurath and Tarski contains forty-one letters from Tarski and forty-two from Neurath spanning the period 1930–1939. Three of them were published in 1992 by Rudolf Haller (Haller 1992). In particular, the letters published by Haller are important for historical matters concerning the mutual influences between the Polish logicians and the Vienna Circle (see also Woleński 1989a and Woleński and Köhler 1998). However, I will focus on another aspect of the correspondence having to do with Neurath's objections to Tarskian semantics. The letters of interest for us come after the 1935 Paris meeting and they are hitherto unpublished.

However, I would like to point out that reading the letters devoted to the discussion of the issues of mutual the influence between the Vienna Circle and the Polish logicians, one is already struck by Tarski's extensive familiarity with the philosophical tenets of the Vienna Circle. For instance, in his letter to Neurath dated 7.9.36 the discussion centers around the following four claims by Neurath (made in Neurath 1935), each of which is disputed by Tarski:

(a) The admissibility of sentences about sentences, the possibility to speak unobjectionably about a language, was accepted broadly by the Vienna Circle before Tarski's lectures in Vienna in 1930.

(b) The claim that the Vienna Circle and the Polish logicians arrived at the same time and independently at the claims contained in (a).

(c) The claim according to which sentences, parts of sentences, etc. are physical entities [Gebilde] was discussed by the Vienna Circle before Tarski's 1930 lectures in Vienna. Tarski does not dispute this but points out that in Warsaw this position was held since 1918.

(d) The claim according to which for the goals of the real sciences one can get by with a universal language.

Some of these issues we will have to come back to. In the same letter Tarski added:

I cannot understand why you also continue to regard semantics as "objectionable" although you have nothing to object to Carnap's discussions of "tautology," "analytic," which run parallel and are closely related to it. I have looked rather carefully at your correspondence with Ms Lutman but it has not helped me at all.[5]

---

[4] For the location of the unpublished correspondence quoted in the chapter see the details given after the notes. Unpublished materials from the Carnap archive have a call number that always begin with "RC" followed by a string of numbers, i.e. "RC 102–55–05." All original sources quoted without a call number are from the Neurath Nachlaß.

[5] "Ich kann nicht begreifen, warum Sie auch weiterhin die Semantik als 'bedenklich' halten obgleich sie den parallel laufenden und sehr nahe verwandten Erörterungen Carnaps, die die Begriffe

We learn here that Neurath had been corresponding with Maria Lutman-Kokoszyńska (henceforth, Kokoszyńska), who in the second half of the 1930s was considered to be one of the representatives of the Lvov-Warsaw school on philosophical issues related to semantics.[6]

What were then Neurath's objections to Tarski's theory of truth?[7] Let us follow the correspondence between Neurath and Tarski. Neurath and Tarski had first met in Vienna during Tarski's visit in 1930. Then Neurath visited Warsaw, twice in 1934, and invited Tarski to be part of the so called Prag-Vorkonferenz, which was planned as a preliminary meeting for the Paris conference of 1935. Thus, they had had several occasions to discuss all kinds of issues related to matters of philosophical and logical interest. In any case, apart from an interesting letter (published in Haller 1992) from Tarski in 1930 concerning the Polish scholars involved in philosophy of the exact sciences, most of the correspondence between them until 1935 is taken up by more mundane things. Meanwhile Tarski had arrived in Vienna in January 1935 with a Rockfeller Foundation fellowship. Most of the correspondence during this period (Neurath was already established in Holland) was concerned with the publication of an article by Tarski in the proceedings of the Prag-Vorkonferenz. It is however interesting that Neurath in a letter dated May 2, 1935 concerning the Paris congress writes to Tarski:

I hope you will contribute something that will be of service to EMPIRICISM. I am constantly worrying that some fine day a book "METAPHYSICA MODO LOGISTICA DEMONSTRATA" will appear. And then we will be blamed even for that.[8]

'Tautologie,' 'analytisch' u.s.w. betreffen nichts vorzuwerfen haben. Ich habe Ihre Korrespondenz mit Fr. Lutman ziemlich sorgfältig durchgeschaut, das hat mir aber nichts geholfen" (Tarski to Neurath, 7.IX.36, Neurath Nachlaß).

[6] Maria Lutman-Kokoszyńska was born in 1905. She earned a Ph.D. under Twardowski after having studied philosophy and mathematics in Lvov. Her Ph.D. thesis was finished in 1928 and was on the topic of "General and ambiguous names." A curriculum vitae dated 1961 is found in Carnap's Nachlaß under RC 088–57–07. For a bibliography of Kokoszyńska's works see Zygmunt 2004.

[7] The only contributions I am aware of that treat to a certain extent Neurath's objections to Tarski are Mormann 1999 and Hofman-Grüneberg 1988. There is also a useful discussion in Uebel 1992. While all of them consulted the Neurath-Tarski correpondence, they made very limited use of it and did not refer to the other archival sources I am using.

[8] "Ich hoffe Sie bringen etwas, das auch dem EMPIRISMUS zugute kommt. Ich habe ja stets die Sorge, dass eines schönen Morgens ein Buch erscheinen wird: 'METAPHYSICA MODO LOGISTICA DEMONSTRATA'. Und dann werden wir noch daran schuld sein sollen." (Neurath to Tarski, 2.V.35, Neurath Nachlaß) See also the letter from Neurath to Carnap dated 19.IV.35 (RC 029–09–60) where Neurath gives voice to the same concern (the correspondence Carnap-Neurath was first studied in Hegselmann 1985). A concern similar to the one expressed in the above quote is expressed as late as 1944: "I have the feeling to continue your Logical Syntax period, before you became Tarskisized with Aristotelian flavour, which I detest. I always fear, that you, a calculatory genius, supports [sic] a kind of possible scholasticism who [sic] leads away from scientific empiricism" (Neurath to Carnap, 1.4.44, RC 102–55–05). See also Neurath to Morris dated 18.XI.44.

By the way, to Neurath's request quoted above (letter dated 19.IV.35), Tarski replied through Kokoszyńska. In a letter from Kokoszyńska to Neurath (dated: Paris, 25.VII.35) the discussion is about Tarski's lecture: "Ich bin von Tarski beauftragt, Ihnen mitzuteilen, dass der endgültige Titel seines referates lauten wird "Die Grundlegung der wissenschaftlichen Semantik" Er hat, Ihrem

This already points to a recurrent theme in Neurath, i.e. the fear that the logical formalism might seduce people into metaphysical positions. In a letter from Neurath to Carnap from 1943 (written in English) in which the danger of semantics is at issue, a full genealogy is also provided:

> I am really depressed to see here all the Aristotelian metaphysics in full glint and glamour, bewitching my dear friend Carnap through and through. As often, a formalistic drapery and hangings seduce logically-minded people, as you are very much... It is really stimulating to see how the Roman Catholic Scholasticism finds its way into our logical studies, which have been devoted to empiricism.
>
> The Scholasticism created Brentanoism, Brentano begot Twardowski, Twardowski begot Kotarbiński, Łukasiewicz (you know his direct relations to the Neo-Scholasticism in Poland), both together begot now Tarski etc., and now they are God fathers of OUR Carnap too; in this way Thomas Aquinas enters from another door Chicago ...
>
> (January 15, 1943, RC 102–55–02)[9]

We see then that Neurath's comment in 1935 already contained the seeds of a worry which would not go away.

1935: THE PARIS CONGRESS AND ITS AFTERMATH

The Paris Congress of 1935 (from 15 to 23 September) represents a turning point in the history of the Vienna Circle and in Tarski's career. During his stay in Vienna in 1935, Tarski had had the opportunity to explain to Carnap and Popper his theory of truth (Polish 1933, German 1935) and upon Carnap's insistence he decided to lecture on it in Paris. In addition, other scholars, such as Arne Naess, came to know Tarski's theory reading the galley proofs during the same period.[10] In Paris, Tarski also gave a second lecture on the concept of logical consequence.[11] At the Congress there were important lectures by Carnap and Kokoszyńska that already built on Tarski's theory of truth. While Carnap and Popper became immediate converts to

---

Wünsche gemäss, sein ursprüngliches Thema geändert, um auf dem Kongresse solche Fragen zu behandeln, die von ziemlich prinzipieller Bedeutung für das Wissenschaftsganze sind" (Kokoszyńska to Neurath, 19.IV.35, Neurath Nachlaß). In my opinion Tarski's famous comment on physicalism and semantics in Tarski 1936 is to be seen in light of Tarski's eagerness to please Neurath. I will not however be dealing with this topic. For a recent discussion see Frost-Arnold 2004.

[9] In general, Neurath uses similar formulations with different correspondents. For instance, the lineage Brentano–Twardowski–Łukasiewicz–Tarski is also given in a letter to Hempel dated 20.II.1943: "It is a sad situation that one has now to object to the Aristotelian metaphysics well formalized by Tarski and Carnap. I shall touch this point only but I think another day I shall explain this point in detail. The Scholasticism via Brentano-Twardowski-Łukasiewicz-Tarski appears now within a calculus but I think the calculus may be useful even within empiricism with a different interpretation, but hardly as it stands" (Neurath to Hempel, 20.II.43, Neurath Nachlaß). See also Neurath to Martin Strauss dated 16.I.43 and Neurath to Carnap dated 27.VIII.38. For an exposition of theories of truth from Brentano to Tarski see Woleński and Simons, 1989. For a survey of the Lvov-Warsaw school see Woleński 1989b.

[10] See Naess to Neurath, dated 8.VII.36.

[11] On Tarski's theory of logical consequence and further references see Mancosu 2006.

Tarski's theory of truth, the Paris congress revealed a wide variety of reactions to Tarski's work. In the *Autobiography*, Carnap says:

> To my surprise, there was vehement opposition even on the side of our philosophical friends... Neurath believed that the semantical concept of truth could not be reconciled with a strictly empiricist and anti-metaphysical point of view... I showed that these objections were based on a misunderstanding of the semantical concept of truth, the failure to distinguish between this concept and concepts like certainty, knowledge of truth, complete verification and the like.
>
> (1963, p. 61)

Neurath touches on the topic in correspondence with Tarski (although there were surely discussions in Paris) already on November 26, 1935 discussing the issue of terminology:

> As far as I can see, the terminology you and Dr Lutman propose seems to give rise to all kinds of confusions. You could maybe emphasize what bearing it has. I still think that my suggestion to call true any Encyclopedia singled out at any point, and accordingly to call "true" all the acknowledged sentences we acknowledge as following from it or contained in it and "false" all those we reject, is terminologically less hazardous. But this is, so to speak, more of a pedagogical problem.
>
> I think that your expositions are in general very important for the questions of logical empiricism. Especially the question as to how "propositions" occur among other "things" etc., and also the problem as to how analytic propositions are to be delimited etc. Unfortunately, I will hardly be able to study these questions more closely in the immediate future. But hopefully not too long from now. The proceedings of the congress in which your chapter is to appear, will certainly be very useful to me.[12]

And after having read Tarski's technical article on truth, Neurath wrote:

> I have read the work you were so kind to send me. Though I do not mean to criticize it in the least, I nonetheless want to say it will certainly give rise to confusion. The restrictions you impose on the concept of truth will not be observed and your formulations will be used for all kinds of metaphysical speculations. But this is a sociological comment which as such is not unimportant.
>
> (Neurath to Tarski, March 24, 1936)[13]

---

[12] "So viel ich sehe, scheint die von Ihnen und Dr Lutman vorgeschlagene Terminologie allerlei Verwirrung anzurichten. Vielleicht können sie betonen, welches die Tragweite ist. Ich denke nach wie vor, dass der von mir gemachte Vorschlag, die jeweils ausgezeichnete Enzyklopedie wahr zu nennen und demgemäss alle aus ihr folgenden oder in sie hinzukommenden anerkannten Sätze 'wahr' alle abgelehnten 'falsch', bezüglich des Terminus weniger gefährlich ist. Aber das ist sozusagen mehr ein pädagogisches Problem.
Ich glaube, dass Ihre Darlegungen für die Fragen des logischen Empirismus im allgemeinen sehr wichtig sind. Insbesondere die Frage, wie die 'Sätze' neben anderen 'Dingen' auftreten usw., auch das Problem, wie analytische Sätze abzugrenzen wären, usw. Leider komme ich in der nächsten Zeit kaum dazu diese Fragen genauer zu studieren. Aber hoffentlich nicht in zu ferner Zeit. Da wird mir der Kongressbericht, in dem Ihre Arbeit erscheint sicher viel nützen." (Neurath to Tarski, 26.XI.35, Neurath Nachlaß)

[13] "Ihre so liebenswürdig übersandte Arbeit habe ich gelesen. Ohne damit die geringste interne Kritik üben zu wollen, möchte ich doch sagen, dass sie sicher Verwirrung stiften wird. Die Einschränkungen, die Sie für den Wahrheitsbegriff vorbringen wird man nicht beachten, wohl

We thus see that Neurath feared a metaphysical usage of Tarski's theory, due to an inappropriate extension of its field of validity (from formal languages to ordinary languages) and was also opposed to Tarski's use of the notion of "truth". At the same time, he recommended using the term "true" in connection to talk of acceptability and rejection from the Encyclopedia. In his reply, dated 21.4. 36, Tarski tried to diffuse the issue by claiming that between him and Neurath on the issue of "truth" there were only terminological differences. However, Neurath thought much more was at stake and from his next letter we glimpse at the constellation of elements that were fueling his resistance:

I thank you very much for your reflections on our "truth definitions." Of course there are to begin with only terminological differences but I have the strong impression that in the discussion concerning the domain of the real sciences your intuition slips very easily into metaphysics. One should fully speak one's mind on this issue. I wrote something to Dr Lutman Kokoszyńska about this. When you hold that it is trivial to say that one speaks with the language about the language then I can only rejoin that an essential part of science consists in defending trivialities against errors. From the beginning of the Vienna Circle, for instance, I have fought against Wittgenstein's attempt to introduce a sort of "elucidations" and thus "illegitimate," almost non- or pre-linguistic considerations in order to then speak of the opposition between "the" language and "the" reality, and hence to speak outside the language [. . .] And insofar as your terminological choice suggests objectionable consequences, it has perhaps not come about independently of these consequences. On the one hand one emphasizes that this concept of truth holds only for formalized languages. On the other hand the concept of truth is of practical interest precisely in non formalized domains. For this reason, if one is not simply to get rid of the term, I am in favor of my terminology, for the latter remains applicable also in non formalized domains. By contrast the terminology you and Lutman use leads to bad things when it is applied to non formalized domains.

(Neurath to Tarski, March24, 1936)[14]

aber Ihre Formulierungen als [Beweis] für allerlei metaphysische Spekulationen verwenden. Aber das ist eine soziologische Bemerkung, die deshalb nicht unwichtig sein muss." (Neurath to Tarski, 24.III.36, Neurath Nachlaß)

[14] "Ich danke Ihnen für die Mitteilungen über unsere 'Wahrheitsdefinitionen'. Natürlich liegen zunächst nur terminologische Unterschiede vor, aber ich habe sehr den Eindruck, dass bei der Diskussion auf realwissenschaftlichem Gebiet Ihre Anschauung sehr leicht ins Metaphysische abgleitet. Darüber müsste man sich ausführlich aussprechen. Ich habe einiges darüber an Dr Lutman-Kokoszyńska geschrieben.

Wenn sie meinen, dass es eine Trivialität ist zu sagen, man spreche mit der Sprache über die Sprache, so kann ich darauf nur sagen, dass die Wissenschaft zu einem wesentlichen Teil darin besteht Trivialitäten gegen Irrtümer zu vertreten. Ich habe z.B. vom Beginn des Wiener Kreises an mich gegen die Versuche von WITTGENSTEIN gewehrt eine Art 'Erläuterungen' also 'nichtlegitime', quasi nicht- oder vorsprachliche Betrachtungen einzuführen, um dann über die Gegenüberstellung von 'der' Sprache mit 'der' Wirklichkeit zu reden, also *ausserhalb* der Sprache. Ich glaube, dass die 'Konstatierungen' von Schlick, die Sätze und doch wieder nicht Sätze sind aus dieser WITTGENSTEINSCHEN Metaphysik herzuleiten sind.

Und sofern Ihre terminologische Wahl bedenkliche Konsequenzen nahelegt, ist sie vielleicht nicht ganz unabhängig von diesen Konsequenzen zustandegekommen. Auf der einen Seite wird betont, dass dieser Wahrheitsbegriff nur für formalisierten Sprachen gelte, andererseits ist der Wahrheitsbegriff gerade in nicht formalisierten Bereich von praktischer Bedeutung. Deshalb bin ich, wenn man den Terminus nicht überhaupt fallen lässt mehr für meine Terminologie,

The quotation clearly shows that Neurath envisaged a radical reinterpretation of the term "true", perhaps one so extreme that it could not even be classified as an "explication" of what its meaning had been all along.

In a subsequent letter Neurath gives more details about the Viennese roots (see Frank 1997) of his objection:

> Long before we made contact with Warsaw, there was a disagreement within the Vienna Circle concerning the question as to whether it makes any sense to compare language with "reality" (for instance whether the language is more complex or less complex than the reality or just as complex and so on) from a position, so to speak, outside of both. The rejection of propositions about "the" reality originated with Frank and within the "Circle" in Vienna chiefly from me. The discussion was connected with a second one which concerned the question, whether "propositions about propositions" are meaningful or not. Wittgenstein, Schlick and others—who however defended their viewpoint less rigorously—and Waismann strictly rejected propositions about propositions, so that the discussion about propositions and reality had to be carried out so to speak *outside* of language, in terms of "clarifications" as "ladder" so to speak that one would afterwards throw away.
>
> (Neurath to Tarski, May 7, 1936)[15]

In order to understand what this amounts to we have to step back and look at some of Neurath's previous work and the debate on the nature of truth which had divided the Vienna Circle.

NEURATH AGAINST THE RIGHT WING OF THE CIRCLE

Wittgenstein's *Tractatus* played an important role in the development of the Vienna Circle. When asked by Tarski in October 1935 (see letter from Tarski to Neurath dated 7.IX.36), Carnap (as reported by Tarski) characterized the Wittgensteinian influence as both stimulating and confining. It was stimulating in that Wittgenstein brought attention to the importance of the problems that relate to language,

---

die im nicht formalisierten Bereich verwendbar bleibt. Während die von Ihnen und Lutman verwendete Terminologie im nicht formalisierten Bereich verwendet zu schlimmen Dingen führt. [Ich wüsste sehr gerne, ob Sie mit der Darstellung übereinstimmen, die Rougier in Paris von der Grenzverschiebung gegeben hat, die durch Ihre Thesen zwischen Metaphysik und Wissenschaft erfolgt sei.]" (Neurath to Tarski, 24.IV.36, Neurath Nachlaß)

[15] "Innerhalb des Wiener Kreises war lange bevor der Kontakt mit Warschau aufgenommen wurde ein Gegensatz da, der sich auf die Frage bezog, ob es einen Sinn hat einen Vergleich der Sprache mit der 'Wirklichkeit' (z.B. die Sprache ist komplexer oder weniger komplex als die Wirklichkeit oder ebenso komplex usw.) sozusagen von einem Punkt *ausserhalb beider*, anzustellen.

Die Ablehnung der Sätze über 'die' Wirklichkeit ging von Frank und innerhalb des 'Zirkels' in Wien vor allem von mir aus.

Die Diskussion verknüpfte sich mit einer zweiten, die mit der Frage zusammenhing, ob 'Sätze über Sätze' sinnvoll seien oder nicht. Wittgenstein, Schlick und andere—die aber ihren Standpunkt weniger scharf vertraten—[aus] Waismann lehnten Sätzen über Sätzen strikt ab, so dass die Diskussionen über Sätze und Wirklichkeit sozusagen *ausserhalb* der Sprache vor sich gehen mussten, als 'Erläuterungen', sozusagen als 'Leiter', die man später wegwerfen müsse usw." (Neurath to Tarski, 7.V.36, Neurath Nachlaß)

i.e. the reducibility of philosophical problems to linguistic problems. On the other hand, Wittgenstein disputed and rejected the possibility of speaking about language in a legitimate way. In a preceding quote by Neurath we have already seen how this second aspect of Wittgenstein's position was central to the Vienna Circle discussions as well as the related problem about the relation between language and reality. Wittgenstein had essentially espoused a correspondence theory of truth[16] whereby the truth of a non-tautological statement consists in its being a picture of a fact. Wittgenstein recognized as acceptable only the logical sentences (which are *sinnloss* but not *unsinnig*) and the sentences of science. In this way he was forced to declare even the propositions contained in the *Tractatus* as "explanations," or "elucidations," which have to be thrown away after one has arrived at an understanding of the *Tractatus* just as one can throw away the ladder after one has climbed upon it (*Tractatus* 6.54). Neurath was a relentless opponent of these Wittgensteinian theses. A constant refrain in Neurath is his rejection of anything that smacks of the "absolute": the "World", the "Truth", etc. In his 1931 article "Physicalism," Neurath attacks central tenets of Wittgenstein's conception which were also shared by other members of the Vienna Circle, such as Schlick and Waismann:

> Wittgenstein and others, who admit only scientific statements as 'legitimate', nevertheless also acknowledge 'non-legitimate' formulations as preparatory 'explanations' which later should no longer be used within pure science. Within the framework of these explanations the attempt is also made to construct the scientific language with the help, so to speak, of pre-linguistic means. Here we also find the attempt to confront the language with reality; to use reality to verify whether the language is serviceable. Some of this can be translated into the legitimate language of science, for example, as far as reality is replaced by the totality of other statements with which a new statement is confronted... But much of what Wittgenstein and others say about elucidations and the confrontation of language and reality cannot be maintained if unified science is built on the basis of scientific language from the beginning; scientific language itself is a physical formation whose structure, as physical arrangement (ornament), can be discussed by means of the very same language without contradictions.
>
> (Neurath 1931*a*, pp. 52–3)

This dense passage contains many characteristic themes of Neurath. The conception of language as a physical formation; the rejection of Wittgensteinian "elucidations"; the possibility of speaking about (parts of) the language within a (part of the) language; the rejection of the comparison between language and reality; the replacement of such a comparison by means of a confrontation of a group of statements with another statement.

What science is about, according to Neurath, is making predictions. At the beginning of this process are observation statements (what later came to be called protocols) by means of which one formulates laws, which are instructions for finding predictions that can then be tested by further observation statements. What is peculiar to

---

[16] For an account of Wittgenstein's theory of truth in the *Tractatus* see Newman 2002 and Glock 2006. See also Mulligan, Simons and Smith 1984.

Neurath's position is the claim that even at the level of observation statements we do not compare the statement with reality. Rather it is always a matter of agreement or disagreement between a body of sentences and the sentence being considered:

Thus *statements are always compared with statements*, certainly not with some 'reality', nor with 'things', as the Vienna Circle also thought up till now. This preliminary stage had some idealistic and some realistic elements; these can be completely eliminated if the transition is made to pure unified science... If a statement is made, it is to be confronted with the totality of existing statements. If it agrees with them, it is joined to them; if its does not agree, it is called 'untrue' and rejected; or the existing complex of statements of science is modified so that the new statement can be incorporated; the latter decision is mostly taken with hesitation. *There can be no other concept of 'truth' for science.*

(Neurath 1931*a*, 53)

Thus, in Neurath's account of truth there is no issue of comparing language to reality. Everything is intralinguistic:

Language is essential for science; within language all transformations of science take place, not by confrontation of language with a 'world', a totality of 'things' whose variety language is supposed to reflect. An attempt like that would be metaphysics. *The one scientific language can speak about itself, one part of the language can speak about the other*; it is impossible to turn back behind or before language.

(Neurath 1931*a*, 54)

According to his anti-absolutism, Neurath denies that alongside existing science there exist a "true" science:

Unified science formulates statements, changes them, makes predictions; however, it cannot itself anticipate its future condition. Alongside the present system of statements there is no further *'true' system of statements*. To speak of such, even as a conceptual boundary, does not make any sense.

(Neurath 1931*b*, 61)

The last quote comes from "Sociology and physicalism," where Neurath expounded on the same claims as his previously cited article on physicalism. In the same vein as in the quotes previously given, Neurath remarks:

Science is at times discussed as a system of statements. *Statements are compared with statements*, not with 'experiences', not with a 'world' nor with anything else. All these meaningless *duplications* belong to more or less refined metaphysics and are therefore to be rejected. Each new statement is confronted with the totality of existing statements that have already been harmonized with each other. *A statement is called correct if it can be incorporated* in this totality. What cannot be incorporated is rejected as incorrect.

(Neurath 1931*b*, 66)

It was in "Protocol sentences" (Neurath 1932/33) that some of the implicit consequences of Neurath's claim came fully to the fore. In particular, Neurath defends an antifoundationalist theory of science. Sentences are checked against bodies of sentences for agreement or disagreement. When a conflict is detected a decision is made as to what to alter. Nothing is sacrosanct. Even observation statements, or protocols, can be given up and thus every statement of science is revisable:

> There is no way to establish fully secured, neat protocol statements as starting points of the sciences. There is no tabula rasa. We are like sailors who have to rebuild their ship on the open sea, without even being able to dismantle it in dry-rock and reconstruct it from the best components.
>
> (Neurath 1932/33, p. 92)

> The fate of being discarded may befall even a protocol statement. There is no 'noli me tangere' for any statement.
>
> (Neurath 1932/33, p. 95)

It was against this picture of science and the radical proposal for the use of 'truth' that went along with it that Schlick attacked Neurath in 1934.

SCHLICK, NEURATH, HEMPEL, AND THE DEBATE ON TRUTH IN NEO-POSITIVISM

Schlick found the fallibilist position defended by Neurath, and as of 1932 also by Carnap, unacceptable and published a sharp attack against it in 1934. This led to replies by Neurath and Hempel. In Paris in 1935, Carnap tried to reconcile the two camps but no unity was to be achieved. As of 1935, Schlick retained his foundationalist outlook, Carnap had moved to his semantic stage, and Neurath kept defending his views in the form of an "encyclopedism" (see Uebel 1992).

Schlick's "On the Foundation of Knowledge" (1934) is a rebuttal to what the author saw as the relativism of Neurath and Carnap. Against the conception of protocols as descriptions of special empirical facts, which can always be revised if need be, Schlick introduced the notion of an affirmation (Konstatierung). It is through this notion that Schlick aimed at recovering what he saw as the rationale for introducing protocol sentences in the first place and in doing so he spelled out the connection to the problem of truth:

> The purpose [of introducing protocols] can only be that of science itself, namely to provide a true account of the facts. We think it self-evident that the problem of the foundations of all knowledge is nothing else but the question of the criterion of truth. The term 'protocol propositions' was undoubtedly first introduced so that by means of it certain propositions might be singled out, by whose truth it should then be possible to measure, as if by a yardstick, the truth of all other statements. According to the view described, this yardstick has now turned out to be just as relative as, say, all the standards of measurement in physics. And that view with its consequences has been commended, also, as an eviction of the last remnant of 'absolutism' from philosophy.
>
> But then what do we have left as a criterion of truth? Since we are not to have it that all statements of science are to accord with a specific set of protocol propositions, but rather that all propositions are to accord with all others, where each is regarded as in principle corrigible, truth can consist only in the *mutual agreement of the propositions with one another*.
>
> (Schlick 1934, p. 374)

Thus, Schlick proceeded to characterize Neurath's position as a "coherence theory of truth" in contrast to the older "correspondence" theory of truth. Against Neurath, Schlick argued that the only plausible meaning that "agreement" between propositions can have in such a truth theory is "absence from contradiction". But then any fictional story which is coherent in itself would have as much a right as scientific knowledge:

> Anyone who takes coherence seriously as the sole criterion of truth must consider any fabricated tale to be no less true than a historical report or the propositions in a chemistry textbook, so long as the tale is well enough fashioned to harbour no contradiction anywhere.
>
> (Schlick 1934, p. 376)

According to Schlick it is not consistency with any sort of statements that can provide the criterion of truth but rather lack of contradiction with quite specific statements. These are the "affirmations" ("Here now so and so") and for this kind of consistency, Schlick concludes, "there is nothing to prevent... our employment of the good old phrase 'agreement with reality'." My interest here is not in explicating Schlick's foundationalist viewpoint but only to point out his disagreement with Neurath on the issue of the criterion of truth and his dubbing of Neurath's position as a coherentist position.

Neurath replied to Schlick's article in "Radical Physicalism and the 'Real' World" (1934). Several points of disagreement with Schlick were addressed. Two of them are particularly important for our understanding of Neurath's reaction to Tarskian semantics. The first concerned the accusation that physicalism did not have an unambiguous criterion of truth; the second, that it did not address the relationship between language and reality. On both these points Neurath clarified and reiterated his previous position. Concerning truth he held once again that

> We shall call a statement 'false' if we cannot establish conformity between it and the whole structure of science; we can also reject a protocol sentence unless we prefer to alter the structure of science and make it into a 'true' statement.
>
> (Neurath 1934, p. 102)

The second point concerned the comparison of language and reality:

> The verification of certain content statements consists in examining whether they conform to certain protocol statements; therefore we reject the expression that a statement is compared with 'reality', and the more so, since for us 'reality' is replaced by several totalities of statements that are consistent in themselves but not with each other.
>
> (Neurath 1934, p.102)

While much more would need to be said both about Schlick's and Neurath's positions what we have covered does at least give the sense of the nature of the opposition between these two members of the Vienna Circle. Finally, it should be mentioned that this part of the debate also included an article by Hempel and a few more items by Schlick. Hempel sided with Neurath and Carnap against Schlick and he also characterized Neurath's position as a "restrained coherence" theory and defined this conception of truth as "a sufficient agreement between the system of acknowledged protocol-statements and the logical consequences which may be deduced from the statement and other statements which are already adopted" (Hempel 1935, p. 54).

Schlick responded with "Facts and propositions" (1935) and defended his approach to truth as a comparison between facts and propositions by discussing the example of checking a statement in a travel guide against a fact:

> What on earth could statements express but facts? ... saying that certain black marks in my Baedecker express the fact that a certain cathedral has two spires is a perfectly legitimate empirical assertion.
>
> (Schlick 1935, p. 402)

He admitted that at times one compares a sentence with another sentence but that there are also the cases where "a sentence is compared with the thing of which it speaks" (401) By contrast, Hempel's reply reasserted that Schlick's talk of comparing a statement from the travel guide with reality simply amounted to the comparison of two statements, i.e. the statement in the travel guide and the statement expressing "the result (not the act!) of counting the spires" (Hempel 1935, p. 94). Neurath strongly objected, as we will see, to be classified as a coherence theorist both in print and in correspondence.

In order to clarify that Neurath was not simply contraposing a "coherence" theory of truth to a "correspondentist" one, it might be useful to say a few things about the reasons that led him to deny that he held such a theory. This is also topical, for some of the secondary literature still claims that Neurath held a coherence theory of truth. Despite the fact that it was Schlick who dubbed Neurath as a coherentist, Schlick himself knew that Neurath defended no such theory. In reply to a letter by Carnap, where Carnap pointed out that Neurath does not accept a coherence theory of truth, Schlick wrote (June 5, 1934):

> I have never doubted that he would refuse to count as a follower of the usual coherence theory. However, I just meant to say that the coherence theory follows from his statements, if one is to take them seriously. I assumed that this was not even clear to himself for his thoughts are too unclear.
>
> (RC 029–28–10)[17]

Neurath was incensed at being dubbed a coherentist. He touched upon the topic with several correspondents including Carnap (15.XI.35 (remarks on a preliminary version of Carnap's *Wahrheit und Bewährung* [RC 110–02–01]), 23.XII.35, 27.I.36), Hempel (11.III.35, 18.II.35, 25.III.35, 29.XI.35, 12.XII.35), Kokoszyńska (8.IV.36, 23.IV.36, 3.VI.36), Nagel (26.II.35), Neider (2.IV.35), and Stebbing (9.III.35). Let me quote from the letter to Stebbing (written in tentative English), as Neurath is voicing his thoughts in preparation for the Paris Congress of 1935:

> Mr Schlick and also Mr Hempel use the name "coherence-theory"... All right—*but I fear*, that for English readers this terminus produces psychological associations which make a connection with the modern Idealism in England... The terminus "coherence theory" seems

---

[17] "Ich habe nie daran gezweifelt, dass er es ablehnen würde, als Anhänger der üblichen Kohärenztheorie zu gelten; aber ich wollte auch nur behaupten, dass aus seinen Aeusserungen, wenn man sie erst nimmt, die Kohärenztheorie folge. Ich nahm an, dass ihm das selbst nicht klar sei, weil seine Gedanken zu undeutlich sind." (Schlick to Carnap, 5.VI.34, RC 029–28–10)

more used by Metaphysicians than by Scientists. Is it not so? If I enough know about Bradley, a.s.f. Joachim a.s.f. is the basis: the "coherence" of the total system (a subject adapted to the well known spirit of Laplace) And if I understand is every statement more or less right in proportion to the quantity of the total coherence, which is inherent in the single statement. That means: the coherence theory of the modern English Idealism seems to be founded in the *absolutism* of the total-world-coherence. But my thesis is directly *against* such a absolutism and for a relativism. The science is a parcel of statements without contradiction and founded on Protocol-Statements. It is possible to make variations of all statements, to bring new statements, to reduce the statements. An[d] we have not an approach to an *absolute system of coherence*/the quasi *one and right world*/as the highest judge. And if we see, that we cannot make an confrontation of *our* parcel of statements with *this total and ideal system of coherence* must the idealistic philosopher, as Joachim use the single statements, and the harbour of refuge for the man of totality-coherence is—that seems so—correspondence. That means. For the idealistic philosophers of such type is the "correspondence" theory very *relative* and the "coherence" theory the ideal-type of an absolute theory. But for us is the correspondence-theory/with Atom-Statements and so further/a form of absolutism and the Theory of "Radical Physicalism" a form of relativism... Excuse please this discurs [sic] about terminology. But I wish to collect terms for Paris.

(Neurath to Stebbing, March 9, 1935)

In addition, Neurath objects to the fact that "coherence theory" is too strongly associated in the literature with Neo-idealism and thus with an absolutism he has always rejected. Neurath was later to write to Carnap:

I have never claimed... that truth consists in the agreements between propositions but only in the agreement with a preferred collection of propositions. This "preference" contains all those elements that are essential for a "realistic" conception.

(RC 102–50–01, December 23, 1935)[18]

However, Neurath would also reject a coherence theory in the sense that the mere consistency of the set of sentences would be enough to consider a set of sentences as true. His emphasis on the "preferred" class of statements points at an extra condition determined by pragmatic factors. Notice moreover how Neurath's claims on 'truth' can be at times stated in such a way that he appears to be giving a theory of truth rather than a proposal for an altogether different usage of the term.

Let us now move to Carnap's use of Tarski's theory as a possible mean to bring peace between Neurath and Schlick.

BACK TO THE PARIS CONGRESS

Of course, Tarski did not provide all the details of his theory of truth at the Paris Congress but he emphasized the most general aspects of his strategy. Central to

---

[18] "Niemals habe ich das behauptet... dass nämlich die Wahrheit in der Uebereinstimmung der Sätze besteht, sondern nur, in der Uebereinstimmung mit einer *bevorzugten* Satzmasse. Diese 'Bevorzugung' enthält alle jene Elemente, die für eine 'realistische' Auffassung wesentlich sind." (Neurath to Carnap, RC 102–50–01, 23.XII.1935)

Tarski's informal characterization were formulations of the project that were bound to bother Neurath. Consider, for instance, the definition of semantics:

> The word 'semantics' is used here in a narrower sense than usual. We shall understand by semantics the totality of considerations concerning those concepts which, roughly speaking, express certain connexions between the expressions of a language and the objects and states of affairs referred to by these expressions. As typical examples of semantical concepts we might mention the concept of denotation, satisfaction, and definition...
>
> (Tarski 1936, 401)

As for truth:

> The concept of truth also—and this is not commonly recognized—is to be included here, at least in its classical interpretation, according to which 'true' signifies the same as 'corresponding with reality'.
>
> (Tarski 1936, 401)

After all the background work we have done we can have a better sense of what's at stake (at least for Neurath) in this sentence. The concept of truth Tarski is after is certainly not the one that was proposed by Neurath but rather the one corresponding to the classical conception. It is thus not surprising that Tarski's work could be interpreted, among other things, as a vindication of Schlick's position in the protocol debate.[19] In addition to Tarski, Kokoszyńska also gave a chapter, "Syntax, semantik und Wissenschaftslogik," which certainly disturbed Neurath's anti-absolutist tendencies. Arguing for the need to extend Carnapian Syntax to Tarskian semantics in the analysis of science, Kokoszyńska used as example "the absolute concept of truth":

> Of late, the classical concept of truth, according to which—as one usually says—the truth of a proposition consists in its agreement with reality, has been labeled as the absolute conception of truth. This conception of truth, as is well known, is called the correspondence theory. This theory is to be contrasted with the coherence theory of truth, according to which the truth of a proposition consists in a certain agreement of this proposition with other propositions. Some logical positivists have in the last few years made a transition from a correspondence theory of truth to a coherence theory of truth. In this transition have found expression both the conviction that the absolute conception of truth is an unscientific concept which should be excluded from philosophical investigation and—as it appears—the opinion that it can be replaced by a syntactic one with the same extension and thus be defined in the syntax language.[20]

---

[19] For instance, Rougier in his unsigned introduction to Neurath 1935 contraposes in a note the positions of Carnap and Neurath (he also adds Popper, Poznanski and Wundheiler [1934]) to those of Schlick, Tarski, and Lutman (see Neurath 1935, p. 5). See also the comment reported by Neurath in Neurath 1936, p. 400 where Rougier sees Tarski's theory as a vindication of Schlick's position that sentences and reality can be compared for agreement.

[20] "Als absoluter Wahrheitsbegriff wird letztens der klassische Wahrheitsbegriff bezeichnet, nach dem—wie man zu sagen pflegt—die Wahrheit eines Satzes in seiner Übereinstimmung mit der Wirklichkeit besteht. Man bezeichnet diese Auffassung der Wahrheit bekanntlich als Korrespondenztheorie. Dieser Theorie steht die Kohärenztheorie der Wahrheit gegenüber, nach der die Wahrheit eines Satzes in gewisser Übereinstimmung dieses Satzes mit andern Sätzen

She went on to claim that Tarski's investigations had shown how to treat scientifically the absolute conception of truth. Thus, the notion of "absolute truth" which had been previously taken to be metaphysical could now be seen as part of the logic of science in its post-syntax phase. As a consequence, problems like "How is the real world?" can be shown not to be pseudoproblems but rather to be suitable for scientific analysis.(p.13) We can see why Neurath felt that Tarskian semantics might end up reviving metaphysical issues that he had tried to dispose of once and for all as pseudo-problems. Moreover, he found himself classed as a coherence theorist, something that, as we have already seen, annoyed him very much.[21] However, in his long review of the Paris meeting for the readers of Erkenntnis, Neurath reports on the discussion that followed the talks by Tarski and Kokoszyńska and although reporting on several objections, many of which due to him, he fairly claimed that the talks had found most people in agreement.

An important development during this meeting stemmed from Carnap's application of Tarski's theory of truth to the protocol debate. Carnap began his lecture "Truth and confirmation" by sharply distinguishing two notions:

> The difference between the two concepts 'true' and 'confirmed' ('verified', 'scientifically accepted') is important and yet frequently not sufficiently recognized. 'True' in its customary meaning is a time-independent term; i.e. it is employed without a temporal specification. For example, one cannot say that "such and such a statement is true today (was true yesterday; will be true tomorrow)", but only "the statement is true". 'Confirmed', however, is time-dependent. When we say "such and such a statement is confirmed to a high degree by observations" then we must add: "at such and such a time."
>
> (Carnap 1936, 18, translation Uebel 1992, p. 198)

Carnap diagnosed the source of the equivocation between the two terms in the misgivings logicians had about the concept of truth, due to the antinomies that had emerged from its unrestricted use, which led to an avoidance of the concept. In an interesting letter to Kokoszyńska, Carnap reflects on the situation before the appearance of Tarski's results:

> After partly reading the proofs of Tarski's essay and seeing that he gives a fully correct definition of the concept of truth, I agree with you thoroughly that "true" and the other concepts related to it are to be seen as scientifically sound. My earlier scepticism, and that of other people, concerning this concept was in fact historically justified, inasmuch as no definition was known which was on the one hand formally correct and on the other hand avoided the antinomies. And the theory that employs these concepts, "semantics" in Tarski's sense, seems to

---

besteht. Bei einigen logischen Positivisten lässt sich in den letzten Jahren ein Übergang von einer Korrespondenztheorie der Wahrheit zu einer Kohärenztheorie nachweisen. In diesem Übergange hat teilweise die Überzeugung Ausdruck gefunden, der absolute Wahrheitsbegriff sei ein unwissenschaftlicher Begriff, der aus den philosophischen Untersuchungen ausgeschaltet werden soll, teilweise aber—wie es scheint—auch die Meinung, er liesse sich durch einen umfangsgleichen syntaktischen ersetzen und auf diese Weise in der Syntaxsprache definieren." (Kokoszyńska 1936a, p. 11)

[21] Neurath was unsuccessful in changing this widespread perception. Indeed, Russell 1940 quipped that according to this view of Neurath "empirical truth can be determined by the police." BonJour 1985, p. 213, ascribes to Neurath a notion of coherence as mere consistency.

me to be an important scientific domain. I consider it very deserving on Tarski's part that he opened up this new domain.

(Carnap to Kokoszyńska, July 19, 1935)[22]

This was thus, according to Carnap, the historical reason for the misuse of the term "true" for "confirmed". This usage was of course in conflict with ordinary usage according to which any declarative sentence is either true or false, something that is not the case for the concept of confirmation. Carnap, in his lecture, then points out the new situation created by Tarski's definition of truth, which allows, under certain restrictions, the consistent use of the adjective "true". As a consequence

The term 'true' should no longer be used in the sense of 'confirmed'. We must not expect the definition of truth to furnish a criterion of confirmation such as is thought in epistemological analyses.

(Carnap 1936, p. 19, trans. Uebel 1992, p. 198)

Using these distinctions Carnap outlined the essential tenets of a theory of confirmation distinguishing between direct confirmation, obtained by confronting the statement with observations, and indirect confirmation, obtained by confronting sentences with sentences. And although he pointed out the danger involved in the talk of "comparison" between sentences and facts he also allowed as unobjectionable the idea that sentences can be confronted with observations thereby striking a middle ground between Neurath and Schlick (see Uebel 1992 for more details).

The published version of Carnap's talk was actually the subject of correspondence between Neurath and Carnap. Neurath asked Carnap to present his conception of truth as a "proposal" but Carnap refused. Moreover, Carnap decided only to present his point of view and not to try to characterize the previous debate for he was convinced he could not do this "without upsetting both of you [Neurath and Schlick]" (letter of December 4, 1935). Neurath, in a final desperate attempt, replied by using "scare tactics":

You'll soon see how questionable it is 1. that one has pinned us with the tag of coherence theory... and 2. that Tarski's and Lutman's indeed valuable considerations circulate with the label "true". If you still can, you should choose a different name for it. I cannot conceive of this term ever contributing to clarification but on the contrary that it will constantly create confusion... I just want to say it again really clearly and sternly because I find painful what,

---

[22] "Nachdem ich einige Korrekturbogen von Tarskis Aufsatz gelesen und gesehen habe, dass er für den *Wahrheitsbegriff* eine vollkommen korrekte Definition aufstellt, stimme ich Ihnen durchaus zu, dass 'wahr' und die andern mit ihm zusammenhängenden Begriffe als wissenschaftlich einwandfrei anzusehen sind. Meine und anderer Leute frühere Skepsis gegen diese Begriffe war ja insofern historisch berechtigt, als keine Definition bekannt war, die einerseits formal korrekt war und andrerseits die Antinomien vermied. Und die diese Begriffe verwendende Theorie, die 'Semantik' im Sinn von Tarski scheint mir ein wichtiges Wissenschaftsgebiet zu sein. Ich halte es für sehr verdienstvoll von Tarski, dass er dieses neue Gebiet erschlossen hat." (Carnap to Kokoszyńska, July 19, 1935, RC 088-57-16)

for instance, Rougier said in the conclusion about the shift of the demarcation line in support of metaphysics.

(Neurath to Carnap, December 8, 1935)[23]

Carnap was not to be deterred from his terminological choices concerning "true" although he agreed with Neurath's criticism of Kokoszyńska's terminology of "absolute truth" (Carnap to Neurath, 27.I.36).

We are finally back to Tarski and Neurath.

TARSKI'S REPLY TO NEURATH

With a better understanding of Neurath's background we can now return to Tarski's reply to Neurath. In his letter dated 28.IV. 36, Tarski replied as follows to Neurath's comments:

I completely agree that "to defend trivialities against errors" is an important task of science. I have indeed for this very reason stressed many times that one must always speak in a language about another language—and not outside the language (from the reductive standpoint, just about my entire 'semantics' should be seen as a triviality; this does not upset me in the least). It seems to me that it is a big mistake, when Wittgenstein, Schlick etc. speak of "the" language instead of (a number of) languages; that might be the true source of the Wittgensteinian "metaphysics". Incidentally, all those who speak about the unified language of science with the slogan "Unity of Science" [Einheitswissenschaft] seem to commit the same mistake. We all know—because of arguments from semantics and syntax—that there is strictly speaking no unified language [Einheitssprache] in which science as a whole could be expressed. It is not enough to say that this is just a temporary, imprecise formulation. For, what should then the final, precise formulation be? Kokoszyńska recently held a lecture on the problem of a Unified Science for the local philosophical society and subjected this point to her criticism; an article from her on this subject is forthcoming in Polish.

(Tarski to Neurath, April 28, 1936)[24]

Here it is interesting to point out that Tarski had drawn attention to the danger of speaking of a single language for science already in Paris in 1935 (see Neurath 1936,

---

[23] "Du wirst bald sehn, wie bedenklich es ist 1. dass man uns den Titel Kohärenztheorie angehängt hat... und 2. dass die wirklich wertvollen Betrachtungen von Tarski und Lutman mit den Terminus 'wahr' herumlaufen. Wenn du noch kannst, solltest Du dafür einen anderen Namen verwenden. Ich kann mir nicht denken, dass dieser Terminus je zur Klärung dient, wohl aber, dass er ständig Verwirrung stiften wird... Ich wills nur noch einmal recht nett und ernst Dir gesagt haben, weil ich nur schmerzlich es empfinde, was z.B. Rougier im Schlusswort über die Verschiebung der Demarkationslinie zugunsten der Mataphysik sagte." (Neurath to Carnap, December 8, 1935, RC 102-50-04) On Rougier's comment see Neurath 1936, p. 401.

[24] "Ich bin völlig Ihrer Meinung, daß es eine wichtige Aufgabe der Wissenschaft ist 'Trivialitäten gegen Irrtümer zu vertreten.' Eben deshalb habe ich ja selbst vielmals betont, daß man stets in einer Sprache über eine andere Sprache sprechen muß—und nicht außerhalb der Sprache (vom rein deduktiven Standpunkt aus ist übrigens meine ganze 'Semantik' fast als eine Trivialität anzusehen;

p. 401). Tarski's point here is quite simple. Since the universal language of science would have to be semantically closed it would end up being inconsistent. Thus, one ought to speak about languages (in the plural) and not about a single language. The argument was later developed at length by Kokoszyńska in her "Bemerkungen über die Einheitswissenschaft" (1937/8) which in all likelihood is the printed version of the lecture referred to by Tarski in the previous quote. Tarski continues:

> Now, as far as my "terminological choice" is concerned, I can assure you, firstly, that it came about completely independently of Wittgenstein's metaphysics and, secondly, that it was in no way a "choice". The problem of truth came up very often, especially in the Polish philosophical literature. One was constantly asking (see for instance Kotarbiński's "Elements"), whether it was possible to define and apply the concept of truth unobjectionably, using such and such properties (which I spelled out in my later work). I simply provided a positive solution to this problem and noted that this solution can be extended to other semantic concepts. Like you, I am certain that this will be misused, that a number of philosophers will "overinterpret" this purely logical result in an unacceptable manner. Such is the common destiny of both small and great discoveries in the domain of the exact sciences (at times, one compares the philosophers to the "hyenas of the battle field").
>
> (Tarski to Neurath, April 28, 1936)[25]

Concerning his relationship to metaphysics, here is what he had to say:

> But I must confess to you that even if I do not underestimate your battle against metaphysics (still more from a social than from a scientific point of view), I personally do not live in a constant and panic fear of metaphysics. As I recall, Menger once wrote something witty on the fear of antinomies; it seems to me that one could apply it—mutatis mutandis—to the fear of metaphysics. It is a hopeless task to caution oneself constantly against metaphysics.

---

das ärgert mich nicht im wenigsten). Es ist—wie mir scheint—ein großer Fehler, wenn Wittgenstein, Schlick usw. von 'der' Sprache anstatt von Sprachen (in Mehrzahl) sprechen; das ist vielleicht die echte Quelle der Wittgensteinschen 'Metaphysik'. Nebenbei gesagt, denselben Fehler scheinen auch alle diejenigen zu begehen, die im Zusammenhang mit dem Stichwort 'Einheitswissenschaft' über die Einheitssprache der Wissenschaft reden. Wir wissen ja alle—auf Grund der Erörterungen aus der Semantik und Syntax—, daß es streng genommen keine Einheitssprache gibt, in der die ganze Wissensschaft ausdrückbar wäre. Es genügt nicht zu sagen, daß das nur eine vorläufige, unpräzise Formulierung ist; denn wie soll die endgültige, präzise Formulierung lauten? (Kokoszyńska hatte vor kurzer Zeit einen Vortrag über das Problem der Einheitswissenschaft in der hiesigen Phil. Gesell. und hat u.a. diesen Punkt einer Kritik unterworfen; es soll ein Aufsatz von ihr in der polnischen Sprache darüber erscheinen)." (Tarski to Neurath, 28.IV.36, Neurath Nachlaß)

[25] "Was nun meine 'terminologische Wahl' betrifft, so kann ich Ihnen versichern, daß sie erstens ganz unabhängig von der Wittgensteinschen Metaphysik zustandegekommen ist und daß es zweitens überhaupt keine 'Wahl' war. Das Problem der Wahrheit kam speziell in der polnischen Philosophischen Litteratur sehr oft vor, man hat immer gefragt, ob man den Wahrheitsbegriff mit den und den Eigenschaften (die ich später in meiner Arbeit genau präzisiert habe) in einwandfreier Weise definieren und verwenden kann (vgl. Z.B. die 'Elemente' von Kotarbiński). Ich habe einfach dieses Problem positiv gelöst und habe bemerkt daß sich diese Lösung auf andere semantische Begriffe ausdehnen läßt. Ebenso wie Sie bin ich sicher, daß man daraus verschieden Mißbräuche machen wird, daß verschiedene Philosophen dieses Ergebnis rein logischer Natur in unerläßlicher Weise 'hinausinterpretieren' werden—das ist das gemeinsame Schicksal aller kleineren und größeren Entdeckungen aus dem Bereiche der exakten Wissenschaften (man vergleicht ja manchmal die Philosophen mit den 'Hyänen des Schlachtfeldes')." (Tarski to Neurath, 28.IV.36, Neurath Nachlaß)

This becomes all the clearer to me when I hear, here at home, various attacks on the very metaphysics of the Vienna Circle (going, namely, in your direction and in that of Carnap), when, for instance, Łukasiewicz talks, with respect to the "Logical Syntax", about Carnap's philosophy, philosophizing etc. (in his mouth this has roughly the same sense as 'metaphysics' in yours). What you blame me for on account of the concept of truth, one blames Carnap for on account of the introduction of the terms 'analytic', 'synthetic', etc. ("Regression to the Kantian metaphysics"); and it seems to me that I was even more justified than Carnap to designate as truth the concept that I discuss. In general it is a valuable task to fill old bottles with new wine.

(Tarski to Neurath, April 28, 1936)[26]

Finally, Tarski points out that to be coherent Neurath should also criticize all the formally defined concepts that are central to syntax and semantics (thus most of those found in Carnap's work in the *Logical Syntax of Language*):

Another point in this connection: my concept of truth, you claim, holds only in formalized languages. But on the contrary, the concept of truth is of practical significance precisely in non formalized domains. One can extend this literally to all precise concepts of syntax and semantics (consequence, content, logical and descriptive concepts, etc.): all these concepts can only be related approximately to the non formalized languages (thus to the actual languages of all non formal sciences [Realwissenschaften]): truth here is no exception.

(Tarski to Neurath, April 28, 1936)[27]

The remaining part of the exchange did not add much to this picture and I have already quoted in section two some passages from the later discussion. Neurath was however to pursue the discussion with Kokoszyńska and we now turn to that part of the exchange.

---

[26] "Aber ich muß Ihnen offen gestehen: wenn ich auch Ihren Kampf gegen die Metaphysik keineswegs unterschätze (noch mehr unter sozialem, als unter wissenschaftlichem Gesichtspunkt), so lebe ich persönlich nicht in einer ständigen, panischen Angst vor der Metaphysik. Wie ich erinnere, hat einmal Menger etwas geistreiches über die Furcht vor Antinomien geschrieben; es scheint mir, daß man das alles—mutatis mutandis—auch auf die Angst vor der Metaphysik übertragen könnte. Es ist eine hoffnungslose Aufgabe, sich stets vor dem Vorwurf einer Metaphysik zu warnen. Das wird mir besonders klar, wenn ich hier bei uns verschiedenen Angriffe eben auf die Metaphysik des Wiener Kreises (und zwar Ihrer und Carnapschen Richtung) höre, wenn z. B. Łukasiewicz a propos der 'Logischen Syntax' über Carnaps Philosophie, Philosophieren usw. spricht (das hat in seinem Mund ungefähr denselben Sinn wie in Ihrem "Metaphysik"). Dasselbe, was Sie mir wegen des Wahrheitsbegriff vorwerfen, wirft man Carnap wegen der Einführung der Termini 'analytisch', 'synthetisch' u.s.w. vor ('Rückkehr zu der Kantschen Metaphysik'); und es scheint mir, daß ich im Grunde noch mehr als Carnap berechtigt war den von mir erörterten Begriff als Wahrheit zu bezeichnen. Im allgemeinen ist es eine wertvolle Aufgabe alte Gefässe mit neuem Trunk zu füllen." (Tarski to Neurath, 28.IV.36, Neurath Nachlaß)

[27] "Noch ein Punkt in diesem Zusammenhang: mein Wahrheitsbegriff gelte nur für die formalisierten Sprachen, andrerseits ist der Wahrheitsbegriff gerade im nicht formalisierten Bereich von praktischer Bedeutung. Das kann man wörtlich auf alle präzisen Begriffe der Syntax und Semantik (Konsequenz, Gehalt, logischer und deskriptiver Begriff u.s.w.) übertragen: alle diese Begriffe können nur annährungsweise auf die nicht-formalisierten Sprachen (also auf die aktuellen Sprachen aller Realwissenschaften) bezogen werden; Wahrheit ist hier keine Ausnahme." (Tarski to Neurath, 28.IV.36, Neurath Nachlaß)

NEURATH AND KOKOSZYŃSKA

Let me recall that at the Paris Congress Kokoszyńska had presented a chapter on issues concerning semantics entitled 'Syntax, Semantik und Wissenschaftslogik." During this meeting she was part of the lively discussions on the concept of truth (basically on Tarski's side) and this led to an extensive correspondence with Neurath. The correspondence between Neurath and Kokoszyńska contains nineteen letters from Kokoszyńska and fourteen letters from Neurath. As a consequence of the Paris discussions, Kokoszyńska had promised to send Neurath some reflections on the viability of a 'sociological' definition of truth (her own label), by which name she meant Neurath's distinctive position as opposed to a coherence theory of truth. She had come to make this distinction under pressure from Neurath who, as we have seen, refused to be classified as a coherentist. She apparently sent her comments in the form of a short essay (which I was not able to locate) which accompanied a letter dated 22.III.1936. We can gather the contents of these essay both from Neurath's reply and from later letters by Kokoszyńska. One central argument against the sociological theory of truth was the following. If we consider as a requirement of any theory of truth that it allows a derivation of the instances of the schema " 'p' is true iff p" then the sociological theory should give rise to " 'p' is acknowledged iff p." But herein lies the absurdity of the proposal, for from the fact that a statement 'p' is acknowledged we then would be able to conclude that p and from p that 'p' is acknowledged. In both directions we can come up with innumerable counterexamples.

Neurath replied with a letter (dated 23.IV.36) containing three dense pages of comments. Neurath's letter is a point by point commentary to Kokoszyńska's essay divided into four parts: (1) linguistic use; (2) coherence theory; (3) "acknowledgement" [Anerkennung] theory of truth (sociological definition), and (4) dangers of the Tarski–Lutman conception of truth.

The first part of the letter points out the variety of uses of the word 'truth' in natural language and refers to the empirical work by Arne Naess of the issue.[28] This was meant to undermine the idea that the "semantic" conception had any better right to claim to capture some sort of ordinary concept of truth than the sociological definition proposed by Neurath. Neurath points out that in different circles, with different linguistic practices, what decides the partition between "true" and "false" depends on a criterion [Instanz] against which the partition is decided. In most cases this criterion turns out to be metaphysical and not in harmony with empiricism. In the case of his proposal, the criterion is empirically given as it consists of the sentences accepted by a specific group of human beings at a certain moment in time. The second part of

---

[28] Appeal to Naess' work is also found in a letter from Neurath to Tarski dated 27 Mai, 1937: "Dass meine Bedenken gegen Semantik sich nur auf die Interpretationen beziehen, die im Empirismus in Frage kommen erwähnte ich schon. Ich glaube, wenn Sie an den Wahrheitsbegriff bei Kotarbiński anknüpfen, Lutman von einen üblichen Wahreitsbegriff spricht, so ist etwas zu wenig Vorsorge getroffen die lebendige Diskussion damit erreicht zu haben, denn es gibt sehr viele Auffassungen von Wahrheit, wie NESS [sic] festgestellt hat." For Naess' work see Naess 1936 and 1938.

the letter questions whether anyone at all defends a theory of coherence as defined by Schlick and discussed also by Kokoszyńska. In the same section Neurath gives an overview of how the problem of truth originated in the Vienna Circle and was pursued in connection to the protocol debate. In the third part, Neurath reiterated his position that a statement should be called true if acknowledged at a certain time by a determined group of people under certain circumstances. The objections to Tarski and Kokoszyńska reassert the generic claim about the metaphysical dangers of the conception. In particular Neurath objected to the fact that the starting point of the Tarskian conception is an appeal to the ordinary usage but at the same time the realm of validity of the theory is limited to formal languages and thus it cannot be applied to natural language; however, these restrictions will not be observed, or so Neurath conjectures, and this will lead to metaphysical abuses of Tarski's theory. Finally, Neurath objected to certain formulations by Tarski and Kokoszyńska such as, e.g. "a proposition can be acknowledge without its holding" or "there can be life on Venus without man experiencing it." Against this type of talk Neurath states that he does "not think to be able to include them in the total body of science." The "holding" [zutreffen] of a statement according to Neurath can only be a question of being a recognition by someone. Not accepting this is tantamount to slip into metaphysics. In his summary of the major points of the letter, Neurath wrote:

The "sociological" definition of truth can be upheld, and certain propositions can thus be characterized as true "now" in its sense. The "sociological" definition of truth corresponds to certain elements of the traditional conception. The Tarski–Lutman definition does not correspond to the ordinary usage in any privileged way (historical question). The Tarski–Lutman definition of truth is only applicable within formalized languages. The Tarski–Lutman terminology lures one into applying it to non-formalized languages and to interpret it in an absolute way. The justificatory explanations by Tarski and Lutman on "acknowledged but not holding" immediately seem to entail absolutist elements and seem not to be applicable within whole science neither according to Neurath's conception (Carnap, Hempel, and so on) nor even according to the very conception expressed elsewhere by Tarski and Lutman.[29]

In her reply Kokoszyńska explicitly stated that she could not accept as a theory of truth any theory which would not prove (all the instances of) the T-schema. That is the reason why she rejects the "sociological" theory as a theory of truth. Concerning the limited domain of applicability of Tarski's theory, Kokoszyńska

---

[29] "Die 'soziologische' Wahrheitsdefinition lässt sich aufrechterhalten, so lassen sich in ihrem Sinne gewisse Sätze als 'jetzt' wahr kennzeichnen.
Die 'soziologische' Wahrheitsdefinition entspricht gewissen Elementen traditioneller Auffassung.
Die T.L. Definition entspricht dem Sprachgebrauch nicht in bevorzugten Weise (historische Frage).
Die T.L. Definition ist nur innerhalb formalisierter Sprachen verwendbar.
Die T.L. Terminologie verlockt dazu sie für nicht-formalisierte Sprachen zu verwenden und absolutistisch zu deuten.
Die begründenden Erörterungen von L.K. über 'zwar anerkannt, aber nicht zutreffend' scheinen unmittelbar absolutistische Elemente zu enthalten und innerhalb der Gesamtwissenschaft nicht verwendbar zu sein, weder nach Auffasung von Neurath (Carnap, Hempel, usw.) noch auch nach sonst geäusserten Auffassung von T. and L." (Neurath to Kokoszyńska, 23.IV.36, Neurath Nachlaß)

pointed out that natural science can be formalized (say as in Carnap's language II) and thus Tarski's definition could immediately be applied. However, on this point Kokoszyńska underestimated the roots of Neurath's objection which rested on the idea that natural science is expressed in a great part through natural language and presents vague concepts [Ballungen] which make its full formalization hopeless. This aspect of Neurath's thoughts can be traced back to his anti-Cartesianism (see Mormann 1999).

I will not pursue in detail the remaining letters except to point out a constant tendency on Neurath's part to push Kokoszyńska into claiming (especially by means of suggesting revisions to her forthcoming article in *Erkenntnis*) that there was no contradiction between his views and those defended by Tarski and Kokoszyńska. Eventually, Kokoszyńska reacted firmly against this attempt and wrote the following (6.9.36):

As far as I understand, you want me to describe the situation as if there were no contradiction between the position you have defended so far concerning the classical concept of truth and the thoughts contained in my comments in *Erkenntnis*. But such a contradiction seems to exist after all. The issue is whether one can reliably use a concept which, so to speak, involves talk of an "agreement with reality". You have to some extent completely rejected this concept for you thought that the determination of such an "agreement" would require one to go beyond the framework of language—which is impossible—and you have tried, to a certain extent, to replace this concept by a sociological-syntactic one. It appears now from Tarski's investigations that one can speak of an "agreement between sentences and reality"—and therefore consider it within language—in positing propositions in which not only names of propositions occur but also names of other things. You have nothing to object to positing such propositions except—what affects mainly you—that they are not necessary in empirical sciences. It thus transpires that one can deal adequately precisely with the concepts which you had rejected so far. The contradiction mentioned above seems to lie therein.[30]

Kokoszyńska concluded by saying that Neurath had only made skeptical remarks in print about the classical conception of truth but that he had never treated the topic exhaustively and publicly. Her intention in corresponding with Neurath was to set limits to such skepticism.

---

[30] "Sie wollten—soweit ich verstehe—dass ich die Sachlage so schildere, als ob kein Widerspruch zwischen Ihrer bisherigen Haltung gegenüber dem klassischen Wahrheitsbegriff un[d] den in meinen Erkenntnis-Bemerkungen enthaltenen Gedanken bestehe. Ein solcher Widerspruch scheint aber doch zu bestehen. Es handelte sich ja darum, ob man sich in verlässlicher Weise mit einem Begriffe bedienen kann, in dem sozusagen von einer 'Übereinstimmung mit der Wirklichkeit' die Rede war. Sie haben teilweise diesen Begriff ganz abgelehnt, da Sie meinten, die Feststellung einer solcher 'Übereinstimmung' müsse ein Ausgehen aus den Rahmen der Sprache erfordern/was unmöglich ist/und teilweise haben sie versucht, diesen Begriff durch einen soziologisch-syntaktischen zu ersetzten. Nun zeigt es sich aus den Untersuchungen von Tarski, dass man von jener 'Übereinstimmung zwischen Sätzen und Wirklichkeit' sprechen/also sie innerhalb der Sprache betrachten/kann, indem man nämlich Sätze aufstellt, in denen nicht nur Namen von Sätzen aber auch von anderen Dingen auftreten. Gegen Aufstellung von solchen Sätzen haben Sie nichts einzuwenden ausser—was speziell Sie betrifft—dass sie in empirischen Wissenschaften nicht nötig sind. Es hat sich also gezeigt, dass man eben diese von Ihnen bisher abgelehnten Begriffe in korrekter Weise behandeln kann. Darin scheint mir der vorher erwähnte Widerspruch zu stecken." (Kokoszyńska to Neurath, 6.IX.36, Neurath Nachlaß)

The correspondence with Kokoszyńska is quite lengthy and often repetitive on Neurath's part. However, it does provide a detailed glimpse of the set of issues that were motivating Neurath while at the same time increasing the reader's frustration for the lack of a clear articulation of Neurath's rationale in his criticism of the theory of truth. He did not object formally to the theory nor to its application within formalized languages. He saw the danger of a possible misapplication of the theory of truth by overextending its limits of application and giving rise to metaphysical pseudo-talk of comparison of language and reality. But while Neurath was focusing on these possible dangers he did not focus on the opposite danger, which consisted in using the word "truth" for talking about "acknowledgement," certainly a quite unintuitive move from the point of view of the ordinary usage of the expression "true." The final chapter in this story I want to consider is the private section on semantics which saw Carnap and Neurath on opposite camps at the 1937 "Congrès Descartes" in Paris.

NEURATH VS. CARNAP: PARIS 1937

The archives on Neurath and Carnap contain two documents which, taken together, mark a culminating point of the debate on semantics within the Vienna Circle.[31] On occasion of the Congrès Descartes (Paris, 1937) Carnap, Neurath, and others met for a private discussion on semantics. Among the invited people were Tarski and Kokoszyńska.[32] Both Carnap and Neurath presented written contributions. Neurath's chapter was entitled "The Concept of Truth and Empiricism" (Neurath 1937a) and Carnap's "The Semantical Concept of Truth" (Carnap 1937).

Neurath's contribution is ten pages long and it is dated July 12, 1937. He begins by acknowledging that he should have made clear, already from his 1931 *Scientia* article on physicalism, that he had only intended to make proposals as opposed to presenting dogmas. On the other hand he claims to have individuated clearly, unlike the other participants in the truth debate, the real opponent, i.e. Wittgenstein and those close to him. His proposal then is to delimit a subject of investigation "where we constantly compare sentences with sentences, investigate their logical extent and their systematic position etc. If one analyzes science in this way then one is engaged in what Carnap called the logic of science." Neurath's proposal is to see "how much can be handled *within* the logic of science" (Neurath 1937a, p.1).

Neurath then proceeds to rehearse the origin of the debate on truth with which we are by now familiar, including Wittgenstein's theses on the comparison between language and "the" reality and the idea that verification consists in a reference to the given. Against this type of talk, Neurath had suggested that both sentences and facts

---

[31] In addition to Carnap 1937 and Neurath 1937a there are three additional pieces by Neurath entitled respectively "Fuer Die Privatsitzung, 30 Juli 1937" (1937b); "Diskussion Paris 1937 Neurath-Carnap" (1937c); "Bemerkungen zur Privatdiskussion"(1937d), classified as K.31, K.32 and K.33 in Neurath's Nachlaß.

[32] Tarski reports positively about the discussion in a letter to Popper dated 4.X.37 (Popper-Tarski correspondence, Hoover Institution, Stanford, Box number 354. Folder ID: 8).

(or states of affairs) were types of objects, objects of the sentence type and objects of the non-sentence type. With the help of new sentences one could now talk about these sentences and non-sentences and thus confront sentences about sentences and sentences about non-sentences.

It is the goal of the logic of science to investigate the logical relationship between, among other things, real sentences. Suppose one wants to study the relationship between theory and experiment in behavioral terms. This normally refers to the activity of scientists first in relation to the experimental apparatus and then in their theoretical formulations. The logic of science, he adds, uses the following trick: it expresses by an observation sentence the outcome of the experimental work, say "At location A ice melts at (temperature) −3 degree"; then it compares it with a theoretical statement, say, "Ice melts at (temperatures) greater than 0 degree." It then investigates how much of an incoherence with a given class of statements it would be to use both sentences simultaneously. This is the way to move away from talk of comparison between 'language' and 'reality' or between 'thought' and 'being'.

Neurath then suggests to apply the 'trick' to semantics. This, he claims, he had already suggested in 1935 at the previous meeting in Paris but he had found no adherents. Carnap and Hempel went along with the formulations of Tarski and Kokoszyńska which, he adds, "can become dangerous for empiricism." He then goes on to make his proposal in terms of the "acknowledgement theory": suppose we are given a sentence of the Encyclopedia which describes (structurally) a sentence, say "it snows", by describing the letters composing it. Then this expression is called a "true sentence" if and only if I am given a sentence of the Encyclopedia: it snows. All of this is done within the logic of science and there is no need to use expressions such as "relations between expressions of language and designated objects." Thus, he proposes to investigate how far one can proceed this way in the framework of logical empiricism.

Moving now to more criticism of the Tarski–Kokoszyńska line, he first points out that it might simply be better to use "accepted (in the Encyclopedia)" and "rejected (in the Encyclopedia)" instead of "true" and "false", which are too loaded with meaning. Against Kokoszyńska he objects that she takes for granted that what she calls "the absolute concept of truth" agrees with the ordinary concept of truth. Against this he adduces the investigations by Arne Naess which "show that there are many common concepts of truth." He proposes the same argument against Tarski who is described as the defender of the traditional philosophical concept of truth, as evidenced by his references to Kotarbiński who, adds Neurath, despite his general sympathies with the logical empiricists, on the issues of truth displays the absolutist tendencies of the Brentano school. Neurath objects that it is not the role of a defender of logical empiricism to discuss more closely a plea [Plädoyer] for the traditional concept of truth, until one shows to him the need to apply this concept in his analysis of science. Later in his chapter Neurath added that "one had already seen in 1935 in Paris, how Tarski and Lutman were actually interpreted and probably not without justice, given that both show a certain 'connivence' vis-à-vis the traditional conception" (Neurath 1937a, p.9). In conclusion, Neurath asked Carnap, Hempel, and "the Polish friends" to discuss whether and how far his "proposal" could be carried out and whether they thought that in this way "semantical and related problems could be brought within the logic of science."

Carnap's typescript is entitled "The semantic conception of truth." It is dated 18.7.37 and it is twelve pages long. Carnap begins by listing four theses he would like to propose for discussion:

1. The semantical conception of truth is correct and unobjectionable;
2. It cannot be replaced by merely syntactical method;
3. It is useful and important;
4. It is in agreement with the concept of truth used in ordinary language.

Under (1) Carnap gives an informal description of the legitimacy of introducing a binary relation Bez (x, y) which captures the notion of denotation. Then he claims that in terms of denotation one can define truth. Both denotation and truth are examples of semantical concepts.

In section (2), Carnap addresses directly Neurath's proposal which, in a way, was an attempt to show the eliminability of talk of truth in terms of syntax. Here Carnap shows that this is not possible. Carnap grants that there are cases where sentences which contain semantical concepts (denotation, truth, etc.) can be transformed in purely syntactical sentences (in the technical sense of the *Logical Syntax*). For instance, "the expression '3 + 4' denotes (the number) 7" can be translated in the syntactical sentence " '3 + 4' is logically-synonymous with '7'." The semantical sentences which are translatable in syntactical sentences are called unessentially semantical sentences. The others are called essentially semantical sentences. There are also cases in which semantical concepts are eliminated by translating the sentence into a sentence of the object-language. For instance " 'Paris is a city' is true" can be translated into a sentence of the object language "Paris is a city." Carnap points out that he had given examples of both strategies in the *Logical Syntax*. He then restates Neurath's proposal as: can one always eliminate the semantical concepts? He answers negatively. Carnap explains that what Neurath calls the 'trick' of science is nothing else than the elimination of a non-essential semantical sentence in the syntactical language. However, he disagrees with Neurath when the latter proposes to translate "truth" by "sentence of the Encyclopedia" or "acknowledged." Carnap argues for the difference between 'true' and 'acknowledged' by remarking that in the case of "true" one does not need to give any temporal or pragmatic parameters, which are however necessary in the second case. This is the solution he had already proposed for distinguishing the two concepts in Paris 1935. Consider the sentence A: "the moon has in its dark side a crater which is even greater than the one it has in the visible side." While it can certainly be agreed that "A does not belong in 1937 to the Encyclopedia" (or "A is not scientifically acknowledged"), this is not the case for "A is not true" or its translation "the moon does not have in its dark side a crater which is even greater than the one it has in the visible side." Thus, Carnap concludes that "A does not belong in 1937 to the Encyclopedia" and "A is not true" do not have the same meaning; thus "true" and "(scientifically) acknowledged" (or "sentence of the Encyclopedia," "scientifically accepted", or "scientifically believed") are different concepts.

Moving on to point (3), Carnap expresses his belief that semantical concepts will turn out to be useful and important for epistemological work. As an example he gives a possible analysis of "x knows y" as "x believes y and y is true." Thus, Carnap concludes, "knows" is a semantical concept. A discussion of some examples with "knows" leads Carnap to observe that while in some cases the semantical notions can be eliminated by moving to sentences of the object language (as in the cases we have looked at before), this cannot be done when the sentence of the object language is not referred to by name. Examples would be: "Each sentence . . . " or "There exists a sentence . . . " Similar to semantical concepts are "seeing", "hearing", "perceiving". By contrast, Carnap adds, "believing", "thinking", "dreaming", "meaning", "imagining", are not semantical concepts.

In section (4) on truth and ordinary language, Carnap claims that he is not interested in the concept of truth of the metaphysicians but only that used in ordinary language. Setting aside the iteration of semantical concepts, which leads to antinomies, one can arrive at an unobjectionable concept of truth for ordinary language which has the same degree of clarity as other concepts used in natural language. The argument proceeds by comparing two sentences: B: "It is true that Goethe died in Weimar in 1832" and C: "Goethe died in Weimar in 1832." Carnap holds that the word "true" is used in ordinary language in such a way that B and C are accepted as synonymous:

> A proposition of the form [B] which contains the word 'true' is more rarely used than [C] namely only when it was preceded by questions, doubts or disputes or when, for some other reason, one wishes to express a stronger emotional emphasis . . . But that is just a psychological and not a logical difference. And this is shown by the fact that no one who would be asked to decide between two propositions such as [B] and [C] would accept the first but reject the other or even leave the latter undecided.[33]

Carnap concluded that since the two sentences are recognized by ordinary speakers as logically equivalent and that the semantical theory of truths also treats them as logically equivalent that there is thus agreement between the ordinary usage and the semantic conception.

The concluding section of the chapter gave some practical advice about how to proceed concerning the disagreements that were obviously present in the circle containing the notion of truth. On the side of the semantic concept of truth Carnap mentions the "Chicagoans" (Carnap, Hempel, Helmer), the Polish (Kotarbiński, Tarski, Lutman[-Kokoszyńska]), and in the opposite camp "Neurath and maybe Ness [sic] and others." Given that the debate cannot be immediately resolved Carnap expresses

---

[33] "Einen Satz von der Form A 1b [B], der das Wort 'wahr' enthält, wird man seltener verwenden als A 1c [C], nämlich gewöhnlich nur dann, wenn Frage, Zweifel oder Bestreitung vorausgegangen sind, oder wenn aus sonstigen Gründen eine stärkere emotionelle Betonung der Behauptung zum Ausdruck kommen soll. . . . Aber das ist nur ein psychologischer, kein logischer Unterschied. Das zeigt sich darin, das niemand in der Umgangssprache, dem zwei solche Sätze wie A 1b [B] und A 1c [C] zur Entscheidung vorgelegt werden, den einen akzeptieren wird, den andern aber ablehnen oder auch nur unentschieden lassen wird." (Carnap 1937, p.9, RC 080–52–01)

his conviction that the differences are due to a lack of clarity and misunderstandings that would disappear within a few years. As for the two groups, he gave the following suggestions:

1. Suggestions for the group of those who want to pursue semantics while their approach is empiristic and antimetaphysical. The latter will set up their terminology and formulations so that the delimitation of metaphysical problems always remains as clear as possible. They will do so not only with consideration for themselves but also for their readers. They will also keep in mind the question of the extent to which the semantical propositions are translatable into non-semantical ones; this especially in favor of those in our circles who, for whatever reason, strive to avoid semantical concepts.

2. Suggestions for the group of those who have reservations about the semantical concepts. They will at first temporize and they will not carry out public polemics against semantics as a whole until the further developments let transpire, first, whether or not the work in the domain of semantics is fruitful for science and especially for the general task we have set ourselves of an analysis of science and, second, whether or not the feared danger of slipping back into metaphysics is real. Therefore, they will not characterize semantical concepts as a whole as metaphysical but will only criticize single specific formulations that they might find objectionable especially if they indeed give rise to pseudo-problems.[34]

CODA

While the 1937 Paris Congress marks a culminating point in the debate on semantics among members of the Vienna Circle it was not the end of the story. Indeed, the conflict flared up anew with the publication of Carnap's "Introduction to semantics" (1942) which led to renewed expressions of skepticism and outright dismissal on the

---

[34] "1. Anregung für die Gruppe derer, die Semantik betreiben wollen und dabei empiristisch und antimetaphysisch eingestellt sind. Diese werden sich—nicht nur mit Rücksicht auf sich selbst, sondern vor allem auch auf ihre Leser—ihre Terminologie und Formulierungen möglichst so einrichten, dass die Abgrenzung gegen metaphysische Probleme immer deutlich bleibt. Sie werden ferner auch die Frage im Auge behalten, in welchem Ausmaß semantische Sätze in nicht-semantische übersetzbar sind; dies vor allem zugunsten derjenigen in unsern Kreisen, die—aus was für Gründen immer—sich bemühen, semantische Begriffe zu vermeiden.
2. Anregung für die Gruppe derer, die Bedenken gegen die semantischen Begriffe haben. Diese werden zunächst eine abwartende Haltung einnehmen und nicht in der Oeffentlichkeit schon gegen die Semantik im ganzen polemisieren, bis die weitere Entwicklung erstens erkennen läßt, ob die Arbeit auf dem Gebiet der Semantik für die Wissenschaft und besonders für unsere Gesamtaufgabe der Wissenschaftsanalyse fruchtbar ist oder nicht, und zweitens, ob die gefürchtete Gefahr des Abgleitens in die Metaphysik wirklich eintritt oder nicht. Sie werden also nicht die semantischen Begriffe im ganzen als metaphysisch bezeichenen, sondern nur die einzelnen etwa auftretenden ihnen bedenklich erscheinenden Formulierungen kritisieren, insbesondere, soweit sie etwa tatsächlich zu Scheinproblemen Anlaß geben." (Carnap, 1937, pp. 11–12, RC 080-52-01)

part of Neurath. For instance, on 22.12.42 Neurath wrote: "Of Tarski's metaphysics I do no longer say anything. It is trivial sad. Aristotle redivivus, nothing more." Eventually, Carnap became exasperated with Neurath:

As you can imagine, I am very sorry about the bad impression you got of my book, and that you even think it is a revival of Aristotelian metaphysics. I try to remember the many and sometimes long conversations we had in the past on Semantics. The first was in the train to Paris 1935. Then there was the public discussion at the Pre-Conference at Paris, with you and Ness [sic] on the one side, and Tarski and me on the other side. After these two discussions I remember I had the definite impression that there were no rational arguments left on your side. When Tarski and I showed that your arguments were based on misconceptions concerning the semantical concept of truth you had nothing to reply. What was left, as far as we saw it, were merely your emotional reactions, namely your dislike of the term "truth" and your vague fear that this would finally lead us back to old metaphysics. Later we sometimes had discussions on the same topic in America; but I did not have the impression that we came any step forward towards a mutual understanding, still less to an agreement... In any case, in spite of the disappointing experiences in the past, I am very willing to continue the discussion with you.

(Carnap to Neurath, May 11, 1943, Neurath Nachlaß, English in original)

Needless to say, there was no reconciliation on this issue and the discussion on whether semantics was loaded with metaphysics continued in the correspondence between Carnap and Martin Strauss in the early 1940s.

As we know, Tarski addressed many of the issues we have discussed in his 1944 chapter on truth. That chapter is well known and I need not enter into Tarski's reply to the criticisms that had been raised against semantics. Many of those criticisms go back to Neurath. In particular, section 14 ("Is the semantic conception of truth the "right" one?"), section 16 ("Redundancy of semantic terms—their possible elimination"), section 19 ("Alleged metaphysical elements in semantics"), section 20 ("Applicability of semantics to special empirical sciences"), and section 22 ("Applicability of semantics to the methodology of empirical science") of Tarski's 1944 article directly address, without mentioning him, issues that Neurath had been raising since 1935.

In conclusion, there were two parts to Neurath's criticism of semantics. On the one hand a background set of strongly held beliefs that led Neurath to his own proposal for using "truth" as "acknowledgement". On the other, the more specific criticisms to the semantic conception of truth that Neurath raised in consonance with those deeply held beliefs. The aim of this chapter has been to show how the criticisms to the semantic conception of truth emerge from those background beliefs and to spell out the discussion which emerged as a consequence with Tarski, Kokoszyńska, and Carnap.[35]

---

[35] This task can be seen as complementary to that carried out in Mormann 1999 which is not as detailed on the reconstruction of the debate but takes a broader view of the philosophical positions held by Neurath and traces his opposition to semantics to his anti-Cartesianism. However, I believe that that is only one of the sources of Neurath's objections. We have seen that his anti-Wittgensteinianism was a powerful factor.

One could now ask how coherent are those deeply held beliefs. More specifically, is Neurath's proposal a defensible one? Ideas going back to Neurath's position are often discussed and criticized in the epistemological literature on coherentism (see Pollock and Cruz, 1999, ch. 3, and Bonjour 1985) where however the discussion is on justification/confirmation rather than truth. Indeed, Bonjour 1985 defends a coherentist position of justification but a correspondentist account of truth. Hofman-Grüneberg 1988 for one endeavors to defend a conception of truth inspired by Neurath's position; in addition, there might be other "vindications" of Neurath that build on the rejection of philosophy of science as the methodology of truth attainment and emphasize the pragmatic component of how warrants are obtained and transmitted within scientific practice. Were any of these positions to mark an interesting and coherent approach to the problem of truth then we would have to recognize that behind the, at times frustratingly vague and unclear, objections by Neurath to semantics there was not just a negative drive but an idea which could be turned into a workable alternative.

Archival documents

FROM NEURATH'S ARCHIVE (HAARLEM, KONSTANZ)

Carnap–Neurath correspondence.
Hempel–Neurath correspondence.
Kokoszyńska–Neurath correspondence.
Morris–Neurath correspondence.
Naess–Neurath correspondence.
Neider–Neurath correspondence.
Stebbing–Neurath correspondence.
M. Strauss–Neurath correspondence.
Tarski–Neurath correspondence.

Neurath 1937*a*, Wahrheitsbegriff und Empirismus (Verbemerkungen zu einer Privatdiskussion mit Carnap im Kreis der Pariser Konferenz), Call number: K.30
Neurath 1937*b*, "Fuer Die Privatsitzung, 30 Juli 1937"; Call number: K.31
Neurath 1937*c*, "Diskussion Paris 1937 Neurath-Carnap"; Call number: K.32
Neurath 1937*d*, "Bemerkungen zur Privatdiskussion", Call number: K.33

FROM CARNAP'S ARCHIVE (PITTSBURGH, KONSTANZ)

Carnap, 1937, Ueber den semantischen Wahrheitsbegriff, RC 080–32–01
The correspondence between Carnap and Lutman-Kokoszyńska is classified under RC 088–57. It contains sixteen items.
The correspondence between Carnap and Neurath is found in different parts of the Nachlaß. Refer to the notes for specific call numbers of the correspondence mentioned in this chapter. While quoting from the Carnap–Neurath exchange if I use a source from the Carnap Nachlaß I give the call number beginning with RC. Otherwise, the item comes from the Neurath Nachlaß.
M. Strauss–Carnap correspondence, RC 102–74 and 102–75.
Schlick–Carnap correspondence. RC 029–27 and RC 029–28.

FROM POPPER'S ARCHIVE (HOOVER INSTITUTION, STANFORD)

Popper–Tarski correspondence. Box number 354. Folder ID: 8.

References

Ayer, A. (1977) *A Part of my Life*, Collins, London.
BonJour, L. (1985) *The Structure of Empirical Knowledge*, Harvard University Press, Cambridge.
Carnap, R. (1934) *Logische Syntax der Sprache*, Springer, Vienna. English translation: *Logical Syntax of Language*, Routledge and Kegan Paul, London, 1937.
—— (1936) Wahrheit und Bewährung, *Actes du Congrès International de Philosophie Scientifique*, vol. 4, Paris, Hermann, pp. 18–23.
—— (1937) Ueber den semantischen Wahrheitsbegriff (Carnap Nachlaß, RC 080–32–01).
—— (1942) *Introduction to Semantics*, University of Chicago Press, Chicago.
—— (1963) Intellectual autobiography, in P. A. Schilpp, ed., *The Philosophy of Rudolf Carnap*, (Library of Living Philosophers, vol. 11), Open Court, La Salle.
Cirera, R. (1994) *Carnap and the Vienna Circle*, Rodopi, Amsterdam.
Feferman, A., and S. Feferman (2004) *Alfred Tarski: Life and Logic*, Cambridge University Press, Cambridge.
Frank, P. (1997) *The Law of Causality and its Limits*, Kluwer, Dordrecht.
Frost-Arnold, G. (2004) Was Tarski's theory of truth motivated by physicalism?, *History and Philosophy of Logic*, 25, pp. 265–80.
Glock, H. J. (2006) Truth in the *Tractatus*, *Synthese*, 148, 345–68.
Grundmann, T. (1996) Can science be likened to a well-written fairy tale? A contemporary reply to Schlick's objection to Neurath's coherence theory, in E. Nemeth and F. Stadler, eds., *Encyclopedia and Utopia. The life and work of Otto Neurath (1882–1945)*, Kluwer, Dordrecht, 1996, pp. 127–33.
Haller, R. (1992) Alfred Tarski: Drei Briefe an Otto Neurath, *Grazer Philosophische Studien*, 43, 1–32.
Hegselmann, R. (1985), Die Korrespondenz zwischen Otto Neurath und Rudolf Carnap aus den Jahren 1934 bis 1945—Ein vorläufiger Bericht, in H.-J. Dahms, ed., *Philosophie, Wissenschaft, Aufklärung*, de Gruyter, Berlin, pp. 276–90.
Hempel, G. (1935) On the logical positivists' theory of truth, *Analysis* 2, 49–59; also in *Selected Philosophical Essays*, ed. R. Jeffrey, Cambridge University Press, Cambridge, 1999, pp. 21–5.
—— (1982) Schlick und Neurath: Fundieung vs. Kohärenz in der wissenschaftlichen Erkenntnis, *Grazer Philosophische Studien*, 16/7, 1–18. Translated in *Selected Philosophical Essays*, ed. R. Jeffrey, Cambridge University Press, Cambridge, 1999, pp. 181–98.
Hofman-Grüneberg, F. (1988) *Radikal-empirische Wahrheitstheorie. Eine Studie über Otto Neurath, den Wiener Kreis und das Wahrheitsproblem*, Verlag Hölder-Pichler-Tempsky, Vienna.
Kokoszyńska, M. (1936a) Syntax, semantik und Wissenschaftslogik, *Actes du Congrès International de philosophie scientifique*, vol. III, Paris, Hermann, pp. 9–14.
—— (1936b) Über den absoluten Wahrheitsbegriff und einige andere semantische Begriffe, *Erkenntnis*, VI, 143–65.
—— (1937/8), Bemerkungen über die Einheitswissenschaft, *Erkenntnis*, VII, 325–35.

Mancosu, P. (2005) Harvard 1940–1941: Tarski, Carnap and Quine on a finitist language for mathematics and science, *History and Philosophy of Logic*, 26, 327–57.

—— (2006) Tarski on models and logical consequence, in J. Gray, J. Ferreiros, eds. *The Architecture of Modern Mathematics*, Oxford University Press, pp. 209–37.

—— (2008) Tarski's engagement with philosophy, forthcoming in S. Lapointe *et al.*, eds, *The Golden Age of Polish Philosophy*, Springer.

Mormann, T. (1999) Neurath's opposition to Tarskian semantics, in J. Woleński and E. Köhler, eds., *Alfred Tarski and the Vienna Circle*, Kluwer, pp. 165–78.

Mulligan, K., Simons, P., and Smith, B. (1984) Truth-makers, *Philosophy and Phenomenological Research*, 44, 287–321.

Naess, A. (1936) *Erkenntnis und Wissenschaftliches Verhalten*, Jacob Dybwad, Oslo.

—— (1938) *"Truth" as conceived by those who are not professional philosophers*, Skrifter Videnskaps Akademi, Oslo: Hist. Fil.Kl, 1938, pp. 1–178.

Neurath, O. (1931*a*) Physikalismus, *Scientia*, 50, 417–21; translated in Neurath 1983, pp. 52–7.

—— (1931*b*) *Empirische Soziologie*, Vienna; translated as Neurath 1973.

—— (1932/3) Protokollsätze, *Erkenntnis*, 3, 204–14; translated in Neurath 1983, pp. 91–9.

—— (1934) Radikaler Physikalismus und 'wirkliche Welt', *Erkenntnis*, 4, 346–62; translated in Neurath 1983, pp. 100–14.

—— (1935) *Le Développement du Cercle de Vienne*, Hermann, Paris

—— (1936) Erster Internationaler Kongress für Einheit der Wissenschaft in Paris 1935, *Erkenntnis*, 377–430.

—— (1937*a*) Wahrheitsbegriff und Empirismus (Verbemerkungen zu einer Privatdiskussion mit Carnap im Kreis der Pariser Konferenz), Call number: K.30, Neurath Nachlaß.

—— (1937*b*) "Fuer Die Privatsitzung, 30 Juli 1937"; Call number: K.31, Neurath Nachlaß.

—— (1937*c*) "Diskussion Paris 1937 Neurath-Carnap"; Call number: K.32, Neurath Nachlaß.

—— (1937*d*) "Bemerkungen zur Privatdiskussion", Call number: K.33, Neurath Nachlaß.

—— (1973) *Empiricism and Sociology*, M. Neurath and R. S. Cohen eds., Reidel, Dordrecht.

—— (1981) *Gesammelte philosophische und methodologische Schriften*, Band I, II, edited by R. Haller and H. Rutte, Hölder-Pichler-Tempsky, Wien.

—— (1983) *Philosophical Papers 1913–1946*, R. S. Cohen and M. Neurath, eds., Reidel, Dordrecht.

Newman, A. (2002) *The Correspondence Theory of Truth*, Cambridge University Press, Cambridge.

Oberdan, T. (1993) *Protocols, Truth and Convention*, Rodopi, Amsterdam.

Pollock, J. L., and Cruz, J. (1999) *Contemporary Theories of Knowledge*, 2nd edn, Rowman and Littlefield, Lanham.

Poznanski, E. and A. Wundheiler (1934) Pojecie prawdy na terenie fizyki [The concept of truth in physics], in *Fragmenty Filozoficzne*, Warszawa, pp. 97–143. A partial German translation by Rose Rand is found in Carnap's Nachlaß under call number RC 081–37–01.

Russell, B. (1940) *An Enquiry into Meaning and Truth*, Allen and Unwin, London.

Rutte, H. (1991) Neurath contra Schlick. On the discussion of truth in the Vienna Circle, in T. E. Uebel, ed., *Rediscovering the Forgotten Vienna Circle*, Kluwer, Dordrecht, pp.169–74.

Schlick, M. (1934) Über das Fundament der Erkenntnis, *Erkenntnis*, 4, 79–99; translated in Schlick 1979, pp. 370–87.

Schlick, M. (1935) Facts and propositions, *Analysis*, 2, 65–70. Reprinted in Schlick 1979, pp. 400–4.
—— (1979) *Philosophical Papers*, vol. II, Reidel, Dordrecht.
Tarski. A. (1933) *Pojęcie prawdy w językach nauk dedukcyjnych* (The concept of truth in the languages of deductives sciences), Prace Towarzystwa Naukowego Warszawskiego, Wydział III, no. 34.
—— (1935) Der Wahreitsbegriff in den formalisierten Sprachen, *Studia Philosophica* (Lemberg), 1, 261–405. Reprinted in Tarski 1986, vol. II, pp. 51–198. English translation in Tarski 1956, pp. 152–278.
—— (1936) Grundlegung der wissenschaftlichen Semantik, in *Actes du Congrès International de Philosophie Scientifique*, vol. III, Paris, pp. 1–8. Reprinted in Tarski 1986, vol. II, pp. 259–268. English translation in Tarski 1956, pp. 401–8.
—— (1944) The semantic conception of truth and the foundations of semantics, *Philosophy and Phenomenological Research*, 4, 341–76. Reprinted in Tarski 1986, vol. II, pp. 661–99.
—— (1956) *Logic, Semantics, Metamathematics*, Oxford University Press, Oxford. 2nd edn 1983.
—— (1986) *Collected Papers*, S. Givant and R. McKenzie, eds., vols. I–IV, Birkhäuser, Basel.
Tugendhat, E. (1960) Tarskis semantische Definition der Wahrheit und ihre Stellung innerhalb der Geschichte des Wahrheitsproblems im logischen Positivismus, *Philosophische Rundschau*, 8, 131–59; now in G. Skirbekk, ed., *Wahrheitstheorien*, Suhrkamp, Frankfurt, 1977, pp. 189–223.
Uebel, T. (1992), *Overcoming Logical Positivism from within: the emergence of Neurath's Naturalism from the Vienna Circle's Protocol Debate*, Rodopi Amsterdam.
Woleński, J. (1989*a*) The Lvov-Warsaw School and the Vienna Circle, in K. Szaniawski, ed., *The Vienna Circle and the Lvov-Warsaw School*, Kluwer, Dordrecht, pp. 443–53.
—— (1989*b*) *Logic and Philosophy in the Lvov-Warsaw School*, Kluwer, Dordrecht.
—— (1993) Tarski as a philosopher, in F. Coniglione, R. Poli, and J. Woleński, eds., *Polish Scientific Philosophy*, Rodopi, Amsterdam, pp. 319–38.
—— (1995) On Tarski's background, in J. Hintikka, ed., *From Dedekind to Gödel*, Reidel, Dordrecht.
Woleński, J., and E. Köhler, (1998) *Alfred Tarski and the Vienna Circle*, Kluwer, Dordrecht.
Woleński, J., and P. Simons, (1989) "De Veritate: Austro-Polish contributions to the theory of truth from Brentano to Tarski" in K. Szaniawski, ed., *The Vienna Circle and the Lvov-Warsaw School*, Kluwer, Dordrecht, pp. 391–442.
Zygmunt, J. (2004) Bibliografia prac naukowych Marii Kokoszyńskiej-Lutmanowej, *Filozofia Nauki*, XII, 155–66.

9

Tarski's Nominalism

Greg Frost-Arnold

TEXTUAL SOURCES FOR TARSKI'S NOMINALISM

Alfred Tarski was very reticent to express his views on traditional topics in metaphysics and epistemology in print: scouring his entire corpus for any explicit pronouncements directly concerning such issues yields virtually nothing. Furthermore, his definitions of truth and of related semantic notions—likely the most fruitful Tarskian innovations for philosophers—are, in Tarski's own estimation, independent of the perennially vexed questions of philosophy: "we may accept the semantic conception of truth without giving up any epistemological attitude we may have had; we may remain naïve realists, critical realists, or idealists, empiricists, or metaphysicians—whatever we were before. The semantic conception is completely neutral towards all these issues" (1944, 362).

The sentiment in this quotation is not unusual; as Patrick Suppes notes, Tarski "was extraordinarily cautious and careful about giving any direct philosophical interpretation of his work" in semantics (1988, 81). Furthermore, Frits Staal, a friend and Berkeley colleague of Tarski's, recalls Tarski expressing "[m]ore than once . . . that he did not like to talk much about philosophy because he thought it was like giving an 'after dinner' speech—in other words, it was not rigorous" (Feferman and Feferman 2004, 318). If we take Staal's testimony at face value, it appears that Tarski not only believed his work in semantics to be independent of traditional philosophical concerns, but he also regarded such concerns with some skepticism.

Nonetheless, Tarski did hold definite views on certain perennial philosophical questions. He shared his stance with others in conversation and correspondence, although virtually never in publications. Much of the scattered, fragmentary evidence of Tarski's philosophical stance has been ably collected and analyzed by Jan Woleński (1993; 2002) and Artur Rojszczak (1999; 2002). One aspect of Tarski's philosophical thought that emerges clearly is that Tarski considers himself a nominalist and an

---

Over the last two years, Steve Awodey, Paolo Mancosu, and Laura Ruetsche all helped me grapple with the content of Carnap's dictation notes that form the basis of this chapter. More recently, Karen Frost-Arnold and Doug Patterson helped hammer my ideas about Tarski's nominalism into a more presentable form. Material from the Carnap Collection is quoted by permission of the University of Pittsburgh. All rights reserved.

anti-Platonist (cf. Mycielski 2004, Feferman 1999). An extremely valuable resource in this regard is (Mancosu, forthcoming).[1] An expression of his attitude can be found in an audio tape of a session during the 1965 Association for Symbolic Logic Meeting, in which Tarski describes himself as "an extreme anti-Platonist." He elaborates, with a joke:

> I represent this very [c]rude, naïve kind of anti-Platonism, one thing which I could describe as materialism, or nominalism with some materialistic taint, and it is very difficult for a man to live his whole life with this philosophical attitude, especially if he is a mathematician, especially if for some reason he has a hobby which is called set theory.
>
> (Quoted in Mancosu 2005, 341)

Tarski here makes explicit what other commentators have noted (Mostowski 1967, 81; Woleński 1993, 322): Tarski's ontological views are in tension with his set-theoretical practice, since a central component of modern Platonism about mathematics is the commitment to a literal or realistic construal of set theory, including at least the assertion that sets exist. But what position, exactly, is Tarski staking out with these labels of 'anti-Platonism,' 'materialism,' and 'nominalism'? Because each of these 'ism's has been around for centuries, and had various advocates, their meanings are not univocally determined.

Fortunately, recently uncovered documents in the Rudolf Carnap Collection shed substantial light upon the specific form of Tarski's nominalist, anti-Platonist position. During the 1940–1941 academic year, several eminent analytic philosophers converged upon Harvard: along with Tarski were Rudolf Carnap, Willard Van Orman Quine, Carl Hempel, Nelson Goodman, and, for the fall semester only, Bertrand Russell.[2] We know from Carnap and Quine's autobiographies that this group met repeatedly over the course of the school year, but their published reminiscings[3] are regrettably brief, so little was known about the details of what this group of philosophical heavyweights discussed. Fortunately, Carnap had the lifelong habit of taking dictation notes in Stolze-Schrey secretarial shorthand, and he kept a detailed record of the discussions at these meetings. Paolo Mancosu has recently presented an insightful overview of these notes, and my own view of these matters is indebted to his analysis (Mancosu 2005). A complete edition of Carnap's 1940–1941 notes, with an English translation, will appear shortly (Frost-Arnold, forthcoming).

This chapter aims to answer three related questions about Tarski's 'nominalism with a materialistic taint' through an examination of Carnap's dictation notes. First, what is Tarski's view? Second, what are the rationales or justifications for his view—and are they defensible? Third, how does Tarski attempt to reconcile his nominalist philosophical scruples with mathematics, since mathematics deals with paradigmatically abstract objects, such as numbers and sets, whose rejection is a standard *sine qua non* of current nominalism? This question becomes more pressing if, like Tarski (and

---

[1] Unfortunately, I did not see Mancosu's chapter until the present chapter was completed.

[2] In fact, a large number of leading figures across academic disciplines convened at Harvard that year, many fleeing the war in Europe; they created a dinner and discussion society, calling themselves the 'Science of Science' group (see (Hardcastle 2003) for details).

[3] See (Quine 1985, 149) and (Carnap 1963, 79).

his interlocutors Carnap and Quine), one takes modern natural sciences seriously, since they are suffused with mathematics. I conclude by considering the relationship between Tarski's nominalism and more recent forms of mathematical nominalism.

TARSKI'S CONDITIONS FOR INTELLIGIBILITY [*VERSTÄNDLICHKEIT*]

Carnap's discussion notes from 1940–1941 cover a wide range of topics. The participants were not dedicated to a single issue; rather, they discussed whatever captured their interest that day. Nonetheless, a plurality of Carnap's discussion notes during the spring semester deal with what he and his collaborators call—most briefly—'finitism.' However, this topic is *not* identical with the cluster of claims philosophers usually associate with the label 'finitism'—namely, the mathematical project associated with Hilbert and his school (specifically, investigating which inferences employed in classical mathematics can be re-cast into a finitistically acceptable form). Tarski, Carnap, and Quine were fully aware that Hilbertian finitism, at least in its original incarnation, was long dead as a research program by 1940, when their conversations began. The notion of finitism at issue in these conversations first appears when Tarski proposes rather strict requirements a language must meet in order for it to qualify (in Tarski's eyes) as *verständlich*, that is, understandable or intelligible. It is here that we find the nub of Tarski's anti-Platonism, his 'nominalism with a materialistic taint.'

Tarski's proposed conditions on intelligible languages vary somewhat from meeting to meeting. Carnap records the first version of it as follows.

January 10, 1941.

Tarski, Finitism. Remark in the logic group.

Tarski: I understand at bottom only a language that fulfills the following conditions:

1. Finite number of individuals.
2. Reistic (Kotarbiński): the individuals are physical things.
3. Non-Platonic: Only variables for individuals (things) occur, not for universals (classes etc.)

(RCC 090–16–28)[4]

Three weeks later, Tarski offers a similar, though not identical, characterization of a language he finds completely understandable. (I have placed my construal of the relation between the following characterization and the previous one in square brackets.)

Finitism.

Tarski: I properly understand only a finite language S_1:

---

[4] Tarski: Ich verstehe im Grunde nur eine Sprache die folgende Bedingungen erfüllt:
1. *Finite* Anzahl der Individuen.
2. *Reistisch* (Kotarbiński): Die Individuen sind physikalische Dinge.
3. *Nicht-platonisch*: Es kommen nur Variable für Individuen (Dinge) vor, nicht für Universalien (Klassen usw.)

only individual variables,	[identical to condition 3 above]
whose values are things,	[identical to 2 above]
whose number is not claimed to be infinite (but perhaps also not the opposite)	
	[modified version of 1 above]
Finitely many descriptive predicates.	[new condition]
	(RCC 090–16–25)[5]

Though terse, these conditions are fairly clear, and they flesh out Tarski's 1965 description of his position as 'nominalism with a materialistic taint.' Let us couch Tarski's proposed conditions for an intelligible language in the modern idiom of model theory. (Though Tarski is considered the father of modern model theory, many of its central concepts—including the central notion of 'truth in a structure'—had not taken their canonical form by 1941 (see Hodges 1986).) We begin with the model-theoretic conception of an interpreted language L, which is an ordered triple $<L, M, \rho>$. L carries the syntactic information about the language: what the symbols of the language are, the grammatical category to which each symbol belongs, which strings of symbols qualify as grammatical formulae and which not, etc. The semantic scheme ρ determines the truth-values of a compound expression formed using logical connectives, given the truth-values of its constituents. M is an interpretation or model that fixes signification of the nonlogical constants of L. Specifically, $M = <D, f>$, where D is a nonempty set, and f is an interpretation function which assigns members of D to singular terms, assigns sets of ordered n-tuples $\subseteq D^n$ to n-ary relation symbols, and a member of $D^n \to D$ to each n-ary function symbol. Now let us use this apparatus to rephrase Tarski's conditions. Tarski is describing a certain type of (interpreted) language L that has the following four characteristics, which I will henceforth refer to as Tarski's 'finitist-nominalist' (FN) conditions:

(FN 1) L is first-order.   *[Anti-Platonic requirement]*

In a fully understandable language, one cannot quantify over properties or relations, only over individuals. Note that in the original formulation of this requirement, Tarski says that 'variables occur only for individuals, not for universals (classes etc.).' The parenthetical end of that claim is surprising, for it is possible to treat classes as individuals, as is done in first-order set theory. So Tarski's first formulation actually outstrips what I have called (FN 1), since it would rule out first-order set theory, whereas (FN 1) as stated does not. However, this discrepancy is not of great importance for identifying the content of the concept of *Verständlichkeit* for Tarski, because the next condition (FN 2) will rule out 'classes etc.' That is, (FN 2) makes redundant any content of Tarski's earlier formulation of his first condition that goes beyond (FN 1). What is noteworthy is that Tarski considered admitting variables for classes, as well as variables for properties and relations, to constitute Platonism.

---

[5] Tarski: Ich verstehe richtig nur ein endliche Sprache S_1: nur Individuumsvariable, ihre Werte sind Dinge; für deren Anzahl wird nicht Unendlichkeit behauptet (aber vielleicht auch nicht das Gegenteil). Endlich viele deskr[iptive] primitiven Prädikate.

One might be tempted to interpret Tarski here as claiming that any string that contains (the symbolic correlate of) 'For all properties...' is not a grammatical formula of L; it is, after all, natural to think of *being first-order* as a grammatical property. However, Tarski appears *not* to think of his proposed conditions as grammatical restrictions. For immediately following the quotation above, Tarski explains that he is willing to construct proofs using sentences containing higher-order variables according to the rules of a proof calculus. However, he claims that he nonetheless does not truly *understand* these sentences, at least not in a full or unqualified sense. Specifically, Tarski says:

I only 'understand' any other language [i.e., a language that does not meet the conditions on intelligibility] in the way I 'understand' classical mathematics, namely, as a calculus; I know what I can derive from what ... With any higher, 'Platonic' statements in a discussion, I interpret them to myself as statements that a fixed sentence is derivable (or derived) from other sentences.

(RCC 090–16–25)

This quotation shows that (FN1) should be understood as a restriction on what is *intelligible* or *understandable*, not on what is grammatical. The notion of genuine 'understanding,' which a 'calculus'[6] alone cannot guarantee, will be discussed at some length at the end of this section.

(FN 2) D consists of 'physical objects' only.   [*Reistic requirement*]

In the elaboration and discussion of (FN 2) within the notes, numbers are specifically disallowed from the universe of discourse. Furthermore, because of (FN 1), not even Frege's reconstruction of numbers as properties of properties is allowed in a finitist-nominalist language. Taken together, (FN 1) and (FN 2), along with the standard treatment of existence in usual systems of symbolic logic, yield a modern formulation of nominalism.[7] The current canonical charaterization of nominalism is that only concrete things exist; abstract objects of any sort (properties, relations, numbers, propositions, sets, etc.) do not exist. We cannot make any first-order existence claims about abstract objects within a language L whose domain of discourse contains no abstract objects—except to say that such objects do not exist. And higher-order existence claims, according to (FN 1), are unintelligible at best, even if we allow them as grammatical strings of L. Tarski restricts the universe of discourse for a fully intelligible language to 'physical objects' only—but what, exactly, are the physical objects? The discussants do not show much interest in settling upon a specific interpretation.

---

[6] Terminological note: as Tarski uses the term in these notes, 'calculus' basically means a language characterized purely proof-theoretically, i.e. a language given in terms of a grammar (L above), rules of inference, and a (possibly empty) set of uninterpreted axioms.

[7] Another terminological note: in Tarski's intellectual circles circa 1940, 'nominalism' may paradigmatically refer only to the claim that *universals* do not exist, without making any claim about abstracta in general; see e.g. (Kotarbiński 1929, 430). Transposed into the context of formalized languages, such a nominalism would take the form of (FN 1) *only*; (FN 2) would not necessarily be included. This terminological difference with current nominalism could explain why Tarski describes his view as 'nominalism *with a materialistic taint*,' for (FN 1) covers the 'nominalism' part, and (FN 2) corresponds to the 'materialistic taint'.

Three options are proposed: (i) elementary physical particles (electrons etc.), (ii) mereological sums composed of elementary particles (or quanta of energy), so that the objects referred to by the names 'London' and 'Rudolf Carnap' will qualify as physical objects, and (iii) spatial and/or temporal intervals (this suggestion derives from the co-ordinate languages of Carnap's *Logical Syntax*) (RCC 090–16–23).

Tarski's third condition on intelligible languages appears in two distinct forms:

(FN 3a: *restrictive version*) D contains a finite number of members; [*finitism*]

or

(FN 3b: *liberal version*) No assumption is made about the cardinality of D.[8]

As we saw above, the restrictive policy is Tarski's initial proposal, but in later conversations, he clearly leans towards the liberal policy (090–16–04 and –05). In his autobiography, Carnap attributes the restrictive version to Quine and the liberal version to Tarski and himself (1963, 79). Presumably, this restriction is why the discussants call their project the construction of a 'finitist' language—and we see clearly how different Hilbert's finitism is from Tarski's.

The last restriction Tarski proposes for a finitist-nominalist language can be couched as follows:

(FN 4) *L* contains only finitely many descriptive predicates.

Tarski offers no justification for (FN 4), and it never plays an explicit role in subsequent discussions, so I will not dwell upon it further. Presumably, these four finitist-nominalist restrictions do not single out one language uniquely—several different (interpreted) languages could satisfy (FN 1–4). I have labeled the above four conditions 'finitist-nominalist (FN) conditions,' since the first two are nominalist, and the third and fourth finitist; let us call any language satisfying them a 'finitist-nominalist' language.

The FN conditions, as Tarski says, are intended to identify languages that are understandable, *verständlich*. But what does Tarski mean by the word *verständlich* and its cognates? Unfortunately, there is very little relevant information on this question in the 1941 notes. (What little there is will be discussed in the next paragraph.) It is frustrating that the discussion notes lack an account of what intelligibility is, since according to all parties involved, attaining it is the acknowledged central motivation for constructing a finitist-nominalist language. More specifically, the notes fail to explain why a language violating any of Tarski's finitist-nominalist criteria is not (fully) understandable. Why, for example, would a sentence beginning with the apparently understandable phrase 'There exists a property such that...' be unintelligible? The notes from the final day of collaborative work on

---

[8] In 090–16–04, however, Tarski excludes interpreted languages whose domain is empty:
Tarski: I would like to have a system of arithmetic that makes no assumptions about the quantity of numbers at hand, or assumes at most one number (0). Let A_n be the system of those sentences of customary arithmetic which are valid only if there are numbers <n; so A_0 has no numbers, A_1 has only 0, and so forth. Let A_ω be the entirety of customary infinite arithmetic. For the purpose of simplification, we want to exclude A_0, so we assume the existence of at least *one* number.

the finitist-nominalist language highlight how unclear and imprecise the concept of *Verständlichkeit* remains for the participants. Carnap writes:

> We agree the language should be as understandable as possible. But perhaps it is not clear what we properly mean by that. Should we perhaps ask children psychological questions, what the child learns first, or most easily?
>
> (RCC 090–16–05)

So Carnap himself does not know what is meant by *verständlich*, even six months into the project, and the dictation notes show no response to his query from the other participants.

Our interpretive prospects are not hopeless, for there is *some* material in the discussion notes that provides insight into what *verständlich* means for Tarski, Quine, and Carnap. In particular, Tarski clearly contrasts understandable languages with purely formal calculi, which enjoy only a 'second-class' type of intelligibility—if any at all.

> Tarski: I fundamentally understand only a language that fulfills the following conditions: [GF-A: Here are the three finitist-nominalist restrictions; see full quote above] ... I only "understand" any other language in the way I "understand" classical mathematics, namely, as a calculus; I know what I can derive from other [sentences] (rather, I have derived; "derivability" in general is already problematic). With any higher "platonic" statements in a discussion, I interpret them to myself as statements that a fixed sentence is derivable (or derived) from certain other sentences. (He actually believes the following: the assertion of a sentence is interpreted as signifying: this sentence holds in the fixed, presupposed system; and this means: it is derivable from certain basic assumptions.)
>
> (RCC 090–16–28)

That is, there is no genuine understanding of sentences couched in a language not meeting (FN 1–4), such as the language of 'classical mathematics.' The contrast between 'intelligible language' and 'uninterpreted calculus' also appears, albeit more briefly, elsewhere in the discussion notes.[9] Thus (to put the point in the terminology of Carnap's *Logical Syntax*), merely knowing the formation and transformation rules of a calculus does *not* constitute genuine understanding of the language corresponding to that calculus, i.e. the language that that calculus is intended to model formally.[10] That is, understanding what a sentence *s* means is *not* merely knowing which sentences are deducible from *s* and from which sentences *s* is deducible. This basic view has found modern expression in John Searle's Chinese Room thought experiment, which purports to show (*inter alia*) that a computer cannot understand a (natural) language, because a computer's operations are restricted to the realm of syntax (Searle 1980). Regardless of what the detailed content of *Verständlichkeit* might be for Tarski and Carnap, it at least requires that a language be more than an uninterpreted calculus or empty formalism. And uninterpreted calculi and empty formalisms are

---

[9] *Tarski*: For the metalanguage M we naturally use a richer language than S_1 [GF-A: S_1 meets the finitist-nominalist conditions] ... But these semantics in M cannot be considered as providing true understanding, rather only as a calculus with finite rules. (RCC 090–15–25; cf. –04).

[10] Formation rules determine the well-formed formulae or grammatical strings of a calculus; transformation rules are commonly called 'rules of inference.'

usually contrasted with *meaningful* or *interpreted* languages. After all, one cannot understand a sequence of spoken or written characters unless that sequence is meaningful.

It appears we can take Tarski to hold that meaningfulness is a necessary condition for intelligibility. A further piece of evidence corroborating this view comes from the German translation of Tarski's monograph on truth; although Tarski did not translate the entire essay himself from the original Polish, he did assist with the translation (Feferman and Feferman 2004, 99), so the language found in the German version is either approved by Tarski, or his own. The concept of *Verständlichkeit* appears in *Wahrheitsbegriff* in precisely the connection discussed above: the contrast between an uninterpreted or purely formal calculus and an intelligible language.

> It remains perhaps to add that we are not interested here in 'formal' languages and sciences in one special sense of the word 'formal,' namely sciences to the signs of which no meaning [*inhaltlicher Sinn*] is attached. ... We shall always ascribe quite concrete and, for us, intelligible meanings [*verständliche Bedeutungen*] to the signs which in occur in the languages we shall consider.
>
> ([1935] 1983, 167)

Besides the sharp opposition between merely formal and intelligible, meaningful languages found here, certain other aspects of this quotation merit attention. First, Tarski requires that the meanings of constants in the language be 'concrete' [*ganz konkrete*]—and the current characterization of nominalism is the claim that only concrete things exist. Second, there is an apparent discrepancy between Tarski and Carnap insofar as Carnap thinks of *languages* (and expressions within a language) as intelligible or not, whereas the above passage treats *meanings* [*Bedeutungen*] as what is or is not intelligible. In any case, we should perhaps avoid placing too much weight upon word and phrase choice, for although Tarski was a meticulous writer, he was not the primary translator for *Wahrheitsbegriff*, so the choices made in the translation may not always reflect his views with complete faithfulness and precision.

With this admittedly partial characterization of *Verständlichkeit* in place, the next question is: why and how does requiring a language to meet the four finitist-nominalist restrictions guarantee the intelligibility (and thus the meaningfulness) of sentences in that language? Unfortunately, in Carnap's notes, that question is neither asked nor answered directly. However, there is some material in the discussion notes that addresses the broader but related issue of what motivates these restrictions. By examining the rationales supplied for imposing these restrictions, we can see more clearly what benefits Tarski believed conditions (FN 1–4) would bring, and we thereby indirectly gain a better understanding of *Verständlichkeit*, and the project as a whole.

RATIONALES FOR THE FINITIST-NOMINALIST CONDITIONS

The official year of birth for modern Anglophone nominalism is generally taken to be 1947, with Goodman and Quine's *Journal of Symbolic Logic* article "Steps Toward a Constructive Nominalism." In a footnote in that article, the authors acknowledge that the initial impetus and strategy for their nominalist project was proposed in 1940 by Tarski, and discussed with him, Carnap, and the authors (1947, 112). Thus, the discussions at Harvard in 1940–1941 can be seen as an important wellspring of current nominalism. In the previous section, Tarski's finitist-nominalist criteria (FN1–4) were outlined and examined in some detail. A pivotal question that was not addressed then, but will be in this section, is the following: *What justifies these finitist-nominalist criteria?* This section discusses two rationales presented by Tarski, Carnap, and Quine for undertaking the finitist-nominalist project: their hostility towards Platonic metaphysics, and what I will call 'the argument from natural science.'

Overcoming Platonic metaphysics

One rationale for pursuing the finitist-nominalist project that appears—both implicitly and explicitly—in the Harvard discussions is the desire to purge metaphysical elements from discourse intended to be cognitively significant, and from language used for scientific purposes in particular. It is well known that the logical empiricists and their allies (e.g. Russell and Wittgenstein) hold a very negative view of metaphysics. The group of Polish philosophers from which Tarski came, the Lvov-Warsaw School, also tended to share this anti-metaphysical animus (Simons 1993, Woleński 1993), though as a group, they were neither as fervently nor as unanimously anti-metaphysical as their Viennese contemporaries.

The anti-metaphysical drive is closely connected to the notion of *Verständlichkeit* discussed in the previous section. One characterization of metaphysics that is widespread among the logical empiricists and their intellectual kin is the following: if a string of symbolic marks *x* is metaphysical, then *x* is *meaningless*.[11] (The converse does not hold: the string ')bPQ))', which is meaningless in standard formalizations of predicate logic, is not metaphysical.) And presumably, if a given word or sentence is meaningless, then it is not understandable or intelligible. The connection to the finitist-nominalist project is clear: by *modus tollens*, if every word and every sentence in a given language is 'fully understandable,' then there are no metaphysical words or sentences in that language. This argument is never explicitly articulated in the discussion notes; in particular, we never find the conditional claim 'if a string of symbols is meaningless, then that string is not understandable.' Nonetheless, given that this conditional seems patently true (how could one understand meaningless nonsense?), it seems reasonable to connect Carnap, Tarski, and Quine's discussions of intelligibility to their shared aversion to metaphysics *qua* meaningless utterances and inscriptions. And if, as just suggested, intelligibility entails meaningfulness, then the construction of an intelligible language will yield a language free of metaphysics. In short, given the unintelligibility of meaningless discourse, a

---

[11] Precisely this characterization is found in Carnap's "Overcoming Metaphysics through the Logical Analysis of Language," but the same idea is also found in the *Tractatus*, as well as in many of Schlick's and Neurath's writings.

fully intelligible language would also be a language free from metaphysical impurities—and such a connection was perhaps at least implicit in the minds of the Harvard discussants.

But the attacks on metaphysics in Carnap's discussion notes are not merely implicit. There are explicit references to objectionable metaphysical theses as well. Tarski and Quine hold that the adoption of (FN1) and (FN2) would prevent a pernicious slide into a certain kind of metaphysics, which they call 'Platonism,' naming the position after the grandfather of all metaphysicians. Recall that Tarski labels (FN1) (the requirement that variables only range over the individual domain, not over properties or relations) the 'non-Platonic' requirement in the first articulation of his proposal (RCC 090–16–28). The participants do not offer a detailed or precise characterization of Platonism, but it involves at least higher-order logic, and probably (full) set theory. ((FN1) rules out higher-order logic, and adding (FN2) to it rules out (first-order) set theory.) For example, in a discussion of "general set theory," Tarski asserts: "With the higher levels, Platonism begins" (RCC 090–16–09). (What Tarski means by 'the higher levels' is not unequivocally fixed by the context; he may mean only transfinite sets, or he may have in mind anything higher than the ur-elements.) And even earlier, in a discussion with Russell and Carnap, Tarski asserts: "A *Platonism* underlies the higher functional calculus (and so the use of predicate variables, especially higher levels)" (RCC 102–63–09). So, in these records, Tarski associates Platonism most closely with second- and higher-order logic, though set theory (presumably even in its first-order variety) is also caught up in the Platonic viewpoint.

We find a similar conception of Platonism in Quine's contemporaneous writings. In a December 1940 lecture at Harvard (and thus before Tarski introduces (FN1–4)), Quine distinguishes mathematics from logic as follows: " 'logic' = theory of joint denial and quantification," while " 'mathematics' = (Logic+) theory of ∈." Quine then claims that "*mathematics is Platonic, logic is not*" (RCC 102–63–04). Why should the set-membership relation introduce Platonic commitments? Quine explains that "there are no logical predicates," but '∈' is a predicate. He then claims:

Predicates first bring *ontological* claims (not because they designate, for they are syncategorematic here, since variables never occur for them; rather): because a predicate takes certain objects as values for the argument variable; so *e.g.* '∈' *demands classes*, universals; thus *mathematics is Platonic, logic is not.*" (RCC 102-63-04).

That is, if there are any true statements of the form '$P \in Q$,' then there must be at least one class (provided '∈' is given the intended interpretation). For Quine, accepting the existence of at least one class is tantamount to accepting Platonism. This position is related to, but stronger than, the one he published a year earlier in "Designation and Existence," for according to the position Quine outlines there, a nominalist could hold '$P \in Q$' to be true, *provided* the nominalist does not *ineliminably quantify* over the Q-position. As a historical matter, Quine harbored suspicions of set theory even before Tarski proposes constructing a finitist-nominalist language, so Quine was presumably a receptive audience for Tarski's proposal. In short, first-order set theory is Platonic, along this line of thinking, because it forces us to admit the (ineliminable) existence of classes as values of variables.

So why do Tarski and Quine also suspect higher-order logic of being metaphysics —even when the domain of discourse consists solely of (concrete) individuals? In a letter to Carnap dated May 1943, reflecting on the Harvard discussions, Quine writes:

> I argued, supported by Tarski, that there remains a kernel of technical meaning in the old controversy about [the] reality or irreality of universals, and that in this respect we find ourselves on the side of the Platonists insofar as we hold to the full non-finitistic logic. Such an orientation seems unsatisfactory as an end-point in philosophical analysis, given the hard-headed, anti-mystical temper which all of us share.
>
> (Creath 1990, 295)

Presumably, the 'kernel of technical meaning in the old controversy' is composed of two decisions: (i) whether (contra (FN 1)) to adopt a higher-order logic, and (ii) whether (contra (FN 2)) to allow non-physical individuals into the domain of quantification (as discussed in the previous paragraph). For Quine, by this point in his career, a commitment to higher-order logics brings in its wake a commitment to the 'reality of universals.' Why? In his "Designation and Existence," published a year before the Cambridge discussions, Quine articulates his famous dictum "[t]o be is to be the value of a variable" (1939, 708). In the same article, he uses this dictum to characterize nominalism within the framework of modern logic: a language is nominalist if its variables do not range over any abstract objects.[12] And properties and relations, which are quantified over in second-order logics, are (for Quine and many others) paradigmatically abstract entities.[13] In short, a language is metaphysical if it quantifies over abstract entities; in first-order set theory, those abstracta are sets, and in higher-order logic, those abstracta are properties and relations.

What is missing from both the discussion notes of 1940–1941, as well as from published writings before and after, is an explanation of *why* admitting abstracta (whether they be sets, relations, or anything else) as values of variables is philosophically objectionable. That is, why does allowing the existence of abstracta violate 'the hard-headed, anti-mystical temper' shared by all the Harvard discussants? This becomes more troubling when we note that classifying higher-order logic and/or set theory as metaphysics does not mesh well with the characterization of metaphysics offered elsewhere by the logical empiricists and their allies. Both the explicatum and the explanandum of the term 'metaphysics' vary over time and between different thinkers. Nonetheless, most logical empiricists, most of the time, strongly resist classifying logic and mathematics as metaphysical. (It is a sign of this that special exceptions are made in their accounts of meaning and knowledge to account for logic and mathematics. For example, Wittgenstein's

---

[12] Quine writes: "In realistic languages, variables admit abstract entities as values; in nominalistic languages they do not" (1939, 708).

[13] Quine takes no position on where or how to draw a sharp line between concrete and abstract entities (1939, 708).

distinction in the *Tractatus* between pseudo-propositions that are nonsense [*unsinning*] and those that are senseless [*sinnlos*] places logic and mathematics in a separate category from metaphysics, even though both 'say nothing about the world.') So not only do Tarski and Quine omit an explanation of why classes and relations are metaphysical, but such a view appears to clash with the view of metaphysics presented by many of their philosophical peers.[14]

Returning to Tarski and Quine's view of higher-order logic and set theory, it may very well be that there is no further justification for their animus. In Goodman and Quine's published chapter on nominalism, the authors admit that their "refusal [to countenance abstracta] is based on a philosophical intuition that cannot be justified by appeal to anything more ultimate" (1947, 105). Of course, we should not assume Quine and Goodman speak for Tarski as well; however, if Tarski did attempt to 'justify' his rejection of abstracta 'by appeal to anything more ultimate,' that justification is not recorded in the Harvard discussion notes, and it did not impress Goodman and Quine enough to include it in their article. We today could impute to Tarski (and/or the other participants) some further justification, but such an attribution would be conjectural, given the evidence available in Carnap's dictation notes.

Second rationale for the FN restrictions: natural science

Another type of justification Tarski and Quine offer for pursuing the finitist-nominalist project can be called the 'argument from natural science.' The previous rationale supported (FN 1–2) (viz. the language is first-order and its domain contains only physical objects), which we could consider support for *nominalism*; the following, however, is only a justification for *finitism* (FN 3). Tarski begins his defense of (FN 3) with a reasonable assertion: the number of individuals in our world "is perhaps in fact finite" (090–16–25). If the universe does contain only finitely many physical things, and if (FN 1–2) hold, then it immediately follows that D has finitely many members—and this is the restrictive version of (FN 3). If we wish rather to leave open the question of whether the number of physical things is finite or not, and we accept (FN 1–2), then the liberal version of (FN 3) follows. Note that without (FN 1–2) or something similar, (FN 3) becomes much more contentious. As explained previously, (FN 1–2) prevent the two most common ways of introducing mathematical objects into a language, and mathematical infinities are usually considered paradigmatic examples of infinite totalities.

---

[14] However, Russell—who was at Harvard for the Fall 1940 semester, and participated in some discussions—is an exception. He claims that "so long as the cardinal number is inferred from the collections, not constructed in terms of them, its existence must remain in doubt, unless in virtue of a metaphysical postulate *ad hoc*" (1918, 156). The fact that the greatest philosophical luminary of Carnap, Tarski, and Quine's early careers called classes 'fictions' and declared the assumption that numbers exist *ad hoc* metaphysics could have played some indirect role in inclining Tarski, Quine, and others to consider refusing numbers into the universe of discourse *prima facie* plausible or reasonable.

Carnap replies to Tarski's claim by suggesting that there *are* infinities. These come in two varieties: logico-mathematical and physical. The usual mathematical infinities will directly violate the nominalist criteria (FN 1–2). As examples of empirical, physical infinities that will not fall afoul of (FN 1–2), Carnap suggests space and, with more conviction, time. He claims:

> even if the number of subatomic particles is finite, nonetheless the number of events can be assumed to be infinite (not just the number of time-points... but the number of time-points a unit distance away from each other, in other words: infinite length of time.)
>
> (RCC 090–16–24)

Carnap's suggestion to use events or spatiotemporal intervals instead of physical objects for the domain of a language of science obviously violates the letter of (FN 2), but Carnap likely believes it does not violate its spirit—for spatiotemporal events are still part of the natural, physical world, unlike numbers and their kin. (However, Kotarbiński, whom Tarski invokes when he proposes (FN 2), explicitly denies that *events* are acceptable for the reist ([1929] 1966, 432).) So, Carnap is suggesting, if we expand (FN2) to allow the domain to contain not just physical *objects* but rather any entity that is (broadly speaking) part of the physical world, then (FN 3) does not force itself upon us—provided there are an infinite number of events.

Tarski responds to Carnap's challenge in two related ways. The first engages Carnap on his own terms; the second suggests that Carnap's critique has missed part of Tarski's motivation for introducing (FN 3), at least in its mature, liberal version. First, Tarski replies directly to Carnap's suggestion that space and time will provide us with infinities, even if there are only a finite number of physical objects in the universe. Tarski asserts that space and time, contrary to appearances, may actually be finite: "perhaps quantum theory will give up continuity and density" for both space and time, by quantizing both quantities (090–16–23). Furthermore, Tarski says, time and space could both be circular, in which case there would not be an infinite number of finite spatial or temporal intervals. In short, Tarski claims that developments in quantum and relativistic physics may in fact show that space and time are actually finite.

Second, Tarski suggests that arguing that there is actually an infinite quantity somewhere in nature misunderstands the rationale behind the liberal version of (FN 3). Presumably (though Tarski does not state this fully explicitly), we should not assume the number of physical things in the world is infinite, because this is an *empirical* matter. What Tarski does say is the following: "we want to build the structure of the language so that this possibility [viz. that the number of physical things is finite] is not excluded from the beginning" (090–16–23). The basic idea is simple: the form of the language we use to describe the empirical world should not prejudge the number of entities in the universe, and Tarski's scheme happily leaves this question open. Put otherwise, 'How many spatial or temporal intervals are there?' is just as empirical a question as 'How many subatomic particles are there?' If one accepts (FN 2), and if one also wishes to incorporate at least full (first-order) arithmetic into one's language, then one would be committed to an infinite number of physical objects. To couch Tarski's worry in Carnapian terms: how many entities there are in the universe—as well as the topological structure of (actual) space and time—are *synthetic* matters, and by adopting (FN 3), we prevent them from becoming analytic ones.[15] That is, answers to questions about the number of things in the universe or the topology of space and time should be determined by the structure of the world, not by the structure of the language used to describe that world.

A FINITIST-NOMINALIST LANGUAGE OF SCIENCE AND MATHEMATICS

The demand for a 'total language of science'

In further formulations of the group's project, an additional condition is placed on the language(s) they are attempting to construct. They wish to formulate a language that simultaneously meets Tarski's conditions on intelligibility, and is also rich enough to use for investigating the logic of science, including metalinguistic investigations of classical analysis and set theory (they sometimes call such a language a "nucleus language").

Jan. 31, 1941

Conversation with Tarski and Quine on Finitism

. . .

We together: So now a problem: What part S of M [the metalanguage of science and mathematics] can we take as a kind of nucleus, so that 1.) S is understood in a definite sense by us, and 2.) S suffices for the formulation of the syntax of all of M, so far as is necessary for science, in order to handle the syntax and semantics of the complete language of science.

(090–16–25)

Similar sentiments are expressed a few months later, as the discussants' year together drew to a close.

June 18, 1941

Final conversation about the nucleus-language, with Tarski, Quine, Goodman, and Hempel; *June 6 '41*

Summary of what was said previously: The nucleus language should serve as the syntax-language for the construction of the complete language of science (including classical mathematics, physics, etc.). The language of science thereby receives a piecewise interpretation, since the n. l. [nucleus language] is assumed to be understandable.

(090–16–05)

---

[15] Carnap's conception of analytic truth during this time is as follows: a sentence s is an analytic in language L if and only if the truth-value of s is fixed merely by specifying L. The 'specification' of L involves a list of semantic rules, such as: 'Socrates' designates (in English) Socrates.

On the one hand, the finitist-nominalist conditions place restrictions on an interpreted language's richness; this condition, on the other, restricts a language's poverty.

Carnap, Tarski, and Quine realize it may not be possible to construct a language that simultaneously satisfies this criterion as well as (FN 1–4). For immediately following the first of the two quotations above, we find:

1. It must be investigated, if and how far the *poor nucleus* (i.e. the finite language S_1) is sufficient here. If it is, then that would certainly be the happiest solution. If it is not, then two paths must be investigated:

2a. How can we justify the *rich nucleus* (i.e., infinite arithmetic S_2)? I.e., in what sense can we perhaps say that we really understand it? If we do, then we can certainly set up the rules of the calculus M with it.

2b. If S_1 does not suffice to reach classical mathematics, couldn't one perhaps nevertheless adopt S_1 and perhaps show that classical mathematics is not really necessary for the application of science in life? Perhaps we can set up, on the basis of S_1, a calculus for a fragment of mathematics that suffices for all practical purposes (i.e., not something just for everyday purposes, but also for the most complicated tasks of technology).

(090–16–25)

In short, they suspect that a metalinguistic analysis of classical mathematics and physics may require a richer language than what the finitist-nominalist criteria allow. Further, if that suspicion is borne out, then either this richer language must be shown to be understandable, or the weaker mathematics sanctioned in finitist-nominalist languages must be shown to be sufficient to deal with all sophisticated practical applications.[16] Unfortunately, we are not informed whether this new condition trumps the finitist-nominalist conditions or not. That is, the notes do not contain Tarski's answer to the following question: if the mathematics recoverable in languages meeting (FN 1–4) is insufficient for practical purposes, then what should be discarded—the demand for a single, intelligible metalanguage of science, or the finitist-nominalist strictures on intelligibility? Thus it is difficult for us to ascertain the relative importance the participants attach to these competing conditions.

However, none of the participants assert that we should abandon the investigations of those portions of (e.g.) set theory that fail to meet the finitist-nominalist criteria. Set theory can still progress, even if parts of it are not fully intelligible—Tarski suggests that set theory then becomes a purely formal (i.e. uninterpreted) calculus, indicating which sentences can be derived from others, as we saw above (RCC 090–16–28). But that is not a barrier on proving theorems within ZFC. Thus the Tarskian nominalist would presumably allow mathematical practice to continue unabated and unimpeded, while simultaneously maintaining that some or all

---

[16] This last alternative was not usual at the time. Wittgenstein writes: "In life a mathematical proposition is never what we want. Rather, we use mathematical propositions *only* in order to infer sentences which do not belong to mathematics from others, which likewise do not belong to mathematics"(*Tractatus* 6.211).

of the claims forwarded in the course of that practice are not, strictly speaking, intelligible. In sum, Carnap, Tarski, and Quine (and occasionally Goodman and Hempel) are attempting to construct a formal language that simultaneously meets the stringent finitist-nominalist constraints (FN 1–4) and is rich enough to serve as a metalanguage for all of scientific discourse, including (at least elementary) mathematics. Since these two conditions pull in opposite directions, this is a difficult goal to reach, as current nominalists can attest. Let us now turn to the 1941 group's attempt to reconstruct arithmetic in a language meeting Tarski's conditions on intelligibility.

Carnap, Tarski, and Quine realize from the outset that one of the most pressing and difficult obstacles facing any attempt to construct a finitist-nominalist language for science will be the treatment of mathematics. Can a language simultaneously meet Tarski's criteria for intelligibility and contain (at least a substantial portion of) the claims of classical mathematics? A sizeable portion of Carnap's notes from the Harvard meetings deals with attempts to answer this question. The discussants focus on the simplest case, *viz.* classical arithmetic. A number of potential pitfalls present themselves: first, what is the content of sentences containing numerals? Can we assert anything with them at all, given that the only elements in our domain of discourse are physical objects? Second, how should we handle numerals that purportedly refer to numbers that are larger than the number of concrete things in the universe? That is, suppose there are exactly one trillion physical things in the universe; what should we then make of the numeral '1,000,000,000,001' and sentences containing it? Finally, what theorems and proofs of classical arithmetic are lost (or possibly lost) in a language that meets the finitist-nominalist conditions? I shall deal with each of these questions in turn.

Number

As seen previously, one of Tarski's requirements for a language to be understandable is that abstract entities are not allowed to serve as denotata of names. So in such a language, the numeral '7' cannot name a natural number, considered as a basic individual object—for the natural numbers (understood as individual objects) are excluded from the domain of discourse. And as mentioned above, since a FN language must also be first-order, Frege's reconstruction of numbers as second-order properties is forbidden as well. But Tarski, Carnap, and Quine want this language to include, at the very least, substantial portions of arithmetic. So how do they reinterpret numerals under this linguistic regime?

Tarski's strategy for introducing ordinal numbers[17] is the following: "Numbers can be used in a finite realm, in that we think of the ordered things, and by the numerals we understand the corresponding things" (090–16–25). Virtually the

---

[17] The group discusses cardinal number *very* briefly in (RCC 090–16–25): "One can also ascribe a cardinal number to a class. Quine: E.g. by the introduction of '($\exists 3x$) . . .' as an abbreviation for '($\exists x$)($\exists y$)($\exists z$)($\sim$ .. = ..&)." I presume that what Quine intends is the following: $\exists 3x \phi x \equiv \exists xyz(x \neq y \land y \neq z \land z \neq x \land \phi x \land \phi y \land \phi z)$.

same[18] proposal is outlined in the part of Carnap's autobiography that deals briefly with the year at Harvard:

> To reconcile arithmetic with the nominalistic requirement, we considered among others the method of representing natural numbers by the observable objects themselves, which were supposed to be ordered in a sequence; thus no abstract entities would be involved.
>
> (1963, 79)

Let us illustrate this idea with a concrete example. Suppose, in our domain of physical things that have been 'ordered in a sequence,' Tom is the eighth thing, John is the fourth, and Harry the eleventh. (Assume the numeral '0' is assigned to the first thing.) Then the arithmetical assertion '7 + 3 = 10' is (put roughly) re-interpreted as 'Tom + John = Harry.' Put model-theoretically, the interpretation function f of an interpreted language meeting the finitist-nominalist requirements assigns to numerals of L objects in D: $f('7') = $ Tom, $f('3') = $ John, and $f('10') = $ Harry. (Arithmetical signs such as '+' are defined via the axiomatization of Peano Arithmetic in Carnap's *Logical Syntax*, §14 and §20.)

This heterodox view of ordinal numbers raises serious questions. First, from whence does the sequential order of the physical objects spring? The ordering is imposed, it appears, by stipulation; Tarski says: "we want the (perhaps finitely many) things of the world ordered in some arbitrary way" (090–16–23). We may assign *any* member of the domain of physical things to the numeral '0,' and we may choose any other member of the domain to be its successor, to be assigned the numeral '1.' The sentence '0 + 1 = 0' will come out *false* under any such stipulation, regardless of which physical objects we choose to 'stand in' for the numbers 0 and 1 (assuming the cardinality of D is greater than one). The relation *is a successor of* need not reflect anything 'in the order of things,' spatial, temporal, or otherwise. There is a second worry about this proposal to reinterpret our usual number-language under a finitist-nominalist regime. Let us suppose that the sentence 'Tom has brown hair' is true. Then, since the name 'Tom' and the numeral '7' both name the same object (model-theoretically, the interpretation function assigns both individual constants the same value), it appears that the sentence '7 has brown hair' will be true. But numbers cannot be brunettes. So this finitist-nominalist interpretation of numerals will make true many assertions about numbers that, intuitively, we do not want to come out true. Tarski, Quine, and Carnap do not even consider this problem (at least, there is no record of it in Carnap's discussion notes).

Technical refinements could perhaps avoid at least some of these unwanted truths. One such refinement is suggested in (Field 2001, 214–15). His basic idea, couched in our terms, is the following. Note that the ordering of physical objects of D should be viewed as arbitrary, and that alternative orderings of the elements of D would still respect the truths of classical arithmetic captured in the original model. We can finesse this fact to eliminate certain unwanted truths: while '7' may be assigned to a

---

[18] The only significant difference is that Carnap claims that the things are "observable." This is most likely a 'mis-remembering' by Carnap, not part of the original proposal.

brunette under one assignment of physical objects to numerals, it will be assigned to a blonde under another, and to various hairless physical objects under other assignments. However, under *all* these assignments, '7 + 3 = 10' is true. This suggests the following refinement to Carnap, Tarski, and Quine's proposal to re-interpret numerals in a finitist-nominalist language: a (mathematical) sentence ϕ is true (in L) if and only if ϕ is true for all assignments of physical-object-values to numerals (satisfying certain conditions: for example, we do not want to include the assignment in which all numerals are assigned to a single object in D). This line of thought shows that perhaps there is a way to interpret '7' that meets (FN 1–4) and certifies substantial portions of arithmetic as true, without committing us to the truth of sentences such as '7 has brown hair.' Of course, if the number of objects in the physical universe is finite, then this proposal will not capture all of standard arithmetic.

Interpreting numerals that are 'too large'

Now we come to a problem concerning mathematics in finitist-nominalist languages that Tarski, Quine, and Carnap *did* recognize and address themselves. Suppose there are only k items in the universe. Carnap poses the question: "How should we interpret [*deuten*]" the number expressions $k + 1, k + 2, \ldots$, "for which there is no further thing there?" (090–16–06). Initially, the group considers three options (employing the usual notation that x' is the successor of x):

(a) $k' = k'' = \ldots = k$,
(b) $k' = k'' = \ldots = 0$,
(c) $k' = 0, k'' = 0', \ldots$

In each of these three cases, at least one of the Peano axioms is violated. If (a) is adopted, then there exist two numbers (i.e. there exist two *physical objects*) that have the same successor; if (b) or (c) is adopted, then the object assigned to '0' will be a successor. None of these options is especially palatable, since none captures the truths of classical arithmetic substantially better than the others. A surprisingly large portion of the discussion notes is devoted to working through proposed solutions to this problem. Strategies other than (a)–(c) are also considered: Carnap suggests identifying numbers with *sequences* of objects instead of objects, so that there is no 'last element' forced upon us; however, Quine and Tarski consider sequences (at least in the form needed to recover full classical arithmetic) just as problematic as sets, numbers, and other abstracta.

There is another problem reconstructing mathematics under a finitist-nominalist regime: as Tarski notes, under these conceptions of number "many propositions of arithmetic cannot be proved in this language, since we do not know how many numbers there are" (RCC 090–16–25). That is, suppose that we do not know how many physical objects there are in the material universe; this ignorance will be reflected formally in a refusal to allow any assumptions about the cardinality of the domain of L. Then, there will be arithmetical assertions we can prove under classical arithmetic, but are unprovable in a finitist-nominalist language. If we allow ourselves *no* assumption about the cardinality of the domain (or just the assumption that at least one object exists, as Tarski suggests), we cannot even prove that '$1 + 1 = 0$' is false.[19] So not only are 'intuitively true' arithmetical sentences declared false in this language, but chunks of previously provable assertions are no longer susceptible of proof. There were other suggestions for how to deal with numbers that are 'too large' in a finitist-nominalist regime besides (a)–(c); however, none meet with substantially more approval from the discussants.

CONCLUSION: NOMINALISM THEN AND NOW

There is still a live philosophical debate surrounding nominalism; the most-discussed recent defense of nominalism is likely Hartry Field's *Science without Numbers* (1980). As mentioned earlier, the project that Field and other current nominalists see themselves pursuing appeared in public in Goodman and Quine's "Steps toward a Constructive Nominalism"—and the Harvard discussions of 1940–1941 prompted this article. To conclude this chapter, I will briefly examine how these discussions, the 'grandfather' of the present debates, relate to current arguments *pro* and *contra* nominalism.

Although Goodman and Quine say, in their seminal article, that their "refusal" to admit the existence of abstracta "is based on a philosophical intuition that cannot be justified by appeal to anything more ultimate," modern nominalists *do* provide a justification for this refusal. The leading modern argument for nominalism can be cast as a simple syllogism, whose major premise is a concise statement of the causal theory of knowledge.

P1. We can only have knowledge of things causally related (or relatable) to us.
P2. Numbers and other abstracta are not causally related (or relatable) to us. Therefore, we cannot have knowledge of numbers or other abstracta.

Both premises have been challenged by various philosophers. More criticism has been leveled at the first, presumably because many philosophers consider 'standing outside' the causal order a defining feature of an abstract object. Another variant of this syllogism replaces P1 with a statement of the causal theory of reference:

P1$_R$. We can only *successfully refer to* things causally related (or relatable) to us.

The conclusion is modified accordingly: we cannot *refer to* numbers or other abstracta. And presumably we cannot say much of significance about items to which we cannot successfully refer. Broadly speaking, neither the causal nor the referential form of this syllogism seems to have won large numbers of converts to the nominalist cause—at least in part because causal theories of knowledge and reference are not overwhelmingly popular nowadays. Causal theories of knowledge and reference did not appear in an explicit, fully-fledged form until the 1960s and 1970s, so it is not surprising that Carnap's 1940–1941 notes do not contain explicit statements of the

---

[19] For example, if we assume (b) or (c) (the discussants' eventual favored choice), then if D contains exactly two elements, then '$1 + 1 = 0$' will be true.

views expressed in P1 and P1$_R$. However, Tarski (and his sympathizers) would perhaps not *deny* that we only know about or refer to entities that are somehow causally connected or connectible with us. After all, if something is a physical object, then (with some exceptions[20]) it *is* causally connectible to us; and if something is an abstract object, then it is not causally connectible to us. Though a causal theory of reference does not appear in Carnap's discussion notes, a related notion may be at work implicitly in the collaborators' minds.

The primary objection to nominalism today is the 'indispensability argument,' usually attributed to Quine and Putnam. Shortly after the co-authored 1947 *Journal of Symbolic Logic* chapter appeared, Quine rejected nominalism. (Goodman did not.) Hilary Putnam and the post-nominalist Quine argued for the existence of mathematical abstracta on the grounds that relinquishing such abstracta would force us to relinquish much of modern science, and that we should be unwilling to pay that price for maintaining nominalist scruples. Current nominalist projects, such as Hartry Field's, usually consist of 'reconstructive' projects that attempt to rebut the indispensability argument. In such nominalist projects, a certain field of natural science is recast in a form that does not appeal to any 'abstract' entities. Field claims that if empirical science can be reconstructed nominalistically, then belief in mathematical objects becomes "unjustifiable dogma" (1980, 9). The literature on the indispensability argument is vast, and I will not comment upon its current status.

Given that the indispensability argument did not appear as such until around 1970, an explicit desire to rebut it cannot be a motivation for undertaking the 1941 project. However, there *is* a precursor of the modern indispensability argument in Carnap's notes. We encountered it above, when discussing the lower bound on the poverty of a finitist-nominalist language's expressive power; the relevant section is reproduced here:

If S$_1$ [the finitist-nominalist language] does not suffice to reach classical mathematics, couldn't one perhaps nevertheless adopt S$_1$ and perhaps show that classical mathematics is not really necessary for the application of science to life? Perhaps we can set up, on the basis of S$_1$, a calculus for a fragment of mathematics that suffices for all practical purposes (i.e. not something just for everyday purposes, but also for the most complicated tasks of technology).

(RCC 090–16–25)

This is not precisely the program Hartry Field has pursued, but it is similar: in both cases, the aim is to show that a proper subset of mathematics, which is nominalistically acceptable, is sufficient for all applications of mathematics in science. This leads us to an interesting question: what is the relationship between the modern indispensability argument and the Tarski–Carnap–Quine demand that their 'understandable' language be sufficient to express (at least a substantial portion of) mathematics and natural science? To answer this question, we need an explicit statement of the indispensability argument. The following is a typical current formulation, due to Colyvan:

---

[20] For example, the laws of physics prohibit me from having causal contact with events outside my past light-cone.

1. We ought to have ontological commitment to all and only those entities that are indispensable to our best scientific theories.
2. Mathematical entities are indispensable to our best scientific theories.
Therefore:
3. We ought to have ontological commitment to mathematical entities.

(2001, 11)

The 1941 project differs from the modern one primarily in the first premise: there is no normative claim concerning ontological commitments explicitly forwarded in the discussion notes. Nothing in the texts decisively rules out attributing this position to the participants as an implicit belief, but this (potentially anachronistic) interpretation is certainly not forced upon us, either. Instead, we can view the discussants as replacing normative-*cum*-ontological issues with the goal of a unified language of science. Whether failure of a language to meet that aim, regardless of the language's other merits, automatically disqualifies it in the discussants' eyes is not discussed in Carnap's notes, as mentioned above. We know that Quine, a decade after the Harvard discussions, opts for disqualification: his grounds for eventually repudiating nominalism are that we cannot recover (in a natural way)[21] a sufficient amount of mathematics to do science if we abide by nominalist strictures. Tarski's description of himself as a 'nominalist with a materialistic taint' in the 1965 ASL meeting suggests that he did not follow Quine.

References

Archives of Scientific Philosophy, Rudolf Carnap Collection, University of Pittsburgh. All rights reserved.
Carnap, Rudolf ([1934] 1937) *Logical Syntax of Language*. La Salle: Open Court Press.
—— (1963) "Intellectual Autobiography," in *Philosophy of Rudolf Carnap*, Paul Schilpp, (ed.). La Salle: Open Court Press.
Colyvan, Mark (2001) *The Indispensability of Mathematics*. Oxford: Oxford University Press.
Creath, Richard (1990) *Dear Carnap, Dear Van*. Berkeley: University of California Press.
Feferman, Anita and Feferman, Solomon (2004) *Alfred Tarski: Life and Logic*. Cambridge: Cambridge University Press.
Feferman, Solomon (1999) "Tarski and Gödel: Between the Lines," in J. Woleński and E. Kohler, eds., *Alfred Tarski and the Vienna Circle*. Boston: Kluwer.
Field, Hartry (1980) *Science without Numbers: A Defense of Nominalism*. Princeton: Princeton University Press.
—— (2001) "Quine and the Correspondence Theory," in *Truth and the Absence of Fact*. New York: Oxford University Press.
Frost-Arnold, Greg (forthcoming) *Carnap, Tarski, and Quine's Year Together: Conversations on Logic, Science, and Mathematics*. La Salle: Open Court Press.

---

[21] Quine sometimes included this parenthetical condition: see (Mancosu, forthcoming, the final paragraph of §3).

Hardcastle, Gary (2003) "Debabelizing Science: The Harvard Science of Science Discussion Group, 1940–41," in *Logical Empiricism in North America*, Gary Hardcastle and Alan Richardson (eds) Minneapolis: University of Minnesota Press, 170–96.

Hodges, Wilfred (1986) "Truth in a Structure," *Proceedings of the Aristotelian Society* 86, 131–51.

Kotarbiński, Tadeusz ([1929] 1966) *Gnosiology. The Scientific Approach to the Theory of Knowledge*, O. Wojtasiewicz (trans). Oxford: Pergamon.

Mancosu, Paolo (2005) "Harvard 1940–41: Tarski, Carnap and Quine on a Finitistic Language of Mathematics for Science," *History and Philosophy of Logic* 26, 327–57.

—— (forthcoming) "From Harvard to Amersfoort: Quine and Tarski on Nominalism," *Oxford Studies in Metaphysics*, vol. 3, Dean Zimmerman (ed.). Oxford University Press.

Mostowski, Andrej (1967) "Alfred Tarski," in *The Encyclopedia of Philosophy, Vol. 8*, P. Edwards (ed.). New York: Macmillan, 77–81.

Mycielski, Jan (2004) "On the Tension between Tarski's Nominalism and his Model Theory," *Annals of Pure and Applied Logic* 126, 215–24.

Quine, W. V. O. (1939) "Designation and Existence," *Journal of Philosophy* 36, 701–9.

—— (1985) *The Time of My Life: An Autobiography*. Cambridge, MA: MIT Press.

Quine, W. V. O. and Goodman, Nelson (1947) "Steps Toward a Constructive Nominalism," *Journal of Symbolic Logic* 12, 105–22.

Rojszczak, Artur (1999) "Why Should a Physical Object Take on the Role of Truth-Bearer?" in J. Woleński and E. Kohler (eds.). *Alfred Tarski and the Vienna Circle*. Boston: Kluwer.

—— (2002) "Philosophical Background and Philosophical Content of the Semantic Definition of Truth," *Erkenntnis* 56, 29–62

Russell, Bertrand (1918) *Mysticism and Logic and Other Essays*. London: Longmans Green.

Simons, Peter (1993) "Nominalism in Poland," in *Polish Scientific Philosophy*, Francisco Coniglione, Roberto Poli, and Jan Woleński (eds.). Amsterdam: Rodopi.

Searle, John (1980) "Minds, Brains, and Programs," *Behavioral and Brain Sciences* 3, 417–24.

Suppes, Patrick (1988) "Philosophical Implications of Tarski's Theory of Truth," *Journal of Symbolic Logic* 53, 80–91.

Tarski, Alfred (1944) "The Semantic Conception of Truth and the Foundations of Semantics," *Philosophy and Phenomenological Research* 4, 341–76.

—— (1983) *Logic, Semantics, Metamathematics*. 2nd edn, J. H. Woodger (trans). Indianapolis: Hackett.

Woleński, Jan (1993) "Tarski as Philosopher," in F. Coniglione, R. Poli, and J. Woleński (eds.). *Polish Scientific Philosophy: The Lvov-Warsaw School*. Amsterdam: Rodopi, 319–38.

—— (2002) "From Intentionality to Formal Semantics (From Twardowski to Tarski)," *Erkenntnis* 56, 9–27.

10

Truth, Meaning, and Translation

Panu Raatikainen

Philosopher's judgments on the philosophical value of Tarski's contributions to the theory of truth have varied. For example Karl Popper, Rudolf Carnap, and Donald Davidson have, in their different ways, celebrated Tarski's achievements and have been enthusiastic about their philosophical relevance. Hilary Putnam, on the other hand, pronounces that '[a]s a philosophical account of truth, Tarski's theory fails as badly as it is possible for an account to fail'. Putnam has several alleged reasons for his dissatisfaction,[1] but one of them, the one I call the modal objection (cf. Raatikainen 2003), has been particularly influential. In fact, very similar objections have been presented over and over again in the literature. Already in 1954, Arthur Pap had criticized Tarski's account with a similar argument (Pap 1954). Moreover, both Scott Soames (1984) and John Etchemendy (1988) use, with an explicit reference to Putnam, similar modal arguments in relation to Tarski. Richard Heck (1997), too, shows some sympathy for such considerations. Simon Blackburn (1984, ch. 8) has put forward a related argument against Tarski. Recently, Marian David has criticized Tarski's truth definition with an analogous argument as well (David 2004, pp. 389–90).[2]

This line of argument is thus apparently one of the most influential critiques of Tarski. It is certainly worthy of serious attention. Nevertheless, I shall argue that, given closer scrutiny, it does not present such an acute problem for the Tarskian approach to truth as many philosophers think. But I also believe that it is important to understand clearly why this is so. Moreover, I think that a careful consideration of the issue illuminates certain important but somewhat neglected aspects of the Tarskian approach.

THE MODAL OBJECTION

The basic idea of the modal objection is simple enough: Instances of T-schema such as

'Snow is white' is true if and only if snow is white

I am very grateful to Douglas Patterson for his valuable comments on an earlier version of this chapter.

[1] See Putnam 1960, 1983, 1985, 1988. For criticism, see Raatikainen, 2003.
[2] As Halbach (2001) has pointed out, analogous arguments have been presented also by Lewy, Strawson, Church, and Quine, though not always directly as a criticism of Tarski.

are, in Tarski's approach, logical consequences of the truth definition and thus necessarily true; but certainly it would have been possible, so the argument goes, that 'snow' denoted, say, grass, in which case it would have been false that 'snow is white' is true if and only if snow is white. In other words, surely the sentence ' "Snow is white" is true if and only if snow is white' is a contingent, empirical claim whose truth value depends on what the expressions of the object language mean, not a necessary truth, as Tarski's approach entails. So, it is concluded, there must be something deeply wrong with Tarski's approach. In what follows, I shall focus mainly on Putnam's version of the modal objection, for Putnam has developed the argument in certain respects further than others, and considering those further developments allows one to clarify some interesting additional issues. I think that to the extent that Putnam's arguments can be rebutted, this should suffice also for the other variants of the modal objection.

In his much-cited 'Comparison of Something with Something Else' (Putnam 1985: see also Putnam 1988), Putnam begins his modal objection by considering the following instance of T-schema:

(1) (For any sentence X) If X is spelled S-N-O-W-SPACE-I-S-SPACE-W-H-I-T-E, then X is true in L if and only if snow is white.

Putnam then presents his objection: 'Since [(1)][3] is a theorem of logic in meta-L (if we accept the definition—given by Tarski—of "true-in-L"), since no axioms are needed for the proof of [(1)] except axioms of logic and axioms about spelling, [(1)] holds in all possible worlds.[4] In particular, since no assumptions about the use of expressions of L are used in the proof of [(1)], [(1)] holds true in worlds in which the sentence "Snow is white" does not mean that snow is white' (Putnam 1985, p. 333). Putnam concludes: 'The property to which Tarski gives the name "True-in-L" is a property that the sentence "Snow is white" has in every possible world in which snow is white, *including worlds in which what it means is that snow is green*... A property that the sentence "Snow is white" would have (as long as snow is white) no matter how we might use or understand that sentence isn't even doubtfully or dubiously "close" to the property of truth. It just isn't truth at all' (Putnam 1985, p. 333).

---

[3] I have changed Putnam's numbering.

[4] Putnam's claim is exaggerated: in the standard cases, where there are infinitely many sentences, at least a weak subsystem of the second-order arithmetic such as ACA—and not just logic—is needed for the truth-definition and the derivation of T-sentences from it. However, as the great majority of philosophers apparently think that theorems of arithmetic also are necessary and *a priori*, and this is the crucial matter here, I shall not make more about this.

Thus let us assume that the *metatheory* does not contain any non-logical axioms except arithmetical axioms, or axioms of the theory of concatenation (or syntax), which amounts to the same (Quine 1946, for example, shows that elementary arithmetic and the elementary theory of concatenation are equivalent). The *metalanguage*, on the other hand, may and often must contain other sorts of non-logical expressions, such as 'green' 'moon', 'round', 'Earth' etc. in our examples; the point is that there are no non-logico-arithmetical axioms governing them. Under these assumptions, T-sentences are just definitional abbreviations of certain theorems of arithmetic, and thus, according to the standard view, indeed necessarily true and *a priori* knowable. Had the metatheory other sorts (e.g. contingent or empirical) of axioms, being a consequence of a definition would not make a theorem anything more than contingent.

John Etchemendy (1988), although reluctant to accept Putnam's most colorful conclusions, says that they are based on a 'sound observation': 'Tarski's definition does not provide an analysis of one important notion of truth' (p. 60, fn 8). More generally, he concludes that 'the theory of truth that results from a Tarskian definition of truth ... cannot possibly illuminate the semantic properties of object language' (Etchemendy 1988, p. 56). The reason Etchemendy gives for these claims is just the modal objection.[5]

CONVENTION T AND TRANSLATION

In order to evaluate the modal objection properly, one needs to take a closer look at Tarski's criterion of material adequacy, that is, his famous *Convention T*. It may be formulated as follows (cf. Tarski 1935, pp. 187–8):

A formally correct definition will be called *an adequate definition of truth* if it has the following consequences:

(a) all sentences

(T) X is true if and only if p,

where 'X' is a structural-descriptive name of a sentence S of the object language L and 'p' is a *translation* of that sentence S into the metalanguage ML.

(b) for all X, if X is true, then X is a sentence of the object language L.

The reference to translation in (a) is important, although is often ignored, presumably because the more popular texts by Tarski (e.g. Tarski 1944) deal only with the case where the object language is assumed to be a (proper) part of the metalanguage (as in the standard example ' "Snow is white" is true if and only if snow is white'); but it is essential to recognize that in this case it is tacitly assumed that the translation from the object language to the metalanguage is the trivial 'homophonic' one. If, on the other hand, one changes the interpretation of the symbols of the object language (with the result, say, that 'white' denotes green), the translation is no longer homophonic and must be made explicit. In his seminal chapter on the concept of truth, Tarski was quite clear about these matters:

We take the scheme [x is a true sentence if and only if p] and replace the symbol 'x' in it with the name of the given sentence, and 'p' by its translation into the metalanguage.

(Tarski 1935, p. 187)

Instances of the schema (T) are nowadays often called T-sentences. As far as I know, this talk of T-sentences originated with Davidson (1973*a*, *b*). Note then that *if*, in a sentence of the form 'X is true if and only if p', either:

(i) 'X' is not a structural-descriptive name of S; or
(ii) 'p' is not a translation of S,

---

[5] For Etchemendy's version of the modal argument, see Etchemendy 1988, pp. 56–7, 60–1.

then the equivalence '*X* is true if and only if *p*' does *not* count as an instance of T-schema, in other words, it is *not* a T-sentence.[6] Consequently, if one changes the interpretation of the symbols of the object language L, a former T-sentence may not be an instance of T-schema any more. That is, properly understood, Convention T necessarily requires that the relations between the object language L and the metalanguage ML be fixed (and remain constant). Let us try to see in more detail why this is so.

THE OBJECT LANGUAGE AS AN INTERPRETED LANGUAGE

As Tarski always insisted, truth can be only defined (because of paradoxes and Tarski's undefinability theorem) for a particular formalized language at a time. Moreover, for Tarski the 'formalized languages'[7] whose truth is under consideration always had to be interpreted languages,[8] as he repeatedly emphasized:

> It remains perhaps to add that we are not interested here in 'formal' languages and sciences in one special sense of the word 'formal', namely sciences to the signs and expressions of which no meaning is attached. For such sciences the problem here discussed has no relevance, it is not even meaningful. We shall always ascribe quite concrete and, for us, intelligible meanings to the signs which occur in the languages we shall consider.
>
> (Tarski 1935, pp. 166–7)

Furthermore, this was, for Tarski, not just an accidental philosophical opinion;[9] rather, it was an essential part of Tarski's whole approach to truth that the meanings of the object language must be fixed. Only so could a truth definition (applied to sentences) make any sense at all:

> For several reasons it appears most convenient to apply the term 'true' to sentences, and we shall follow this course. [footnote omitted]
>
> Consequently, we must always relate the notion of truth, like that of a sentence, to a specific language; for it is obvious that the same expression which is a true sentence in one language can be false or meaningless in another.
>
> (Tarski 1944, p. 342)

---

[6] There is much unclarity and confusion on this matter in the literature. Thus one often counts sentences such as ' "Snow is white" is true iff the moon is made out of cheese' as T-sentences, and talks about false T-sentences. But such sentences simply are not T-sentences. I think one should call them, e.g. alleged or apparent T-sentences, or T-like sentences (as Lepore and Ludwig 2005 do), in order to clearly distinguish them from the genuine T-sentences.

[7] One may also note that the title of the Polish original of 'The concept of truth in formalized languages' did not even speak about formalized languages, but translates in fact as 'The concept of truth in the languages of deductive sciences'.

[8] To be sure, certain characterizations of 'formalized languages' by Tarski are quite misleading and confusing, e.g., when he writes that formalized languages 'can be roughly characterized as artificially constructed languages in which the sense of every expression is uniquely determined by its form' (Tarski 1935, pp. 165–6).

[9] Apparently Tarski originally accepted this idea by accepting his teacher's Leśniewski's 'intuitionistic formalism', according to which all languages, including formal ones, are already interpreted (this was considered not to be an obstacle for their formalization). But Tarski still held this view much later (still in 1969), when he otherwise had distanced himself quite a lot from Leśniewski's philosophical ideas.

We shall also have to specify the language whose sentences we are concerned with; this is necessary if only for the reason that a string of sounds or signs, which is a true or a false sentence but at any rate meaningful sentence in one language, may be a meaningless expression in another.

(Tarski 1969, p. 64)

... the concept of truth essentially depends, as regards both extension and content, upon the language to which it is applied. We can only meaningfully say of an expression that it is true or not if we treat this expression as a part of a concrete language. As soon as the discussion concerns more than one language the expression 'true sentence' ceases to be unambiguous. If we are to avoid this ambiguity we must replace it by the relative term 'a true sentence with respect to the given language'.

(Tarski 1935, p. 263)

Therefore, it is necessary in Tarski's setting to focus on an interpreted language with constant meanings. If one varies the interpretation of the symbols of the object language L, the language changes to a different language L'; and (because one can define a truth predicate only for a particular language—an interpreted language—at a time) a former truth definition (true-in-L) is not a truth definition for this latter language L'; a former T-sentence does not count any more as a T-sentence (because T-sentences are defined only relative to a particular truth definition), and wholly different sentences become instances of T-schema—e.g. assuming that 'white' denoted (in-L') green, one should now have 'The sentence "Snow is white" is true-in-L' if and only if snow is green'.

All this is in stark contrast to the way formal languages are viewed in mature model theory, even though Tarski also importantly influenced the development of the latter. That is, in model theory, a language L is a completely uninterpreted and syntactic formal language. An L-structure W is defined as a pair (D, I), consisting of the domain D and the interpretation function I. The latter maps the non-logical symbols of L to elements of D (that is, the function I maps individual constants to elements of D, predicates to subsets of D, etc.).[10] In changing the structure, one varies the interpretation, but the language L remains the same.

Let us note in passing that the interpretation function I establishes a link between the object language and a domain of extra-linguistic objects, and hence is a semantical concept in Tarski's sense (see also below); hence, it would be somewhat problematic to presuppose it in the Tarskian definition of truth,[11] which should not according to Tarski presuppose any semantical notions; the meanings of the object language must thus get fixed in some other way. Accordingly, it is important not to conflate Tarski's philosophical project of defining truth *simpliciter*, and the model-theoretic notion of truth-in-a-model defined in the above setting; their different understanding of what a language consists of is particularly relevant. However, all too often these are not

---

[10] Obviously, there are different ways to formulate these ideas, but in practice they are equivalent to the one presented here.

[11] Though, it is of course perfectly acceptable in its proper context, in model theory, whose aims are quite different.

clearly distinguished, and many misunderstandings derive from this. In particular, I suspect that such a conflation partly explains the popularity and attractiveness of the modal objection.

To recap, Tarski's approach to defining truth proceeds in certain order: First, an interpreted language equipped already with its meanings is chosen as the object language. Second, one presents a definition of the truth predicate for this particular interpreted language. The truth predicate defined is relative to this language and its interpretation. Finally, one shows that the definition is materially adequate by deriving T-sentences, which are doubly relative to the interpretation of the object language. *As an expression of German* (understood as an interpreted language), 'weiss' necessarily means (means-in-German) what it does, namely white, and the same holds for all other expressions. As we have seen, if 'weiss' denoted green, or 'schnee' denoted grass, for example, the language would not be German any more. The identity of a language, in Tarski's setting, essentially depends on meanings of its expressions. Consequently, the equivalence ' "Schnee ist weiss" is true-in-German if and only if snow is white' is, and should be, necessary, for the truth predicate is tied to the particular interpreted language (cf. Milne 1997).

Let us now reconsider the modal objection. It is certainly true that expressions can change their meaning, and that the language could have so evolved that, for example, 'white' would denote green. However, from the Tarskian point of view, that language would no longer be English or, in short, L (as an interpreted language supplied with its meanings)—even if it were syntactically identical with L. Call this latter language L'. Even in such a possible world, it would nevertheless be true that 'white' denotes-in-L white, and that 'Snow is white' is true-in-L, if and only if, snow is white. It would only be the case that 'white' denotes-in-L' green, and that 'Snow is white' is true-in-L', if and only if snow is green. In other words, ' "Snow is white" is true-in-L, if and only if, snow is white' is indeed true in every possible world and thus necessary.

In sum, Tarski's definition of truth does, *pace* Putnam, depend in a sense also on the meaning and not only on the spelling. Namely, meaning is built into the Tarskian approach via interpretation of the object language. So it seems that Putnam's modal objection can be effectively rebutted by pointing out that there is an illegitimate change of object language in the midst of the argument. Many of the critical replies to Putnam have indeed made this point (see e.g. García-Carpintero 1996, Fernandez Moreno 1992, 1997, Niiniluoto 1994, Halbach 2001, Woleński 2001), and as far as it goes, this reply is, I think, on the correct lines.

THE IDENTIFICATION OF THE OBJECT LANGUAGE

The whole issue is not, however, that easy to bypass, for Putnam is in fact aware of this 'language change reply'—as it might be called—and he has a further objection to this line of reply—an objection of which most of his critics seem to be ignorant. In *Representation and Reality* (Putnam 1988), Putnam reports how he raised the modal objection in a conversation with Carnap in the early 1950s: he complained that it isn't a logical truth that the (German) word 'Schnee' refers to the substance snow, nor is it a logical truth that the sentence 'Schnee ist weiss' is true in German if and only if snow is white. Carnap's reply was, Putnam recalls, that everything depends on the way the name of the language—'German' or whatever—is defined. '[I]n philosophy, Carnap urged, we should treat languages as abstract objects, and they should be identified (their names should be defined) by their semantical rules. When "German" is defined as "the language with such and such semantical rules", it is logically necessary that the truth condition for the sentence "Schnee ist weiss" in German is that snow is white' (Putnam 1988, p. 63). Putnam tells us that he was not satisfied, but did not continue the argument: 'What I thought but did not say was: And, pray, what semantical concepts will you use to state these "semantical rules"? And how will those concepts be defined?' (Putnam 1988, p. 63). Putnam then goes on to argue in some detail that if one attempts thus to define a language, one needs to appeal to the concept of truth, and that this would make the language change reply circular (Putnam 1988, pp. 63–5).

Carnap apparently thought that languages should be identified (their names should be defined) by their semantical rules, and it may be that this is begging the question.[12] But be that as it may, it is important to note that this is not Tarski's view. Tarski explicitly points out the difference here between his own approach and that of Carnap, according to which we regard 'the specification of *conditions* under which sentences of a language are *true* as an essential part of the description of this language' (see Tarski 1944, p. 373, note 24; my emphasis). For Tarski, on the other hand, the interpreted object language is instead specified simply through its metalinguistic translation (see e.g. Tarski 1935, pp. 170–1; cf. Fernandez Moreno 1992, 1997; Milne 1997, Feferman 2004). In accordance, Tarski described the metalanguage in the following ways:

...the metalanguage contains both an individual name and a *translation* of every expression (and in particular of every sentence) of the language studied...

(Tarski 1935, p. 172; my italics)

...to every sentence of the language...there corresponds in the metalanguage not only a name of this sentence of the structural-descriptive kind, but also a sentence having the *same meaning*.

(Tarski 1935, p. 187; my italics)

However, one could point out that Tarski's approach still assumes the notion of *meaning,* in the disguise of translation, or the sameness of meaning. Does this mean that, at the end of the day, Tarski fails to achieve his expressed aim, that is, to define truth without assuming any *semantical* concepts? It has been frequently suggested that this is indeed the case. However, this is not necessarily so. In order to see this, we need to recall what Tarski meant by 'semantical'. Tarski's paradigmatic examples of

---

[12] But see Fernandez Moreno 1997.

semantical concepts were satisfaction, denotation, truth, and definability (see Tarski 1935, pp. 164, 193–4; 1936, p. 401). He explained his understanding of 'semantical concept' as follows:

> A characteristic feature of the semantical concepts is that they give expression to certain relations between the expressions of language and the objects about which these expressions speak, or that by means of such relations they characterize certain classes of expressions or other objects.
>
> (Tarski 1935, p. 252)

In contrast, I submit that it is possible to view translation, in this context, as a purely syntactic, effective mapping between two languages, without assuming any relations between either language and objects about which they speak. Translation, so viewed, is *not* a semantical concept in Tarski's sense, and does not presuppose truth or related notions (most importantly, satisfaction, by means of which the others can be defined).[13] Hence, it seems to be, after all, admissible for Tarski to presuppose such a notion of translation in his approach without begging the question (cf. Milne 1997; see also below).

To conclude, Putnam's contention that defining the interpretation of the object language necessarily requires the notion of truth for that language is unproven, and the modal objection can indeed be disarmed—without begging the question—by recognizing that in the Tarskian approach, the object language, as an interpreted language with the meanings of its terms and hence their translations into the metalanguage held fixed, must remain constant and is not to be varied.

A CLOSER LOOK AT THE TARSKIAN TRUTH DEFINITION

Let us now look in more detail, with a particular example, on how exactly Tarski himself specifies the meanings of the object language expressions and gives a truth definition. That one can derive the instances of T-schema in the metatheory is due to careful stage-setting; specifically, as Field (1972) has emphasized, the Tarskian definitions of satisfaction and truth are based on prior definitions of denotation for individual constants and of application for predicate constants—in short, of *primitive denotation*.[14]

---

[13] It must be granted that that issue is not absolutely crystal clear. For example, in 1944 Tarski wrote: 'Within theoretical semantics we can define and study some further notions, whose intuitive content is more involved and whose semantic origin is less obvious; we have in mind, for instance, the important notions of *consequence, synonymity,* and *meaning*'. He adds (fn. 20) that all those notions can be defined in terms of satisfaction; and refers to Carnap (1942) for the definition of synonymity. Doesn't this passage undermine my conclusion in the text? I am inclined to that not. First, Tarski seems to be talking here about intralinguistic synonymity between two expressions of the object language L, and not about interlinguistic synonymity (translation) between L and ML. Second, Tarski only says that it is possible to define synonymity in terms of satisfaction; he does not state that it cannot be fixed in any other way. Third, he is here referring more to Carnap's work than to his own.

[14] For simplicity, I assume that L does not contain function symbols and that it only has monadic predicates.

For example, let us assume that the object language L is a (semi-formal) fragment of German. A Tarskian definition of denotation for names then takes the form of a list:

$Denotes_L(x, y) \leftrightarrow$

$$[(x = \ulcorner Frankreich \urcorner \wedge y = France) \vee$$
$$(x = \ulcorner Deutschland \urcorner \wedge y = Germany) \vee$$
$$\vdots$$
$$(x = \ulcorner K\ddot{o}ln \urcorner \wedge y = Cologne)].$$

Note that the number of primitive proper names is finite; consequently, denotation for names can be explicitly defined in the metalanguage; i.e. $Denotes_L(x, y)$ can always be eliminated, and one can use the right-hand side of the equivalence, which is a formula of the unextended metalanguage (assumed to contain no semantical concepts), instead. An analogous definition can be given for predicates:

$Applies_L(x, y) \leftrightarrow$

$$[(x = \ulcorner Stadt \urcorner \wedge City(y)) \vee$$
$$(x = \ulcorner Staat \urcorner \wedge State(y)) \vee$$
$$\vdots$$
$$(x = \ulcorner Rund \urcorner \wedge Round(y))].$$

This is how Tarski in practice fixes the interpretation of the object language (more exactly, the interpretation of its primitive non-logical symbols). Surely such a list-like explicit definition, which makes primitive denotation eliminable, does not presuppose any semantical notions. This should remove any remaining doubts as to whether Tarski could nail down the meanings of expressions of the object language without leaning on semantical concepts. In fact, denotation and application could be subsumed under a more general notion of satisfaction (see Tarski 1935, pp. 190, 194), but for expository purposes, it is useful to present them separately as above. (A list-like characterization of primitive denotation such as above may strike one as disappointingly shallow philosophically, and one may sympathize Field's (1972) demand for a more substantial account of denotation, but there is, logically speaking, nothing in principle wrong in Tarski's approach—it is not in any way question-begging or circular.)

The recursive definitions of satisfaction and truth are familiar (For simplicity, let us assume that the object language L contains, as logical constants, only $\sim$ (negation), & (conjunction), and E (existential quantifier)). I shall use $\sim$, &, E, for the object language symbols, and $\neg, \wedge, \exists$ for the respective metalanguage symbols (and I assume that the metalanguage has also $\vee, \rightarrow, \leftrightarrow$, and $\forall$). A and B are formulas of L, n is a name in L and P is a predicate in L. σ, τ are infinite sequences of objects, and $\sigma(j)$ ($\tau(j)$) is the jth member of the sequence σ (of the sequence τ).

$Satisfies_L(\sigma, x) \leftrightarrow$

$[(x = \ulcorner P(n) \urcorner \wedge (\exists y) \, (Denotes_L(\ulcorner n \urcorner, y) \wedge Applies_L(\ulcorner P \urcorner, y)) \vee$
$(x = \ulcorner P(v_j) \urcorner \wedge Applies_L(\ulcorner P \urcorner, \sigma(j))) \vee$
$(x = \ulcorner A \& B \urcorner \wedge Satisfies_L(\sigma, \ulcorner A \urcorner) \wedge Satisfies_L(\sigma, \ulcorner B \urcorner)) \vee$
$(x = \ulcorner \sim A \urcorner \wedge \neg Satisfies_L(\sigma, \ulcorner A \urcorner)) \vee$
$(x = \ulcorner (Ex_i)A \urcorner \wedge (\exists \tau)[(\forall j)(j \neq i \rightarrow \tau(j) = \sigma(j)) \wedge Satisfies_L(\tau, \ulcorner A \urcorner)]].$

Note that this is not an explicit but a recursive definition, for $Satisfies_L$ occurs also in the right-hand side of the equivalence. It is, however, possible to turn it to an explicit definition, with a help of a little bit of set theory.[15] The definition of truth is then simple:

$True_L(x) \leftrightarrow [x \text{ is a closed formula} \wedge (\forall \sigma)(Satisfies_L(\sigma, x))$.

All these definitions at place, one can then see that all the instances of T-schema, such as:

$[True_L(\ulcorner Stadt(K\"oln) \urcorner) \leftrightarrow City(Cologne))],$

can be derived in the metatheory.

DIFFERENT INTERPRETATIONS OF SEMANTICAL DEFINITIONS

Now just to what extent such T-sentences are either true by definition and necessary, or contingent (the question at stake in the modal objection), is certainly parasitic to the modal status of what I shall call D-sentences and A-sentences. That is, by D-sentences, I mean sentences such as:

$(\forall x)[Denotes_L(\ulcorner Mond \urcorner, x) \leftrightarrow x = \text{the moon}],$

and by A-sentences, analogously, sentences such as

$(\forall x)[Applies_L(\ulcorner Rund \urcorner, x) \leftrightarrow Round(x)].$

Note that just like T-sentences, all D- and A-sentences are, in the Tarskian approach, provable theorems in the metatheory (given the definitions) and apparently necessarily true (assuming that the metatheory contains only arithmetical or set-theoretical axioms as its non-logical axioms; cf. note 4). The fundamental question concerns the modal status of such sentences; the modal objection could now be rephrased as the complaint that it is certainly a contingent empirical fact that, e.g. 'Mond' denotes moon in German, and not a necessary truth as Tarski's approach seems to entail. The detour through T-sentences is really redundant and makes the issue unnecessarily complex and opaque.

Now it is true that such D- and A-sentences come out as 'true by definition' in the approach that Tarski's takes to primitive denotation, and are provable in the metatheory, because *Denotes* and *Applies* can be explicitly defined. However, we have seen

---

[15] Or, alternatively, one can transform it to an axiomatic theory. This is relevant in what follows.

above that this is, after all, exactly how it should indeed be. The two-part definition of primitive denotation is constitutive for L as an interpreted language, and D- and A-sentences are immediate consequences of these definitions. Although it is obviously not necessary that 'Mond', as a mere string of symbols and viewed purely formally and syntactically, denotes moon, it is nevertheless the case that as a word of the interpreted language L, it necessarily denotes the moon.

One can look at the definition of primitive denotation in two different ways.[16] First, one may take the definition as purely *stipulative*, such that it defines the artificial language L as an abstract entity under consideration. From this perspective, there is nothing external for the definition to be right or wrong about. However, one may alternatively be interested in an actual, concrete natural language, e.g. German, or rather a suitable formalizable fragment of such a language, and attempt to capture by a definition the pre-existing denotation relation[17] of that language in the metalanguage.[18] The definition aims to be *usage reporting*. From this perspective, one may well conclude in some case that definition or not, it has got the facts wrong. If the definition contained, for example, as its part the clause

$$(x = \ulcorner München \urcorner \wedge y = Munster),$$

one would have all the reasons to protest that it just isn't the case in German that the denotation of 'München' is Munster—'München' denotes Munich—and to revise the definition. Surely, nothing in the formal definition itself dictates how to view it, but it is certainly possible to take the latter attitude towards the definition (cf. Davidson 1990).

At this point, it is illuminating to recall Carnap's distinction between pure and descriptive semantics (see Carnap 1942, pp. 11–15). *Descriptive semantics* is concerned with historically given natural languages, such as German, and is based on empirical investigation. *Pure semantics,* on the other hand, is analysis of semantical systems with artificial languages which are stipulatively defined. It is entirely analytic and without factual content. 'Here we lay down definitions for certain concepts, usually in the form of rules, and study the analytic consequences of these definitions. In choosing the rules we are entirely free' (Carnap 1942, p. 13). And we have seen that according to Carnap, in philosophy one must confine oneself to pure semantics. For Carnap, pure and descriptive semantics seem to be largely independent projects.

Tarski made an analogous distinction between *descriptive* and *theoretical semantics* (Tarski 1944, p. 365). By 'descriptive semantics', he refers to the totality of investigations on semantic relations which occur in a natural language. Apparently by 'theoretical semantics' Tarski means kind of study he is himself pursuing. Fernandez Moreno (1997) suggests that theoretical semantics as undertood by Tarski corresponds to pure semantics in the sense of Carnap. However, I find this slightly

---

[16] For more about the difference between stipulative and usage reporting (or lexical) definitions, see, e.g. Belnap 1993.
[17] More exactly, its restriction to the relevant fragment.
[18] Obviously, the way I have developed the truth definition above is already inclined towards this interpretation.

problematic, or at least misleading. Carnap apparently viewed (in pure semantics) the definitions of semantical relations as purely stipulative, that is, thought that such definitions stipulatively define the language in question, and are analytically true of it. The language here is an artificial, formal language—an abstract object arbitrarily defined by the stipulations that govern its semantical relations.[19]

So what about Tarski? It is true that Tarski constantly insisted that colloquial languages give rise to semantical paradoxes, and that truth can only consistently be defined for a formalized language. This has led many to assume that Tarski, just like Carnap, wanted to restrict his 'theoretical semantics' exclusively to artificial formal languages—that it is not at all applicable to the real-life natural languages. The case with Tarski is, however, more complicated than that. We have seen above that formalized or not, the languages under consideration must, for Tarski, be 'concrete' and already interpreted, in other words, must come already equipped with 'concrete' meaning. This alone makes them quite different from the artificial formal languages as usually understood. Tarski also thought that his semantical tools can be applied to restricted languages of various special sciences, say, of chemistry—as long as they do not contain semantical vocabulary.

Moreover, Tarski suggests that theoretical semantics is, after all, applicable to natural languages, if 'only with certain approximation' (Tarski 1944, p. 365). That is, 'the approximation consists in replacing a natural language (or a portion of it in which we are interested) by one whose structure is exactly specified, and which diverges from the given language "as little as possible"' (Tarski 1944, p. 347). Similarly, Tarski writes, 'if we translate into colloquial language any definition of a true sentence which has been constructed for some formalized language, we obtain a fragmentary definition of truth which embraces a wider or narrower category of sentences' (Tarski 1935, p. 165, fn. 2). In fact, Tarski at one point emphasized that by 'formalized languages', he 'does not have in mind anything essentially opposed to natural languages;' and he continues: 'On the contrary, the only formalized languages that seem to be of real interest are those which are fragments of natural languages (fragments provided with complete vocabularies and precise syntactical rules) or those which can at least be adequately translated into natural languages' (Tarski 1969, p. 68).

For Tarski, the main problem with colloquial languages was that they are semantically closed,[20] for it is this aspect of them that leads to antinomies. However, suitable (semantically open) fragments of natural language, with sufficiently specified grammar, were wholly acceptable for him as object languages for truth definitions. Tarski had only complaints against natural language taken in its entirety (cf. Woleński 1993). Tarski himself described his view of theoretical and descriptive semantics thus:

The relation between theoretical and descriptive semantics is analogous to that between pure and applied mathematics, or perhaps to that between theoretical and empirical physics; the

---

[19] Whether this is a completely fair interpretation of Carnap's views I am not sure—it may well be an oversimplified account (in any case, his later thoughts about explication suggest a more sophisticated view). However, this does not really matter; my aim here is to argue that Tarski did not hold the view I describe here—whether or not this is exactly the overall view of historical Carnap.

[20] Or, more accurately, that they purport to be semantically closed (see Patterson 2006).

role of formalized languages in semantics can be roughly compared to that of isolated systems in physics.

(Tarski 1944, p. 365)

As a consequence of all the above, it seems as if Tarski was, unlike Carnap, inclined to view the definitions of semantical relations as usage reporting. That is, Tarski was inclined to the think that his definitions ultimately attempt to capture the actual semantical relations to the world of (fragments of) existing natural languages, rather than being merely stipulative specifications of artificial formal languages. (Such languages, of course, can certainly still play a role in the usage-reporting project.)

ON TRUTH DEFINITIONS AND TRUTH THEORIES

If one slightly relaxes Tarski's requirement that we do not use *any* semantical concepts in the truth definition, instead of explicitly defining primitive denotation one can add $Denotes_L$ and $Applies_L$ as new primitive predicates to the metalanguage, and then extend the metatheory with all D- and A-sentences as axioms governing them. One can then either explicitly define satisfaction and truth (assuming some set theory) in terms of primitive denotation, *or* add $True_L(x)$ and $Satisfies_L(x,y)$ as additional primitive predicates and turn the relevant definitions to axioms governing them; the exact details do not matter here, where we are interested primarily in primitive denotation. The result is a *theory* of primitive denotation and truth, not a definition, and the D- and A-sentences are axioms of the theory. From this perspective, it is easier (than with definitions) to look at the theory as attempting to describe the actual denotation relations of the real target language, here German, and one can consider the axioms as having, in a sense, *empirical* content—exactly what, in part, the advocates of the modal objection demand. The suggested axiom $(\forall x)[Denotes_L(\ulcorner München \urcorner, x) \leftrightarrow x = Munster]$, for example, would then be, even if an axiom, just a false hypothesis which should be revised, if the object language is supposed be (a fragment of) German.

But isn't it essential to the Tarskian approach to be able to explicitly define all semantical concepts? Does not giving up this requirement reopen the threat of paradoxes? And did not Tarski himself expressly oppose axiomatic theories of truth? These are good and natural questions to ask. However, I think that they suggest a bit of an oversimplified picture of Tarski's view. It is true that from the beginning, Tarski announces the intention explicitly to define truth without using any semantical concepts, and it is also true that he eventually succeeds in doing so. Moreover, the possibility of explicitly defining truth in a logico-mathematical metatheory with no semantical concepts certainly removes any worries of the possibility of antinomy. However, it seems to be a mistake to assume that for Tarski, the primary solution to paradoxes is and has to be the requirement of explicit definability of the semantical concepts (in contrast to what, e.g. Soames (1984, 1999) and Etchemendy (1988) seem to suggest). Rather, for Tarski, the real source of paradoxes was the universality or the semantical closedness of a language, and accordingly, the principal solution is the clear distinction between the object language and the metalanguage (cf. Heck 1997). Whether or not one is able, and prefers, to give explicit definitions is a further issue.[21]

Moreover, the consistency of the above axiomatic theory of primitive denotation is guaranteed, for it can be easily shown to be a conservative extension of the original, unextended metatheory; therefore, no paradoxes can possibly threaten it. Hence there is little reason to resist such a move, and it is indeed difficult to see any reason why Tarski would have doubted the consistency of this theory—given that the separation of the object language and the metalanguage is clearly respected. In fact, even the full axiomatic theory of truth and satisfaction is likewise a conservative extension of a suitable unproblematic metatheory.[22]

It must be granted that there are some passages in Tarski where he contrasts the axiomatic approach with the definitional approach, and makes some critical remarks on the former (see Tarski 1936, pp. 405–6, cf. 1935, pp. 257–8). One problem Tarski mentions is the question whether the axiomatic semantical theory is consistent. However, in the approach we have just discussed this is not at all a problem; the consistency of the theory is guaranteed. Furthermore, Tarski complains that an axiomatic theory would be 'highly incomplete', and that 'the choice of axioms always has a rather accidental character'. But if we look closer what Tarski really says, it becomes apparent that he has in mind first and foremost the weak theory which consists in mere T-sentences, and possible *ad hoc* extensions of this theory (Tarski 1935, pp. 257–8). The reasons he gives do not thus seem to count against just any kind of axiomatic theory of truth. Consequently, it seems that Tarski would not have had any strong reasons to object to an axiomatic theory such as one described above, which is in effect just Tarski's definitions transformed to an axiomatic theory. It is really just a different way of looking at Tarski's truth definition, and does not bring with it anything essentially new. Moreover, arguably Tarski himself was well aware of the possibility of such a transformation of his truth definition into a theory (cf. Heck 1997).

In sum, it is possible, without betraying the spirit of Tarski's project, to transform the Tarskian truth definition to an axiomatic theory, which can be interpreted to have empirical content. However, this does not mean that the relevant axioms and theorems are contingent. They still are constitutive and essential for the language in question. Perhaps they could be taken as another example of necessary truths that are knowable only *a posteriori*.

---

[21] If, however, one takes seriously Tarski's once declared requirement of physicalistic acceptability of the semantic notions, the need of explicit definability may be more acute. However, I am inclined to think that physicalism was not really that essential to Tarski's project; the only context where he talks about it (Tarski 1936) was a popular presentation of his work for an audience with many logical positivists there. See also Frost-Arnold 2004.

[22] Not object theory. Assuming that the object language has at most the expressive power of the language of first-order arithmetic (of course, it may have nothing to do with arithmetic or mathematics), the weak subsystem of second order arithmetic ACA is sufficient for most purposes. The full axiomatic theory of truth over the language of first-order arithmetic, which allows induction scheme to be applied also to formulas which contain truth predicate, is equiconsistent with ACA.

References

Belnap, N. D. (1993) 'On Rigorous Definitions', *Philosophical Studies* 72, 115–46.
Blackburn, S. (1984) *Spreading the Word*, Oxford University Press, Oxford.
Carnap, R. (1942) *Introduction to Semantics*, Harvard University Press, Cambridge, MA.
David, Marian (2004) 'Theories of Truth', in I. Niiniluoto, M. Sintonen and J. Woleński (eds.), *The Handbook of Epistemology*, Kluwer, Dordrecht, 331–413.
Davidson, D. (1973a) 'In Defence of Convention T', reprinted in D. Davidson, *Inquiries into Truth and Interpretation*. Oxford University Press, Oxford, 65–75.
—— (1973b) 'Radical Interpretation', reprinted in D. Davidson, *Inquiries into Truth and Interpretation*. Oxford University Press, Oxford, 125–39.
—— (1990) 'The Structure and Content of Truth', *Journal of Philosophy* 87, 279–328.
Etchemendy, J. (1988) 'Tarski on Truth and Logical Consequence', *Journal of Symbolic Logic* 53, 51–79.
Feferman, S. (2004) 'Tarski's Conceptual Analysis of Semantical Notions', in A. Benmakhlouf (ed.), *Sémantique et épistémologie*, Editions Le Fennec, Casablanca [distrib. J. Vrin, Paris], 79–108.
Fernandez Moreno, L. (1992) 'Putnam, Tarski, Carnap und die Warheit', *Gräzer philosophische Studien* 43, 33–44.
—— (1997) 'Truth in Pure Semantics: A Reply to Putnam', *Sorites* 8, June 1997, 15–23.
Field, H. (1972) 'Tarski's Theory of Truth', *Journal of Philosophy* 69, 347–75.
Frost-Arnold, G. (2004) 'Was Tarski's Theory of Truth Motivated by Physicalism?', *History and Philosophy of Logic* 25, 265–80.
García-Carpintero, M. (1996) 'What is a Tarskian Definition of Truth?', *Philosophical Studies* 82, 113–44.
Halbach, V. (2001) 'How Innocent is Deflationism?', *Synthese* 126, 167–94.
Heck Jr, R. (1997) 'Tarski, Truth and Semantics', *Philosophical Review* 106, 533–54.
Lepore, E. and K. Ludwig. (2005) *Donald Davidson: Meaning, Truth, Language, and Reality*, Oxford University Press, Oxford.
Milne, P. (1997) 'Tarski on Truth and Its Definition', in T. Childers, P. Kolář and V. Svoboda (eds.), *Logica '96: Proceedings of the 10th International Symposium*, Filosofia, Prague, 189–210.
Niiniluoto, I. (1994) 'Defending Tarski against his Critics', in B. Twardowski and J. Woleński (eds.) *Sixty Years of Tarski's Definition of Truth*, Philed, Warsaw, 48–68.
Pap, A. (1954) 'Propositions, Sentences, and the Semantic Definition of Truth', *Theoria* 20, 23–35.
Patterson, D. (2006) 'Tarski, The Liar and Inconsistent Languages', *Monist* 89, 150–77.
Putnam, H. (1960) 'Do True Assertions Correspond to Reality?', in H. Putnam, *Mind, Language and Reality, Philosophical Papers* Vol. 2, Cambridge University Press, Cambridge, 70–84.
—— (1983) 'On Truth', in L. Cauman *et al.* (eds.) *How Many Questions? Essays in Honour of Sidney Morgenbersser*, Hackett, Indianapolis, 35–56; page references to the reprint in H. Putnam, *Words and Life* (ed. J. Conant) Harvard University Press, Harvard, 1994, 315–29.
—— (1985) 'Comparison of Something with Something Else', *New Literary History*, 17, 61–79; page references to the reprint in H. Putnam, *Words and Life* (ed. J. Conant) Harvard University Press, Harvard, 1994, 330–50.
Putnam, H. (1988) *Representation and Reality*, MIT Press, Cambridge.

Quine, W. V. (1946) 'Concatenation as a Basis for Arithmetic', *Journal of Symbolic Logic*, vol. 11, 105–14.
Raatikainen, P. (2003) 'More on Putnam and Tarski', *Synthese* 135, 37–47.
Soames, S. (1984) 'What is a Theory of Truth?', *Journal of Philosophy* 81, 411–29.
—— (1999) *Understanding Truth*, Oxford University Press, Oxford.
Tarski, A. (1935) 'The Concept of Truth in Formalized Languages', in A. Tarski: *Logic, Sematics, Metamathematics* (2nd edn) J. Corcoran (ed.), Hackett, Indianapolis, 1983, 152–278.
—— (1936) 'The Establishment of Scientific Semantics' in A. Tarski: *Logic, Semantics, Metamathematics* (2nd edn) J. Corcoran (ed.), Hackett, Indianapolis, 1983, 401–8.
—— (1944) 'The Semantic Conception of Truth and the Foundations of Semantics', *Philosophy and Phenomenological Research* 4, 341–76.
—— (1969) 'Truth and Proof', *Scientific American* 220 (June 1969), 63–77.
Woleński, J. (1993) 'Tarski as a Philosopher', in F. Coniglione *et al.* (eds.) *Polish Scientific Philosophy: The Lvov-Warsaw School,* Rodopi, Amsterdam, 319–38.
—— (2001) 'In Defense of the Semantic Definition of Truth', *Synthese* 126, 67–90.

11

Reflections on Consequence

John Etchemendy

In The *Concept of Logical Consequence* (CLC) [13], I presented an extended argument that the standard, Tarskian analysis of logical consequence and logical truth is wrong. In the years since its publication, over a score of authors have written reviews, articles, or portions of books criticizing various arguments I gave in the book.* Nearly all have presented what the authors considered devastating replies to some or all of my arguments. Many of the replies are very thoughtful and contain much with which I entirely agree. Other authors misunderstood crucial parts of my argument, no doubt because I expressed them poorly. But all in all, the attention the book has received is gratifying. My only regret is that due to an onerous administrative appointment at my university, I was unable to reply to individual articles as they came out.

This chapter is not meant to be a "reply to my critics." Such a reply would be of very little interest to any one reader, inasmuch as the critics themselves disagree so sharply on fundamental points, and so the lines of criticism are often at odds with one another. Instead, the chapter is meant to be a rethinking of my overall argument in light of what I have learned from the various critiques, in particular what I have learned about ways in which CLC was confusing, incomplete, or otherwise misleading. Where appropriate, I indicate in footnotes how points made in this chapter relate to specific criticism that has appeared in print.

Let me say at the outset that I still believe that all of the significant points made in the book are essentially correct. Indeed, I am confident that most readers will not need my help answering some of the criticism that has been offered in the literature. A fair amount of that criticism has centered on historical questions about Tarski's 1936 chapter "On the Concept of Logical Consequence" [37], which I took as the philosophical locus of the standard analysis. Tarski's short chapter is remarkably puzzling in many ways, but as I said in the book, and as most commentators agree, the important issue is not what Tarski was thinking when he wrote the chapter, but whether the account he proposed is correct. Though I still hold to the reading of Tarski's chapter

---

* A partial list is contained in the references at the end of this chapter. Some of these were reactions to my earlier articles [11] and [12] (plus private correspondence) which covered some of the points discussed at greater length in CLC. I apologize to any authors whose work I have inadvertently overlooked.

described in the book, the historical debate is more a distraction than a useful guide to the important issues surrounding the account.

The articles and reviews have, however, convinced me that I made at least two serious mistakes in writing CLC: one a sin of commission, the other a sin of omission. The sin of commission was that I simply included too much, trying to anticipate objections, rationales, and modifications that might be raised in defense of Tarski's account. This is all to the good in one sense, but in another made it hard to see the forest for the trees. My overall argument, the "forest," can be summed up quite succinctly. Tarski's analysis involves a simple, conceptual mistake: confusing the symptoms of logical consequence with their cause. Once we see this conceptual mistake, the extensional adequacy of the account is not only brought into question—itself a serious problem given the role the semantic definition of consequence is meant to play—but turns out on examination to be at least as problematic as the conceptual adequacy of the analysis. To put it bluntly, the account fails both conceptually and, in most applications—in fact in virtually all applications—extensionally as well. That is the forest. In this chapter I'll try to fill in just enough of the trees to make the justification of these claims clear.

Fixing the sin of omission, unfortunately, pulls in the opposite direction. In the book I intentionally avoided discussing my own views on consequence and model-theoretic semantics. At the time, I thought it better not to muddy the discussion with both a negative argument directed at Tarski's analysis and a positive argument for an alternative view. Since the defect in the analysis is entirely independent of my positive views, I did not want readers to imagine that the positive views, with which they might disagree, were somehow part and parcel of the critique.

I see now that this strategy was a mistake. First, several authors have criticized the book by proposing views of model-theoretic semantics remarkably similar to my own. The fact that they consider their proposed accounts of model theory criticism of my arguments suggests that they have seriously misconstrued the target of my critique. In particular, the critique is not aimed at model-theoretic techniques, properly understood, nor at the view that logical consequence is a fundamentally semantic, not syntactic, notion. Second, I now realize that without the positive side of the story, readers were legitimately puzzled about the overall significance of my arguments. If we acknowledge that the Tarskian analysis is wrong, does it mean that large tracts of accepted logical practice must be abandoned? Do many of the main technical results in the field suddenly lose their intuitive or philosophical significance? In fact, I think the significance is quite the opposite. Recognizing the flaw in the standard analysis has a liberating effect: it allows us to give a sensible account of much work that does not fit neatly into the picture that results from the flawed analysis, and opens up new areas of legitimate study that seem precluded by the analysis. Although I suggested as much at the end of my book, it is clear that the assertion alone was not enough to give readers a clear understanding of why this might be so.

My second goal in this chapter, then, is to provide the missing, positive account that I should have included in CLC. By this I don't mean I will propose a competing analysis of logical consequence: this project will have to wait until later. But I will sketch what I consider the proper understanding of model-theoretic semantics and its relation to the pretheoretic notions of logical consequence and logical truth. I claim that model theory, properly understood, does not yield an analysis of the logical properties, but presupposes them. This is not to say that a model-theoretic semantics for a particular language does not illuminate the logical properties of the target language. It does, but the illumination results not from an analysis of the basic logical notions, but rather from an entirely different characteristic of the model-theoretic technique, a characteristic which I will isolate and explain.

In order to add the positive account without obscuring the main argument even more—adding more trees, so to speak—I devote the first two sections of this chapter to my criticism of Tarski's analysis, and postpone discussing my positive views on logic, model theory, and the consequence relation to later sections. I am convinced that the point of view described in the latter two sections is in fact widely held, though not widely discussed, and so make no claims for the originality or novelty of that view. Still, I emphasize once again that my criticism of Tarski's account does not rely on the acceptance of that point of view.

CONCEPTUAL ADEQUACY OF TARSKI'S ACCOUNT

Let me begin by briefly recounting Tarski's analysis and sketching the main objections I raise in the book. For the reasons explained, I won't try to repeat the detailed arguments presented there, but will simply suggest their flavor and encourage the reader to go back to the original if the current summary is unsatisfying or if it seems I've overlooked an obvious point.

Tarski proposes a reductive analysis of the logical properties. The analysis purports to reduce logical truth to ordinary truth (or satisfaction) of an associated generalization (or open formula). Similarly, it reduces logical consequence to material consequence plus generalization: an argument is logically valid, according to the analysis, if every argument in an associated class of arguments preserves truth, where by "preserves truth" we mean simply that it has one or more false premises or a true conclusion. The analysis ensures that a logically true sentence is true because it is an instance of the associated generalization, and so could not be false without falsifying the generalization. It ensures that a logically valid argument preserves truth—has a false premise or true conclusion—because the argument is a member of its associated class.

I'll say more in a moment about how we get from a sentence to its associated generalization, or from an argument to its associated class of arguments. But for now let me note the remarkable appeal of such a reductive analysis. Surrounding the intuitive concepts of logical consequence and logical truth are a host of vague and philosophically difficult notions—notions like necessity, certitude, *a prioricity*, and so forth. Among the characteristics claimed for logically valid arguments are the following: If an argument is logically valid, then the truth of its conclusion follows necessarily from the truth of the premises. From our knowledge of the premises we can establish, without further investigation, that the conclusion is true as well. The information expressed by the premises justifies the claim made by the conclusion. And so forth. These may be vague and ill-understood features of valid inference, but they are the characteristics that give logic its *raison d'être*. They are why logicians have studied the consequence relation for over two thousand years.

In spite of the importance of these characteristics, we needn't be happy about the vagueness. And that is why the reductive analysis of consequence is so attractive. Tarski shows us, if he is right, how the logical properties can be reduced to the well-understood notion of truth, plus whatever is involved in specifying the associated generalization or class. Like magic, the vague and obscure notions that sit at the core of our discipline simply disappear. No wonder we find Tarski's account so appealing: if it works, it allows us to set aside a breathtaking number of philosophical issues. Who could not want the account to succeed?

So how does Tarski characterize the associated generalization and the associated class of arguments? Let's consider arguments first. According to Tarski's account, the associated class consists (roughly speaking) of all arguments displaying a similar "logical form," where logical form is defined by the appearance of the so-called "logical constants," plus the pattern of the remaining expressions. So, for example, assuming *if . . . then* is the only logical constant in the following argument:

If Tarski was right, then Etchemendy is wrong.
Tarski was right.
So, Etchemendy is wrong.

the associated class consists of all arguments displaying the following form:

If P then Q
P
So, Q

What Tarski means by "all arguments" of the displayed form is not just arguments actually expressible in the language, but all arguments expressible in sufficiently similar languages. Exactly what is meant by this isn't crucial for our purposes. But a more modern way of expressing the analysis is to stick with the original argument and ask whether it preserves truth (in the actual world) regardless of how we interpret the constituent sentences *Tarski was right* and *Etchemendy is wrong*, or alternatively, regardless of how we interpret the names *Tarski* and *Etchemendy*, and the predicates *was right* and *is wrong*. If the argument preserves truth on all interpretations of the non-logical constants, then it is said to be logically valid.

The definition of logical truth is similar. We hold fixed the logical constants in the sentence, and quantify away the contribution of the remaining expressions. For example, the sentence:

If Tarski was right, then Tarski was right

is logically true, according to the analysis, because every instance of the form *If P then P* (or *If R(a) then R(a)*) is simply true. Or in modern parlance, the displayed sentence is logically true because it remains true however we interpret the expressions *Tarski* and *was right*.

I have already mentioned the great attraction of the reductive analysis: the fact that it replaces a host of obscure modal or epistemic notions with the vastly clearer notion of truth. But we can now add to that attractiveness the observation that the account is also quite plausible, at least at first glance. Consider our sample argument above. This is a logically valid argument, and so of course preserves truth—that is, has a false premise or true conclusion. What's more, it is an instance of an argument form, *modus ponens*, all of whose instances preserve truth. Indeed, instances of any of the well-known rules of inference—the Aristotelean syllogisms, universal instantiation, even the omega rule—display these same features. Each such argument is an instance of an argument form all of whose instances preserve truth. The reductive analysis simply takes the natural step of proposing these obvious features of valid rules of inference as *definitive* of the logical consequence relation. This is surely an attractive and plausible proposal.

In CLC, I argued that the reductive analysis fails. In what sense? Roughly speaking in any sense that would give it philosophical, logical, or foundational interest. In this section, I begin with the easy part: isolating the conceptual flaw in the account. Suppose we ask whether the account captures, directly or indirectly, any of the intuitive characteristics of the consequence relation mentioned above. Does it guarantee, or for that matter give us any reason to expect, that inferences that qualify as valid according to the analysis will have any of the modal, epistemic, semantic, or informational characteristics ascribed to logically valid arguments? My concern here is not with the extension of the account, which I will discuss at length in the next section, but with its conceptual adequacy: with the question of whether there is any *conceptual* assurance that arguments declared valid by the account will in fact be genuinely so.

Perhaps not surprisingly, the answer is *no*. This is simply a generalization of an observation first made by Wittgenstein when Russell entertained a similar, reductive account of logical truth. I will make the point in my own words, rather than Wittgenstein's, and about logical consequence rather than logical truth, since that is the important concept. The crucial point is this. The property of being logically valid cannot simply consist in membership in a class of truth preserving arguments, however that class may be specified. For if membership in such a class were all there were to logical consequence, valid arguments would have none of the characteristics described above. They would, for example, be epistemically impotent when it comes to justifying a conclusion. Any uncertainty about the conclusion of an argument whose premises we know to be true would translate directly into uncertainty about whether the argument is valid. All we could ever conclude upon encountering an argument with true premises would be that either the conclusion is true or the argument is invalid. For if its conclusion turned out to be false, the associated class would have a non-truth-preserving instance, and so the argument would not be logically valid. Logical validity cannot guarantee the truth of a conclusion if validity itself depends on that self-same truth.

It might help to look at an analogous, but obviously faulty definition of consequence where the same problem arises. Suppose we defined a logically valid argument as an argument that simply preserves truth, that is, has either a false premise or a true conclusion. Then of course any "valid" argument with true premises would have to have a true conclusion, since otherwise it would be invalid. Nevertheless this definition of validity misses the crucial feature of genuine consequence: *the* fact that we can draw conclusions about sentences whose truth values we do not antecedently know, based on our knowledge of other sentences that logically imply them. If logical validity were nothing more than truth preservation, then our knowledge that the premises of an argument are true would only tell us that either the conclusion of the argument is true (and hence the argument "valid") or the conclusion is false (and hence the argument invalid). Whatever enables us actually to infer consequences from our knowledge of other claims has simply dropped out of the account.

I claim that Tarski's reductive analysis of consequence, though certainly more involved, suffers from precisely the same conceptual omission. The fact that validity is tied to a larger class of arguments does not help, nor does the appeal to logical constants as a means of specifying that class. Indeed the problem remains no matter how narrowly one construes the term "logical constant," no matter how general the resulting argument forms. Indeed, consider again the instance of *modus ponens* shown above. Surely, if anything is a logical constant, *if . . . then* must be. But even here, if the logical validity of the argument came down to nothing more than the fact that every instance of the illustrated form preserves truth, then the truth of the premises could never be used to establish the truth of the conclusion. The conclusion might be true and the argument valid, or the conclusion false and the argument invalid. The consequence relation, as characterized by the reductive account, involves nothing that would incline us toward one of these possibilities rather than the other.

Of course, this problem does not infect genuinely valid arguments, like the instance of *modus ponens* illustrated above. Indeed, that's the point. The crucial feature of *modus ponens* is that we can recognize that all of its instances preserve truth without knowing the specific truth values of the sundry instances. My own view is that we recognize this by virtue of the meaning of the expression *if . . . then* and our knowledge of how the remaining constituents can contribute to the truth values of the premises and conclusion. This gives us an independent guarantee—independent, that is, of the actual truth values of premises and conclusions—that all the instances of the argument form preserve truth. This independent guarantee, and only this independent guarantee, is what enables us to infer that a conclusion is true on the basis of the truth of the premises.

It is obvious that in the absence of such a guarantee, we would not have a logically valid argument, regardless of which expressions we considered logical constants and regardless of the truth preservation of its instances. Suppose we have an argument form all of whose instances preserve truth, just as the reductive account requires, but suppose that the only way to recognize this is, so to speak, serially—by individually ascertaining the truth values of the premises and conclusions of its instances. In other words, suppose there is no independently recognizable guarantee of truth preservation, as there is with *modus ponens*, only the brute fact that the instances preserve truth. Would an instance of this argument form be logically valid? Clearly not. For example, we could never come to know the conclusion of such an argument in virtue of our knowledge of its premises. Indeed, the premises would provide no justification whatsoever for a belief in the conclusion. For, by hypothesis, knowing the specific truth value of the conclusion in question would be a prerequisite to recognizing the "validity" of the argument.[1]

We can and do recognize that all instances of *modus ponens* preserve truth, and we do this without having the foggiest idea of the actual truth values of most of the sentences that make up those instances. The characteristic that enables us to do this, at least in the case of simple valid forms, is clearly the essential feature of logical consequence. For an argument *with* this characteristic can be used to extend our knowledge: we can know antecedently that the argument preserves truth, subsequently discover that its premises are true, and thereby infer that its conclusion is true as well. By contrast, an argument form *without* this characteristic could never be so deployed.

I said a moment ago that the conceptual problem with the reductive account applies regardless of how tightly constrained our notion of a logical constant. Let me expand on this a bit. When presented with the omission just described, people often reply that the reductive account is not defective, but rather incomplete. What is required is a careful characterization of the logical constants. This characterization, it is thought, will explain why when these and only these expressions are held fixed, the only arguments that can possibly satisfy the reductive definition are those that display the required guarantee of truth preservation, that is, those that are genuinely valid. In other words the crucial guarantee, it is thought, flows jointly from the truth preservation of the associated arguments plus certain special features of the logical constants. Of course, until we see what those features are, this is little more than an article of faith, though an article of faith that has sustained many a supporter of the reductive analysis.

But this article of faith is simply false. It is not hard to prove that there are no features of the logical constants capable of providing this assurance, at least on the assumption that the truth-functional connectives are logical constants. To see this, consider the following argument form:

$$P(a) \wedge Q(a)$$
$$\neg P(b)$$
$$\text{So, } P(c) \rightarrow Q(c)$$

This is obviously not logically valid: from premises of the indicated forms we are in no way justified in inferring the corresponding conclusion. Now it happens that there are non-truth-preserving instances of this argument form. But notice that this is not guaranteed by any features, global or local, of the truth-functional connectives appearing in the argument. For there would not be any non-truth-preserving

---

[1] Graham Priest [25] accuses me of confusing the "epistemic order" with the "definitional order," drawing an analogy with the notion of computability and the Church/Turing analyses of that notion. As Priest correctly points out, we might know that a function is effectively computable without knowing that it is Turing computable. But Priest has misunderstood my argument, the point of which is that it is possible for an argument that is not in fact logically valid—one that has none of the epistemic or other characteristics of a valid argument—to satisfy Tarski's definition. The point has nothing to do with knowing whether Tarski's definition applies, but rather with the characteristics (or lack of characteristics) of the arguments to which the definition could in fact apply. The right analogy would be if we could show that it is possible for functions that are not effectively computable to be Turing computable (which of course we can't).

instances if the world contained only two objects, or if all the objects in the world fell into two indistinguishable types. And yet the truth-functional connectives would presumably still have whatever features we thought definitive of the logical constants. This shows that the conceptual flaw in the reductive analysis will never be corrected by specifying characteristics of the logical constants, at least on the assumption that these characteristics are enduring features of the truth-functional connectives. *The source of the guarantee observed in genuinely valid arguments is not the truth preservation of their instances plus special features of the logical constants.*

No selection of logical constants rules out the possibility discussed four paragraphs back: arguments that satisfy the reductive definition due to the "brute fact" that their instances preserve truth, but which do not display the guarantee of truth preservation that makes an argument genuinely valid. No matter how we characterize the logical constants, Tarski's definition provides no assurance that every argument that satisfies the definition will be logically valid.

Let me summarize. Genuinely valid arguments carry with them an independent guarantee of truth preservation, a guarantee that can be recognized antecedent to our knowledge of the actual truth values of their premises and conclusion. Now what is important for present purposes is not how we diagnose this crucial guarantee. I've indicated that I think it emerges from semantic characteristics of the language, but others may disagree. Kant, if I remember correctly, attributed it to the *a priori* structure of the understanding; others seem content to appeal to a primitive notion of logical necessity. But in any event, what is important for now is only that we recognize the following two points. The first is that without such an independent guarantee of truth preservation, logical consequence would be a completely flaccid relation. It would be impossible to use logically valid arguments to extend our knowledge, to justify the truth of a conclusion, or to prove that a given theorem follows from accepted axioms. The second point is that the reductive analysis just omits the guarantee, attempting to replace it with that which the guarantee is a guarantee *of*. We ignore whatever it is that assures us that every instance of a logically valid form preserves truth, and say that logical validity simply *consists* in every instance of the given form preserving truth. It is like confusing the symptoms with the disease, effects with their cause: understandable confusions, but confusions nonetheless.

I take both of these points to be undeniable, the first about the consequence relation, the second about the reductive analysis. When you think carefully about these two points, you will see that they show that none of the central characteristics of the consequence relation—whether modal, epistemic, semantic, or informational—are captured by Tarski's analysis. Since much of my discussion in CLC focused on the standard modal characterization, many commentators have not quite understood this point. For example, Timothy Smiley interprets my book as an attack on the "nonmodal aspect" of Tarski's definition. Smiley goes on to say:

> A debate is called for, but it will be more fruitful if it asks for what purposes necessity is an essential ingredient of consequence. For example, someone who does not endorse Aristotle's doctrine of proof and episteme may well be content with proofs that establish the bare truth of theorems, and it is not obvious that this requires a modal relation of consequence. [36]

My reply should be obvious: Fine, jettison all talk of modalities.[2] Concentrate on nothing more than the fact that the consequence relation allows us to establish the truth of sentences based on the truth of others. But that is precisely the problem, precisely the characteristic that Tarski's definition ignores. If the consequence relation involved nothing more than what the reductive definition maintains, then the relation could never be used to "establish the bare truth of theorems." When we encounter a new inference of any specified form, our sole guarantee would be that it either preserves truth or constitutes a counterexample in virtue of which the argument form is invalid. This guarantee, the only one Tarski's definition offers, can never establish the truth of anything.

This was, as I said, the easy part. It is clear that Tarski's definition tries to reduce a "cause"—the logical consequence relation—to its "symptoms," the truth preservation that the consequence relation guarantees. And it is equally clear that this guarantee of truth preservation is the essential feature of logical consequence, the feature that makes it possible to infer the conclusion of a valid argument from its premises. In short, the reductive analysis omits the single most important characteristic of the consequence relation. Let's consider what follows from this fact.

EXTENSIONAL ADEQUACY OF TARSKI'S ACCOUNT

We might summarize the observations of the preceding section with a simple slogan: All of the instances of *modus ponens* preserve truth because it is a logically valid argument form. This is true. What is false is that *modus ponens* is logically valid *simply because* its instances preserve truth. What follows from the fact that Tarski's definition is based on the latter, faulty assumption?

First and most obvious, it follows that we have no assurance that the reductive account will yield the correct extension. When we apply the definition to an arbitrary language, choosing some subset of its expressions as logical constants, we have no blanket assurance that the arguments declared logically valid in fact *are* logically valid. Let me call this the question of *overgeneration*. There is an equally problematic question of *undergeneration*—are there any logically valid arguments in the language which, on the given selection of logical constants, are not declared valid—but let me set this second issue aside for the moment.

In CLC, I discuss at length how, when, and where the reductive account overgenerates. Obviously, applications of the account will overgenerate if there are argument forms all of whose instances in fact preserve truth, yet which do not provide the guarantee of truth preservation required of logically valid arguments. Intuitively, this can happen if the truth preservation is an upshot of facts that have nothing to do with

---

[2] Note, by the way, that my own explanation of the validity of *modus ponens* is semantic, not modal. I am no particular fan of modal characterizations of consequence, and thought I had made that relatively clear in CLC. In spite of that, some commentators seem to have concluded that I identify logical truth with necessary truth and logical consequence with necessary truth preservation. Nothing could be further from the truth.

logic, the consequence relation, or anything plausibly related to it.[3] As I explain in the book, this happens whenever the language, stripped of the meanings of the non-logical constants, remains relatively expressive, or if the world is relatively homogeneous, or both. This is not hard to see. If the expressions we've chosen as logical constants are sufficiently expressive, then it will always be possible to come up with argument forms whose instances uniformly preserve truth in spite of the fact that they are not logically valid: we need only find a non-logical generalization that is expressible in these terms and cast it into an argument form. The more homogeneous the world, the easier this task becomes, since fewer expressive resources are required.

So when does Tarski's account work, in the sense of not overgenerating? Well, we can say with confidence that the account works when applied to the language of propositional logic, treating the truth-functional connectives ($\neg$, $\wedge$, $\vee$, $\rightarrow$, etc.) as the sole logical constants. This is an exceedingly weak language, and so long as the world provides us with an adequate supply of true and false propositions, the only argument forms whose instances universally preserve truth are those whose truth preservation is guaranteed by the meanings of the chosen connectives. They are all, in other words, genuinely valid arguments. Mind you, it is easy to introduce a new "logical constant" for which this is not the case, for example the operator $\odot$ discussed in chapter 9 of CLC. But if we treat the truth functions as the only logical constants, this application of the account successfully avoids the problem of overgeneration. This is not because, even here, the reductive account is testing for the right thing, but only because the truth-functional operators are expressively weak and the world sufficiently heterogeneous.

What happens when we move to the language of first-order logic? To put it mildly, things get complicated. If we apply Tarski's unmodified definition to such a language, adding the first-order quantifiers and identity predicate to our list of logical constants, then the account overgenerates right and left. For one thing, the quantifiers and identity predicate allow us to express many numerical truths that are substantive, non-logical claims about the world. For example, we can express the fact that there are more than three billion objects using a sentence that contains only the quantifiers, the identity predicate, and the truth-functional operators. Call this sentence β. According to the reductive account, β is a logical consequence of any premises whatsoever, since any argument with this conclusion preserves truth. What is happening here is exactly what we predicted two paragraphs back: the chosen logical constants are now sufficiently expressive that the basic conceptual flaw in the account manifests itself in concrete, extensional errors. Every instance of the argument form:

$$P$$
$$So, \beta$$

---

[3] I am being intentionally vague here, but only to maintain an ecumenical stance about the source of the consequence relation. Those who share my view that the relation emerges from the semantic characteristics of the language can replace the vague phrase "facts that have nothing to do with logic" with "facts that have nothing to do with the semantics of the language." Others can make corresponding replacements, depending on their views of consequence. In what follows, I will continue to speak of "non-logical" or "substantive" facts in this way, leaving it to the reader to make appropriate substitutions.

preserves truth, in spite of the fact that these arguments do not display the required guarantee that would justify an inference from premise to conclusion. The fact that the argument form contains only expressions traditionally considered logical constants is no protection against the fundamental mistake made by the reductive account: taking the symptoms of the genuine consequence relation as definitive of the cause. The above argument form is just one of many examples of the misdiagnoses that result; I refer the reader to CLC for additional examples and a more extensive discussion of this phenomenon.

This sort of example was used by Wittgenstein to convince Russell of the flaw in the reductive account. Modern applications of the analysis avoid this embarrassment by adding, without explanation or rationale, a new twist. Rather than say that any argument form all of whose instances preserve truth is logically valid, we require additionally that the instances preserve truth *in every (actual but possibly restricted) domain of quantification*. Since there are domains of quantification containing fewer than three billion objects, we thereby dodge these particularly blatant instances of overgeneration.

In CLC, I argue that this new twist is at best unmotivated and at worst inconsistent with the original, reductive account.[4] Be that as it may, it is not worth repeating those arguments here, for the revised account is subject to the same fundamental flaw as the original reduction. It is still possible for an argument form to have only truth preserving instances—instances that preserve truth in every existing domain—without being logically valid, without having the guarantee of truth preservation needed to support an inference from premises to conclusion. In the book I emphasize this in various ways, including pointing out the peculiar position of the finitist who, if the reductive account were correct, would be forced to accept as logically valid many first-order arguments that obviously are not. For example, if there is a largest domain, then any inference whose conclusion asserts that there are no more objects than the cardinality of this domain will be incorrectly declared logically valid. Similarly, the inference from the claim that a relation is transitive and irreflexive to the claim that the relation has a "least" or "greatest" element will be declared logically valid if all domains are finite, even if there is no largest domain.

Many people have misunderstood the point of this argument, which is not directed at the finitist, nor meant to show that Tarski's analysis presupposes the axiom of infinity—perhaps an interesting point, but not an objection I would consider significant. Rather, it shows that even the modified account, incorporating varying domains of quantification, suffers from the same conceptual flaw described in the last section. It still provides no conceptual assurance—whether due to the truth preservation of instances, characteristics of the logical constants,

---

[4] Giving this argument in detail would be impossible in an article, but let me rule out one motivation that may spring to mind. We might reason that although the universe is the size it is, it could have been larger or smaller, and varying the domain is meant to take account of these possibilities. But consideration of the possible size of the universe is completely irrelevant to the reductive account, which is based on the assumption that we can reduce logical consequence to facts about how the world actually is, not how it could have been. Surprisingly enough, many commentators have missed this basic point.

or the variation in the domain—that all arguments which satisfy the definition are actually valid. The symptoms of consequence on which Tarski's account is based are not a reliable indicator of genuine consequence, even when we vary the domain of quantification.

This is an important realization, even if you believe, as do I, that finitism is false; indeed, even if you believe that finitism is necessarily false.[5] For although the availability of infinite domains may assure us that the specific examples mentioned two paragraphs back are not mistakenly declared valid, it does nothing to assure us that there aren't other arguments that are. It is clear that the output of even the modified account depends on facts, such as the size of the universe, that have no bearing on the logical consequence relation. What is not clear is whether any such facts expressible in the first-order language cause this application of the account to overgenerate. This uncertainty is a direct consequence of the conceptual flaw already discussed, and the flaw applies with full force to the modified definition.

Assuming there are no finitists among us, and assuming as well a reasonably powerful set theory, it turns out that the modified reductive account does not overgenerate when applied to the first-order language. Now if you understand that I've said so far, a question should immediately come to mind: How do we know that this application of the account does not overgenerate? This is not an idle question. If we apply even the modified account to first-order languages, and limit ourselves to considerations internal to the Tarskian definition of consequence, *there is absolutely no way to determine whether all of the arguments declared logically valid are in fact logically valid.* There is no way to rule out the possibility that general, extralogical facts expressible using the first-order quantifiers, identity and the truth-functional connectives give rise to truth preserving argument forms that are not logically valid, that display the symptoms of validity but not the underlying cause. Perhaps these facts are more complex cardinality claims similar to β, but whose truth is not blocked by the trick of varying the domain of quantification. Or perhaps there are obscure algebraic or set-theoretic facts that are true in every domain, but not because they are logically true. If we think we are assured that this application of the account does not overgenerate—whether on general philosophical grounds or because of any characteristics of the expressions we've chosen as logical constants—we are simply fooling ourselves.

---

[5] Vann McGee [21, 23] has replied to my argument by claiming that finitism is not simply false but necessarily false, since mathematical objects like pure sets exist necessarily. I am not sure how to assess the truth of this claim, though I suspect I agree with it. Still, it does not weaken my argument. When we apply Tarski's account to a particular language we make a host of decisions about the kinds of objects we will take as legitimate interpretations of the non-logical constants, for example whether predicates are interpreted by properties or arbitrary sets, and whether individual constants may refer to abstract objects as well as concrete ones. When we make the standard decisions on these matters—using sets to interpret predicates, and so forth—it follows from McGee's (entirely reasonable) assumptions about mathematical truth that an application of the account will indeed have the extension it has necessarily. When the extension is right, it will be necessarily right; when it is wrong, it will be necessarily wrong. If we accept McGee's assumption, then, the problem is that Tarski's account provides no general assurance that such an application of the account will be necessarily right rather than necessarily wrong. As we'll see, it is sometimes the former, sometimes the latter.

Though it is not an idle question, it does have an answer. In fact we can prove that this particular application of the account does not overgenerate by appealing to an entirely different tool for studying the consequence relation: a system of deduction. Now it is generally accepted that deductive techniques do not provide an analysis of the logical consequence relation. Nevertheless, it is possible to set down a simple collection of deductive rules whose repeated application will never permit us to prove a sentence that is not a genuine consequence of the assumed premises. How is this possible? First, we set out a handful of argument forms whose instances are all logically valid, that is, whose instances all display the requisite guarantee of truth preservation. Obviously, we choose forms like *modus ponens*, while avoiding those like "from any sentence, infer β," since the former do, but the latter do not, display the requisite guarantee. Second, we observe that the logical consequence relation is transitive, and hence repeated application of the primitive valid rules can never lead to a conclusion that is not a genuine consequence of the original premises.

What these two points show is that the careful application of deductive techniques allows us to design systems that are recognizably sound, systems we can be sure do not "overgenerate." And with first-order logic, it happens that we can use such a system to prove that the (modified) Tarskian account does not overgenerate, either. This follows from the so-called "completeness" theorem for first-order logic. The theorem assures us that any argument declared valid by the (modified) Tarskian account is provable in the deductive system, and hence is sure to be logically valid, thanks to the intuitive soundness of that system.[6] Seen in this light, the theorem is actually misnamed, for its import is to transfer our assurance of the soundness of one characterization of consequence, the deductive system, to another characterization of consequence, the Tarskian definition, whose "soundness" we can never independently ascertain.[7]

What I have just argued is that the application of the (modified) reductive account to first-order languages can be proven correct. Or rather, I've argued that we can prove this application does not overgenerate, since I've set aside for the moment the issue of undergeneration. But if you understand the argument, you will begin to see why I claim that the faulty analysis has little philosophical, logical, or foundational interest. The common mythology is that the Tarskian definition is important because we have an independent, conceptual assurance of its extensional adequacy, and this allows us, among other things, to prove the extensional adequacy of other characterizations of consequence, such as our system of deduction. But once we recognize the conceptual

---

[6] A more cautious and correct statement would be that the completeness theorem shows that if there are infinite domains, and if the presupposed axioms of set theory hold, then the modified reductive account does not overgenerate. The theorem obviously provides no assurance for the finitist, for if the finitist is right, this application of the account demonstrably overgenerates.

[7] This is closely related to Kreisel's construal of the completeness theorem in [20]. The difference is that Kreisel accepts the reductive account of logical consequence, but is worried about the fact that standard applications of it survey only domains that are sets, while we often use the first-order language to talk about proper classes of objects. But even if we included proper classes among the domains surveyed, we would still have to worry about the possibility of overgeneration. The completeness theorem assuages both worries, Kreisel's and mine.

flaw in Tarski's account, we see that it is not, contrary to mythology, in better shape than our deductive characterization of consequence. Quite to the contrary, the deductive techniques are actually in better shape: as we have seen, the careful application of these techniques can at least give us a characterization of consequence that is recognizably sound. This is more than we can say for the Tarskian definition of consequence, where our assurance of "soundness" is entirely derivative from the deductive system.

We can emphasize this point by asking what we can conclude in cases where we have an intuitively sound deductive system and a Tarskian definition of consequence, but the completeness theorem fails. In these cases, the Tarskian definition asserts the logical validity of arguments that go beyond what the deductive system can prove. The standard mythology would have it that in such cases the deductive system is incomplete. But this presupposes that the Tarskian definition is guaranteed not to overgenerate. Not only do we have no such guarantee, once we appreciate the flaw in the reductive analysis, we see that it predictably will overgenerate. So in the absence of a completeness theorem, our only legitimate conclusion is that *either* the deductive system is incomplete, *or* the Tarskian definition has overgenerated, *or possibly both*.

We have already seen a simple example of this. If we apply the unmodified Tarskian definition to a first-order language, we will not be able to prove completeness, since the standard deductive system cannot prove sentences like β from arbitrary premises. But this does not show that the deductive system is incomplete, but rather that this application of the reductive definition overgenerates: β is not a logical truth, and in fact should not be provable from random premises.

A more interesting example is second-order logic, where the problem appears even in the modified account. If we apply Tarski's account of consequence to this language in the most natural way, treating both first- and second-order quantifiers as logical constants, then the resulting consequence relation extends well beyond any intuitively sound deductive system for the language. But what can we conclude from this? Can we infer that there is no complete deductive system for second-order languages, that any candidate system leaves some *genuinely* valid arguments unprovable? Or is the problem that the Tarskian definition of consequence overgenerates when applied to these languages, declaring sentences logically true and arguments logically valid that in fact are not? Or perhaps both?

The answer is that we can't really tell, at least not based on Tarski's reductive definition of consequence. It is well known that when the Tarskian definition is applied to second-order languages, certain highly abstract set-theoretic claims are declared logically true. For example, we can easily formulate sentences containing only identity, truth functions, and first- and second-order quantifiers that are true in all domains if and only if the Continuum Hypothesis is true, and other sentences that are true in all domains if and only if the Continuum Hypothesis is false. Let CH and $\neg CH$ be representative sentences of this sort.[8] If the Continuum Hypothesis is

---

[8] For example, we can take CH to be the second-order sentence that says there are no subsets of the domain larger than **N** and smaller than **R**. Since "larger than **N**" and "smaller than **R**" are definable in (full) second-order logic, this sentence will be true in all domains just in case the Continuum Hypothesis is true. Similarly, we can take ¬CH to be the sentence that asserts that if

true, then *CH* will be declared, by the reductive account, a logical consequence of any sentence whatsoever. If the Continuum Hypothesis is false, the latter sentence will have a similar fate. But from an intuitive standpoint it seems that neither the argument form:

$$P$$
$$So,\ CH$$

nor the argument form:

$$P$$
$$So,\ \neg CH$$

displays the guarantee of truth preservation required of logically valid arguments.[9] It does not seem that we are logically justified in concluding CH from a random premise, even if the Continuum Hypothesis happens to be true (perhaps unbeknownst to us). A more reasonable hypothesis is that the identification of the symptoms of consequence with genuine consequence here fails, thanks to the expressive power of this language. All of the instances of one of these argument forms will indeed preserve truth, but not because it is logically valid.[10]

As I said, the wayward behavior of the reductive analysis when applied to second-order languages is well known. Some philosophers, including Quine, have concluded from examples of this sort that *second-order logic is not logic*. This conclusion has to count as one of the more surprising and implausible conclusions of recent philosophy. After all, second-order languages, like all languages, have a logical consequence relation. Some inferences employing the expressive devices of these languages are logically valid, and others are not. True, the consequence relation for these languages may be vague or underspecified, depending on the vagueness or underspecification of the expressions that make up the language, and perhaps also because of the vagueness of our understanding of the consequence relation itself. But the idea that studying the logic of these languages is somehow not the business of logic is hardly a supportable conclusion. If we are convinced that the above argument forms are not logically valid—certainly a reasonable position—then we should simply conclude that this

---

the domain is at least the size of **R**, then there are such intermediate-sized subsets. One of these is true in all structures, though we don't know which.

[9] I should mention, in case it is not obvious, that the fact that I've used argument forms with trivial premises is not significant. In both this and the first-order case I could give examples in which the forms of the premises are significant. The examples I've chosen are just easier to describe.

[10] Much more could be said about second-order logic than I can say here. For example, there are various ways to modify the interpretation of the second-order quantifiers to decrease the expressive power of the language. We can, for instance, construe them in the manner of so-called "weak" second-order logic, or perhaps as plural quantifiers (appropriately generalized to handle relation variables). And as the expressive power of the chosen logical constants decreases, so too will the instances of overgeneration, for reasons I have already explained. But this does not affect the point made in the text. When the quantifiers are interpreted as quantifying over all subsets of the domain—surely a possible interpretation, and probably the most natural—the problems discussed here unavoidably arise. All that matters is that this is a possible interpretation of second-order quantification; whether there are also weaker interpretations is irrelevant.

is a case where Tarski's analysis overgenerates, something that we know, for conceptual reasons, is bound to happen. The symptoms of consequence—truth preserving instances—are not reliable indicators of the sought-after cause: genuine logical consequence.

So far, I have only discussed the problem of overgeneration, but the reductive account can undergenerate as well. Most obviously, it will fail to detect logically valid arguments if the validity of those arguments depends on expressions not in the chosen collection of logical constants. To take a simple example, suppose we apply the account to an interpreted propositional language, treating only the truth-functional connectives as logical constants. Although all the arguments identified as valid in this case are genuinely so, there may well be valid arguments that are overlooked. For example, suppose the following argument is expressible in the language (where *Triangle(a)* asserts that *a* is a triangle):

> *Triangle(a)*
> $a = b$
> So, *Triangle(b)*

This argument would not be declared valid for obvious reasons: its validity depends on the meaning of the identity predicate, which is not among the expressions we've treated as logical constants, plus the fact that *Triangle* is an extensional predicate. Similarly, if the language contains quantifiers, but we do not treat them as logical constants, most of the interesting arguments of first-order logic will be judged invalid. None of this, of course, is the least bit surprising.

The real problem with undergeneration arises when a language contains expressions that figure into valid argument forms, but which we cannot treat as logical constants in Tarski's account for fear of the opposite problem: overgeneration. For recall that the more expressive the list of logical constants, the more likely it is that the reductive account will overgenerate. But how often does this occur? In the book, I give a very simple, but artificial example of a language in which no selection of logical constants characterizes what intuitively seems the right set of logically valid arguments. But let's avoid artificial examples and jump headlong into a controversial one.

Suppose the language in question contains a binary predicate, say $\cong$, that asserts that two objects are identical in shape. Then it seems at least arguably the case that the conclusion of the following argument follows logically from its two premises, much like our previous example:

> *Triangle(a)*
> $a \cong b$
> So, *Triangle(b)*

Surely, this conclusion must be true provided the premises are true: indeed, its truth preservation is guaranteed by the meanings of the constituent expressions. One could even argue that it is formally valid, since it holds for any *a* and *b*, and even holds when we replace one shape predicate with another. Unfortunately, if we treat enough predicates as logical constants to validate this (and similar) arguments, Tarski's account is sure to overgenerate. Contingent facts about the shapes of objects in the universe will result in arguments that are declared logically valid, but which do not display the guarantee of truth preservation that seems evident in our chosen example.[11]

As I said, this is a controversial example, but it is worth mentioning for a couple of reasons. For one thing, it is not obvious that this example is all that different from the previous argument, which is universally acknowledged to be logically valid. Numerical identity justifies substitution of individual constants, so long as the predicates involved are extensional. Identity of shape would seem to justify similar substitutions, albeit within a more narrow class of predicates. Given the similarity of these inferences, it is hard to see why they should be treated differently. Of course, most philosophers have been raised, under the influence of Quine, to say that the former inference is an instance of logical consequence, while the latter is something quite different: "analytic consequence," or something of the sort. But the idea that the justification underlying the first inference is different in kind from the second is supported by nothing more than the fact that the reductive account of consequence can be made to work in the first case but not in the second. Given the flaw in the reductive account, this is hardly a persuasive consideration.

Still, philosophers are extremely wary of any mention of the meaning of predicates—with the exception of identity, which receives special dispensation. So are there other cases where this problem occurs, where any selection of logical constants either overgenerates or undergenerates? Well, how about first-order logic? Before modifying the reductive account, this was precisely the situation we were in. If we include the standard collection of logical constants, then sentences like β turn up as logical truths. But if we delete any of these from the list, many obviously valid arguments are not so declared. And how about second-order logic? Once again we have the identical problem, only this time varying the domain doesn't come to the rescue. If we include the second-order quantifiers among the logical constants, then claims like CH (or $\neg CH$) are declared logically true. But if we exclude them, many intuitively valid arguments are judged invalid.

Many logicians and philosophers react to my conceptual critique of Tarski's account by retreating to an extensional stance, saying the only thing that really matters is that the analysis be extensionally correct. I have no doubt that, when push comes to shove, Tarski would have said the same thing—as, in fact, would I. So let's try to assess the "material adequacy" of the definition, as Tarski would have put it. What can we say about the account from a purely extensional standpoint? We can say that it is an unqualified success in one case: propositional languages in which the atomic sentences are logically independent. *But that is about all we can say without adding significant caveats.* With a first-order language, the analysis fails unless we add an important modification whose consistency with the original analysis has yet to

---

[11] For instance, suppose $P(x)$ is a shape predicate satisfied by only finitely many objects, say n. Then $\neg P(a)$ will be declared a logical consequence of any collection of premises that imply that N objects not equal to a satisfy $P(x)$. Just as we noted earlier that in the first-order case we must assume an infinite universe or the (modified) Tarskian account will overgenerate, here we would, for a start, have to hope that every shape predicate is actually satisfied by infinitely many objects. Again, note that appeals to "possible objects" and "possible satisfaction" are completely irrelevant to the reductive account.

be explained.[12] Adding this modification avoids some obvious extensional errors in first-order languages, but does not help when we move, for example, to second-order languages.

The issue of extensional adequacy is even more troubling than this tiny survey indicates. So far, I have focused on three rather similar languages: propositional, first- and second-order logic. When we widen the scope of our survey, the Tarskian analysis becomes increasingly implausible. For example, much of the most interesting work in logic during the past thirty years has grown out of so-called "index" or "possible world" semantics, pioneered by Saul Kripke, Stig Kanger, and others. This work includes modal logic, epistemic logic, temporal logic, deontic logic, the logic of indexicals, and so forth. Yet in none of these cases does the consequence relation studied admit of a plausible Tarskian characterization.

This deserves emphasis, since it is on the one hand so obvious, yet on the other, so consistently overlooked. If we were to follow Tarski's lead, the way to study the logic of, say, knowledge and belief, would be to treat these operators as logical constants and consider the argument forms whose instances, purely as a matter of fact, preserve truth. But this way madness lies. Contingent, but perfectly general facts about knowledge and belief, perhaps of the depressing sort studied by Kahneman and Tversky, would be enshrined as logically valid arguments of epistemic logic. For example, suppose that the inference from φ to ψ is a particularly subtle fallacy of first-order logic. Then to decide whether $Bel_\alpha(\psi)$ is a logical consequence of $Bel_\alpha(\varphi)$ we would have to find out if anyone—any actually existing believer—saw through the fallacy, that is, believed an instance of φ but not the corresponding instance of ψ. If so, it would not be a logical consequence; if not, it would. Similarly, to decide whether $Bel_\alpha(\varphi)$ logically implies $\exists x(x \neq a \wedge Bel_x(\varphi))$, we would have to find out if any propositions

---

[12] Greg Ray [27] argues that Tarski originally intended to employ varying domains, presumably to prove that the feature is consistent with the original account. But Ray's interpretation is inconsistent with the motivations Tarski gives for the reductive account, with Tarski's explicit description of the account, and with the consequences that he expressly draws from it. Yet even if Tarski himself were assuming varying domains without telling his readers, which he clearly was not, an explanation would be needed for what is in fact a radical departure from the core idea of the reductive analysis. In what is surely one of the more interesting examples of defensive zeal, Ray claims that Tarski is wrong about one of the simplest consequences of the account (that logical consequence reduces to material consequence when all expressions are treated as logical constants), since this does not accord with the account Ray tries to impose on the article. Ray then goes on to accuse me of presenting an "invalid argument," which (we find out only in a footnote) is invalid because it takes Tarski at his word about this obvious consequence of the account [27, p. 648]. Why, according to Ray, would Tarski actually intend an account that is at such variance with his explicit description, with his express motivation, and with the consequences he draws from it? Because, Ray says (following Wilfrid Hodges [18] and Gila Sher [34]), he was addressing the chapter to philosophers. It is interesting that Tarski should have been concerned that his philosophical audience would not understand the clause "and you must vary the domain of quantification," though he assumes they will follow his discussions of omega incompleteness, Gödel's theorems, satisfaction, and so forth. Is this concept really so difficult? It is even more interesting that he would provide a motivation at odds with his "real" account, and go on to draw consequences that follow from the stated account but not from the "real" account. One wonders if Tarski could have said anything to convince these commentators that he actually said what he meant. I urge readers interested in this exegetical issue to read Tarski's article; nothing I could add would provide a more convincing refutation.

are, as a matter of fact, believed by one and only one person. My guess is not, but that would not make this a logically valid inference, as Tarski's analysis implies. And of course we'd also have to settle the question of whether any propositions are believed by a thousand people but not a thousand and one, or a thousand and one but not a thousand and two. Since there are only finitely many believers, the answer would eventually be dubbed a logically valid inference, according to the reductive account. To take yet another example, we'd also have to determine whether anyone believes that there are more than three but fewer than seven things in the universe, for if not (and I suspect not), the negation of this claim would be a logical truth. And so forth. As I said, this way madness lies.

Similar issues arise when we try to apply the reductive analysis to any of the other languages mentioned, from modal logic to the logic of indexicals. These applications immediately involve us in a host of empirical or quasi-empirical questions similar to those mentioned for belief. No one knows, or has ever tried to find out, what the actual extension of the Tarskian consequence relation would be in any of these cases. I leave it as an exercise for the reader to try out any of these applications to see why.[13]

This is not, of course, how these logics are investigated. Kripke semantics, in its many variations, bears no relation to the reductive account of logical consequence given by Tarski. To be sure, in Kripke semantics we use semantic techniques pioneered by Tarski to define the relation of truth in a structure, and we define logically true sentences to be those that come out true in every structure. But this vague similarity is as far as the resemblance goes. We conduct no investigations of which sentences involving knowledge and belief (or necessity and possibility, or "I," "here," and "now") are actually true—that is, true of actual knowers and believers out there in the actual world. Yet such issues would be an essential part of these investigations if Tarski's reductive analysis were correct. The answer, of course, is that in characterizing the logic of these languages we are doing something quite different. I will return to what that is later. For now, what is important to recognize is that the reductive analysis is not used in studying these languages, and would not work if it were.

Tarski's reductive definition of consequence works for propositional languages with logically independent atomic sentences. It can be made to work, with some significant tinkering, for first-order languages (again, with logically independent predicates and functions), and certain close relatives of these.[14] But it fails as soon as we add

---

[13] One might think that of all these applications, modal logic would be the one most likely to succeed. But even here, we are immediately embroiled in substantive issues expressible using the modal operators that one does not ordinarily consider part of modal logic. For example, we would have to decide (or discover) whether there are any properties which one object has necessarily, but which another object has contingently. If not, then $\forall x(\Box Px \vee \Box \neg Px)$ will be a (Tarskian) consequence of $\exists y \Box Py$. This is not a logical consequence in any modal logic I am familiar with. Again, as with epistemic logic, there are a host of similar examples.

[14] The "close relatives" I am referring to are first-order languages supplemented with various numerical quantifiers. This is obvious in the case of quantifiers already definable in first-order logic, but interestingly, the account produces plausible results when we add the quantifier *there exist uncountably many* and even, I believe, the quantifier *there exist infinitely many*. Of course, in all of these cases we have to employ the modified reductive account, in which we vary the domain of quantification; the unmodified account gets the extension radically wrong.

any logically interesting expressions that go significantly beyond the truth-functional connectives, first-order quantifiers, and identity. It fails if those expressions are predicates and relations; it fails if they are non-truth-functional sentential operators; it fails if they are higher-order quantifiers. In all cases, it fails for exactly the reason explained in the first section: having uniformly truth preserving instances is no guarantee of logical validity. I think it is clear—even from a narrowly extensional standpoint—that the reductive account of consequence is a failure.

CONSEQUENCES OF TARSKI'S ACCOUNT

The conclusions of the last paragraph may seem a damning indictment of Tarski's account of logical consequence. But in an odd quirk of intellectual history, these conclusions are actually embraced by the most ardent defenders of the reductive analysis. When you accept Tarski's analysis as capturing the essence of logical consequence, its haphazard behavior on most choices of logical constants gives rise to a seemingly important issue. That issue is sometimes raised with the question "What are the *genuine* logical constants?"; sometimes under the rubric "What is logic?"

What is really being asked here—though the defenders of the reductive account would never phrase it this way—is simply this: when does the Tarskian analysis get the extension of the consequence relation right? And not surprisingly, most of the answers we find in the literature are roughly the same as mine. What are the genuine logical constants? The truth-functional connectives, first-order quantifiers and identity, plus or minus epsilon. What is logic? First-order logic, give or take a bit.

I have already intimated what I consider the correct—indeed the obvious—answer to the question "What is logic?" Any language, regardless of its expressive devices, gives rise to a consequence relation, a relation that supports inferences from sentences in the language to other sentences in the language. The study of this relation is the study of the logic of that language. When Carnap, Kanger, and Kripke studied languages with modal operators, they were doing logic. When Hintikka applied similar techniques to epistemic notions, he was doing logic. When Kaplan investigated indexicals, he was doing logic. Second-order logic is logic (though the Continuum Hypothesis may not be). And these are logic not in a derivative or secondary or lesser sense: they are studying precisely the same thing we study in first-order logic, though in languages with additional expressive resources. Logic is not limited, *de jure*, by the expressive power of the devices in the language, as the reductive account unavoidably implies, though it may be limited, *de facto*, by the clarity of those devices and the availability of techniques to study them.

This is why the issue of the adequacy of the reductive analysis is important. It is not simply an abstract question about a piece of philosophical analysis. Accepting the faulty account leads to an extraordinarily limiting view of the appropriate subject matter of logic. Consider an analogy. Suppose in the early days of chemistry a technique had been developed that worked reasonably well in classifying inorganic compounds. Suppose further that this technique had been taken as definitive of the subject matter of chemistry: chemistry was just the study of those compounds that could be classified using the technique. But suppose the technique simply failed when applied to organic compounds, and as a consequence organic chemistry was declared "not chemistry." No doubt this would have impeded the development of organic chemistry, though I'm sure it would not have stopped it completely. Organic chemists would have pushed ahead, recognizing the importance of their work regardless of what it was called.

In many ways, this is similar to what has happened in logic. Lip service is given to the Tarskian analysis of logical consequence, and to the extremely narrow view of logic that it implies. But much of the interesting work in logic is done outside the confines of that view. I have pointed to some important examples that are well known among philosophers and logicians, but these are only the tip of the iceberg. Let me gesture toward two additional examples of a very different sort. Both fall, by my lights, squarely within the legitimate boundaries of logic; neither admits of a Tarskian analysis.

I said above that any language gives rise to a consequence relation, and that this relation is a legitimate subject of logical investigation. I actually believe this is true of any well-defined system of representation, whether it takes the form of a traditional language or not. A good example is a database. A database stores information in a systematic format, and it is often an extremely important question whether a given piece of information is a consequence—yes, a logical consequence—of the information the database contains. A good deal of work has gone into the study of such questions, and recently into issues that arise when dealing with information contained in heterogeneous databases, where the same information may be represented in very different forms. For instance, one database may contain a field recording an individual's date of birth; another may record the person's age in years at the time of entry, along with a record of when the entry was made; a third might simply indicate whether the individual was a minor when the record was created. The information stored in any one of these databases bears a host of logical relations to the others. They can be inconsistent with one another; one may logically imply information that allows us to update another; and so forth.[15]

When characterizing the logic of a database, or of a collection of heterogeneous databases, the one thing we cannot ignore is the specific structure and interpretations of the various fields. Indeed, it is rarely the case that anything like the traditional logical constants are found as components of a database. The logical constants are often used in query languages designed for accessing information in a database, but almost never in the database itself. The important logical issues that arise here are not amenable to a Tarskian analysis. In fact the issues bear more relation to what Quine would disparage as "analyticity" or "analytic consequence." But this is just Quine's

---

[15] Logicians who are loath to give serious consideration to representational systems other than traditional languages are inclined to say that databases are models. This is simply a confusion. Models are abstract, set-theoretic entities which we use to characterize the semantics of a system of representation. Databases, in contrast, are full-fledged representations. They have a semantics; they can be true or false, accurate or inaccurate; they bear logical relations both to other databases and to sentences in more traditional languages. It happens that they have what I have elsewhere [5] called a homomorphic semantics, but this does not make them models.

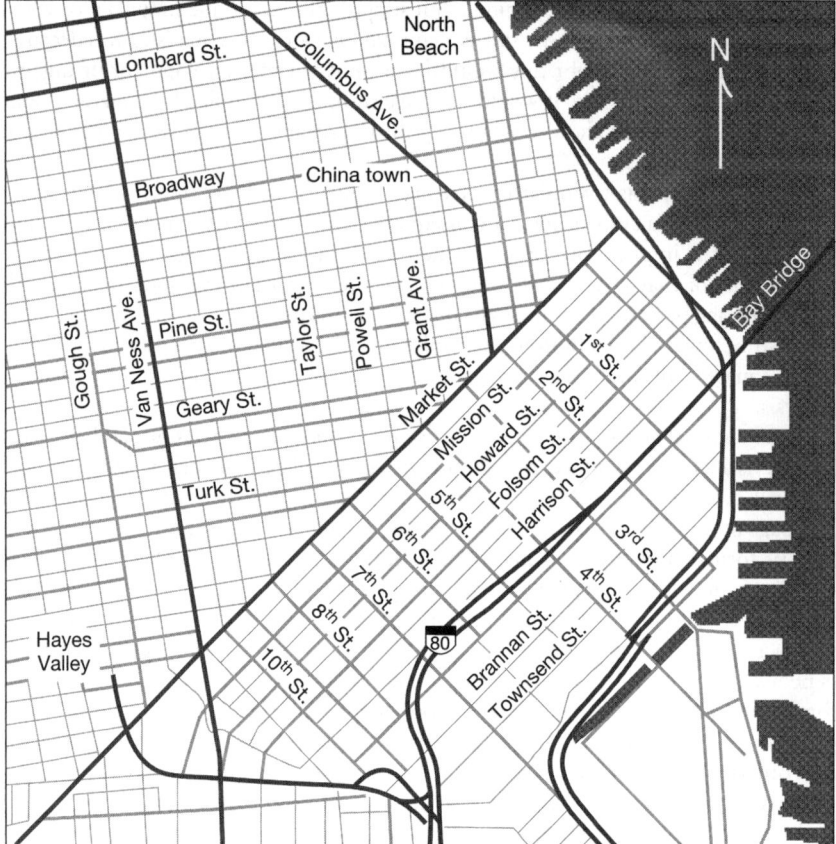

Figure 11.1. A map of downtown San Francisco.

way of marking the artificial boundary that results from the reductive account of consequence.

The second example is closer to home—my home, at any rate. For many years, Jon Barwise and I, along with many students and colleagues, have studied the logic of various forms of graphical and diagrammatic representation.[16] We are particularly interested in what we call heterogeneous reasoning, reasoning that involves information provided in multiple forms. A simple example of such reasoning is the following. Suppose you are given two pieces of information: the map of San Francisco shown in Figure 11.1 and the assertion "The Old San Francisco Mint is at the corner of Mission and Fifth Streets." Here, now, is a quiz. Which of the following sentences follows

---

[16] See for example Barwise and Etchemendy [3, 4, 5, 6] and the papers collected in Allwein and Barwise [1].

logically from the information you've been given: "The Old Mint is south of Chinatown," or "The Old Mint is east of Golden Gate Park." If you know San Francisco, you may realize that both of these assertions are true. But whether you know San Francisco or not, you can see that only the first is a consequence of the information provided.

It takes only a moment's thought to appreciate how hopeless the Tarskian account of consequence is when applied to this sort of inference. What features of the map would be our candidate logical constants? At least when we discussed inferences involving predicates, non-truth-functional operators, or second-order quantifiers, our only problem was that the reductive account gets the wrong extension. With the present example, there is no clear way even to begin applying the analysis. Yet I dare say that inferences of this sort, and the logic that underlies them, are far more common in everyday life than those studied in first-order logic.

Much more could be said about this example, particularly about the implausibility of replacing such inferences with inferences characterizable in first-order logic, but that would take us away from the basic point. That point is this. Tarski's analysis of consequence is based on a simple mistake: the identification of the symptoms of consequence with their cause. When we accept this identification, based on the fact that the symptoms and cause happen to be coextensive in a tiny collection of languages with very limited expressive resources, we risk missing the greater part of logic. Taken seriously, the analysis would rule out any reasonable treatment of modal, epistemic, deontic, or temporal logic. Taken seriously, it precludes the systematic study of the logic of predicates and relations, the logic of noun phrases, and the logic of at least some quantifiers. Finally, taken seriously it rules out the logical investigation of representational systems that take forms other than that of a traditional language, both those that have been around since before recorded history, like maps and diagrams, and those of more recent origin, like computer databases.

MODEL THEORY AND THE MODELING PERSPECTIVE

In the years since Tarski published his article on logical consequence, model theory has become one of the dominant disciplines in mathematical logic. The history of model theory is complex, and includes much work that predates both Tarski's article on consequence and his seminal monograph on truth. Still, there is little question that model theory in its present form owes more to Tarski's work than to the work of any other single individual.

It is important to understand that my rejection of the reductive analysis of logical consequence is not an attack on model theory or model-theoretic semantics *per se*, but rather on a particular view of these techniques. In this section, I will try to make clear what I consider the proper understanding of model-theoretic techniques for studying the logic of a language. My explanation will, of necessity, be fairly dense, but I hope it is sufficient for those already familiar with standard applications of model theory. Readers not interested in model-theoretic semantics should feel free to skip to the final section.

When we give a model-theoretic semantics for a language, we characterize a class of set-theoretic objects alternatively called *structures*, *interpretations*, or *models* for the language. Once these are described, we use Tarski's semantic techniques to define a relation between these structures and the sentences of the language, a relation known as truth in a structure and usually written $\mathfrak{A} \models \varphi$ for structures $\mathfrak{A}$ and sentences φ. A sentence is said to be logically true if it is true in all structures; a sentence φ is said to be a logical consequence of a set Σ of sentences if φ is true in every structure in which the members of Σ are all true, that is, if it preserves truth in every structure.

In CLC I described how the model theory for propositional and first-order languages can be seen as a more or less direct outgrowth of Tarski's reductive account of logical consequence. I devote a chapter of the book to this explanation, but I can describe the gist of it in a paragraph or two. The structures for these languages are quite simple. For the most part, they simply assign objects of a semantically appropriate type to certain expressions of the language. The expressions are the atomic sentences of propositional languages and the predicates, functions, and individual constants of first-order languages. These are, of course, exactly the expressions that in these languages are traditionally considered non-logical constants. Now if we think of the non-logical constants as a special kind of variable and structures as assignments of values to these variables, the relation of truth in a structure is nothing more than the ordinary satisfaction relation: not truth in a structure, but satisfaction of an open formula by actual objects in the actual world.[17] Thus, on this view, a sentence φ is logically true just in case the universal closure $\forall u_1 \ldots \forall u_n \varphi$ that explicitly quantifies the special variables is simply true, just as Tarski's reductive analysis would have it.

When model-theory is seen through this lens, structures are often called *interpretations* of the language, since the assignment of a value to a non-logical constant can equally well be thought of as assigning an interpretation to the expression. If the expressions of the language have antecedent interpretations, the structure that assigns each non-logical constant its actual semantic value is called the *intended* interpretation of the language. The model-theoretic characterization of logical truth would, using this nomenclature, go like this: A sentence is logically true if it is true (in the

---

[17] Dale Jacquette [19], Graham Priest [25], Gerhard Schurz [30, 31], and Gila Sher [34] all think that Tarski was assuming, or should have been assuming, that structures contain both existing and non-existing (merely possible, or perhaps even impossible) objects. I'm not sure which is more difficult to accept, the idea that we can build structures out of non-existent objects or the idea that Tarski had this in mind. Structures are built from actual objects, whether concrete or abstract, and the truth values had by sentences in those structures are determined by the actual properties and relations of those objects. I'm not sure how to make sense of the envisioned alternative. Are structures containing only non-existent objects actual, or are they too non-existent? If the former, this is a truly remarkable set-theoretic feat; if the latter, then do we have to revise Tarski's definition to quantify over all existing and non-existing structures? In any event, once we decide to appeal to non-existent objects, we forsake the principle benefit of the reductive account, the elucidation of a philosophically difficult notion by means of concepts that are significantly clearer and easier to understand. I do not deny, by the way, that we can use actual objects to represent alternative possibilities: I will say more about this in a moment.

actual world) no matter how the non-logical constants are interpreted. Again, the parallel with Tarski's reductive account should be apparent.

We have seen that the reductive analysis of the logical properties is mistaken. The same mistake is inherited by the interpretation of model-theoretic semantics just described. But this is a problem with the described interpretation, not with model theory itself. For there is an alternative view that makes perfectly good sense of model-theoretic practice—much better sense, in fact, than the Tarskian view. In CLC I called this alternative representational semantics and briefly described it in order to distinguish it from the Tarskian perspective. But I clearly did not say enough to forestall confusion, so let me try to rectify that here.

The guiding idea of the representational view of model theory is simple, and in fact widely held, though not widely articulated. The idea is this. The set-theoretic structures that we construct in giving a model-theoretic semantics are meant to be mathematical models of logically possible ways the world, or relevant portions of the world, might be or might have been. They are mathematical models in a sense quite similar to the mathematical models used to study, say, the possible effects of carbon dioxide in the atmosphere, only they are used to study semantic phenomena, not atmospheric, and specifically to characterize how variations in the world affect the truth values of sentences in the language under investigation. The main difference is that in model theory we generally use discrete mathematics rather than the continuous mathematics used in physical modeling, though one can easily devise languages where discrete tools do not suffice. I called this view of model theory "representational" because the set-theoretic structures are seen as full-fledged representations: models of the world.

I will say more in a moment to add texture to the representational perspective, but for now let me finish this simple, initial sketch. According to the representational view, our goal in constructing a semantics is to devise a class of models that represents all logically possible ways the world might be that are relevant to the truth or falsity of sentences in the language, and to define a relation of truth in a model that satisfies the following constraint: a sentence φ should be true in model A if and only if φ would be true if the world were as depicted by A, that is, if A were an accurate model. The models are designed to represent the world in a particularly straightforward way, and this is important. Any individual model represents a logically possible configuration of the world and any two (non-isomorphic)[18] models are logically incompatible: at most one can be accurate. But jointly, they are meant to represent all of the possibilities relevant to the truth values of sentences in the language. In other words, if we've designed our semantics right, the models impose an exhaustive partition on the possible circumstances that could influence the truth of our sentences. Because of this, it is a trivial consequence that sentences which are true in every model are logically

---

[18] Throughout this section, when I speak of "two" models, I presuppose the modifier "nonisomorphic." If I had more room to discuss representational semantics, I would explain how in certain semantics, non-isomorphic models can be representationally equivalent, that is, represent exactly the same possible circumstances. I will ignore this fact here, since it is irrelevant to the present issues. I also set aside issues that arise when we allow partial models in the semantics, or when we compare models from different semantics.

true, and arguments that preserve truth in every model are logically valid, at least if the representational criteria are genuinely satisfied.[19] Note, however, that this does not give us an analysis of the logical properties, since the logical notions are presupposed from the very start, in the criteria by which we assess our class of models. I will return to elaborate on this point later.

This is only a very rough sketch of the representational perspective. Let me add some texture to the sketch before discussing it in detail. When we study the logic of a language, we are generally interested in the logic of only some of its expressions. In propositional logic we are interested in the truth-functional connectives; in first-order logic we additionally focus on the quantifiers and identity; in modal logic we add necessity and possibility; in epistemic logic, knowledge and belief; and so forth. Because we focus on only some of the expressions in the language, the semantics of the remaining expressions can be treated differently from those whose logic we aim to explicate. I will say we treat them "categorematically" for lack of a better term.

When we are not focusing on the logic of a particular expression or category of expressions, we need not model the specific semantic behavior of that expression. It is enough to characterize the minimal semantic behavior common to expressions of the same semantic category. This is what I mean by the categorematic treatment of an expression. In propositional logic, any sentence with no truth-functional structure is simply treated as providing a truth value, true or false, to the larger sentences of which it is a constituent. This is not to say there is no interesting or even logically relevant semantic behavior among these sentences, but only that we are not attending to it at present. Similarly, in first-order logic, where we are not concerned with the logic of predicates, a model can simply assign an arbitrary set to represent the semantic contribution of a monadic predicate, since all monadic predicates, whatever the details of their semantic behavior, will have some extension or other.

Note that we needn't treat all members of a given category categorematically just because we treat some that way. For example, in studying the logic of indexicals, Kaplan focuses on a handful of singular terms, "I," "here," and "now," while treating all other singular terms categorematically. In propositional logic, we treat sentences with no truth-functional structure categorematically, while giving sentences built using the truth-functional connectives a more detailed semantic treatment. In first-order logic, we characterize the specific semantics of the identity predicate, but treat the remaining binary predicates categorematically. Note also that an expression treated categorematically in one semantics may be given a detailed semantic treatment in another. Nothing special hangs on the choice

---

[19] I will set aside the important question of how we know our models actually depict every relevant possibility. Merely intending our semantics in this way is not sufficient, since limitations of our modeling techniques may rule out the depiction of certain possibilities, despite the best of intentions. This is arguably the case in the standard semantics for first-order logic, for example, where no models have proper classes for domains. Similarly, if we built our domains out of hereditarily finite sets we would have no model depicting an infinite universe. These are not problems with representational semantics per se, but with our choice of modeling techniques. Analogous problems arise in mathematical models of physical phenomena.

of whether to treat an expression categorematically or not. It simply depends on which expressions we wish to focus on. This too is an important point that I will come back to later.

Now the fact that we treat expressions categorematically does not change the fundamentally representational perspective of the semantics we construct, though it can give rise to some confusion. What is confusing is that the categorematic treatment of certain expressions ignores any specific meanings these expressions may have, if in fact they were drawn from an antecedently interpreted language. For example, if we start with the language of elementary arithmetic or a first-order language containing the predicates $\cong$ and *Triangle* from our earlier example, there will be models that do not represent genuine possibilities—that is, possibilities consistent with the antecedent meanings of these expressions. For instance, there will be models in which an object a is in the extension assigned to *Triangle*, an object b is not in that extension, and yet the pair $\langle a, b \rangle$ is in the extension assigned to $\cong$. But this is hardly surprising or problematic, given that we have only characterized the minimal semantic behavior shared by all expressions of the respective categories. These models represent relevant possibilities for some expressions of those categories, though perhaps not all. This is analogous to what happens when, in a propositional semantics, we assign the truth value *false* to the atomic sentence $a = a$. This truth value assignment represents a genuine possibility for *some* atomic sentences, and since we are treating such sentences categorematically, the specific meaning of $a = a$ is irrelevant to the semantics.

How does the categorematic treatment of certain expressions affect the representational view of model theory? Well, not much at all. Where earlier we said that every model is meant to represent a logically possible configuration of the world, we now need to add a qualification to handle expressions treated categorematically. Each model is meant to represent a semantically relevant circumstance for at least *some* expressions in those categories—or, if you will, for *some* interpretations of the expressions so treated. This is a minor change, but it is potentially confusing because there are now two dimensions of variation. Structures still represent possible circumstances—this is the important dimension of variation—but our decision to treat some expressions categorematically introduces a second dimension, since these expressions receive uniform treatment in spite of potentially significant variations in their meanings. Thus in propositional logic, we afford $a = a$ the same treatment as *Triangle*(a), despite the fact that the first may express a logical truth, while the second may express a contingent claim about the world.

From the representational standpoint, there is only one significant effect of the categorematic treatment of expressions. If we do not treat any expressions categorematically and our semantics meets the representational guidelines—that is, if every logically possible circumstance is represented by some model and our definition of truth in a model is correct—then we can be sure that *all and only* logically valid arguments of the language will be declared valid by the semantics. But suppose we are dealing with an antecedently interpreted language, and yet treat some expressions categorematically. Then it is always possible that our semantics will not declare some genuinely valid arguments valid, namely those whose logical validity depends on specific meanings our semantics ignores. Again, there are no surprises here. In propositional logic, the following argument is not revealed to be logically valid:

> *Triangle*(*a*)
> *a* = *b*
> So, *Triangle*(*b*)

But the validity of this argument emerges as soon as we give a more detailed semantics, one that does not simply treat the sentences categorematically.[20]

Now it is important to see that representational semantics is not simply a minor redescription of Tarski's reductive analysis of consequence, or of "interpretational semantics," as I called the view of model theory that emerges from that analysis. There are many ways in which they differ, but the most important is that they impose different and conflicting criteria of adequacy on the semantics. What this means is that a semantics that is acceptable from the representational stance may be completely unacceptable from the reductive stance, and conversely, one that satisfies all criteria from the reductive perspective may be entirely inadequate from the representational. It is like the difference between billiards and pool: there are obvious similarities, to be sure, but pretty quickly the difference in rules (and the presence or absence of pockets) can no longer be ignored.

Let's look at a few examples, since in fact the criteria diverge almost immediately. For example, I earlier alluded to the fact that on the reductive analysis of consequence, it is hard to understand why in the semantics for first-order languages we vary the domain of quantification. Certainly, there are languages with restricted quantifiers, and even languages (such as English) in which the restrictions may be determined contextually from one use to the next. But suppose we are interested in the consequence relation for a language in which ∀ really means *"for all,"* a language in which this expression quantifies over everything that happens to exist. From the reductive perspective, if we treat this expression as a logical constant, we should fix its meaning and survey the argument forms whose instances uniformly preserve truth, for example any argument with β as its conclusion. Replacing the unrestricted quantifier with various restricted quantifiers, quantifiers that do not quantify over everything, seems clearly inconsistent with the stated goal of

---

[20] In CLC I discussed the representational view of model-theoretic semantics in some detail, and described how it differs from the Tarskian perspective. But I did not elaborate on the common practice of treating certain expressions categorematically, since I assumed it would be clear how the practice fits into the representational perspective. But this clearly confused some commentators. For example, Gila Sher [34] claims that no one interprets model theory as a representational semantics, but her evidence is simply the categorematic treatment of names and predicates in the standard first-order semantics. I think, on the contrary, that it is patently obvious that many logicians and most philosophers (including Sher) adopt a fundamentally representational stance, though they may be unclear how different this is from the Tarskian analysis of consequence. My reasons for saying this will become clear later, since most model-theoretic semantics can only be understood representationally. Another commentator, Manuel García-Carpintero, gives an excellent and thoughtful analysis of the intuitions underlying the representational semantics of first-order languages in [14]. My main disagreement with García-Carpintero is his assumption that this was what Tarski had in mind in his analysis.

determining the unrestricted quantifier's logic. When we view the same issue from the representational perspective, however, the inconsistency immediately goes away. First-order structures, viewed as representations of the world, should of course have different domains: this is simply our way of representing the fact that, although the world is the size it is, this could have been different. The same feature that seems a straightforward violation of the reductive account is in fact demanded by the representational perspective.

A more illuminating example, or collection of examples, are the various Kripke semantics mentioned earlier. Let me describe these in representational terms, since I know of no other way to view them. In a Kripke semantics, a structure consists of a set of indices, I, plus a relation R on I that specifies whether one index is "accessible" from another. Each index is associated with what is in effect a first-order structure, specifying a domain, extensions of the predicates, and so forth. The members of I are for heuristic reasons called *possible worlds* (though they are in fact simply set theoretic objects of some sort), and one of them, sometimes denoted @, is singled out as the *actual world*. According to the heuristic, the first-order structure associated with a given index represents the non-modal facts of the possible world corresponding to that index. Thus the first-order structure assigned to the index @ represents the non-modal facts in the actual world. The remaining members of I, along with the accessibility relation R, are simply an ingenious way of representing modal (or epistemic, deontic, temporal, etc.) facts about the world. For example, if a particular state of affairs holds at an index i accessible from @, this represents that this is a possible (though perhaps not actual) state of affairs. In other words, an entire Kripke structure represents a world—a single world—replete with both modal and non-modal facts.[21]

The crucial feature of a Kripke semantics is that for any logically possible configuration of the world, including both modal and non-modal facts (or epistemic and non-epistemic facts, etc.), there will be a Kripke structure representing that configuration. To hark back to our discussion of epistemic logic, there will be structures representing worlds in which $Bel_\alpha(\psi)$ is true whenever $Bel_\alpha(\varphi)$ is true, but also worlds in which this is not the case, however subtle the fallacious inference from φ to ψ may be. There will be worlds in which no one believes there are more than three but fewer than seven objects, but also worlds in which some people do. And so forth. What this means is that any sentence that comes out true in every Kripke structure must be true regardless of how the semantically relevant circumstances—in this case, epistemic and non-epistemic facts—happen to shake out.[22]

---

[21] Many people mistakenly believe that Kripke semantics commits us to the existence of possible worlds of some sort or other. This is just a confusion resulting from a simplistic view of the technique used in the semantics to represent modal (or epistemic, etc.) facts. The semantics is neutral about the issue of whether there are possible worlds in any ontologically significant sense; it simply uses the heuristic as a technique for representing various alternative modal facts.

[22] Similarly, returning to our example from footnote 13, there are Kripke structures in which $\exists y \Box Py$ is true but $\forall x(\Box Px \lor \Box \neg Px)$ is false, and others in which this is not the case, showing that the latter is logically independent of the former, in contrast to what the Tarskian account will say about such cases.

Kripke semantics obviously satisfies the guidelines for a representational semantics. Indeed, the great contribution of the semantics is that it gives us a remarkably flexible way to represent facts that play determining roles in the truth or falsity of sentences in a wide range of languages. But there is no sensible way to construe it as an application of Tarski's reductive account of consequence, as should be clear from our earlier discussion of these languages. To apply the reductive account, we would have to hold fixed the meanings of the operators in question and determine which arguments preserve truth—in the actual world—under various interpretations of the non-logical constants. A Kripke semantics does nothing even vaguely resembling this, and so by these criteria would have to be judged an out and out failure.

Once we appreciate how far removed representational semantics is from the reductive account of consequence, it becomes clear that there is no language, indeed no system of representation, whose logic cannot in principle be studied using model-theoretic techniques. For example, suppose we are interested in studying the logic of color predicates, perhaps in the context of a first-order language. Clearly the traditional first-order semantics, in which all predicates except identity are treated categorematically, would have to be supplemented. But it is not hard to see how the supplement might go. One simple option would be to assign to color predicates appropriate regions in a color space, and then have models map (some or all) objects in their domain to random points in color space. This would give us a more detailed representation of the range of logically possible circumstances relevant to the truth values of sentences involving these predicates. Naturally, important questions would arise in constructing such a semantics, but it is clear enough how it would be done. Again, the resulting semantics, like Kripke semantics, would bear no relation to Tarski's reductive account of consequence.

To take another example, I mentioned in the last section that it is unclear how we would even begin to apply the Tarskian analysis of consequence to diagrammatic forms of representation, since these are so different, both syntactically and semantically, from traditional languages. But this does not mean that a *representational* semantics is difficult to construct for these forms of representation. As long as we can devise model-theoretic techniques for representing circumstances relevant to the truth or falsity, accuracy or inaccuracy, of these types of representation, nothing prevents us from studying their logic as well. Such semantic accounts are no more difficult to provide than a model-theoretic semantics for traditional languages.[23]

Let me conclude this discussion by returning to a couple of issues touched upon earlier. The first is the difference between the categorematic treatment of expressions in a representational semantics and the distinction between logical and non-logical constants central to Tarski's reductive analysis. I have already mentioned one respect in which these are very different. When we accept Tarski's analysis, it becomes an extremely important question which expressions are the legitimate logical constants, for choosing the wrong ones will yield a radically incorrect consequence relation. In representational semantics, in contrast, the decision to treat certain expressions

---

[23] See for example Barwise and Etchemendy [5], Shin [35], and the papers collected in Allwein and Barwise [1].

categorematically is entirely arbitrary, depending only on whether we are interested in the logic of those expressions. Of course, we will sometimes treat expressions categorematically for important practical reasons—for example, we may not have a clue how to give a detailed treatment of their semantics—but there is nothing logically or philosophically significant about this choice. A second difference is that in Tarski's analysis, we genuinely hold fixed the meaning, both intension and extension, of the chosen logical constants. To treat *believes'* as a logical constant, we must survey actual believers and actual beliefs. In representational semantics, even the behavior of expressions not treated categorematically enjoys more flexibility. The fact that models are simply representations of semantically relevant circumstances allows us to survey alternative extensions of these expressions—not different meanings, but different ways the world might be. This is as it should be: we are studying the logic of these expressions, not general facts that may be expressed using them. Epistemic logic is not psychology, modal logic is not metaphysics, and second-order logic is not set theory.

Several commentators on CLC have argued that Tarski had in mind something like representational semantics when proposing his analysis, but this question should be finally laid to rest by the observations of the preceding paragraph. The fact that Tarski saw the choice of logical constants to be a crucial step in applying his analysis, the fact that he explicitly points out that the choice is not arbitrary, and finally his acknowledgment that logical consequence reduces to material consequence when all expressions are treated as logical constants, show that he could not have had in mind representational semantics. If we are engaged in representational semantics, none of this is even remotely the case.[24]

---

[24] Gila Sher, though she claims no one views model-theoretic semantics representationally, goes on to propose an interpretation of Tarski's analysis that looks suspiciously like a representational semantics with the categorematic treatment of names and predicates. But Sher continues to claim that the choice of logical constants is crucial, for reasons that are obscure. She begins by emphasizing the importance of the notion of formality: "Necessity... is by itself a problematic notion, but formality can be viewed as a modifier of necessity: not all necessary consequences are logical, only *formal*-and-necessary (or *formally* necessary) consequences are. The key to understanding logical consequence is, thus, formality" [34, p. 672]. This may seem a promising start, despite the rather abrupt "thus." But Sher goes on to describe a notion of formality which, among other things, implies that the formal rules of modal and epistemic logic are not, contrary to appearances, formal in the required sense. Predictably, it turns out that Sher's formality requirement is only satisfied by first-order logic and minor variants.

What is the relationship between formality and necessity that justifies Sher's "thus"? Sher explains it this way: "[The concern about non-logical generalizations] does not apply to my conception, where logical consequence is reducible not to just any kind of generality, but to a special kind of generality, namely, *formal* and *necessary* generality. Speaking in terms of models: Suppose there is an accidental property H, of all models for a given language. The notion of model is defined within some background theory, T, based on its notion of 'formal structure.' If T is an adequate theory of formal structure, then T includes the theorem 'Some formal structure A does not possess the property H' and, in accordance with this theorem, the apparatus of models defined in T will include a model representing a formal structure in which H does not hold" [34, p. 681].

Sher is saying here that if there is an accidental or non-logical feature that holds of all models (say the Continuum Hypothesis or the absence of proper classes among the models' domains), then you simply need a background theory of formality that says the feature does not really hold of all models. But this is simply nonsense. Every class of models has such features—indeed infinitely many such features—including the models used in first-order logic. For example, if the Continuum

The second issue is the issue of analysis itself. If Tarski's analysis worked, it would be a genuine analysis, in the sense that it characterizes the logical properties in terms that do not presuppose those very same properties. As I mentioned earlier, representational semantics gives us no such analysis, since the logical notions are used to assess the class of models devised for the semantics. Each model is meant to depict a logical possibility; no two are logically consistent; and the "sum" of the models is logically necessary—that is, every semantically relevant possibility is represented. This has a consequence that some may find disappointing, though it should hardly be surprising. We cannot look to model-theoretic semantics to answer the most basic foundational issues in logic. For example, if we have serious doubts about whether the principle of excluded middle is a logical truth, the classical semantics for propositional languages will not provide an answer. For the same intuitions that suggest that it is a logical truth are used in defining the class of structures—truth value assignments—that are employed by this semantics.

Does this mean that model-theoretic semantics, construed representationally, provides no illumination about the logical properties of the languages studied? Not by any means. We can see how it illuminates these properties both concretely and abstractly. Concretely, a well-designed semantics shows us how the truth values of sentences in the target language vary as the non-linguistic facts represented by the structures vary, and accordingly explains persistent patterns of truth values that emerge due to the semantics of these sentences. Logical truth and logical consequence are just two such persistent patterns. Naturally, the explanation is only as clear as the semantics, and in particular relies on a clear understanding of how the structures used in the semantics are meant to represent possible circumstances. If, so to speak, the "semantics" of our models is obscure, this will detract from or even negate the explanatory power of the model-theoretic semantics. But assuming a clear understanding of the states of affairs depicted by our models, the semantics shows precisely how the logic of the language arises from the meanings of its constituent expressions, modulo any basic logical assumptions incorporated into the models themselves. For example, the classical semantics for propositional logic may not provide a fully grounded explanation of the principle of excluded middle, but it does explain why, given this basic assumption, a complex sentence like $\neg(P \wedge (\neg P \vee (Q \wedge R))) \vee Q$ is necessarily true. This provides illumination of a very real sort.

There is another, more abstract way to describe this illumination. In a model-theoretic semantics, although the class of models is itself a representational system,

---

Hypothesis is true (or false), that will have an effect even on what first-order models there are; similarly, in standard first-order model theory there are no proper classes among the domains; and so forth. Having a theory of formal structure that says these features are not there doesn't help, it simply means your theory is false. Or, to put the point another way, if an adequate theory of formal structure must be able to prove that no such features hold of the class of models, then there can be no such (true) theory, any more than there can be a true theory proving that two plus two is five.

Any class of structures will have features that are not logically necessary. The crucial question is not whether there *are* such features, but whether the features are expressible using the chosen logical constants, and hence whether they have an impact on the extension of the reductive definition. The only way to prove that they don't have such an impact is by means of a completeness theorem, as explained earlier.

it is a system with a particularly simple logic—in fact the simplest logic possible: no model is logically true or logically false, no model follows logically from another, and so forth. I will say that the system of models is logically "transparent," since there are no non-trivial consequence relations between representations in the system. Thus in a representational semantics we describe the logical properties of a logically *complex* system of representation in terms of the logical properties of a transparent system of representation. We show why, for example, *Triangle(b)* is a consequence of the premises *Triangle(a)* and $a = b$ by characterizing the semantics of these sentences in terms of a class of representations in which there are no non-trivial consequence relations of this sort. Since we are presenting the semantics of our target language in terms of another representational system, this is the best we can possibly do.[25]

CONCLUDING PHILOSOPHICAL POSTSCRIPT

The reductive analysis of consequence is by no means a silly or trivially mistaken account. On the contrary, it is both attractive and plausible: attractive, because it promises to eliminate a host of obscure notions at the core of logic in favor of the vastly clearer notion of truth; plausible, because the definition is based on features that are indeed important characteristics of logically valid arguments. This is why the account has been put forward repeatedly, not only by Tarski, but in slightly different forms by Bolzano, Russell, Quine, and others. Yet in spite of its plausibility and attractiveness, the account is wrong: the identified features are not what underlie logical consequence, but merely symptomatic of the genuine relation. Sometimes these symptoms are coextensive with the cause, but more often they are not.

The main problem with the account, however, has nothing to do with subtle philosophical issues, but rather with the wide-ranging consequences of accepting the faulty analysis. I have discussed the consequences for logic if we take the reductive account seriously earlier. I also noted that the account is only given lip service among many working logicians, who of necessity abandon the analysis in order to study the logic of languages with expressive resources that go much beyond propositional or first-order logic. But the analysis has also had a significant impact in philosophy proper, perhaps even more so than in logic itself. Let me conclude this chapter with some very brief remarks about the influence of the account on work in the philosophies of logic, mathematics, and language.

The influence of the reductive account has been most direct in the philosophy of logic, where the analysis provides the field with one of its principal problems. A great deal of effort has been devoted to the question of which expressions are "genuine" logical constants, and precisely what features make them so. This of course is an extremely important question if the reductive analysis is correct. After all, if the expression *if...then* turns out not to be a logical constant, then according to the reductive account *modus ponens* is not a logically valid argument form. On the other

---

[25] For a more extensive discussion of ways in which model-theoretic semantics illuminates the semantics of a language, see Barwise and Etchemendy [2] and Etchemendy [11].

hand, if *believes* or *same shape* or *is red* turn out to be logical constants, then logic becomes, in effect, an empirical discipline. Sentences like:

No one believes there are more than three but fewer than seven objects.

will then qualify as truths of logic. Once we accept the reductive account, the problem of the logical constants appears to hold the key to the difference between genuinely valid inference and inference that obviously is not. With stakes this high, this becomes a philosophical issue that demands attention.

Wittgenstein is well known for his claim that the problems of philosophy arise out of fundamental confusions, and that their proper solutions lie in clarifying those antecedent confusions. Personally, I think that this is not at all true of most philosophical problems. But the problem of the logical constants, and the closely related question of what is logic, are clear examples of Wittgenstein's claim. The problem arises for no other reason than our acceptance of an incorrect account of logical consequence. When we fail to distinguish the symptoms of consequence from genuine consequence, we are bound to get faulty results. The idea that these results are due to the correctness or incorrectness of our selection of logical constants is simply a misdiagnosis of what went wrong. Any expressive device—predicates, adverbs, indexicals, quantifiers—can in principle affect the logical properties of a language, can give rise to arguments that are guaranteed to preserve truth in virtue of the way those devices work. The expressions traditionally singled out in the argument forms studied by Aristotle or Boole or Frege or Gentzen are simply expressions whose logic is particularly clear, interesting, and widely applicable. These traditional constants will no doubt share many properties, as will any finite collection of expressions, but the idea that the properties they share are somehow *definitive* or *determinative* of logic is based on a confusion.

The reductive account of consequence has had an equally extensive influence in the philosophy of mathematics, though the influence is more diffuse. Most of the influence comes via the claim that logic is identical to first-order logic. We have seen why this view seems inevitable given the reductive account of consequence: as soon as we venture very far from the expressive resources of first-order logic, the resulting "logical consequence" relation bears little or no relation to logic—not due to the real logic of these expressions, but due to the faulty analysis of consequence. The identification (or misidentification) of logic with first-order logic has important consequences for how we understand the nature of mathematics and mathematical truth. To take the most obvious example, the logicist claim that arithmetical truth is reducible to logical truth is clearly false if we accept this identification. But it is arguably true when we consider the logic of languages containing more powerful expressive devices. Of course the conclusion that the reduction is possible in a more powerful language may not carry with it some of the epistemological benefits envisioned by the early logicists—there is no getting around Gödel's incompleteness theorems—but it may nonetheless provide illumination about the nature of mathematical truth. I do not pretend to have solutions to the longstanding debate inspired by the logicist's claim. But it is obvious that sorting these issues out requires a reasonably clear understanding of logical truth and logical consequence, not one based on an analysis incapable of dealing with more powerful logics.

A very different example is our understanding of geometrical reasoning, which has been hampered by the absence of any account of valid reasoning that involves diagrammatic or other non-linguistic forms of representation. As long as we adhere to the reductive account of logical consequence, we will never make progress understanding this sort of reasoning, for reasons already discussed. This point in fact applies much more widely than the philosophy of mathematics. Logic is in part a service discipline, providing precise, idealized models of valid and invalid reasoning, models which in turn help us describe and understand the process of rational investigation, whether in mathematics, the sciences, or everyday life. To the extent that the models we develop fail to address important types of deductive reasoning, we make the task of philosophers investigating those domains correspondingly difficult. Since the most highly developed model of deductive reasoning is that provided by first-order logic, philosophers naturally try to model, say, scientific reasoning by applying the notions derived from this theory. But it is likely that, as in the case of geometrical reasoning, the first-order model is inadequate to capture significant portions of the deductive reasoning that takes place in these disciplines.

I have already alluded to the impact of the reductive account in the philosophy of language. Here, the influence flows largely from the work of Quine. Quine and his followers have long disparaged the notions of analytic truth and analytic consequence, arguing that it is impossible to sensibly distinguish analytic truths from deeply held empirical beliefs, that the distinction is simply an unfounded "dogma" of empiricism. But most followers of Quine pull their punches when it comes to logical truth and logical consequence: these notions, unlike analyticity, can be clearly and definitively characterized by means of the reductive analysis, or so the Quinean would like to believe. This allows them to assume the legitimacy of logic—or at any rate, first-order logic—while denying the legitimacy of any appeal to the analytic/synthetic distinction, or to a full-bodied conception of meaning.

Needless to say, a detailed analysis of Quinean epistemology and philosophy of language is far beyond the scope of the present article. But it should be clear that the distinction between analyticity and first-order logic, or between first-order logic and more powerful logics, is brought into question once we recognize the defect in the reductive analysis. Quine's attack on analyticity applies equally to the notions of logical truth and logical consequence (as Quine himself sometimes acknowledges). My own view is that the attack fails in both cases, but a Quinean can consistently maintain that it succeeds in both, that the concept of logical consequence, like that of analyticity, is an unfounded dogma of empiricism. Whether he would be willing to accept the model of rationality that emerges—wherein the web of belief comes to resemble a disconnected pile of sand—is an open question. But these are questions that must be postponed until later.

References

[1] Allwein, Gerard, and Jon Barwise, eds. *Logical Reasoning with Diagrams*. Oxford University Press, 1996.

[2] Barwise, Jon, and John Etchemendy. "Model-theoretic Semantics." In *Foundations of Cognitive Science*, Michael Posner, ed., MIT Press, 1989, 207–43.

[3] —— "Visual Information and Valid Reasoning." In *Visualization in Mathematics*, Walter Zimmermann and Stephen Cunningham, eds., Mathematical Association of America, 1991, 9–24. Reprinted in [1].

[4] —— "Hyperproof: Logical Reasoning with Diagrams." In *Proceedings of the 1992 AAAI Spring Symposium on Diagrammatic Reasoning*, AAAI, 1992, 80–4. Reprinted in *Reasoning with Diagrammatic Representations*, AAAI Press, 1994.

[5] —— "Heterogeneous Logic." In *Diagrammatic Reasoning: Cognitive and Computational Perspectives*, Janice Glasgow, N. Hari Narayanan and B. Chandrasekaran, eds., MIT Press, 1995, 211–34.

[6] —— "Computers, Visualization, and the Nature of Reasoning." In *The Digital Phoenix: How Computers are Changing Philosophy*, T. W. Bynum and James H. Moor, eds., Blackwell, 1998, 93–116.

[7] Blanchette, Patricia. "Models and Modality." *Synthese* 124 (2000): 45–72.

[8] Chihara, Charles. *The Worlds of Possibility*. Oxford University Press, 1998.

[9] —— "Tarski's Thesis and the Ontology of Mathematics." In *The Philosophy of Mathematics Today*, Matthias Schirn, ed., Clarendon Press, 1998, 157–72.

[10] Curtis, Gary. "Review of *The Concept of Logical Consequence*, by John Etchemendy." *Nous* 28 (1994): 132–5.

[11] Etchemendy, John. "Models, Semantics and Logical Truth." *Linguistics and Philosophy* 11 (1988): 91–106.

[12] —— "Tarski on Truth and Logical Consequence." *Journal of Symbolic Logic* 53 (1988): 51–79.

[13] —— The Concept of Logical Consequence. Harvard University Press, 1990. Reissued by CSLI Publications and Cambridge University Press, 1999.

[14] García-Carpintero Sánchez-Miguel, Manuel. "The Grounds of the Model-Theoretic Account of the Logical Properties." *Notre Dame Journal of Formal Logic* 34 (1993): 107–31.

[15] Gómez-Torrente, Mario. "Tarski on Logical Consequence." *Notre Dame Journal of Formal Logic* 37 (1996): 125–51.

[16] Hansen, William. "The Concept of Logical Consequence." *Philosophical Review* 106 (1997): 365–409.

[17] Hart, W. D. "Review of *The Concept of Logical Consequence*, by John Etchemendy." *Philosophical Quarterly* 41 (1991): 488–93.

[18] Hodges, Wilfrid. "Truth in a Structure." *Proceedings of the Aristotelean Society* 86 (1986): 135–51.

[19] Jacquette, Dale. "Tarski's Quantificational Semantics and Meinongian Object Theory Domains." *Pacific Philosophical Quarterly* 75 (1994): 88–107.

[20] Kreisel, Georg. "Informal Rigour and Completeness Proofs." Reprinted in *Problems in the Philosophy of Mathematics*, Imre Lakatos, ed., North Holland, 1969, 138–71.

[21] McGee, Vann. "Review of Etchemendy, The Concept of Logical Consequence." *Journal of Symbolic Logic* 57 (1992): 254–5.

[22] —— "Two Problems with Tarski's Theory of Consequence." *Proceedings of the Aristotelean Society* 92 (1992): 273–92.

[23] —— "*The Concept of Logical Consequence* and the Concept of Logical Consequence." San Francisco: American Philosophical Association, Pacific Division Meeting, March 1993.

[24] O'Hair, Greg. "Logical Consequence and Model-Theoretic Consequence." *Logique et Analyse* 35 (1992): 239–49.
[25] Priest, Graham. "Etchemendy and Logical Consequence." *Canadian Journal of Philosophy* 25 (1995): 283–92.
[26] Ray, Greg. "On the Possibility of a Privileged Class of Logical Terms." *Philosophical Studies* 81 (1996): 303–13.
[27] —— "Logical Consequence: A Defense of Tarski." *Journal of Philosophical Logic* 25 (1996): 617–77.
[28] Read, Stephen. "Formal and Material Consequence." *Journal of Philosophical Logic* 23 (1994): 247–65.
[29] Sagüillo, José. "Logical Consequence Revisited." *Bulletin of Symbolic Logic* 3 (1997): 216–41.
[30] Schurz, Gerhard. "Logical Truth: Comments on Etchemendy's Critique of Tarski." In *Sixty Years of Tarski's Definition of Truth*, B. Twardowski and J. Woleński, eds., Philed, Kraków, 1994, 78–95.
[31] —— "Tarski and Carnap on Logical Truth—or: What is Genuine Logic?" In *Alfred Tarski and the Vienna Circle*, J. Woleński and E. Köhler, eds., Kluwer, 1998.
[32] Shapiro, Stewart. "Logical Consequence: Models and Modality." In *The Philosophy of Mathematics Today*, Matthias Schirn, ed., Clarendon Press, 1998, 131–56.
[33] Sher, Gila. *The Bounds of Logic: A Generalized Viewpoint*. MIT Press, 1991.
[34] —— "Did Tarski Commit 'Tarski's Fallacy'?" *Journal of Symbolic Logic* 61 (1996): 653–86.
[35] Shin, Sun-Joo. *The Logical Status of Diagrams*. Cambridge University Press, 1994.
[36] Smiley, Timothy. "Consequence, Conceptions Of." In *The Routledge Encyclopedia of Philosophy*, Edward Craig, ed., Routledge, 1998.
[37] Tarski, Alfred. "O pojciu wynikania logicznego." *Prezglad Filozoficzny* 39 (1936): 97–112. Translated as "On the Concept of Logical Consequence" in [38], 409–20.
[38] —— *Logic, Semantics, Metamathematics*. Oxford University Press, 1956.

12
Tarski's Thesis

Gila Sher

"Tarski's Thesis" is the claim that a certain invariance condition can serve as our criterion of logicality. My goal in this chapter is to explain the thesis, provide it with a philosophical justification, and respond to three recent criticisms due to Solomon Feferman.

CRITERION OF LOGICALITY

In a 1966 lecture, "What are the Logical Notions?", Tarski's proposed the following criterion of logicality:

Invariance under Permutation: A notion is logical iff it is invariant under all permutations of the individuals in the "world" (or universe of discourse).[1]

By "notions" Tarski understood not linguistic or conceptual entities but objects of the kind referred to by such entities, i.e., objects in the world, including individuals, properties (sets), relations, and functions. "World" he understood as including both physical and mathematical objects and as forming a type-theoretic hierarchy, based on *Principia Mathematica* or a similar theory. In the present context it will sometimes be convenient to view objects as operators (characteristic functions representing them) and use standard set theory with urelements rather than *Principia Mathematica* as our background theory.

By centering his attention on objects or operators (worldly entities) rather than constants (linguistic entities) Tarski follows the precedent of the Boolean, truth-functional definition of logical connectives in propositional logic. This definition

---

I started writing this chapter while visiting the philosophy department at the University of Santiago de Compostela in Spain. I would like to thank the participants in my philosophy of logic seminar, and in particular Concha Martinez, José Miguel Sagüillo, and Luis Villegas-Forero for stimulating conversations on issues related to this chapter. I am also thankful to members of the LOGOS group in Barcelona and to the participants in the conference "Foundational Issues in Logic: Logical Consequence and Logical Constants Revisited" for their feedback. Thanks to Denis Bonnay for his discussion and written comments, and to Peter Sher for his comments.

[1] Paraphrase of Tarski (1966): 149.

identifies logical connectives with certain objects, namely, Boolean truth functions, and it is these objects, rather the names or descriptions used to refer to them, that are said to capture the idea of logicality on the propositional level. One advantage of the objectual route is that it avoids complications arising from the vagaries of linguistic usage.[2] Another advantage is the existence of a richer, more precise, and more sophisticated machinery for talking about operators than about constants.

Before examining Tarski's specific criterion, let us consider the idea of a general criterion of logicality independently of its content. What is the purpose of such a criterion? What would a systematic principle that demarcates the logical from the non-logical (not just on the level of propositional connectives but also on the level of quantifiers and other non-propositional operators) accomplish? The answer, I believe, is this: First, it would bring an end to the current practice of an ad-hoc, utterly uninformative, definition-by-enumeration of the logical operators other than connectives. Second, it would solve a serious problem that threatens to undermine Tarski's model-theoretic definition of logical consequence, and with it the entire field of logical semantics. Furthermore, such a principle would considerably deepen our understanding of the nature of logic, expand our ability to approach logic critically, create a fertile domain of mathematical investigations, help solve outstanding problems in linguistic semantics, and perhaps make other contributions as well, e.g., explain the relationship between the concept of logicality and other central philosophical concepts, explain logic's relation to neighboring fields (both within and outside philosophy), and so on.

One would have expected Tarski to motivate his criterion by the problem that threatened his own definition of "logical consequence," and whose full import he recognized and brought to our attention (Tarski 1936), namely, the problem that the definition's adequacy depended on the existence of an adequate criterion of logicality. At the time Tarski worried that such a criterion would never be found (in which case his definition would be forever unjustified), and this naturally leads us to expect that his 1966 lecture was intended to assuage those worries.

However, judging from what Tarski explicitly said (and did not say) in his 1966 lecture, his route to the criterion of logicality was completely divorced from his early concerns.[3] Instead, Tarski arrived at this criterion based on general considerations concerning the demarcation of fields of knowledge. His starting point was Klein's demarcation of geometrical fields based on their invariance properties. Klein suggested that each geometric field could be characterized by the invariance condition satisfied by its notions. This condition had the form:

Geometric Invariance: Geometric notion O is invariant under all 1–1 transformations of the geometrical space onto itself which preserve X.

---

[2] See Sher (2003). A similar advantage accrues to the objectual, model-theoretic definition of logical consequence as opposed to the linguistic, substitutional definition of this concept. (See Tarski 1936 and Sher 1996a.)

[3] One of the things that Tarski explicitly said (p. 145) is that he was *not* interested in the problem of logical consequence (or, as he put it, logical truth) in that lecture.

By strengthening X we restrict the transformations taken into account, getting more specific geometrical notions; by weakening X we increase the transformations taken into account, getting more general notions. Thus, if X is the requirement that the ratio of distances between points be preserved, the class of notions satisfying Geometric Invariance is the class of *Euclidean* notions. By strengthening X to the requirement that actual distances between points be preserved we obtain a characterization of narrower geometric notions, namely those applicable to *rigid bodies* (which don't change their *size* under movements or transformations); and by weakening X to the requirement that (I will express by saying that) openness (open sets) be preserved, we obtain a characterization of very broad geometric notions, namely the *topological* notions. Now, Tarski asked: What would happen if we weakened X as much as possible, i.e., if we set no requirements on the transformations taken into account? Then, we would get the condition.

General Invariance: Notion O is invariant under all 1–1 transformations of space, or the universe of discourse, or the "world" onto itself (or under all permutations of the "world").

This invariance condition takes *all* 1–1 transformations into account and, as a result, characterizes our most general notions. What is the science which studies these notions? Tarski suggested that this science is *logic*. Logic deals with our most general notions, notions which are invariant under all 1–1 transformations of the world onto itself.

Today, we usually adopt a slightly different version of Tarski's criterion. In fact, Tarski's (1966) lecture remained unknown for many years, and the current version is historically traced to Lindström's (1966) generalization of Mostowski (1957). This version invokes "isomorphisms" (or "bijections") instead of "permutations" (or "transformations") and refers to a totality "structures" rather than a to single, universal, "world." One way to formulate this criterion is:

Invariance under Isomorphism: An operator O is logical iff it is invariant under all isomorphisms of its argument-structures

where:

(i) A *structure* is an m-tuple, m ≥ 1, whose first element is a universe, A (i.e., a non-empty set of objects treated as individuals, that is, as objects lacking inner structure), and whose other elements (if any) are set-theoretic constructs of elements of A.

(ii) Two structures, $\langle A, \beta_1, \ldots, \beta_n \rangle$ and $\langle A', \beta'_1, \ldots, \beta'_k \rangle$, are *isomorphic*—$\langle A, \beta_1, \ldots, \beta_n \rangle \cong \langle A', \beta'_1, \ldots, \beta'_k \rangle$—iff n = k and there is a bijection f from A to A' such that for every $1 \leq i \leq n$, β'_i is the image of β_i under f.

(iii) An operator O represents an *object of a given type*—an individual, a property of individuals, an n-place relation of individuals (n > 1), an n-place function from individuals to an individual, a property of properties of individuals (i.e., a monadic first-order quantifier), a relation of properties of individuals (i.e., a relational first-order quantifier), a property of relations of individuals (i.e., a polyadic quantifier), etc.—and specifies its extension (or constitution) in each universe.

Specifically:

- An operator representing an individual a assigns to each universe A a 0-place function whose fixed value is a if $a \in A$, and which is treated in some conventional manner otherwise.
- An operator representing a first-order property assigns to each universe A a function from all members of A to a truth-value (which, provisionally, we assume is T or F).
- An operator representing an n-place first-order relation ($n > 1$) assigns to each universe A a function from all n-tuples of members of A to {T, F}.
- An operator representing a first-order monadic quantifier assigns to each universe A a function from all subsets of A to {T, F}.
- An operator representing a first-order binary relational quantifier assigns to each universe A a function from all pairs of subsets of A to {T, F}.
- An operator representing a first-order polyadic quantifier (of the simplest type) assigns to each universe A a function from all binary relations on A to {T, F}.

Etc.

(iv) If O is an operator whose arguments are of types $t_1, \ldots, t_n$,[4] A is a universe and $\beta_1, \ldots, \beta_n$ are constructs of elements of A of types $t_1, \ldots, t_n$ respectively, then $\beta_1, \ldots, \beta_n$ are arguments of O in A (or $\langle \beta_1, \ldots, \beta_n \rangle$ is an argument of O in A) and $\langle A, \beta_1, \ldots, \beta_n \rangle$ is an *argument-structure* of O.

For example:

(a) The first-order property "is red" is represented by an operator, R, which for every universe A is assigned a function, $R_A: A \to \{T, F\}$, such that for any $a \in A$, $R_A(a) = T$ iff a is red. (Its argument-structures are structures $\langle A, a \rangle$, where A is a universe and $a \in A$.)

(b) The first-order identity relation is represented by an operator, =, which for every universe A is assigned a function, $=_A: A \times A \to \{T, F\}$, such that for any $a, b \in A$, $=_A(a, b) = T$ iff $a = b$. (Its argument-structures are structures $\langle A, a, b \rangle$ where $a, b \in A$.)

(c) The first-order existential quantifier is represented by an operator, ∃, such that $\exists_A: P(A) \to \{T, F\}$, and for every $B \subseteq A$: $\exists_A(B) = T$ iff B is not empty.[5] (Its argument-structures are structures $\langle A, B \rangle$ where $B \subseteq A$.)

(d) The first-order monadic cardinality quantifiers, "There are exactly κ things such that," where κ is any cardinal, finite or infinite, are represented by operators, K, of the same kind as ∃, and such that for every $B \subseteq A$: $K_A(B) = T$ iff the cardinality of B—$|B|$—is κ. (Their argument-structures are the same as those of ∃.)

(e) The first-order monadic quantifier "It is a property of humans" is represented by an operator H of the same kind as ∃, and such that for every $B \subseteq A$: $H_A(B) = T$

[4] Types of arguments are the same as types of operators. (See (iii) above.)
[5] $P(A)$ is the power set of A.

iff all the members of B are humans. (Its argument-structures are the same as those of ∃.)

(f) The first-order polyadic quantifier "Is a well-ordering" is represented by an operator W such that $W_A: P(A \times A) \to \{T, F\}$, and for every $R \subseteq A \times A$: $W_A(R) = T$ iff R well-orders A. (Its argument-structures are structures $\langle A, R \rangle$ where $R \subseteq A \times A$.)

And so on.

We now define:

An n-place operator O is invariant under all isomorphisms of its argument-structures
iff
for any of its argument-structures, $\langle A, \beta_1, \ldots, \beta_n \rangle$ and $\langle A', \beta'_1, \ldots, \beta'_n \rangle$: if $\langle A, \beta_1, \ldots, \beta_n \rangle \cong \langle A', \beta'_1, \ldots, \beta'_n \rangle$, then $O_A(\beta_1, \ldots, \beta_n) = O_{A'}(\beta'_1, \ldots, \beta'_n)$.

It is easy to see that all the standard logical operators—e.g., (b) and (c), as well as the logical connectives when considered as objectual operators[6]—are logical according to this criterion, and that all blatantly non-logical operators—operators like (a) and (e)—are not. But the Invariance-under-Isomorphism criterion is a substantive criterion that does not just repeat what we think of as logical prior to a systematic, theoretical reflection. Quantifiers like the infinitistic (d)'s and (f) are also logical. Other non-standard logical operators include the uncountability quantifier and the monadic and relational "most."[7] In general, mathematical operators as they appear in first-order theories—e.g., the first-order set-membership operator ($\in$)—are not logical, but when raised to a higher order—e.g., the second-order set-membership operator ($\in$)—they are logical.[8]

---

[6] e.g., the logical connective "&" when considered as an objectual operator (as when it appears in an open formula of the form "Bx and Cx") is represented by an operator $\cap^2$ such that $\cap^2_A: P(A) \times P(A) \to P(A)$ and for every $B, C \subseteq A$: $\cap_A(B, C) =$ the intersection of B and C. For the sake of determining its logicality we represent this functional quantifier by the relational quantifier $\cap^3$ such that $\cap^3_A: P(A) \times P(A) \times P(A) \to \{T, F\}$, and for any $B, C, D \subseteq A$: $\cap^3_A(B,C,D) = T$ iff D is the intersection of B and C. (Its argument-structures are structures $\langle A, B, C, D \rangle$ where $B, C, D \subseteq A$.)

[7] These are defined as follows:

(i) The first-order monadic quantifier "There are uncountably many" is represented by an operator, U, of the same kind as ∃, and such that for every $B \subseteq A$: $U_A(B) = T$ iff B is uncountable. (Its argument-structures are the same as those of ∃.)

(ii) The first-order monadic quantifier "Most" (as in "Most things are B") is represented by an operator, M^1, of the same kind as ∃, and such that for every $B \subseteq A$: $M^1_A(B) = T$ iff $|B| > |A - B|$. (Its argument-structures are the same as those of ∃.)

(iii) The first-order relational quantifier "Most" (as in "Most B's are C's") is represented by an operator M^2 such that $M^2_A: P(A) \times P(A) \to \{T, F\}$, and for every $B, C \subseteq A$: $M^2_A(B,C) = T$ iff $|B \cap C| > |B - C|$. (Its argument-structures are structures $\langle A, B, C \rangle$ where $B, C \subseteq A$.)

[8] These operators are defined as follows:

(i) The first-order membership relation is represented by an operator, $\in$, of the same type as =, such that for any a, b in A, $\in_A(a, b) = T$ iff b is a set and a is a member of b. (Its argument-structures are the same as those of =.)

What about the logical connectives considered, as they usually are, as propositional operators? There are two ways to deal with propositional connectives: either we expand the notion of structure so that the Invariance-under-Isomorphism criterion applies to such operators, or we give a disjunctive criterion of logicality, dealing with propositional and objectual operators separately. Not surprisingly mathematicians (e.g., Tarski and Lindström) have opted for the former, but as a philosopher I prefer the latter. I think that the philosophical idea underlying logicality is realized on different levels of abstraction for the two types of operator, and to signal this difference I define:

Logicality: An operator is logical iff it either satisfies the Truth-Functionality criterion for propositional operators or it satisfies the Invariance-under-Isomorphism criterion for objectual operators.

Leaving the relation between Truth-Functionality and Invariance-under-Isomorphism aside for a moment, our next question is: What is the philosophical meaning of the Invariance-under-Isomorphism criterion?

PHILOSOPHICAL SIGNIFICANCE OF INVARIANCE-UNDER-ISOMORPHISM

The idea that logic is characterized by an invariance condition—i.e., by the things it does not distinguish between—has a long history. Kant, for example, says that "[general logic] treats of understanding without any regard to difference in the objects to which the understanding may be directed" (1781/7: A52/B76), and Frege says that "[p]ure logic ... disregard[s] the particular characteristics of objects" (1879: 5). But this trait can be construed in different ways, and two philosophical construals of Invariance-under-Isomorphism are: (a) generality (Tarski 1966), and (b) formality (Sher 1991).

Generality

In proposing his logicality criterion Tarski continually emphasized the fact that notions invariant under more transformations are more general than notions invariant under fewer[9] transformations. Thus, in geometry, we have more transformations preserving the ratio of distances between points than transformations preserving the actual distances between them, and more transformations preserving openness than transformations preserving the ratio of distances. Accordingly, notions invariant

---

(ii) The second-order membership relation is represented by an operator, $\in$, which for every universe A is assigned a function, $\in_A$: $A \times P(A) \to \{T, F\}$, such that for any a in A and B included in or equal to A, $\in_A(a, B) = T$ iff a is a member of B. (Its argument-structures are structures $\langle A, a, B \rangle$ where a is in A and B is included in or is equal to A.)

[9] "Fewer" here means "proper subset" rather than "smaller cardinality," as when we say that the set of odd positive integers has fewer elements than the set of positive integers.

under transformations preserving openness are more general than those invariant under transformations preserving the ratio of distances, and the latter are more general than notions invariant under transformations preserving actual distances.

To obtain *the most general* notions we renounce all restrictive conditions on the transformations partaking in the invariance condition. And invariance under all (bisective) transformations characterizes the *logical notions*. The distinctive mark of logicality, on this conception, is thus *utmost generality*, and this trait is captured by the Invariance-under-Isomorphism (or permutation) criterion.

Thus Tarski says:

> Now suppose we continue this idea, and consider still wider classes of transformations. In the *extreme case*, we would consider the class of *all* one-one transformations of the space, or universe of discourse, or 'world', onto itself. What will be the science which deals with the notions invariant under this widest class of transformations? Here we will have very few notions, all of a very general character. I suggest that they are the logical notions.
>
> (1966: 149; my underline)

It is natural to associate *utmost generality* with another characteristic feature of logic, *topic neutrality*, and this seems to strengthen the plausibility of interpreting Invariance-under Isomorphism as maximal generality.

But does Invariance-under-Isomorphism yield *the* most general notions? In "logicality and Invariance" (2006) Denis Bonnay challenges the identification of Invariance-under-Isomorphism with maximal generality:

> The [interpretation] in terms of generality rests on the assumption that invariance under the biggest class of transformations yields maximal generality. The idea is that the group of all permutations is as "big" as one might wish, because in that case the transformations do not respect any extra-structure, such as *e.g.*, the topological structure of the space. Let us have a closer look at this idea. Permutation invariance just says that as soon as there is an automorphism linking ⟨M, A⟩ and ⟨M, A′⟩, a quantifier Q acting on M has to give A and A′ the same value. On the one hand, this is indeed liberal, because no further structure beyond the extensions A and A′ on M is taken into account. But on the other hand, this is quite demanding: for ⟨M, A⟩ and ⟨M, A′⟩ to be similar from a logical point of view, they have to share exactly the same structure—they have to be isomorphic. Now there are a lot of other concepts of similarity between structures which are used in model theory and in algebra which are far less demanding. Instead of requiring the structure to be fully preserved, they lower the requirement to some kind of approximate preservation. Why should we refrain from resorting to these other concepts? To sum up, even if one grants that generality is a good way to approach logicality, there is no evidence that the class of all permutations is the best applicant for the job.
>
> (Bonnay 2008: 38)

Bonnay's point is well taken. In the extreme case we can remove all constraints on the functions involved, requiring logical operators to be invariant under *all functions* (from argument-structures to argument-structures of a given kind) *whatsoever*. This would give us the utmost general notions (in one reasonable sense of the word), but these notions would have very little to do with what we think of as logic. All the standard logical notions would fail this criterion, and the notions that would satisfy it would be such notions as: "is an individual," "is a property of individuals," "is an n-place relation of individuals (n > 1)," "is a property of properties of individuals," etc. Logic, according to this characterization, would be a theory of *semantic types*, not a theory of *inference* (or *transmission of truth*) as we intend it to be. I conclude that: (a) Invariance-under-Isomorphism does not mean utmost generality, and (b) if we want to preserve any semblance to what we intuitively mean by logic, we cannot regard utmost generality, or for that matter topic neutrality, as *the* mark of logic.

Formality

On my interpretation (Sher 1991 and elsewhere), the Invariance-under-Isomorphism criterion is a criterion of *formality* or *structurality*: isomorphic structures are formally identical; identity-up-to-isomorphism is formal identity. The basic idea is that logic is a theory of reasoning based on the formal (structural) laws governing our thinking on the one hand and reality on the other, and the Invariance-under-Isomorphism criterion says that to be formal is to treat isomorphic structures as the same structures. Formal operators do not distinguish between isomorphic arguments (or rather between isomorphic argument-structures, since some formal features of arguments depend on the formal traits of the underlying universe).

The view that Invariance-under-Isomorphism captures the concept of formality (or structurality) is well-known from the philosophy of mathematics. Structuralists, in particular, view mathematics as the science of structure (or formal structure), and Invariance-under-Isomorphism as a mark of structurality. The Invariance-under-Isomorphism criterion characterizes logic as a theory of formal or structural inference, inference based on the laws governing formal or structural operators.

What is the relation between logic and mathematics under this interpretation? I will attend to this question in the next section, but in the meantime let me say that on the "formalist" conception of logic, logic and mathematics are interconnected theories, approaching the same topic, *the formal*, from different, yet interrelated, perspectives. Mathematics investigates the laws of formal structure; logic applies these laws in general reasoning. Logic includes mathematics, raised to a higher-order, so it can be applied in inference in general. The idea is that formal operators—union, intersection, complementation, non-emptiness, majority ("most"), finiteness, and others—are applicable to structures of objects studied in all areas of knowledge, and therefore inferences based on the laws governing them are valid in all areas.

This universal applicability of the formal operators explains logic's generality and topic neutrality. Logic does not distinguish between different topics of discourse since the formal laws governing the behavior of individuals, properties, and relations in different areas are the same. (In all areas individuals are identical to themselves, the union of non-empty properties is non-empty, etc.) Their differences concern something other than these formal laws, and logic abstracts from such differences. Comparing the two characterizations of logic associated with the Invariance-under-Isomorphism criterion, then, we can say that the *formality* of logic *ensures* its *generality* (not absolute generality, but a very high degree of generality), while the *generality* of logic *does not ensure* its *formality*. This is but one advantage of taking formality rather than generality as the mark of logic. In the remainder of this chapter I will assume Invariance-under-Isomorphism characterizes logicality as formality.

PHILOSOPHICAL JUSTIFICATION OF THE INVARIANCE-UNDER-ISOMORPHISM CRITERION

Now that we have a basic understanding of the Invariance-under-Isomorphism criterion, our next task is to provide a philosophical justification for this criterion. I think it is quite clear that this criterion satisfies the first methodological desideratum mentioned above, namely, systematicity and informativeness (i.e., a genuine principle of logicality as opposed to a definition by enumeration).[10] But it also satisfies the other desiderata. For example, it has opened new areas of research in mathematics and linguistics and helped solve standing problems in both disciplines.[11] Here, however, I would like to focus on substantive philosophical points that support this criterion, i.e., give it what may be called "a foundational justification." By this I mean showing how the philosophical conception of logic associated with this criterion—namely, the "formalist" conception briefly delineated in the last section—is capable of providing a foundation for logic largely due to its association with this criterion.

Methodological quandary: holistic vs. foundationalist foundation

In thinking about a foundation for logic most of us think in foundationalist terms: we think that the only way to establish logic is by using epistemic resources that are more basic than those produced by logic itself. And this leads us to a pessimistic conclusion: since no sufficiently rich branch of knowledge is more basic than logic, there is no possibility of establishing logic; a foundation for, or a justification of, logic is in principle impossible. The source of the problem, it is easy to see, is the foundationalist conception of the foundation (justification, grounding) relation as intuitively *strongly ordered*. In the ideal case, foundationalism requires that our entire system of knowledge be ordered by an anti-reflexive partial-ordering, that this ordering have an absolute base consisting of minimal (initial, atomic) elements, and that each non-minimal element

---

[10] Among other things, the Invariance-under-Isomorphism criterion does for objectual logical operators what the Boolean, truth-functional criterion did for propositional logical operators, namely, provide a complete, precise, systematic definition, fleshing out their structure, and explaining how they "work." (How they work is best shown by a "constructive" or "bottom-top" definition of logical operators. Such a definition is formulated in Sher 1991, Ch. 4, and is informally described in Sher 1996*b*.)

[11] For example, it has led to the development of "model-theoretic logic" and "generalized-quantifier theory." Some remarkable results of these new fields are Lindström's characterization of (standard) first-order logic, Keisler's completeness proof for first-order logic with the quantifier "uncountably many," the solution to the problem of determiners in linguistic semantics, and the theories of polyadic and branching quantifiers in natural language. The literature here is enormous. For a small sampling see Keisler (1970), Lindström (1974), Barwise and Feferman (1985), Higginbotham and May (1981), Barwise and Cooper (1981), Keenan and Stavi (1986), Van Benthem (1983 and 1989), Westerståhl (1985 and 1987), Keenan (1987), and Sher (1991, chs. 2, 4, and 5).

in the system be connected to each minimal element grounding it by a finite chain. This central feature of the foundationalist method is its Achilles heel; due to it, foundationalism has, in principle, no resources for grounding the basic constituents of knowledge—the disciplines constituting the lowest echelon in the foundationalist hierarchy. In particular, foundationalism is incapable of providing a foundation for logic. As a basic branch of knowledge, logic can partake in the foundation of other sciences, but no science (or combination of sciences) can provide a foundation for logic. Having postulated (i) that any resource for founding logic must be more basic than the resources produced by logic itself, and (ii) that there are no (or not enough) resources more basic than those produced by logic, foundationalism is incapable of founding logic.

In view of these considerations, it is clear that a foundation for logic must be *holistic*. I will not be able to explain in great detail the idea of a holistic foundation, or *foundational holism*, here. (For an extended discussion see Sher 2006.) But a few points have to be made:

- Foundational holism would provide a foundation for logic in the sense of describing its basic mechanisms, justifying the definitions of central meta-logical concepts, solving standing problems in the philosophy of logic, identifying constraints on logic, elucidating the relation between logicality and related concepts, sorting out and accounting for the distinctive characteristics of logic, explaining logic's role in our system of knowledge, throwing light on the relation between logic and mathematics, providing critical tools for detecting errors and making improvements in logical theory, etc.

- Foundational holism is not coherentist. It requires that knowledge be grounded in reality; in fact, it strengthens foundationalism by requiring that every branch of knowledge be grounded in reality. But, being holistic, it permits us to use all the resources available to us in providing such a grounding.

- Foundational holism does not require an absolute, infallible foundation, but it requires a solid foundation.

- Foundational holism requires a theoretical, and not just an intuitive, grounding of logic.

- Foundational holism rejects vicious circularity, but not circularity per se. Which circularity is vicious is determined by holistic methods.

Having made these methodological points, I will proceed to show how the formalist conception of logic accomplishes some of the foundationalist tasks mentioned above due to its association with the Invariance-under-Isomorphism criterion.

A. *Explanation of logic's connection to truth*. It is commonplace to say that as a theory of logical truth (truth of a certain kind) and of logical consequence (transmission of truth of a certain kind) logic is intimately connected with truth. But what, exactly, are the nature of this connection and its constraints on logic? Let us start with general theoretical considerations.

Assuming the classical idea that truth importantly involves some correspondence relation between truth-bearers and reality, let us consider two truth-bearers, S_1 and

S_2, whose truth-conditions straightforwardly and paradigmatically exemplify this idea. (For the sake of simplicity, let us further assume that S_1 and S_2 are distinct and non-synonymous.) Now, suppose that according to some logical theory, L, S_2 is a logical consequence of S_1. In symbols:

(1) (*Level of Logic*) $S_1 \models^{12} S_2$.

Further suppose that S_1 is true. Then (1) says that the truth of S_1 extends to, or is transmitted to, or is preserved by, S_2:

(2) (*Level of Language*) $T(S_1) \longrightarrow T(S_2)$.

((1) says something stronger than that, but let us attend to the weaker claim first.)

Let $\mathfrak{S}_1$ and $\mathfrak{S}_2$ be the situations that have to be realized for S_1 and S_2 to be true and that would guarantee their truth were they to be realized. Figuratively:

(3) (*Level of Language*) $T(S_1) \quad\quad T(S_2)$.
    $\quad\quad\quad\quad\quad\quad\quad\quad\quad \Updownarrow \quad\quad\quad \Updownarrow$
    (*Level of World*) $\mathfrak{S}_1 \quad\quad\quad \mathfrak{S}_2$.

Now, suppose that in the world $\mathfrak{S}_1$ is the case but $\mathfrak{S}_2$ is not. (In the extreme case, $\mathfrak{S}_1$ rules out $\mathfrak{S}_2$.) I.e.,

(4) (*Level of World*) $\langle \mathfrak{S}_1, \text{not } \mathfrak{S}_2 \rangle$ (in the extreme case : $\mathfrak{S}_1 \Rightarrow \text{not } \mathfrak{S}_2$).

Then, our logical theory is wrong. No matter what L says, S_2 is *not* a logical consequence of S_1:

(5) (*Level of Logic*) $S_1 \not\models S_2$.

Logic, indeed, is constrained by truth more deeply than the above consideration suggests. Suppose that in the world both $\mathfrak{S}_1$ and $\mathfrak{S}_2$ are the case, but $\mathfrak{S}_1$ being the case does not require $\mathfrak{S}_2$ being the case:

(6) (*Level of World*) $\langle \mathfrak{S}_1, \mathfrak{S}_2, \quad \mathfrak{S}_1 \not\Rightarrow \mathfrak{S}_2 \rangle$.

Then, again:

(7) (*Level of Language*) $S_1 \not\models S_2$.

Now, an adequate criterion of logicality has to explain, or be incorporated in an account that explains, this alethic constraint on logic. The formalist interpretation of the Invariance-under-Isomorphism criterion delineated in the last section is embedded in a "formalist" account of logic that does just that. We can sum up its main points as follows:[13]

(i) The logical constituents of truth-bearers—especially, their *logical constants*—represent formal properties, relations, and functions, where formality is interpreted as Invariance-under-Isomorphism.

---

[12] '$\models$' is a symbol of an unspecified kind for logical consequence. "$S_1 \models S_2$" reads: "S_1 logically implies S_2."

[13] For a more detailed account see Sher (1991) and related papers.

(ii) The logical form of truth-bearers is obtained by holding their logical constants fixed and treating their non-logical constants as variable.

(iii) Corresponding to a truth-bearer S is a situation, $\mathfrak{S}$, that would make S true if it were to be realized; corresponding to the logical form of S is the formal skeleton of $\mathfrak{S}$, which contains those parameters of $\mathfrak{S}$ which correspond to its logical constituents. For example, corresponding to "Something is white and round" is a structure, $\langle A, B, C \rangle$ where A is the intended universe of discourse, B is the collection of white things in A, C is the collection of round things in A, and the intersection of B and C is not empty. The formal skeleton of $\mathfrak{S}$ contains the formal parameters of $\mathfrak{S}$ corresponding to the logical constants of S, namely: intersection and non-emptiness (a cardinality parameter).

(iv) Logical consequence is a relation between truth-bearers which represents a universal formal law connecting the situation corresponding to the "premise" truth-bearers to the one corresponding to the "conclusion" truth-bearer. Alternatively, logical consequence correspond to, and is largely due to, a law connecting the formal skeleton of the "premise" situations to the formal skeleton of the "conclusion" situation. This law is universal in the sense that it holds in all formally possible situations, or in all possible formal-structures. For example:

Something is white

is a logical-consequence of

Something is white *and* round

because it is a formal law that whenever an intersection of two subsets is not empty, the first of these subsets is not empty; it is not a logical consequence of

Something is white *or* round

because it is not a formal law that whenever a union of two subsets is not empty, the first of these subsets is not empty.

If we regard formal laws as formally necessary, we can concisely represent the present conception of logical consequence thus:

Level of Language: S_1 *logically implies* S_2.
                                  ⇕
Level of World: $\mathfrak{S}_1$ *formally necessitates* $\mathfrak{S}_2$.

(v) In contemporary (Tarskian) semantics we represent the formally possible situations vis-à-vis a given language by the totality of *models* for that language. Universal formal laws are represented by *regularities across all models*.

(vi) This explains the standard (Tarskian) semantic definition of logical consequence: S is a logical-consequence of K iff S is true in all models (i.e., formally possible situations) in which all the members of K are true, i.e., when S is a logical consequence of K, this is due to some formal law connecting the situations corresponding to S and K.

Note: This account assumes a background theory of formal structure, used to formulate the logicality criterion, delineate the totality of formally possible situations represented by models, determine the laws governing them (i.e., the formal laws underlying logical consequence), etc. The appeal to such a background theory is licensed by the holistic methodology of the account. This is an important point that is easy to miss. Indeed, it is so common to associate the foundational goal with the foundationalist method and the holistic method with the renunciation of this goal, that many philosophers evaluate *any* foundational proposal based on foundationalist standards.[14]

The holistic approach enables us to maneuver the limitations of the background theory in a rational and effective manner. Consider, for example, the incompleteness phenomenon. The holist reasons that in the same way that we are forced to use incomplete mathematical background-theories in providing a foundation for physics, so we are forced to use incomplete mathematical background-theories in providing a foundation for logic. To the extent that no perfect, complete theory of formal structure is available (temporarily or in principle) but some quite advanced theories do exist, we rely on the best background theory we can find, and avow ignorance with respect to those cases of logical truth and consequence that this theory cannot handle.

Yet the holist can still hold on to the classical concept of truth, i.e., ensure that there is a fact of the matter about how, say, the "continuum quantifier" behaves. This he does by using a *complete* version of his chosen background theory—specifically, the theory of some *model* of his theory of formal structure—to determine *facts* about the behavior of logical operators and the laws governing them, and an *incomplete* axiomatization to derive whatever *knowledge* he can have of those facts. Truth is anchored in a complete (if inaccessible) theory of formal structure, knowledge—in an effectively axiomatized, hence accessible (if incomplete) version of that theory. The facts are as they are, but knowledge is in principle limited.

Explanation of logic's role in knowledge and its place in our system of knowledge

The formalist account of logic enables us to explain the role played by logic in our system of knowledge (and to the extent that this explanation is compelling, it is also

---

[14] In the case of logic, one relevant example is Etchemendy (this volume). Etchemendy thinks that due to the incompleteness of any reasonable background theory of formal structure we cannot establish the formal necessity of Tarskian consequences. (See his criticism of Sher 1996*a* in fn. 24 of Chapter 11). This claim is right for a foundationalist, who requires absolute certainty and intuitive completeness (hence also technical completeness) of a putative foundation, but not for a holist who, contesting the appropriateness of such demands, allows the background theory of formal structure, like all human theories, to be short of perfect (and technically incomplete). I will presently attend to the incompleteness problem in a little more detail, but the point here is that while Etchemendy himself is, for all I know, a holist, he applies foundationalist standards to the foundational claims in Sher (1996*a*), despite the fact that they are explicitly offered as holistic claims. He seems not to appreciate the possibility of a holistic foundation, or foundational holism.

supported by it). According to this explanation logic plays a dual role in knowledge: first, it sets general constraints on what counts as knowledge, and second, it creates useful tools for expanding and correcting our knowledge. Let us consider the latter first:

Expansion of knowledge

Being finite and relatively short-living creatures, we cannot hope to establish all our knowledge directly but have to resort to such indirect means as *inference* to obtain a considerable portion of our knowledge. In inference we use our knowledge of the relations between objects or situations plus some knowledge of these objects or situations to obtain new knowledge which, as inferred knowledge, does not require independent verification. For example, if we have knowledge about the chemical constitution of objects and the relations between chemical structures, we can use this knowledge to obtain new knowledge about objects. But while chemical laws enable us to expand our knowledge in a small number of areas, formal laws enable us to expand it in all areas. Given that formal features of objects are constantly referred to in all discourse—one cannot talk about anything without saying that certain objects are in the *complement* or *intersection* of certain properties, that certain properties are *non-empty*, or *universal*, or have κ *objects* falling under them, etc.—we can use our knowledge of these features to develop a wholesale method of expanding our knowledge. Logic, on this conception, utilizes our knowledge of the formal behavior of objects to formulate rules of inference that sanction our movement from what we know to what (prior to this movement) we did not know. Knowledge of some formal laws may be more useful for expanding our overall knowledge than knowledge of others, so it might be useful to build limited logical systems geared to those features. But in principle logic can provide us with rules for expanding our knowledge based on any laws governing the formal behavior of objects.

Constraints on knowledge

Due to the prevalence of formal features of objects and our constant reference to such features in discourse and theorizing, the threat of formal errors in our system of knowledge looms large. But due to the fact that the formal does not distinguish between different domains of knowledge, it is possible to take care of such errors in "one fell swoop", so to speak, i.e., in a way that protects all (or most) fields of knowledge at once. This opportunity is seized by logic. Logic builds into our language rules that prevent us from making errors pertaining to the (law-governed) formal behavior of objects in any area. For example, by telling us that statements of the form "$\Phi a \ \& \sim \Phi a$" are false (or that a combination of statements of the form "Φa" and "$\sim \Phi a$" is inconsistent) logic prevents us from making certain errors concerning the behavior of objects under the complementarity operation (in any field). By telling us that inferences of the form "$(\forall x)(\exists y)\Phi xy$; therefore $(\exists y)(\forall x)\Phi xy$" are invalid, logic prevents us from assuming certain symmetries exist where they do not. And so on.

Solution to Tarski's problem

In his 1936 chapter, "On the Concept of Logical Consequence," Tarski sought a definition of "logical consequence" that would satisfy two intuitive constraints:

> Certain considerations of an intuitive nature will form our starting-point. Consider any class K of sentences and a sentence X which follows from the sentences of this class. From an intuitive standpoint it can never happen that both the class K consists only of true sentences and the sentence X is false. Moreover, since we are concerned here with the concept of logical, i.e., *formal*, consequence, and thus with a relation which is to be uniquely determined by the form of the sentences between which it holds, this relation cannot be influenced in any way by empirical knowledge, and in particular by knowledge of the objects to which the sentence X or the sentences of the class K refer. The consequence relation cannot be affected by replacing the designations of the objects referred to in these sentences by the designations of any other objects. The two circumstances just indicated, ... seem to be very characteristic and essential for the proper concept of consequence.
>
> (Tarski 1936: 414–15)

Based on these considerations Tarski formulated his semantic definition of "logical consequence":

> *The sentence X follows logically from the sentences of the class K if and only if every model of the class K is also a model of the sentence X.*
>
> (Ibid.: 417)

Is this an adequate definition? Does it satisfy the intuitive constraints? At first Tarski gave a positive answer:

> It seems to me that everyone who understands the content of the above definition must admit that it agrees quite well with common usage. This becomes still clearer from its various consequences. In particular, it can be proved, on the basis of this definition, that every consequence of true sentences must be true, and also that the consequence relation which holds between given sentences is completely independent of the sense of the extra-logical constants which occur in these sentences.
>
> (Ibid.)

But soon he qualified his answer:

> I am not at all of the opinion that in the result of the above discussion the problem of a materially adequate definition of the concept of consequence has been completely solved. On the contrary, I still see several open questions, ... one of which—perhaps the most important—I shall point out here.
>
> (Ibid.: 418)

This question was the demarcation of logical constants:

> Underlying our whole construction is the division of all terms of the language discussed into logical and extra-logical. This division is certainly not quite arbitrary. If, for example, we were to include among the extra-logical signs the implication sign, or the universal quantifier, then our definition of the concept of consequence would lead to results which obviously contradict ordinary usage. On the other hand, no objective grounds are known to me which permit us

to draw a sharp boundary between the two groups of terms. It seems to be possible to include among logical terms some which are usually regarded by logicians as extra-logical without running into consequences which stand in sharp contrast to ordinary usage.

(Ibid.: 418–19)

These qualifications led Tarski to conclude his chapter on a skeptical note:

Further research will doubtless greatly clarify the problem which interests us. Perhaps it will be possible to find important objective arguments which will enable us to justify the traditional boundary between logical and extra-logical expressions. But I also consider it to be quite possible that investigations will bring no positive results in this direction, so that we shall be compelled to regard such concepts as 'logical consequence' and ... 'tautology' as relative concepts which must, on each occasion, be related to a definite, although in greater or less degree arbitrary, division of terms into logical and extra-logical.

(Ibid.: 420)

The Invariance-under-Isomorphism criterion offers a positive solution to Tarski's problem. It offers a demarcation of logical operators under which Tarski's definition of logical consequence can be shown to satisfy the intuitive constraints. To see how, consider the following:

1. Tarski set two intuitive constraints on an adequate definition of logical consequence:

    (C1) Necessity: When X follows logically from K, X follows necessarily from K.

    (C2) Formality: When X follows logically from K, X follows formally from K.

2. Regardless of what Tarski himself understood by necessity, if we show that his definition satisfies a robust standard of necessity, we will have shown that it satisfies whatever weaker standard he might have had in mind.

3. Formality can be interpreted both syntactically and semantically. Philosophers often think of formality syntactically, but the key to vindicating Tarski's definition is to think of it semantically.

4. Tarski himself offers the key to a semantic interpretation of formality:

    [As a formal relation, logical consequence] cannot be influenced in any way by ... knowledge of the objects to which the sentence X or the sentences of the class K refer. The consequence relation cannot be affected by replacing the designations of the objects referred to in these sentences by the designations of any other objects.

    (Ibid.: 414–15, cited above)

5. This paragraph suggests that the formal is characterized by its inability to distinguish the identity of objects in a given universe of discourse. This is an invariance characterization: formal relations are invariant under replacements of objects. Now, if we interpret "replacement" as "1–1 and onto transformation or mapping," and "replacement of objects" as "replacement of objects of all types induced by replacement of the individuals in a given universe of discourse," then we get the Invariance-under-Isomorphism criterion of logicality.

6. Under this criterion all logical operators are derivable from mathematical operators by raising them to a higher order (as we have seen on p. 307), and in this sense they are essentially mathematical.

7. But the laws governing mathematical operators are intuitively formal and necessary (where this necessity is an especially strong kind of necessity, stronger than biological, physical, and even metaphysical necessity). Therefore, if logical consequence is due to the formal (or mathematical) laws governing the logical operators, logical consequences are formal and (strongly) necessary.

8. Now, on a formalist reading Tarski's definition does satisfy the antecedent of this conditional. The totality of models represents the totality of formal possibilities; logical consequences preserve truth across all models; they do so due to the logical structure of the sentences involved; this logical structure reflects the formal skeleton of the situations described by those sentences; therefore the preservation of truth is due to connections that hold between the formal skeletons of the situations involved in all formal possibilities; and formal connections persisting through the totality of formal possibilities are laws of formal structure. It follows that consequences satisfying Tarski's definition are formal and necessary, as required by the intuitive constraints (however strong the necessity constraint is taken to be).[15]

Explanation of the distinctive characteristics of logic

Logic is often characterized by its basicness, generality, topic-neutrality, necessity, formality, strong normative force, certainty, a-priority, and/or analyticity. While, as foundational holists, we reject the purported analyticity of logic and qualify its a-priority, we can explain its other characteristics (including quasi-apriority) based on the Invariance-under-Isomorphism criterion, i.e., explain why the laws of logic and its consequences are as basic, general, topic-neutral, formal, strongly normative, and highly certain as they appear to us to be, and to what degree they are a-priori.

We have already seen how the Invariance-under-Isomorphism criterion, either alone or together with other elements of the formalist account, explains the *formality*,

---

[15] In defending the adequacy of Tarski's definition it may seem that we have to confront Etchemendy's 1990 challenge to it, but in fact we don't. Etchemendy considers two conceptions of logic: the so-called representational and interpretational conceptions. But the formalist conception of logic offered here (and in Sher 1991) falls under neither category. Since Etchemendy's criticisms center on features of those conceptions that are not shared by the present conception, his criticisms do not concern us here. This includes his claim that the problem of logical constants is a "red herring." Etchemendy regards the problem of logical constants as a red herring not because he thinks logical constants do not pose a genuine problem to Tarski's definition, but because he thinks that Tarski's definition is plagued by other problems as well and merely solving the logical constants problem will not by itself establish its adequacy. However, the additional problems Etchemendy alludes to are specific to the interpretational construal of logic and do not arise on the formalist construal. Therefore, on that construal the problem of logical constants, far from being a "red herring," is the main obstacle to the adequacy of Tarski's definition. For a fuller critique of Etchemendy's (1990) see Sher (1996*a*).

generality, topic-neutrality, and *necessity* of logic. Let us, then, turn to the other characteristics.

Basicness and *strong normative force*. Logic is intuitively more basic than other disciplines. The grounding of geography, biology, and chemistry involves establishing their *logical* consistency, i.e., establishing that their laws obey the laws of logic, but the grounding of logic does not involve establishing that its laws obey the laws of geography, biology, and chemistry. This gap is related to a gap in the normative force of logic and other disciplines. Chemistry, biology, and geography have to attend to the strictures of logic, but logic need not attend to their strictures. Logic has normative authority over these disciplines, but not vice versa. The Invariance-under-Isomorphism criterion explains why this is so: Since chemical properties are not preserved under isomorphisms, logic has a stronger invariance property than chemistry. As a result, logic does not distinguish chemical differences between objects and is not subject to the laws governing chemical properties. But chemistry does distinguish formal differences between objects; for example, it distinguishes between *one* atom and *two* atoms. So chemistry is subject to the laws of formal structure. For example, chemistry is bound by the law

$$(\exists!2x)\Phi \supset \sim(\exists!3x)\Phi,$$

as in

$(\exists!2x)$ x *is a Hydrogen atom in water molecule w* $\supset \sim(\exists!3x)$ x *is a Hydrogen atom in w*.

And the same holds for most other disciplines. (The case of mathematics will be discussed separately below.)

Certainty and quasi-a-priority. Logic has a relatively high degree of certainty, not in the sense that we are less likely to make errors in applying the logical laws than other laws, but in the sense that the logical laws themselves are unlikely to be refuted by our empirical discoveries. The Invariance-under-Isomorphism criterion explains why logic is immune to refutation in this sense. Since most of our empirical discoveries do not concern the formal regularities in the behavior of objects—i.e., regularities governing features of objects that are invariant under isomorphism—logic is not affected by most of these discoveries, and in this sense it is resistant to refutation and, furthermore, *a-priori-like*. Now, if formal laws were completely immune to discoveries having any empirical element, then logic would be strictly a-priori. But holism allows a certain degree of interconnection between all disciplines, hence on the holistic approach logic is only quasi-a-priori. What kind of empirical discoveries could affect logic? Empirical discoveries affect logic only in very rare cases, and therefore we have no ready examples, but one challenge to classical logic did come from physics (Birkhoff and von Neumann 1936), and by extrapolating from it we could arrive at a possible scenario in which empirical discoveries would affect logic. Suppose we discover that in some region of reality (e.g., the quantum region) objects or states behave in a way that is radically different from what we have observed elsewhere, and we have good reasons to believe that it concerns the basic formal behavior of objects (states, properties). For example, suppose we have good reasons to believe that their behavior is deeply non-Boolean. Then this discovery would pose serious questions to classical logic.

Explanation of the relation between logic and mathematics

Ever since Frege, logic and mathematics have been treated as closely related disciplines whose relation requires an explanation. And one of the least noted, but methodologically most important achievements, of Frege's logicism was the enormous economy it brought to the philosophical tasks of explaining the nature of logic and mathematics and providing them with a foundation. By reducing mathematics to logic, logicism reduced two mysteries to one. Instead of having to explain both the nature of logic and the nature of mathematics we now had to explain only the nature of logic; and instead of the monumental task of constructing both a foundation for logic and a foundation for mathematics, we had the more manageable task of constructing a foundation only for logic. However, the search for a foundation for logic (independently of mathematics) led to nowhere. The most influential attempt to construct an account of logic that would complement logicism—Carnap's conventionalism—has by and large been discarded, and this, together with the almost unanimous rejection of logicism itself, has left us, once again, with the extremely difficult task of providing an explanatory account and a foundation both for logic and for mathematics.

The formalist account of logic, with its Invariance-under-Isomorphism criterion of logicality, offers an explanation of the relation between logic and mathematics that has the same methodological advantage as Frege's explanation without having its shortcomings. Like Frege's account, it reduces the two fields to one, hence the two foundational tasks to one. But this time it is logic that is reduced to mathematics rather than mathematics to logic. Or, alternatively, both logic and mathematics are reduced to the formal. Mathematics, in this account, builds a theory of formal structure, and logic provides a method of inference based on this theory. I will call the new approach "mathematicism." If *logicism* is the view that mathematics has a logical foundation, *mathematicism* is the view that logic has a mathematical foundation. But there is a considerable methodological advantage to mathematicism over logicism. While today we have no promising foundational account of logic not centered on mathematics, we do have a number of promising foundational accounts of mathematics not centered on logic; for example, the Platonist account, the naturalist account, and the structuralist account. It is true that these accounts assume logic in the background, but since mathematicism seeks to give a *holistic* foundation for logic, this does not pose a special difficulty. Logic does not stand at the center of any of these accounts, therefore the circularity involved is (at least prima facie) not vicious.

But the current situation is even more felicitous. Not only are several accounts of mathematics compatible with logical formalism, one of these accounts, the *structuralist* account, is very close to it in spirit. This is reflected in the fact that mathematical structuralism and logical formalism share the same identity criterion: invariance under isomorphism. Invariance under isomorphism is the identity criterion of logical operators according to logical formalism, and it is also the identity criterion, or at least an identity-criterion of choice, of mathematical structures according to mathematical structuralism. Thus, Shapiro says:

> No matter how it is to be articulated, structuralism depends on a notion of two systems that exemplify the "same" structure. That is its point.... [W]e ... need to articulate a relation among systems that amounts to "have the same structure".
>
> There are several relations that will do for this.... The first is *isomorphism*, a common (and respectable) mathematical notion.... Informally, it is sometimes said that isomorphism "preserves structure".
>
> (Shapiro 1997: 90–1; my underline)
>
> A purported implicit definition characterizes at most one structure if it is *categorical*—if any two models of it are isomorphic to each other.
>
> (Ibid.: 13; my underline)
>
> Because isomorphism . . . [is an] equivalence relation . . . one can informally take a structure to be an isomorphism type.
>
> (Ibid.: 92; my underline)

Indeed, it would be just as appropriate to call our account of logic "logical structuralism" as to call it "logical formalism" (and to call the structuralist account of mathematics "mathematical formalism" as to call it "mathematical structuralism").[16]

Furthermore, we can achieve the same methodological goal without reducing either discipline to the other, namely, by tracing both mathematics and logic to the same root, i.e., *the formal (structural)*. Analytically, logic and mathematics develop in tandem from a basic engagement with the formal (the structural). We can represent their joint development along something like the following lines: In stage 1, we develop a rudimentary logic-mathematics which studies some very basic formal operators, say complementation, union, intersection, and inclusion. Based on this knowledge we develop, in stage 2, a logical framework for theories in general, and using it we develop a more sophisticated mathematical theory of formal structure (say, naive set theory). Based on this theory we develop, in stage 3, a more sophisticated logical framework, say the logical framework of standard first-order logic with its standard logical operators ($\exists$, $\forall$, $=$, and the truth-functional connectives). And using this framework we develop, in stage 4, a more advanced mathematical theory of formal structure (say, axiomatic set theory). In stage 5 we use this advanced theory to develop a criterion of logicality (for example, the Invariance-under-Isomorphism criterion) and a semantic definition of logical consequence (for

---

[16] My misgivings about "structuralism" is that there are many kinds of structure, not all mathematical or logical (for example, physical or biological structures which are not preserved under isomorphisms). To distinguish mathematical and logical structures from other structures I call them "formal." But "formalism" has unwanted connotations of its own, namely, Hilbertian formalism. Once we make clear, however, that our use of "formal" is semantic, this association should dissolve. In Sher (2001) I used "formal-structural" for the formalist account of logic so as to signal both its affinity to the structuralist account of mathematics and its difference from Hilbert's formalism.

example, Tarski's model-theoretic definition), and based on these, an expanded logical framework—say, so-called generalized first-order logic (or standard second-order logic). And this process may continue: using this enriched logic we may arrive at a still more powerful mathematics and, based on it, perhaps a stronger logic. And so on.

To deal with the formal in logic and in mathematics we operate on different levels. In mathematics we construe the formal as (for the most part) lower-order, in logic we construe it as (for the most part) higher-order. Take, for example, the notion of number or the notions of union, intersection, and complementation. In axiomatic arithmetic numbers are individuals, but the numerical quantifiers are operators on properties; in axiomatic set-theory union, intersection, and complementation are operations on individuals, but in logic they are operators on properties (or propositions). As studied in mathematics, these notions do not satisfy the Invariance-under-Isomorphism criterion, but as studied in logic they do. And the same holds for other formal notions: for example, the membership relation of axiomatic set theory is not logical, but the membership relation of higher-order logic is. (A more nuanced version of this account would say that the mathematician treats some mathematical concepts as non-logical and others as logical. The number theorist, for example, treats numbers as non-logical entities, but the background mathematical concepts he uses to talk, and formulate questions, about numbers—e.g., the concept of set-membership—as logical.)

Tarski's take on the philosophical ramifications of the new logicality criterion for the relation between logic and mathematics is different from mine:

> The question is often asked whether mathematics is a part of logic. Here we are interested in only one aspect of this problem, whether mathematical notions are logical notions, and not, for example, in whether mathematical truths are logical truths, which is outside our domain of discussion. Since it is now well known that the whole of mathematics can be constructed within set theory, or the theory of classes, the problem reduces to the following one: Are set-theoretical notions logical notions or not? Again, since it is known that all usual set-theoretical notions can be defined in terms of one, the notion of belonging, or the membership relation, the final form of our question is whether the membership relation is a logical one in the sense of my suggestion. The answer will seem disappointing. For we can develop set theory, the theory of the membership relation, in such a way that the answer to this question is affirmative, or we can proceed in such a say that the answer is negative. So the answer is: 'As you wish!'.

(Tarski 1966: 151–2)

In my view, the new logicality criterion leads to a more intricate and interesting answer to this question. It suggests that there is a division of labor between logic and mathematics, one that leads to different practices in the two disciplines. Logic and mathematics approach the formal from two different, though complementary, perspectives, and therein lie both their similarities and their differences. Mathematics seeks to discover formal laws, logic seeks to implement them; mathematics is interested in the formal as it concerns objects, logic is interested in the formal as it concerns thought or language. And our cognitive capacities are such that discovery is best systematized in terms of individuals and their properties, implementation—in terms of properties and relations and especially in terms of properties and relations of properties and relations. The formal is differently represented in logic and in mathematics, but at bottom it is the same in both. (For additional points and a slightly different perspective, see Sher 1991, chs. 3 and 6.)

Tools for justifying logic's claims and detecting its errors

By using mathematical truth as a basis for logical truth, we are licensed to use mathematics, and indirectly, the tools used to justify it and detect its errors, as a tool for justifying and detecting errors in logic. For example, to the extent that mathematical or rational intuition is a tool for justifying mathematical assertions, it is also a tool for justifying the supervening logical assertions. Or to the extent that sometimes (if rarely) physical discoveries have formal ramifications, they can be used to corroborate or throw doubt on logical assertions. Or to the extent that a new claim, or an old conjecture, is proved in mathematics, we can use it to justify a logical rule of proof or a logical inference. For example, the newly discovered proof of Fermat's Last Theorem justifies all the hitherto unjustified logical rules of inference of the form:

$$(\exists!k^n x)\Phi, (\exists!l^n x)\Psi, (\forall x)(\Phi \equiv \sim\Psi); \text{ therefore } \sim(\exists!m^n x)(\Phi \vee \Psi),$$

where $n > 2$ and $k, l, m > 0$. Or if we find compelling reasons for including the Continuum Hypothesis or its negation in our theory of formal structure, we can use them to justify either the logical inference

$$(\exists!2^{\aleph_0} x)\Phi; \text{ therefore } (\exists!\aleph_1 x)\Phi$$

or the logical inference

$$(\exists!2^{\aleph_0} x)\Phi; \text{ therefore } \sim(\exists!\aleph_1 x)\Phi.$$

And so on.

These are some of the foundational advantages of the Invariance-under-Isomorphism criterion and the formalist theory of logic within which it is offered.

It should be noted that the Invariance-under-Isomorphism criterion also contributes to a *critical approach to the philosophy of logic*. The prevalent philosophy of logic today adheres to the so-called *"first-order thesis"*[17] which says that standard first-order logic is the whole of logic. Very few systematic or theoretical grounds have been adduced in support of this thesis, and for the most part it has been accepted without serious argument. The Invariance-under-Isomorphism criterion challenges this thesis on several grounds. For one thing, it challenges one of the few theoretical arguments used to support it, namely, the argument from completeness (Quine 1970). Investigations connected with this criterion (e.g., Keisler 1970) have proved that standard first-order logic is definitely not the strongest (extensional) logic which

---

[17] This epithet is due to Barwise (1985) who made similar points to those I am about to make.

has the virtue of being complete; stronger first-order logics—for example, first-order logic with the added logical quantifier "there are uncountably many"—are also complete. More importantly, the Invariance-under-Isomorphism criterion demonstrates that a systematic, theoretical, philosophically anchored, highly explanatory, mathematically rich, and linguistically fruitful criterion of logicality is possible. In so doing it sets a new, higher standard of justification for theses concerning the scope of logic, a standard that, as far as I can judge, has not been met by any of the known justifications of the first-order thesis.

Our final task before turning to Feferman's criticisms is to show how the Invariance-under-Isomorphism criterion for objectual logical operators relates to the Boolean, truth-functional criterion for propositional logical operators (logical connectives).

Invariance-under-Isomorphism and Truth-Functionality. In making statements we usually work with two types of structures—objectual structures and propositional structures, and we use two types of operators—objectual operators and propositional operators. Thus, in making a statement of the form

$$\sim(\exists x)(Bx \;\&\; \sim Cx)$$

we first consider an objectual structure with two properties, B and C; then, working with the objectual operator $\sim$ (the objectual correlate of the propositional operator $\sim$, namely, complementation), we focus our attention on B and the complement of C; next, working with the objectual operator & ($\cap$) we shift our attention to the intersection of B and the complement of C; then, working with the objectual operator $\exists$, we consider the possibility that this intersection is not empty; and finally, thinking in propositional terms and using the propositional operator $\sim$, we say that this possibility is not realized: nothing is both a B and a non-C.

Now, if we commonly use operators of two types, objectual and propositional, each defined in terms of the corresponding structure, then we need two (albeit coordinated) criteria of logicality, each formulated in terms of the relevant structure.[18] Invariance-under-Isomorphism is a criterion of logicality for objectual operators, and Truth-Functionality is a criterion of logicality for propositional operators. How are they connected? The formalist answer is that the same idea—*formality*—lies at the bottom of both criteria, and the same technical device—*invariance under "isomorphism"*—is used in both, but with respect to different structures:

(I) *An objectual operator is logical iff it is invariant under all isomorphisms of its argument-structures, which are objectual.*

(II) *A propositional operator is logical iff it is invariant under all isomorphisms* of its argument-structures, which are propositional.*

---

[18] And we also need two related alethic predicates, "satisfaction" and "truth"—the former applying to open formulas whose operators, if any, are all objectual, and whose definition accordingly refers to objectual structures; the latter applying to closed formulas (sentences) whose new operators (i.e., those added to the operators of their open sub-formulas), if any, are all propositional, and whose definition accordingly refers to propositional structures.

Tarski's Thesis

What is an isomorphism* of propositional structures? When are two propositional structures formally the same? Well, formality in the domain of propositions is, on the classical approach (tentatively adopted here) *preservation of Boolean structure*. And the Boolean features of propositional structures are a generalization of the Boolean features of objectual structures. The basic parameter in this generalization is *binary structure* or *complementarity*, which is common to both objectual and propositional structures, and we can arrive from the objectual form of this parameter to its propositional form in three steps that, in the simplest case, can be described as follows:

(i) Objectual step:

Given an object a in a universe A and a set of objects or a property B in A, there are exactly two possibilities with respect to a, exactly one of which is realized: a is a B, a is a $\overline{B}$ (complement of B in A), the latter being equivalent to: a is not a B (in A).

(ii) Situational step:

Given the situation s in which a is a B (in A), there are exactly two possibilities with respect to s, exactly one of which is realized: s is the case, s is not the case (not-being-the-case being the complement of being-the-case).

(iii) Propositional step:

Given a proposition p corresponding to s, there are exactly two possibilities with respect to p, exactly one of which is realized: p is true, p is false (false being the complement of true).

These steps connect objectual structures to propositional structures and form a bond between the logicality criterion of objectual operators and the logicality criterion of propositional operators: an operator, objectual or propositional, is logical iff it does not distinguish the non-formal features of its argument-structures. Since the generalization from objectual to propositional structures is such that the only formal feature of a proposition is its binary value (truth or falsity), a propositional operator is logical iff it is invariance under 1–1 mappings of propositions which transfer each proposition into a proposition with the same binary value (i.e., its truth value).

Technically, we can define:

(a) A *propositional structure* is as an $n + 1$-tuple $\langle P, p_1, \ldots, p_n \rangle$, where P is the set of all propositions of a given language and $p_1, \ldots p_n$ are elements of P.
(b) An *argument-structure* for a k-place propositional operator is a propositional structure of length $k + 1$.
(c) Two propositional structures $\langle P, p_1, \ldots, p_n \rangle$ and $\langle P, p'_1, \ldots, p'_m \rangle$ are isomorphic* iff $n = m$ and there is a truth-bijection from P to P, i.e., a 1–1 truth-preserving function f from P onto P such that for every $1 \leq i \leq n$, p'_i is the image of p_i under f.

Truth-functionality is thus (classical) formality on the propositional level.[19]

We are now ready to consider Feferman's criticisms.

FEFERMAN'S CRITICISMS

The Invariance-under-Isomorphism criterion is a substantive criterion, and as such it invites substantive criticisms. In "Logic, Logics and Logicism" (1999), Solomon Feferman offers three substantive criticisms of the claim that this criterion is a necessary and sufficient criterion of logicality (referred to as "the Tarski–Sher thesis"). Feferman formulates the criterion (in terms sanctioned by a certain definability result due to McGee 1996) as follows:

An operation O across domains is a logical operation according to the Tarski–Sher thesis if and only if for each cardinal $\kappa \neq 0$ there is a formula ϕ_κ of $L_{\infty,\infty}$ which describes the action of O on domains of cardinality κ.

(Feferman 1999: 37)

Here, however, I will continue to employ our earlier terminology in discussing his criticisms.

Feferman criticizes the Tarski–Sher thesis on three counts:

1. "*The thesis assimilates logic to mathematics, more specifically to set theory*" (ibid., my italics).

Elaboration:

The first [point], I think, speaks for itself, ... but it will evidently depend on one's gut feelings about the nature of logic as to whether this is considered reasonable or not. For Sher, to take one example, this is no problem. Indeed, she avers that "the bounds of logic, on my view, are the bounds of mathematical reasoning. Any higher-order mathematical predicate or relation can function as a logical term, provided it is introduced in the right way into the syntactic-semantic apparatus of first order logic." ([Sher 1991], pp. xii–xiii) What that "right way" is for her is spelled out in a series of syntactic/semantic conditions... (*[ibid.]*, pp. 54–5) ... .[Although these conditions restrict us to] logical operation[s] ... of type-level at most 2 ... [this] is not set-theoretically restrictive.... In particular, we can express the Continuum Hypothesis and many other substantial mathematical propositions as logically determinate statements on the Tarski–Sher thesis.... But in so far as ... the thesis requires the existence of set theoretical entities of a special kind, or at least of their determinate properties, it is evident that we have thereby transcended logic as the arena of universal notions independent of "what there is".

(Feferman 1999: 37–8)

2. "*The set-theoretical notions involved in explaining the semantics [of the background language] are not robust.*" (ibid.: 37; my italics)

---

[19] There are of course more familiar ways to construe isomorphism*; for example, using structures whose distinguished elements are truth-values. But I was looking for a construal that would be philosophically transparent, regardless of its familiarity or elegance.

Elaboration:

Point 2 is in a way subsidiary to point 1. The notion of "robustness" for set-theoretical concepts is vague, but the idea is that if logical notions are at all to be explicated set-theoretically, they should have the same meaning independent of the exact extent of the set-theoretical universe. For example, they should give equivalent results in the constructible sets and in forcing-generic extensions. Gödel's well-known concept of absoluteness provides a necessary criterion for such notions and, when applied to [the kind of operators considered by the Tarski–Sher thesis] considerably restricts those that meet this test. For example, the quantifier "there exist uncountably many x" would not be logical according to this restriction, since the property of being countable is not absolute.

(Ibid.: 38)

Feferman, however, qualifies his support of the absoluteness criterion somewhat:

One should be aware that the notion of absoluteness is itself relative and is sensitive to a background set theory, hence again to the question of what entities exist.

(Ibid.)

3. "No natural explanation is given by [the Tarski–Sher Thesis] of what constitutes the *same logical operation over arbitrary basic domains.*" (Ibid.: 37; my italics)

Elaboration:

It seems to me there is a sense in which the usual operations of the first-order predicate calculus have the *same meaning* independent of the domain of individuals over which they are applied. This characteristic is *not* captured by invariance under bijections. As McGee puts it "The Tarski–Sher thesis does not require that there be any connections among the ways a logical operation acts on domains of different sizes. Thus, it would permit a logical connective which acts like disjunction when the size of the domain is an even successor cardinal, like conjunction when the size of the domain is an odd successor cardinal, and like a biconditional at limits." (McGee 1996: 577)

(Feferman 1999: 38)

For Feferman, this point is more compelling than the other two:

For me, point 3 is perhaps the strongest reason for rejecting the Tarski–Sher thesis, at least as it stands

(Ibid.)

But his objection concerns only the sufficiency part of the Tarski–Sher thesis:

I agree completely [that] the Tarski–Sher thesis [is] a necessary condition for an operation to count as logical.

(Ibid., inversed sentence structure)

Still, it is a clear and strong criticism:

I . . . believe that if there is to be an explication of the notion of a logical operation in semantic terms, it has to be one which shows how the way an operation behaves when applied over one domain M_0 connects naturally with how it behaves over any other domain M'_0.

(Ibid.: 38–9)

As "a first step in that direction" Feferman proposes a revision of the Invariance-under-Isomorphism criterion. The revision consists in replacing "Isomorphism" by "Homomorphism," the resulting concept of logical operation being that of a "homomorphism invariant operation" (ibid.: 39). I will examine Feferman's proposal below, but first let me consider his criticisms.

CONSIDERATION OF FEFERMAN'S CRITICISMS

I will begin by putting Feferman's criticisms in a proper perspective. There are a few significant points of similarity between Feferman's approach to the issues in question and mine (I prefer not to speculate about Tarski). First, Feferman does not question either the need for a criterion of logicality or the appropriateness of the semantic method for such a criterion. Second, Feferman regards the issue of logicality as a foundational issue, and is not averse to the pursuit of foundational studies. (On the contrary; Feferman has been extensively engaged in important foundational work, two examples of which are Feferman 1993*a* and Feferman and Hellman 2000.) Furthermore, Feferman's approach is neither logicist nor Platonist, conventionalist, intuitionist, or indispensabilist, but he is seeking a new approach to the foundations of logic and mathematics. (See, e.g., Feferman 1984, 1993*a*, and 1993*b*.) Finally, Feferman, as noted above, accepts the Invariance-under-Isomorphism criterion, as it stands, as a *necessary* condition on logicality, and his own proposal for a *sufficient* condition involves only a limited revision of that criterion. In light of these observations, I think it is reasonable to view Feferman's criticism as a restricted, internal criticism, rather than a full-scale external criticism. Nevertheless, this is a veritable criticism that requires careful consideration.

Assimilation of logic to mathematics

A disagreement between a mathematician and a philosopher on the relation between logic and mathematics, such as that between Feferman and myself, was anticipated by Tarski:

[T]he two possible answers [to the question whether mathematics is separate from logic] correspond to two different types of mind. A monistic conception of logic, set theory, and mathematics ... appeals, I think, to a fundamental tendency of modern philosophers. Mathematicians, on the other hand, would be disappointed to hear that mathematics, which they consider the highest discipline in the world, is a part of something so trivial as logic; and they therefore prefer a development of set theory in which set-theoretical notions are not logical notions.

(Tarski 1966: 153)[20]

---

[20] The ellipses and square-bracket formulations are partly intended to neutralize Tarski's tendency to identify the impact of the Invariance-under-Isomorphism/Permutation criterion on logic and mathematics with logicism. As pointed out earlier in this chapter, the new criterion leads to the "mathematization" of logic rather than to the "logicization" of mathematics. Although

But I think there is more to Feferman's position than a certain type of mind or a reverential attitude toward mathematics. In my view (as outlined above), mathematicians have a solid reason for regarding mathematics as dealing with non-logical notions, namely: their task. Their task (or one of their main tasks) is to discover and systematize the laws governing formal structures rather than apply these laws in discourse and reasoning. And the natural way for humans to study the laws governing a certain kind of structure is to construe these structures as structures of basic elements (of some kinds), i.e., in the case of formal structures, as structures of elements that do not satisfy the Invariance-under-Isomorphism criterion. But the two construals of formal objects do not conflict. To see this more clearly, let us draw an analogy to the conception of numbers in mathematical structuralism.

From the structuralist point of view there is no difference between studying the laws of arithmetic by studying a certain system of numbers or the corresponding system of sets. But to study the arithmetical laws the mathematician is best served by choosing some specific entities to work with, be they numbers or sets. From the point of view of the working number theorist, then, arithmetic is a theory of a specific kind of objects, but that does not conflict with the philosophical claim, reached by abstraction and generalization, that numbers are mere *places in a structure*, whose occupants' identity is immaterial.

In the case of formal operators, the notions mathematicians work with are, for the most part, lower-order, non-logical notions, while the notions logicians work with are, for the most part, logical notions, obtained from lower-order, non-logical, mathematical notions by "raising" them to a higher-order. It is this raising that captures their nature as formal or structural elements, and the laws governing them as laws of formal structure. Together, these two perspectives systematize our idea of formality.[21]

So we see that Feferman's justified claim that there are significant differences between logic and mathematics *is in fact satisfied* by the Invariance-under-Isomorphism criterion, especially on the formalist interpretation I have given to it here and in Sher (1991).

Feferman's criticism, however, raises other issues as well, some directly, others indirectly. One issue it raises indirectly is the role of common-sense intuition, or "gut feelings," in determining the relation between logic and mathematics. On this issue, I am afraid, we are in disagreement, since in my view the relation between logic and mathematics has very little to do with gut feelings. It is true that in approaching this issue, and in various stages of pursuing it, we use everyday intuition. But once we approach it as a theoretical issue, as we do when we construct a rigorous criterion of

either way mathematics and logic are one, the direction of reduction is philosophically significant: logicism attributes to mathematics the properties usually associated with logic, while mathematicism attributes to logic the properties usually associated with mathematics.

[21] An analogy with equivalence classes in mathematics might be helpful here. In some cases a given idea is better expressed by an equivalence class than by any of its constituent classes. But an equivalence class could not express this idea without its constituent classes, which are generally not equivalence classes, exemplifying it. In that sense, there is a division of labor between equivalence- and non-equivalence-classes in expressing that idea. This, indeed, is a natural way to understand mathematical structuralism as well.

logicality and develop a systematic account of logic to go with it, the role of gut-feelings becomes very limited. In fact, Feferman himself regards foundational studies as having a largely *theoretical* role: namely, "*conceptual clarification; interpretation [and] reduction... of problematic concepts and principles; organizational... foundations; and reflective expansion of concepts and principles*" (Feferman 1993*b*: 106). As such they are entitled to results that conflicts with some of our "gut feelings."

Since the Invariance-under-Isomorphism criterion, combined with the formalist account of logic, offers an informative and systematic account of the concept of logical operator, solves serious conceptual problems (e.g., with the definition of logical consequence), explains the relation between logic and truth, elucidates the role of logic in our system of knowledge, critically establishes many of the intuitive attributes of logic, and offers a substantive and methodologically economical account of the relation between logic and mathematics, it should not be judged based on "gut feeling."

Another issue raised by Feferman's criticism is ontological commitment. Feferman upholds the traditional view that logic, unlike mathematics, should have no ontological commitments. By assimilating logic to mathematics, he claims, the Invariance-under-Isomorphism criterion burdens it with considerable ontic commitments. By this Feferman means one of two things: (i) the fact that we resort to a set theoretical background language carries with it ontological commitments to sets; (ii) the enormous expressive power of the logic sanctioned by that criterion carries commitments to many ontologically-laden set theoretic theses. Clearly (i) is common to standard first-order logic and the logic sanctioned by our criterion. So let us turn to (ii). Consider the sentence:

$$(\exists! 2^{\aleph_0} x) x = x \equiv (\exists! \aleph_1 x) x = x,$$

This is a well-formed sentence of the logic sanctioned by the Invariance-under-Isomorphism criterion, but for its truth-value in uncountable models to be determined, logic must be committed either to the continuum hypothesis (CH) or to its negation (~CH). Does this saddle logic with the same ontological commitments as those of mathematics?

To get a first inkling of the difference between logical and mathematical commitments, consider the difference between the way the logical CH and the mathematical CH behave under negation. (This is a theme known from comparisons of first- and second-order CH; see, e.g., Shapiro 1991.) Let us call the mathematical statement expressing CH "CH_M" and the logical statement expressing CH "CH_L". Then, whereas ~CH is captured by "~CH_M", it is not captured by "~CH_L". "~CH_M" can be added to set theory as an axiom without rendering set-theory inconsistent. But " ~CH_L" cannot be a logical law, since logic—both standard logic and the logic sanctioned by the Invariance-under-Isomorphism criterion—has countable models (in which CH is trivially satisfied), and these would prevent it from being true in all models.

The main point is that while mathematics has direct ontological commitments, logic's ontological commitments are for the most part indirect. Aside from a few direct technical commitments—for example, a commitment to the existence of at least one individual (given the technical requirement that a model have a non-empty universe)—logic has only indirect ontological commitments, namely, commitments through its background theory of formal structure. And even these commitments are not existential in the usual sense; rather, they are commitments to the *formal possibility* of existence. Thus, as an axiom within (mathematical) set theory, Infinity says that an infinite set actually exists, but as an axiom within a background theory for logic, it says that an infinite structure of objects is *formally possible*.[22]

Non-robust logical notions

Feferman notes that many of the logical operators sanctioned by the Invariance-under-Isomorphism criterion are not "robust" and argues that only "robust" operators should be classified as logical. The word "robust" can be interpreted in many ways, but Feferman has a specific interpretation in mind: for an operator defined in set-theoretical terms to be robust is to have "the same meaning independently of the exact extent of the set-theoretical universe" (cited above). And this idea, Feferman suggests, is captured by the set-theoretical concept of "absoluteness": to be robust is to be "absolute" (in the set-theoretic sense). The set-theoretic concept of *absoluteness* was introduced by Gödel in the course of proving the relative consistency of the Axiom of Choice and the Generalized Continuum Hypothesis. His proofs involved the claim that in the "constructible universe" $V = L$ (i.e., L exhausts the whole universe of sets). And to establish this claim Gödel used absoluteness results, whose basic concept can be defined as follows:

A formula $\Phi(x_1, \ldots x_n)$ *is absolute* from a transitive class M to a transitive subclass N *iff* $\forall x_1 \ldots x_n [x_1, \ldots, x_n \in N \supset (N \models \Phi \equiv M \models \Phi)]$.

Gödel was especially interested in formulas which are absolute from V to L, and in particular, in the fact that the operation of forming all the "constructible" (definable) subsets of a given set is absolute from V to L. (See Gödel 1940 and discussion in Solovay 1990.) But the concept of absoluteness has been generalized in various ways, leading to many new applications.[23]

From the point of view of Feferman's criticism of the Invariance-under-Isomorphism criterion, the most relevant feature of the absoluteness requirement is that it does not allow operators to change their meaning by expansion or contraction of a given universe. This requirement renders "finite" an absolute operator (relative to ZFC) but "uncountable" not. A subset of the universe that satisfies "is finite" in a smaller

---

[22] I briefly discussed this matter in Sher (1996a: 682) where I pointed out additional references.
   Note: I do not mean to say that including CH, for example, as an axiom of our background theory of logic does not actually commit that theory to provide the bijections needed to secure the fact that the size of the continuum is aleph-1. What I mean to say is that if CH is included as an axiom of this theory, then *it represents a formal law whose scope is the totality of formally possible structures of objects*. If we include CH in this theory, we are of course actually committed (not "possibly committed") to the existence of the requisite bijections (or of something else that will do the same job). I would like to thank Denis Bonnay for raising this issue.

[23] See, e.g., Burgess (1977), Väänänen (1985), and Tourlakis (2003).

model of set theory also satisfies it in a larger model (and vice versa, assuming it is included in the smaller model), but a subset of the universe that satisfies "is uncountable" in a small model (for example, a Löwenheim-Skolem model) does not satisfy it in a standard model. Accordingly, the quantifier "finitely many" is absolute, but "uncountably many" is not. But both quantifiers are logical according to the Invariance-under-Isomorphism criterion. Therefore, this criterion must be rejected, or so Feferman says.

In responding to Feferman's second criticism, I will first show that this criticism is weaker than it may seem to be, and then I will question the relevance of absoluteness to logicality.

(A) First, it should be pointed out that Feferman's criticism is directed at an *artifact* of a particular background theory we use to formulate the Invariance-under-Isomorphism criterion, but the idea underlying this criterion is not wedded to this, or any other background theory. In particular, the conception of logicality as formality, and even the conception of formality as invariance under 1–1 replacements of individuals, is not inherently connected to a particular set-theoretical language for which the question of "absoluteness" arises.

But even assuming this background language, Feferman's criticism is weaker than it may seem to be. Whereas in one sense the operator "uncountably many" changes its meaning from universe to universe, in another, more relevant sense, it does not. Let me explain. Clearly, as defined in a first-order set-theory—call it "T"—the predicate "x is uncountable" is satisfied by some countable set (i.e., an individual b to which countably many individuals a stand in the relation "x is a member of y") in some model of T. But the quantifier "there are uncountably many" is not satisfied by any countable set (a collection of countably many individuals) in any model of a first-order logical system in which it serves as a logical quantifier. To see this, the reader has to know how such a logical system is constructed, and this is something I have not discussed here. (A relevant discussion appears in chapter 3 of Sher 1991.) But let me try to explain the general principle underlying this claim briefly.

Consider the following:

In a first-order set-theory, T_1, we cannot see that the predicate "x is uncountable" (of T_1) is satisfied by a countable set in some model of T_1. To see that it is, we have to go to another theory, T_2, which is at least as strong (in the relevant sense) as T_1 and in which we can truly say that the formula "x is uncountable" of T_1 is satisfied by some countable[24] set in some model of T_1. Intuitively, from the point of view of T_2 the T_1-predicate "x is uncountable" is not robust, but from the point of view of T_1 it is.

Now, it is an essential feature of a logic L that the following are all done on the same level of discourse, or within the same background theory—call it "T1": (i) the definitions of the logical constants of L, (ii) the definitions of the operators corresponding

---

[24] i.e., countable from the point of view of T_2.

to them, (iii) the definition of the models of L, (iv) the definition of "true in a model," (v) the definition of "countable" and "uncountable," (vi) the definition of robustness, etc. From the above considerations it follows that from the point of view of T1 the logical quantifier "there are uncountably many" of L has a fixed meaning and as such is robust. This is expressed by the fact that (from the point of view of T1) the Löwenheim–Skolem theorem does not hold the logical constants of L: "(For uncountably many x) x = x" has no countable models. (Of course, from the point of view of yet another theory, T2, T1 itself may be subject to the Löwenheim–Skolem theorem. But from the point of view of T2, "robustness" may have a non-standard meaning as well, as Feferman noted.)

(B) Absoluteness is an interesting and in certain respects a desirable property, but should we restrict our concept of logicality to operators satisfying this property? To put things in perspective, there are many interesting and desirable properties we don't restrict our concepts to. Take, for example, *decidability*. Decidability is an interesting and desirable property of logics, yet we do not restrict ourselves to decidable logics. The price of setting decidability as an upper boundary on our concept of logic is simply too high. Clearly, sentential logic or even monadic standard first-order logic is too narrow to exhaust our concept of logic or even to serve as a working logic for mathematics. Or consider *completeness*. Completeness is a desirable property of theories. But we would have to remove most of mathematics from the realm of axiomatized first-order theories if we were to require that only complete axiomatizations be permitted in that realm. Saying that *generally* only complete theories are genuine theories would be even more absurd.

The question arises whether the same does not hold for absoluteness. It clearly does in some cases. For example, we cannot restrict set theory to absolute concepts, since this would involve omitting many of its most basic concepts, e.g., the concept of cardinality. But does it hold in the case of logic?

Let us first see how the formalist conception of logic answers this question. From the point of view of this conception, logic requires a background theory of formal structure, and it is an open question what the best theory of formal structure is. In principle we are looking for the most economical theory that is sufficiently strong to account for formal structures in a comprehensive manner. Two more economical candidates than ZFC are ZFC + (V = L) and Feferman's predicative system, but there are other candidates as well, and the jury is still out on what the best available theory is. However, absoluteness per se is not a reasonable constraint on a theory of formal structure, since *a property is absolute iff it is insensitive to a certain formal difference between universes* (namely, the difference between *larger* universes and *smaller* universes *included* in them). This means that a theory that admits only absolute notions neglects some formal differences between objects, and as such is not an acceptable theory of *formal structure*.

These considerations show that: (i) the fact that absoluteness is desirable for some purposes does not mean that it is appropriate for the purpose of constructing a criterion of logicality; and (ii) to accept absoluteness as a constraint on logicality we need to renounce the formalist conception of logic and the associated justification of our criterion of logicality. I would be very interested to examine a philosophical conception of logic that fits in with the absoluteness requirement and offers a foundational justification of a concept of logicality satisfying it. As far as I know, none is available yet.

Operators with "non-uniform meaning", "split identity", or "unnatural behavior"

Feferman's main objection to the Invariance-under-Isomorphism criterion is that it sanctions logical operators lacking a unified identity, or a natural connection between the way they behave in different universes, or (when we consider the terms denoting them) the same meaning in different universes. Such operators can behave one way over universes of cardinality κ and another way over universes of cardinality $\lambda (\neq \kappa)$, i.e., their meaning, or identity, depends on the size of the universe, and there is no natural connection between the way they behave in universes of size κ and universes of size λ.

Before considering Feferman's criticism, it would be instructive to note that his particular example of such an operator is in fact *not* countenanced by my version of the "Tarski–Sher" thesis. Feferman's example is that of a *propositional connective*, O, "which acts like disjunction when the size of the domain is an even successor cardinal, like conjunction when the size of the domain is an odd successor cardinal, and like a biconditional at limits" (cited above). I agree with Feferman's claim that O is not a proper logical operator, but not with his reason for claiming so. Propositional connectives should not depend on the size of the universe (of individuals) because this has nothing to do with *truth-functionality*. The problem with O, as I see it, is that as a *propositional* operator it should not take into account *universes of individuals* at all. And in my version of the logicality criterion propositional operators do not. Propositional operators (connectives) are defined in terms of propositional rather than objectual structures, and propositional structures have a universe of propositions rather than a universe of individuals. Indeed, they take into account only one universe—the universe of all propositions. The operator mentioned in Feferman's example is therefore not logical according to my (version of the) "Tarski–Sher" logicality criterion.[25]

But the phenomenon Feferman talks about is true of other operators satisfying this criterion. Take the objectual operator Q defined by: Given a universe A and a subset B of A,

$Q_A(B) = T$ *iff either* A *is countable and* B $= $ A *or* A *is uncountable and* B *is not empty*.

---

[25] 1. See p. 322–4 above.

2. I should indicate that I had the opportunity to correct Feferman's error when I received a pre-publication copy of McGee (1996) from which this example is taken, but I failed to do so, since in the context of McGee's chapter it seemed an insignificant point. In the present context, however, it is more significant.

Then Q behaves like ∀ in countable universes and like ∃ in uncountable universes.[26] Is the fact that Q has this "split" identity a good reason for refusing to count it as a logical operator? In answering this question I will make a few points:

(A) To the extent that we refuse to count Q as a logical operator because it is an "unnatural" operator, it should be noted that numerous unnatural objects (properties, relations, functions) are widely accepted in other fields. Feferman himself (2000) brings numerous examples of what he calls "monstrous" or "pathological" objects that are generally accepted by mathematicians (he included).

Indeed, even in standard logic there are many "unnatural" operators, including logical operators of "split" identity or meaning, operators which do not seem "to have the same meaning" or "be the same operators" in different settings. Two examples would suffice:

(a) A 132-place propositional connective, C, such that:

(i) C behaves like a 132-place Conjunction in rows with 0–23 T's,
(ii) C behaves like a 132-place Disjunction in rows with 24–79 T's, and
(iii) C behaves like the Majority Connective in all other rows.

(b) A quantifier Q*, *definable in standard first-order logic*, such that:

(i) Q* behaves like "All" (∀) in universes of cardinality <101,
(ii) Q* behaves like "Some" (∃) in universes of cardinality 101–745, and
(iii) Q* behaves like "None" (∼∃) in universes of all other cardinalities.

Intuitively, the identity (or meaning) of these operators is no less "split" than that of Q, yet they are accepted as legitimate logical operators by most logicians and philosophers (including Feferman, I am sure). Why, then, should we discriminate against Q? In what way is Q *less natural*, or its identity or meaning *more "split"*, than those of C and Q*, which we all accept as legitimate logical operators?

(B) My point is not that there is no value or interest in a specific concept of "natural operator" (or "natural connection between an operator's behavior in different universes"), but such a concept has nothing much to do with our idea of *logicality*.

We may wish to distinguish "natural" logical operators from "unnatural" logical operators or "natural" operators in general from "unnatural" operators in general,

---

[26] This is not the only kind of operator that exhibits this phenomenon. In connection with my claim that propositional connectives are *not* sensitive to the size of objectual universes, I would like to clarify that some logical operators *defined in terms* of objectual functions corresponding to the logical connectives (union, intersection, etc.) are sensitive to the size of universes of individuals, but *these operators themselves do not correspond to any propositional connective*. Thus, the functional operator F, defined, for an objectual universe A and subsets B and C of A, by:

(i) $F_A(B, C) = B \cap C$ if A is countable;
(ii) $F_A(B, C) = B \cup C$ if A is uncountable

is a logical operator according to the Invariance-under-Isomorphism criterion, but it is not the objectual equivalent of *any* propositional connective.

in the same way that we may wish to distinguish "natural" functions from "unnatural" functions or "natural" relations in general from "unnatural" relations in general. But just as the latter would not undermine, or force us to change, our criterion of a functionality (of a relation being functional), so the former would not undermine, or force us to change, our criterion of a logicality (of an operator being logical).

We could impose on ourselves a "naturalness" constraint in choosing a logical system to work with, but this would be a separate constraint from the "logicality" constraint we would impose on such a system.

(C) Finally, there is a strong unity (or uniformity) to the concept of *logical operator* delineated by the Invariance-under-Isomorphism criterion and a clear concept of *same logical operator* associated with it. Both are generated by our interpretation of this criterion as a criterion of *formality:* All and only formal operators are logical, and each logical operator describes one way in which an operator takes into account some formal features of a given situation. Thus all logical operators are *unified* in being formal, and a logical operator is *the same* in different universes iff there is some formal pattern of objects-having-properties-and-standing-in-relations-within-situations that its trajectory through the different universes represents. Since the size of the universe is a basic formal feature of objectual situations, it is—and should be—a central parameter of some objectual formal operators.

FEFERMAN'S CRITERION OF LOGICALITY

Feferman's criterion of logicality is proposed "as a first step" in the "direction" of showing "how the way an operation behaves when applied over one domain M_0 connects naturally with how it behaves over any other domain M_0'" (Feferman 1999: 38–9). It is obtained from the Invariance-under-Isomorphism criterion by replacing "isomorphism" by "homomorphism" (or what is sometimes called "strong homomorphism"), i.e., by replacing the requirement that logical operators be invariant under any *1–1* and *onto* transformations of structures by the requirement that they be invariant under any *onto* transformations of structures. We can formulate Feferman's criterion as follows:

Invariance-under-Homomorphism: *An operator O is logical iff it is invariant under all homomorphisms of its argument-structures,*

where:

(i) A structure, $\langle A, \beta_1, \ldots, \beta_n \rangle$, is *homomorphic* to a structure $\langle A', \beta_1', \ldots, \beta_k' \rangle$ iff $n = k$ and there is a surjection f from A to A' such that for every $1 \leq i \leq n$, β_i' is the image of β_i under f,

(ii) An n-place operator O is invariant under all homomorphisms of its argument-structures iff for any of its argument-structures, $\langle A, \beta_1, \ldots, \beta_n \rangle$ and $\langle A', \beta_1', \ldots, \beta_n' \rangle$: if $\langle A, \beta_1, \ldots, \beta_n \rangle$ is homomorphic to $\langle A', \beta_1', \ldots, \beta_n' \rangle$, then $O_A(\beta_1, \ldots, \beta_n) = O_{A'}(\beta_1', \ldots, \beta_n')$.

The effects on the *generality* and *formality* of our concept of logicality are: (a) Since every bijection is a surjection but not vice versa, there are more surjections than bijections, and Invariance-under-Homomorphisms (surjections) is invariance under more transformations of structures. As a result, the concept of logicality associated with the new criterion is more *general* than that associated with the old criterion. All logical operators under the former are logical under the latter, but not vice versa. The new criterion, however, does not render logic maximally *general* (it does not require logical operators to be invariant under all transformations whatsoever); therefore the concept of logicality associated with it cannot be fully explained or justified in terms of *generality*. (b) Since surjections overlook certain gaps in size between their domain-universe and their range-universe (mapping larger universes into smaller ones), invariance under surjections does not respect an important *formal* difference between structures, namely, difference in size or cardinality. As a result, the new criterion leads to a concept of logicality that parts ways with that of *formality*, making the explanation and justification of the old criterion inaccessible to it.

The new criterion, however, may be thought to satisfy Feferman's requirement that logical operators "behave in the same way in all universes." Intuitively, a homomorphism is a mapping h such that the distinguished elements of the smaller structure are obtained from those of the larger one by '"shrinking along' h" (ibid.: 39), and this "shrinking" explains the sense in which an operator preserves its identity when moving from larger universes to a smaller ones.

Is *Invariance-under-Homomorphism* a reasonable criterion of logicality? To help us answer this question let us point at a few significant examples of operators that do and do not satisfy it.

(a) *Isomorphism-invariant operators that are also homomorphism-invariant:*

(i) The operators corresponding to the logical connectives.

(ii) The existential and universal quantifiers.

(iii) The quantifier "is well-founded" (whose arguments, in any given universe, are the binary relations on that universe).

(b) *Isomorphism-invariant operators that are* not *homomorphism-invariant:*

(i) The (standard) Identity relation.

(ii) Cardinality quantifiers (including finite-cardinality quantifiers like "There are exactly 5" and infinite-cardinality quantifiers like "There are uncountably many").

(iii) The monadic quantifier "Most" (as in "Most things are B").

(iv) Quantifiers that behave like one familiar quantifier in universes of certain cardinalities and like a different familiar quantifier in universes of other cardinalities (for example, Q of the last section).

These examples suggest that the Invariance-under-Homomorphism criterion gives rise to a "hybrid" logic. This logic coincides neither with standard first-order logic nor with our "formal" logic, yet it is not intermediate between the two either, since in certain ways it is weaker than standard first-order logic. In particular, neither the identity relation (=) nor the finite-cardinality quantifiers ("There are at-least/exactly/at-most n things such that") of standard first-order logic satisfy it. At the same time it is stronger than standard first-order logic since it is satisfied by such non-standard quantifiers as the well-foundedness quantifier.

Feferman does not fully embrace the Invariance-under-Homomorphism criterion as a criterion of logicality. Rather, having "been moving more and more to the position that the classical first-order predicate logic has a privileged role in our thought" (ibid.: 32), he is looking for ways to adjust it so it classifies all and only the standard logical operators as logical. His investigations first lead to an adjustment that, assuming Invariance-under-Homomorphism is so formulated as to apply to objectual operators only, could be expressed by:

Adjusted Invariance-under-Homomorphism criterion (I):

A first-order operator is logical iff it is:

either (i) a monadic quantifier satisfying the Invariance-under-Homomorphism criterion,

or (ii) a truth-functional connective,

or (iii) an operator definable from logical operators within the λ-calculus.

By formulating the Invariance-under-Homomorphisms criterion in such a way that it applies to propositional connectives as well, however, Feferman obtains a more unified version of this adjusted criterion:

Adjusted Invariance-under-Homomorphism criterion (II):

A first-order operator is logical iff it is:

either (i) a monadic quantifier satisfying the Invariance-under-Homomorphism criterion,

or (ii) a propositional operator (monadic or not monadic) satisfying this criterion,

or (iii) an operator definable from logical operators within the λ-calculus.

The adjusted criterion differs from the original Invariance-under-Homomorphism criterion in setting a *type restriction* on logical quantifiers: only *monadic* first-order quantifiers—quantifiers of the type O(B), where B is a subset of a given universe—and not first-order quantifiers of any other type—i.e., relational or polyadic quantifiers—are logical. That is, only monadic quantifiers are subject to the Invariance-under-Homomorphism test. (Linguistically, this restricts us to quantifiers of the form "(Qx)Px", ruling out in advance, i.e., prior to applying the Invariance-under-Homomorphism criterion, all relational quantifiers (e.g., "Most^2," as in "Most B's are C's") and polyadic quantifiers (e.g., "Is a well-ordering").)

This restriction yields *almost* the desired result: all and only the logical operators of standard first-order logic *without identity* are logical.

What about Identity? In considering this question Feferman says:

It is undeniable that the relation of identity has a "universal", accepted, and stable logic (at least in the presence of totally defined predicates and functions, as is usual in PC with =), and that argues for giving it a distinguished rule in logic even if it should not turn out to be logical on its own under some cross-domain invariance criterion, such as under homomorphisms.

(Ibid.: 44)

To include identity as a logical operator we can simply *postulate* that it is, closing logical operators under definability as before. We thus get the third version of the adjusted criterion:

Adjusted Invariance-under-Homomorphism criterion (III):
A first-order operator is logical iff it is:
either (i) a monadic quantifier satisfying the Invariance-under-Homomorphism criterion,
or (ii) a propositional connective satisfying this criterion,
or (iii) the identity relation,
or (iv) an operator definable from logical operators within the λ-calculus.

This criterion classifies identity and the finite-cardinality quantifiers as logical, thus providing a characterization of the standard first-order logical operators as logical.

Is either the Invariance-under-Homomorphism criterion or the Adjusted Invariance-under-Homomorphism criterion (in any of its versions) an adequate criterion of logicality?

Van Benthem (2002) and Bonnay (2008) point out that the Invariance-under-Homomorphism criterion is subject to Feferman's first two criticisms—assimilation of logic to mathematics and non-robust logical operators—and as such is inadequate from his own perspective. I would add that by affirming the logicality of the finite-cardinality quantifiers—including "split identity/meaning" finite-cardinality quantifiers (those whose behavior in universes of different sizes is "unnaturally connected") —Feferman's third adjusted criterion also violates the third criticism. Finally, Bonnay (2008) criticizes the *ad hoc* nature of Feferman's restriction of logical quantifiers to monadic ones in the adjusted versions of his criterion.[27]

Most of these criticisms, however, do not speak against Feferman's criteria from my point of view, since the "weaknesses" they talk about are no weaknesses at all from my perspective. The one exception is the *ad hocness criticism*, which points to what, in my view, is the main challenge to any criterion of logicality, namely, a solid philosophical justification, which is missing from Feferman's discussion, and indeed not even attempted by him. That such a justification needs pursuing is also Feferman's view of the matter:

Whether that [i.e., the notion of a logical operation as "definable from homomorphism-invariant monadic operations"] (or any other invariance notion) can be justified on fundamental conceptual grounds is . . . in need of pursuit.

(Feferman 1999: 32)

---

[27] Feferman's tries to justify this restriction linguistically, by appealing to a linguistic conjecture which says that most non-monadic quantifiers used in natural language are "lifted" in one way or another from monadic quantifiers (Keenan and Westerståhl, 1997). But this conjecture is restricted to natural-language applications, is not strictly universal, is (at least as of now) unsubstantiated, and assumes the logicality of monadic quantifiers that Feferman rejects. More importantly, it is not clear that linguistic support of a logical-philosophical restriction is of much relevance.

This is a good note on which to end. I must add, however, that there exist other serious proposals for revision of the Invariance-under-Isomorphism criterion. These include Peacocke (1976), McCarthy (1981), MacFarlane (1991), Bonnay (2008), and Casanova's (2007), and they each require a careful consideration.

References

Barwise, J. (1985) "Model-Theoretic Logics: Background and Aims". In Barwise and Feferman 1985: 3–23.
Barwise, J. and R. Cooper (1981) "Generalized Quantifiers and Natural Language". *Linguistics and Philosophy* 4: 159–219.
Barwise, J. and S. Feferman, eds. (1985) *Model-Theoretic Logics*, New York: Springer-Verlag.
Birkhoff, G. and J. von Neumann (1936) "The Logic of Quantum Mechanics". *Annals of Mathematics* 37: 823–43.
Bonnay, D. (2008) "Logicality and Invariance". *The Bulletin of Symbolic Logic* 14: 29–68.
Burgess, J. P. (1977) "Forcing". In *Handbook of Mathematical Logic*, ed. J. Barwise. Amsterdam: Elsevier 403–52.
Casanovas, E. (2007) "Logical Operations and Invariance". *Journal of Philosophical Logic* 36(1): 33–60.
Etchemendy, J. (1990) *The Concept of Logical Consequence*. Cambridge: Harvard.
—— (2008) "Reflections on Consequence". This volume: 263–99.
Feferman, S. (1984) "Foundational Ways". In Feferman 1998: 94–104.
—— (1993*a*) "Working Foundations—'91". In Feferman 1998: 105–24.
—— (1993*b*) "Why a Little Bit Goes a Long Way: Logical Foundations of Scientifically Applicable Mathematics". In Feferman 1998: 284–98.
—— (1998) *In Light of Logic*. New York: Oxford.
—— (1999) "Logic, Logics, and Logicism". *Notre-Dame Journal of Formal Logic* 40: 31–54.
—— (2000) "Mathematical Intuition vs. Mathematical Monsters". *Synthese* 125: 317–22.
Feferman, S. and G. Hellman (2000) "Challenges to Predicative Foundations of Arithmetic". In *Between Logic and Intuition—Essays in Honor of Charles Parsons*, eds. G. Sher and R. Tieszen. Cambridge: Cambridge University Press: 317–38.
Feferman, S. *et al.* eds. (1990) *Kurt Gödel: Collected Works*, vol. II. New York: Oxford University Press.
Frege, G. (1879) "Begriffsschrift, a Formula Language, Modeled upon That of Arithmetic, for Pure Thought". In *From Frege to Gödel: A Source Book in Mathematical Logic, 1879–1931*, ed. J. van Heijenoort. Cambridge: Harvard University Press, 1967: 1–82.
Gödel, K. (1940) "The Consistency of the Axiom of Choice and of the Generalized Continuum Hypothesis with the Axioms of Set Theory". In Feferman *et al.* 1990: 33–101.
Higginbotham, J. and R. May (1981) "Questions, Quantifiers and Crossing". *Linguistic Review* 1: 41–79.
Kant, I. (1781/7) *Critique of Pure Reason*. 1st and 2nd edns. Tr. N. Kemp Smith. London: Macmillan, 1929.
Keenan, E. L. (1987) "Unreducible *n*-ary Quantifiers in Natural Language". In *Generalized Quantifiers: Linguistic and Logical Approaches*, ed. P. Gärdenfors. Dordrecht: D. Reidel: 109–50.
Keenan, E. L. and J. Stavi (1986/1981) "A Semantic Characterization of Natural Language Determiners". *Linguistics and Philosophy* 9: 253–329.

Keenan, E. L. and D. Westerståhl (1997) "Generalized Quantifiers in Linguistics and Logic". *Handbook of Logic and Language*, eds. J. van Benthem and A. ter Meulen. Amsterdam: Elsevier: 837–93.

Keisler, H. J. (1970) "Logic with the Quantifier 'There Exist Uncountably Many'". *Annals of Mathematical Logic* 1: 1–93.

Lindström, P. (1966) "First Order Predicate Logic with Generalized Quantifiers". *Theoria* 32: 186–95.

—— (1974) "On Characterizing Elementary Logic". *Logical Theory and Semantic Analysis*, ed. S. Stenlund. Dordrecht: D. Reidel: 129–46.

McCarthy, T. (1981) "The Idea of a Logical Constant". *Journal of Philosophy* 78: 499–523.

MacFarlane, J. G. (2000) *What Does it Mean to Say that Logic is Formal?* Ph.D. dissertation, University of Pittsburgh.

McGee, V. (1996) "Logical Operations". *Journal of Philosophical Logic* 25: 567–80.

Mostowski, A. (1957) "On a Generalization of Quantifiers". *Fundamenta Mathematicae* 44: 12–36.

Peacocke, C. (1976) "What Is a Logical Constant?" *Journal of Philosophy* 73: 221–40.

Quine, V. W. (1970/86) *Philosophy of Logic*. 2nd edn. Cambridge: Harvard University Press.

Shapiro, S. (1991) *Foundations without Foundationalism: A Case for Second-Order Logic*. Oxford: Oxford University Press.

—— (1997) *Philosophy of Mathematics: Structure and Ontology*. Oxford: Oxford University Press.

Sher, G. (1991) *The Bounds of Logic: A Generalized Viewpoint*. Cambridge: MIT.

—— (1996a) "Did Tarski Commit 'Tarski's Fallacy'?" *The Journal of Symbolic Logic* 61: 653–86.

—— (1996b) "Semantics and Logic". In *Handbook of Contemporary Semantic Theory*, ed. S. Lappin, 509–35. Oxford: Blackwell.

—— (2001) "The Formal-Structural View of Logical Consequence". *The Philosophical Review* 110: 241–61.

—— (2003) "A Characterization of Logical Constants *Is* Possible". *Theoria* 18: 189–97.

—— (2006) "Epistemic Friction and the Illusion of Foundationalism". Manuscript.

Solovay, R. M. (1990) "Introductory Note to 1938, 1939, 1939a, and 1940". In Feferman *et al.* 1990: 1–25.

Tarski, A. (1936) "On the Concept of Logical Consequence". In *Logic, Semantics, Metamathematics*, tr. J. H. Woodger, 2nd edn, ed. J. Corcoran. Indianapolis: Hacket 1983: 409–20.

—— (1966) "What Are Logical Notions?" *History and Philosophy of Logic* 7 (1986): 143–54.

Tourlakis, G. (2003) *Lectures in Logic and Set Theory*, vol. II. Cambridge: Cambridge University Press.

Väänänen, J. (1985) "Set-Theoretic Definability of Logics". In Barwise and Feferman 1985: 599–643.

van Benthem, J. (1983) "Determiners and Logic". *Linguistics and Philosophy* 6: 447–78.

—— (1989) "Polyadic Quantifiers". *Linguistics and Philosophy* 12: 437–64.

—— (2002) "Logical Constants: The Variable Fortunes of an Elusive Notion". *Reflections on the Foundations of Mathematics: Essays in Honor of Solomon Feferman*, eds. W. Sieg, R. Sommer, and C. Talcott. Association for Symbolic Logic: 420–40.

Westerståhl, D. (1985) "Logical Constants in Quantifier Languages". *Linguistics and Philosophy* 8: 387–413.

—— (1987) "Branching Generalized Quantifiers and Natural Language". In *Generalized Quantifiers: Linguistic and Logical Approaches*, ed. P. Gärdenfors. Dordrecht: D. Reidel.

13

Are There Model-Theoretic Logical Truths that are not Logically True?

Mario Gómez-Torrente

The question in the title is similar to other important unsettled questions prompted by attempted mathematical characterizations of pretheoretical notions, by coextensionality "theses" about them.[1] Church and Turing's thesis that a function is computable iff it is recursive[2] gives rise to the question "are there computable functions that are not recursive?" The thesis, especially associated with Stephen Cook, that a natural problem has a feasible algorithm iff it has a polynomial-time algorithm[3] gives rise to the question "are there natural problems having a feasible algorithm that do not have a polynomial-time algorithm?"[4] But there are remarkable dissimilarities too. A central one is that the notion of "model-theoretic logical truth" is a notion relative to (at least) a choice of a set of formalized languages, of a set of logical constants, of a notion of model, and of a notion of truth in a model, while the notions of recursiveness and of a polynomial-time algorithm are not relative to anything in any such conspicuous way. Nevertheless, arguments against particular coextensionality theses, involving fully relativized notions of model-theoretic logical truth, will be significant

---

Parts of this chapter were presented in workshops at the University of California at Irvine (2002), the Universidade Nova de Lisboa (2003), the University of Melbourne (2004), and the Universidad de Santiago de Compostela (2006), and as lectures at the Universidad Nacional Autónoma de México (2006) and the Universidade Federal do Rio de Janeiro (2006). I thank the audiences at these events for very helpful comments and criticism. Special thanks to Rodrigo Bacellar, Bill Hanson, Øystein Linnebo, Gila Sher, and two anonymous readers.

[1] Here by "pretheoretical" of course I don't mean "previous to any theoretical activity"; in this sense there could hardly be pretheoretical notions of computability, of a feasible algorithm, or of logical truth. What I mean is "previous to the theoretical activity of mathematical characterization."

[2] See Church (1936) and Turing (1936/37). Of course, the notion of recursiveness is provably coextensional with the notions of lambda-conversion and Turing-machine computability used respectively in the enunciation of the original theses of Church and Turing.

[3] See especially Cook (1991); see also Cook (1971).

[4] The three converse questions have some interest too, but it is widely conceded that properly understood they must have a negative answer. In the case of logical truth, it is widely conceded that, provided one constructs one's model theory with sufficient care, there cannot be pretheoretical logical truths that are false in some model.

provided the theses are *prima facie* reasonable and/or have been actually proposed by logicians.[5]

One common way of arguing against coextensionality theses involving fully relativized notions of model-theoretic logical truth has been by arguing that some model-theoretic logical truths in the relevant sense are not necessary or are not *a priori*, for under most pretheoretical conceptions of logical truth a logical truth must be necessary and *a priori*. One important fact in the background of this chapter is that the process of formalization leaves unclear, though perhaps not undetermined, the answers to some questions one must answer before one can ask if an interpreted sentence of a formalized language is necessary or *a priori*. This is not in itself a weakness of formalization, since the main aim of formalization is to obtain formal sentences which, unlike their correlates in natural language, have a truth-conditional content made absolutely precise by stipulation; in this process, features of the natural language sentences relevant to their modal character and their epistemology are simply abstracted from. But then formalization does have the result that some questions about the modal and epistemic status of a model-theoretic account of logical truth do not have a clear, or perhaps even a determinate answer. One aim of this chapter is to emphasize this often neglected fact. The chapter's main aim, however, is to argue that, once one lays open some natural or at least plausible ideas about the modal character and the epistemology of the classical first-order quantifiers, some *prima facie* reasonable coextensionality theses are false or at least must be somewhat qualified.

In particular I will argue that, given those ideas, the specific coextensionality thesis put forward by Tarski, the main proponent of the model-theoretic method for the mathematical characterization of logical truth, doesn't hold even though it is *prima facie* reasonable. In order to get to this conclusion, I will first enunciate and distinguish a number of coextensionality theses that sound Tarskian somehow, and I will offer a quick evaluation of each (in Section I). For each of these theses I will claim that either it is weaker or stronger than Tarski's thesis. In the course of this examination of theses I will survey some previous critiques of model-theoretic characterizations of logical truth, and will find them unsatisfactory. In Section II I will state the thesis that most deserves the name 'Tarski's thesis', and I will note that, under natural assumptions, there are model-theoretic logical truths in the sense relevant to Tarski's thesis that are not necessary, and hence not logically true under most conceptions of logical truth. Some of those natural assumptions include assumptions about the modal behavior of the classical first-order quantifiers, that are not part of their explicit classical

---

[5] I share the view, especially associated with Kreisel (1967), that one can give informal but potentially conclusive arguments both against *and* for theses asserting the coextensionality of pretheoretical and theoretical concepts. In this chapter I will be especially concerned with giving tentative arguments *against* certain particular coextensionality theses in the case of model-theoretic logical truth. (Kreisel's own argument *for* one of these coextensionality theses will be mentioned below.) At least until recently, an unKreiselian view has been widespread in the case of the Church-Turing thesis. The view seems to have been widely held that while this thesis can be refuted, it cannot be conclusively argued for, since it relates a theoretical concept and a pretheoretical one that cannot be used in rigorous general reasonings. But a Kreiselian view of the Church-Turing thesis has been urged by Harvey Friedman, Saul Kripke, and others.

extensional model theory. But, as advanced above, some such assumptions need to be made explicit before one can meaningfully ask the question whether a classical quantificational sentence is necessary. And I will argue that the assumptions I will use are the most natural given the classical extensional model theory.

Similarly, one needs assumptions about the epistemology of classical first-order quantifiers before one can ask the question whether a classical quantificational sentence is *a priori*. The final Section III describes some assumptions about the epistemology of classical first-order quantifiers that, though plausible, are potentially more controversial than the assumptions about their modal behavior described in Section II. I note that, under these potentially controversial assumptions, some model-theoretic logical truths in the sense relevant to Tarski's thesis, and even in the sense relevant to some weaker theses described in Section I, are not *a priori*, and hence presumably not logically true. Nevertheless, it appears that those theses need only be slightly qualified in order to free them from the counterexamples I will offer.

I

An especially important coextensionality thesis that Tarski held, but that is clearly not his (strongest) coextensionality thesis, is the following:

(T1) A sentence of a classical propositional/quantificational language is logically true in the pretheoretical sense iff it is true in all classical propositional/quantificational models which (re)interpret its constants (other than its classical propositional/quantificational logical constants).

(T1) is very specific about the class of sentences it talks about: the sentences of classical propositional and quantificational languages, both first- and higher-order. It is also very specific about the class of models it talks about, and about the notion of truth in a model that is at stake, which are just the classical, Tarskian ones:[6] in particular, a propositional model is any assignment of values from the set {Truth, Falsehood} to the propositional letters, and a quantificational model is seen as a sequence composed of a set-domain of quantification built out of existing objects, plus extensions drawn out of this domain for the predicate, function, and individual constant letters of a language in the relevant class. Also, (T1) talks about an absolutely specific set of logical constants, namely: the truth-functional propositional connectives and the

---

[6] There are, of course, some doubts voiced in the literature about whether Tarski's (1936) notion of a quantificational model is what I'm calling 'the classical notion of a quantificational model'. (See, e.g., Etchemendy (1988), Bays (2001), Mancosu (2006). For alternative views see Gómez-Torrente (1996) and Ray (1996).) In my view there are in fact differences between Tarski's notion and the current notion (see e.g. Gómez-Torrente (2000)), but these are not the differences purportedly detected by the doubters. One of these differences is that in (1936) Tarski required, as a precondition for the applicability of his theory of logical consequence, that the domain of an interpretation of a first-order language be denoted by a non-logical predicate of the language, and this convention is not used with the current notion; I will come back to this convention. But it is relatively uncontroversial, at any rate, that at some point Tarski adopted all the now common conventions about models, and that (T1) (and (T1(1)) below) were Tarskian theses.

classical quantifiers of finite order (plus the predicate of identity and/or a predicate of intra-typical membership in some formulations).

Is (T1) true, or is it false? Most people until relatively recently have thought that it must be true (or at least most of those who agree that the higher-order quantifiers are logical constants have thought that it must be true). Both critics and defenders of the model-theoretic approach (including myself in previous work) have tended to agree that variations of Kreisel's (1967) argument put beyond reasonable doubt (T1) as restricted to propositional and first-order logical constants. (See Etchemendy (1990), ch. 11, Hanson (1997), Gómez-Torrente (1998/9).) To be precise, the following is what has been thought to be beyond doubt:

(T1(1)) A sentence of a classical propositional/first-order quantificational language is logically true in the pretheoretical sense iff it is true in all classical propositional/quantificational models which (re)interpret its constants (other than its classical propositional/first-order quantificational logical constants).

In particular, the question whether all model-theoretic logical truths in the sense of (T1(1)) are logically true has been thought to receive a positive answer by the following Kreiselian argument: let S be a propositional or first-order model-theoretic logical truth; then, by the completeness of propositional and first-order logic, S is derivable without premises in a wide array of deductive calculi; and for any of these calculi one can easily check by inspection that they can only yield sentences that strike one as logically true, under a wide variety of pretheoretical conceptions of logical truth. Nevertheless, in Section III we will see a possible qualification to the conclusion of this argument.[7]

Recently several people have given arguments purporting to show that certain higher-order quantificational sentences are true in all classical quantificational models which (re)interpret their constants (other than their classical logical constants) and yet are not logically true. Etchemendy (1990), ch. 8, and McGee (1992) are perhaps the foremost examples of proponents of alleged counterexamples to (T1). The issue is a subtle one, but it seems fair to say that these attempted refutations have not gained anything close to a wide acceptance. McGee's alleged counterexample, in fact, is based on assumptions which go against the received view in set theory. My own view is that Etchemendy's and McGee's alleged counterexamples are unconvincing. (I give a critical discussion of these counterexamples in Gómez-Torrente (1998/9). See also Soames (1999), ch. 4, for specific discussion of Etchemendy.)

The alleged counterexamples and the general arguments supporting (T1) against them are too sophisticated for me to go into them here without digressing excessively.

---

[7] Another qualification is in any case needed in view of sentences such as '$(\exists x)(P(x) \vee \sim P(x))$', which are true in all (non-empty) models but may not be logically true. One can of course relax the convention of not contemplating empty models and prove suitable completeness theorems for the corresponding slightly non-standard calculi and appropriate variations in the notion of truth in a model. (See e.g. Quine (1954).) But one can also adopt a reasonable (yet apparently unexplored) view (described in a later note) according to which sentences such as '$(\exists x)(P(x) \vee \sim P(x))$' are not properly interpreted (or do not express propositions) when no non-empty domain for the quantifiers has been provided.

But what I want to claim for the moment is not that (T1) has not been refuted so far. I want to claim that (T1) is too weak to be called 'Tarski's thesis' (despite being so called both by McGee (1992) and by myself in Gómez-Torrente (1998/9)). (T1) is not the strongest coextensionality thesis Tarski postulated. The reason why it is too weak is that it talks about a very restricted set of logical constants. Tarski was clearly not concerned with the statement of a thesis about a set of logical constants so severely restricted by stipulation (of a list) as the set mentioned in (T1). To be sure, he contemplated the possibility that the notion of a logical constant might be so hopelessly obscure as to make arbitrary any delimitation of the borderline between logical and non-logical constants. But his later work (e.g., Tarski (1966)) shows that he never quite accepted that possibility. Tarski would have been ready to accept that other constants besides the classical logical constants of quantificational languages are logical constants, even assuming that the borderline between logical and non-logical constants may be fuzzy to some extent.

Tarski's thesis was something which, unlike (T1), is reasonably liberal about the class of logical constants it talks about. Let's consider this:

(T2) A sentence of a formal language which possibly extends a classical propositional/quantificational language with new logical constants which are propositional connectives, quantifiers or predicates is logically true in the pretheoretical sense iff it is true in all classical propositional/quantificational models which (re)interpret its constants (other than its logical constants).

(T2) is just like (T1), but it does not restrict itself to any severely limited set of logical constants; consequently, it also does not restrict itself to sentences of classical quantificational languages, but talks about sentences with possibly new logical constants which are propositional connectives, quantifiers, and predicates having the same syntax as their analogues in classical quantificational languages. For example, one of the languages (T2) talks about is a typical quantificational modal language.

A decisive problem with (T2) is that, no matter how one understands the notion of truth in a model that appears in its formulation, and given a natural choice of logical constants, it is obviously false, and it is pretty absurd to think that Tarski might have had something like this in mind. Classical propositional or quantificational models are clearly not appropriate for a theory of the logical properties of non-extensional logical constants, such as '$\Box$' ("it is necessarily the case that"). To make the problem vivid, concentrate on a simple example. Think of a classical propositional language with 'p' as the only propositional letter (and the only non-logical constant), and add to it '$\Box$'. A classical propositional model for this language simply assigns a truth-value to 'p'. So there are just two classical models for this language: M_T which assigns Truth to 'p', and M_F which assigns Falsehood to 'p'. And thus there are just four possible combinations of truth-values in M_T and M_F for the formula '$\Box$p': (1) '$\Box$p' is true both in M_T and in M_F; (2) '$\Box$p' is true in M_T and false in M_F; (3) '$\Box$p' is false in M_T and true in M_F; (4) '$\Box$p' is false both in M_T and in M_F. Given (1) or (2), the sentence '(p $\supset$ $\Box$p)' is true in all classical models, and it is thus a counterexample to (T2), for it is not a pretheoretical logical truth on any reasonable conception of logical truth; given (3) or (4), the sentence '(p ⊃∼ □p)' is true in all classical models, and for the same reason is again a counterexample to (T2).

So Tarski's thesis was something weaker than (T2). One possibility would be to refine (T2) by specifying a finite set of extensional and non-extensional constants we want to make our claim about and by trying to be specific about some corresponding suitable non-classical notions of model plus accompanying notions of truth in a model. For example, consider this:

(T3) A sentence of a classical propositional/quantificational/modal language is logically true in the pretheoretical sense iff it is true in all propositional/quantificational/Kripke models which (re)interpret its constants (other than its classical propositional/quantificational/modal logical constants).

(T3) may sound reasonable and entrenched in logical practice. In this respect it seems to me to be very much like (T1). This is not to say that it has not been criticized in the literature. A prominent example of a criticism of this kind can be derived from Ed Zalta's (1988). Zalta gave a type of sentences which could be seen as counterexamples to (T3) if we accepted his considerations and used them from a certain point of view. (But Zalta did not construct his examples as counterexamples to the adequacy of the model-theoretic method.)[8] Here is perhaps the simplest of his examples (the others are essentially identical).

Think of a propositional modal language which includes a monadic sentential operator 'A' taken as a logical constant meaning "it is actually the case that." For such a language there is a somewhat standard Kripkean definition of model and of truth in a model, and hence also of model-theoretic logical truth as truth in all models. A model for such a language is a quadruple of the kind (W, R, α, V), where W is a set (intuitively, a set of worlds), R a binary relation on W (intuitively, the relation in which two worlds w_1 and w_2 stand when all propositions true in w_2 are possible in w_1), α a member of W (intuitively, the actual world of the model), and V an assignment of truth-values to pairs propositional letter-world. V can be extended to a full assignment of truth-values to every pair formula-world by means of certain well-known recursive satisfaction clauses for the logical constants of the propositional modal language. A sentence is called true in a model of this kind if it is assigned the value Truth in the world α of the model. And a sentence is called logically true in the model-theoretic sense if it is true in all models (or in all models where R verifies a certain property).

The recursive satisfaction clause for formulae of the form 'Aφ' says that they are assigned the value Truth in a world w of a model of this kind if φ is assigned the value Truth in the world α of the model. This means that every formula of the form of

(1) Ap ⊃ p

---

[8] Zalta *identifies* the concepts of logical truth and of model-theoretic logical truth, so it is no surprise that he does not see his considerations as a criticism of any coextensionality thesis. But this view is obviously objectionable, for surely there is a wide conceptual gap between the notions of logical truth and model-theoretic logical truth. (This gap has been forcefully adverted to by Etchemendy (1990), even though Etchemendy's arguments for non-coextensionality seem to me less fortunate.)

will be assigned the value Truth in the world α of any model, since if 'p' is false in α then 'Ap' is false in α. So every formula of the form of (1) is a model-theoretic logical truth given the somewhat standard Kripkean definition of model and of truth in a model for languages containing the modal logical constant 'A'.

Now, a reasonable principle about logically true formulae is that any proposition they may come to express as a result of giving them an interpretation which respects the meanings of the logical constants ought to be necessary. However, if we interpret the letter 'p' by means of a sentence expressing a contingently true proposition, say "Kripke is a philosopher," it appears that the content then expressed by (1) is contingent: in a possible world in which Kripke is not a philosopher it is still true that Kripke actually is a philosopher (because he is a philosopher in *our* world, the actual world), but in that world it is not true that he is a philosopher. This is reflected in the somewhat standard Kripkean model theory of the modal language we are considering, since the necessitation of (1),

(2) $\qquad\qquad\qquad\square(Ap \supset p)$,

is not a model-theoretic logical truth; there is a model such that (1) is not true in some of the worlds of the model possible relative to α, and hence (2) is false at α in this model.[9]

Thus we would supposedly have a model-theoretic logical truth in the sense of (T3), (1), which, when 'p' is interpreted in a suitable way, is not necessary. If we take it as a *datum* that a logical truth must be necessary (and this would seem eminently reasonable under most pretheoretical conceptions of the notion of logical truth), it follows that some model-theoretic logical truths in the sense of (T3) are not real logical truths.

One problem with this alleged counterexample is that it is a bit dubious that the actuality operator 'A' is a good candidate for logicality. But perhaps it is consistent with our vague intuitions about logical constancy to take 'A' to be a logical constant, or at least a potential logical constant. A more serious worry, however, is that although the model-theoretic definition of logical truth for modal languages containing the operator 'A' that Zalta uses is somewhat standard, it is not as entrenched as the corresponding definition for modal languages not containing 'A'. In fact, there is room for choice concerning languages containing 'A' even within the *Kripkean* possible worlds semantics mentioned in (T3). Some theorists have adopted the following alternative. Call now a model a *quintuple* of the kind (W, R, α, β, V), where things are as before with W, R, α and V, and β is a member of W, possibly but not necessarily α. Keep all the recursive satisfaction clauses as before, but say that a sentence is true in a model (W, R, α, β, V) if it is true in β. Then clearly (1) is false in some models of the new kind, simply because it's false at some worlds in some models of the earlier kind.[10]

---

[9] David Kaplan (1977) gave an example very similar to (1) and claimed that it is a contingent logical truth. The example is 'ANp $\supset$ p', where 'N' is a "now" operator with a semantics in tense logic analogous to that of 'A' in modal logic. Zalta's example is less objectionable as a counterexample because 'N' is less clearly a logical constant than 'A'.

[10] For related critical remarks on Zalta, see Hanson (2006).

Zalta's examples would convincingly refute (T3) if they crippled all reasonable Kripkean model theories for languages containing 'A', or simply if they were directed at a more entrenched, or "the" entrenched Kripkean model theory. It would be good, of course, if there were fewer doubts about the choice of logical constants on which Zalta's counterexamples are based. But even if we grant that the choice is right, a suspicion remains because the examples are directed at the model theory of a non-extensional operator with a non-entrenched semantics. There is no standard recipe for constructing the Kripkean semantics of new modal operators we may come up with.[11]

In any case, as happened with (T1), one problem with (T3) and similar theses is that they talk about a very restricted set of logical constants. In fact, as long as one restricts oneself to a smallish finite set of logical constants (as in (T1), (T3), and similar theses), even the set of *extensional* logical constants among them may, for all we know, be always too restricted. And a problem peculiar to (T3), that in any case disqualifies it as a suitable Tarskian thesis, is that it talks about a notion of model (Kripke models) that Tarski simply does not talk about or even adumbrate in his classic chapter of 1936.

To summarize, I think that the thesis that Tarski probably had in mind was something weaker than (T2) but not as specific as (T1) or (T3). Further, Tarski's thesis must have been one that seems reasonable to postulate when one restricts one's attention to classical propositional/quantificational models, for these are the models that Tarski clearly has in mind. Finally, Tarski's thesis must have made a broad claim about the *notion* of a logical constant—not about a set of logical constants characterized by specifying a mere list.

II

The following thesis, (T4), satisfies all these desiderata, and seems to me quite likely to capture the essence of what Tarski had, at least implicitly, in mind:

(T4) A sentence of a formal language which possibly extends a classical propositional/quantificational language with new *extensional* logical constants which are propositional connectives, quantifiers, and predicates is logically true in the pretheoretical sense iff it is true in all classical propositional/quantificational models which (re)interpret its constants (other than its extensional logical constants).

(T4) is just like (T2), but it restricts itself to quantificational languages with extensional logical constants, and tacitly presupposes a natural extension of the classical notion of truth in a model for such languages (see, e.g., the examples below). This was very probably Tarski's intent. There are well-known dismissive remarks of Tarski about the presumable impossibility of giving non-extensional constants "any precise meaning" (Tarski (1935), p. 161); these remarks are in his classic monograph on truth, published one year before his chapter on logical consequence. Besides, it's clear

---

[11] Let me stress, however, that Zalta's example does succeed in refuting a particular coextensionality thesis, regardless of how reasonable or entrenched it may have been.

that if he had had in mind a thesis about non-extensional logical constants, he would not have restricted himself to extensional models such as the classical quantificational models, in view of elementary considerations such as those that show that (T2) is false. Furthermore, (T4) seems to underlie to a good extent the practice of using classical models in the model theory of languages having as logical constants generalized quantifiers, the predicate of identity, and other extensional constants denoting notions invariant under permutations of the quantificational domain.

What does 'extensional' mean, exactly? It's hard to be precise, but the rough idea, which will be enough for my purposes, is this (the extension to the polyadic cases is obvious):

(a) A (monadic) connective © is extensional if, whenever φ and $\varphi\prime$ are formulae which are either both satisfied or both unsatisfied by the same valuation v of the variables at a world w, ©φ is satisfied by v at w iff ©$\varphi\prime$ is satisfied by v at w.[12]

(b) A (monadic) quantifier Q is extensional if, whenever φ and $\varphi\prime$ are formulae which are satisfied at a world w by the same set of valuations differing at most at 'x' from a valuation v, $Qx\varphi$ is satisfied by v at w iff $Qx\varphi\prime$ is satisfied by v at w.

(c) A (monadic) predicate P is extensional if, whenever t and $t\prime$ are terms with the same denotation under a valuation v of the variables at a world w, $P(t)$ is satisfied by v at w iff $P(t\prime)$ is satisfied by v at w.

Now that we know what (T4) means and having claimed that it probably deserves the title 'Tarski's thesis' more than any of the other theses we have considered,[13] we can ask: is (T4) true?

One thing that would seem clear is that (T4) is not trivially false. But there are a number of examples that have been given in the literature which, if they were convincing, would show that (T4) is false in a rather trivial way. Two of these examples are due again to Etchemendy (1990) and McGee (1992). Etchemendy has noted that, if one takes as logical constants the extensional monadic predicates 'P' and 'M', meaning respectively "is or was a president of the U.S. on or before 2005" and "is a male," then the quantificational sentence

(3) $(\forall x)(P(x) \supset M(x))$

is true in all classical quantificational models, since no matter what model we choose (3) will be true in the model. In this case, that means that no matter what (set-sized) quantifier domain of actually existing things we choose, every object in that domain will be either a non-president or a male (there haven't been any female presidents in the actual world). However, (3) is not a logical truth under most conceptions of logical truth, for it is not even necessary. Or, at least, (3) is not necessary if it is interpreted in such a way that it quantifies over any of a wide class of natural ranges

---

[12] Note that, as desired, 'A' is not extensional in this sense.

[13] As mentioned in an earlier note, in (1936) Tarski required the domain of some models to be the denotation of a non-logical predicate of the formal language under consideration. However, he later adopted the now common convention and presumably he stuck otherwise to his views in the 1936 chapter.

for its quantifier; for example, if it ranges over "absolutely everything,"[14] or over the set of humans, etc., then (3) is not intuitively necessary. And hence it is not a pretheoretical logical truth, under most conceptions of logical truth.

McGee has given another purported counterexample (but he has not categorically asserted that it is a counterexample). Take as a logical constant the extensional quantifier '($\exists^{PC}$x)', meaning "there are at least a proper class of x's such that"; then the quantificational sentence

(4) $\qquad\qquad\qquad\sim(\exists^{PC}x)(x = x)$

is true in all classical quantificational models, since no matter what set-sized quantifier domain we choose, the sentence will be true in that domain. And yet (4) is not true for proper class-sized domains, much less necessary; or, at least, (4) is not necessary when the proposition it expresses quantifies over, e.g., "absolutely everything," or over the class of sets, etc. So (4) is not logically true, under most conceptions of logical truth.

A problem for anyone who wants to use these examples against (T4) is that the arguments needed for this must be premised on suspicious choices of expressions as logical constants. There is a patent intuition, I think, that neither 'P' nor 'M' are logical constants, and there is to say the least no clear intuition that '($\exists^{PC}$x)' is a logical constant, so the persuasive force of Etchemendy's and McGee's examples is quite limited. This intuition also vindicates the initial impression that a refutation of (T4) cannot be trivial (or at the very least cannot be as trivial as the refutations that would be provided by those examples). Part of what makes finding a refutation of (T4) nontrivial is that one cannot choose or define just about any constants, pick any non-logical truth involving them and claim that that truth is a model-theoretic logical truth in the sense of (T4) because it is true in all models given that we decide to count the constants in question as logical. If one wants to refute (T4), one must rather find constants about whose logicality there is an independent and reasonably entrenched intuition, and show that some sentence that is not a logical truth is nevertheless a model-theoretic logical truth in the sense of (T4).

As I announced at the beginning, I do think that (T4) is false. But the reason why I think it's false is not trivial (at least not trivial in the just indicated sense in which the reasons provided by the preceding attempted counterexamples would be trivial if they worked). If I'm right that (T4) is false, that will mean that a *prima facie* reasonable, non-trivial strengthening of (T1), that is in all probability the coextensionality thesis held by Tarski, is false, and hence that Tarski's theory of logical truth and logical consequence is defective to some extent.[15]

The way in which I will argue that (T4) is false will be by exhibiting certain quantificational sentences containing only constants that are intuitively logical and extensional, which are true in all classical quantificational models, but which, like Etchemendy's, McGee's, and Zalta's sentences, are *not necessary* (in the sense that

---

[14] In the sense recently elucidated by Tim Williamson (2003).
[15] As noted in Section I, there are strong arguments for its adequacy when its range of application is restricted in the manner specified in (T1(1)) and even in (T1). In Section III, however, we will see a possible qualification even of (T1(1)).

their non-logical vocabulary can be interpreted, and indeed it is typically interpreted, in such a way that the truths then expressed are not necessary). Since under most pretheoretical conceptions of logical truth, interpreted sentences that are not necessary are not logically true, it will follow that (T4) is false under such conceptions.[16]

The examples I will present are based on an assumption about the modal behavior of the classical first-order quantifiers that is not part of their explicit classical model-theoretic semantics, but which is, I think, the most natural assumption one can make. There are basically two kinds of propositions that a formula dominated by a classical universal quantifier (a formula of the form '$(\forall x)\varphi$') might be taken to express, which are reflected in the two most standard semantics for quantificational modal logic. The first kind of proposition is a proposition the content of whose quantifier gets specified when one specifies a quantificational domain, given purely in extension, plus a property which further restricts the range of the quantifier at a particular world. Thus, for example, given this view, in order to specify the content of the quantifier in the proposition expressed by (a use of) the sentence (3), what one has to do is to specify a class of objects, given purely in extension, plus a property. For example, one specifies the set {Franklin Roosevelt, Eleanor Roosevelt, George Bush, Laura Bush} plus the property of being Texan. Assuming that the other expressions have their intuitive meaning, (3) then expresses the proposition that, roughly, "each of the Texans in the set {Franklin Roosevelt, Eleanor Roosevelt, George Bush, Laura Bush} is, if a president, a male." This proposition is true in those worlds where, as in the present one, the Texan presidents in the set {Franklin Roosevelt, Eleanor Roosevelt, George Bush, Laura Bush} are all males. The way this idea is reflected in one of the standard semantics for quantified modal logic is as follows: given a previously specified domain D and a property Π, one says that $(\forall x)\varphi$ is satisfied by a valuation v at a world w iff every valuation u of the variables (with objects of the previously given domain D) which differs from v at most at 'x' and which assigns to 'x' an element of D that has the property Π in w satisfies φ at w. (Typically Π is taken to be the property of existence.) This clause has the effect that with a quantifier '$(\forall x)$' one quantifies in a world w only over the objects from D that have the property Π in w.

The second kind of proposition that a quantificational formula might be taken to express is a proposition the content of whose quantifier gets specified when one specifies simply a quantificational domain, given purely in extension, which constitutes the range of the quantifier at any particular world. Given this view, in order to specify the content of the quantifier in the proposition expressed by (a use of) the sentence (3), what one has to do is simply to specify a class of objects, given purely in extension. For example, one specifies the set {Franklin Roosevelt, Eleanor Roosevelt, George

---

[16] On a minority of pretheoretical conceptions of logical truth, including perhaps Tarski's, a refutation of (T4) that appealed to intuitions about necessity would be dismissed on the grounds of some kind of skepticism about modality. However, I think it's most significant to evaluate (T4) using assumptions shared by the majority of views, and one such is the assumption that any bona fide logical truth must be necessary. Note that (T4) sounds initially plausible regardless of whether one is in the minority of skeptics about modality or in the majority of non-skeptics (just as (T1) has sounded plausible to many people regardless of the details of their pretheoretical conceptions of logical truth).

Bush, Laura Bush} as before, and that's enough. Assuming the other expressions have their intuitive meaning, (3) then expresses the proposition that, roughly, "each of the things in the set {Franklin Roosevelt, Eleanor Roosevelt, George Bush, Laura Bush} is, if a president, a male." This proposition is true in those worlds where, as in the present one, the presidents in the set {Franklin Roosevelt, Eleanor Roosevelt, George Bush, Laura Bush} are all males; and it will be false, for example, in worlds where Eleanor becomes president. The way this idea is reflected in the other standard semantics for quantified modal logic is as follows: $(\forall x)\varphi$ is said to be satisfied at a world w by a valuation v of the variables with objects of the previously given domain D iff every valuation u of the variables with objects of D which differs from v at most at 'x' satisfies φ at w. This clause has the effect that with a quantifier '$(\forall x)$' one quantifies in a world w over all the objects of the previously specified domain D.[17]

Note that under this way of understanding the content of the universal quantifier, when we consider whether, e.g., the quantificational sentence (3) is true at a world, we always ask ourselves whether each of the objects of a previously fixed domain D is either out of the extension of 'P' for that world or in the extension of 'M' for that world. Assuming that these extensions are the sets of presidents and males, respectively, what we always ask ourselves is whether each of *those same objects, the objects of that same domain* D, is either a non-president or a male in the world at issue.[18]

Regardless of what of these two ways of understanding the modal behavior of the classical quantifier we choose, that does not conflict with its status as an intuitive logical constant. Or, in the case of the first way, there is no conflict when the property Π is a logical property like existence. And in fact the two kinds of quantifiers are taken as logical constants in quantificational modal logic. Further, under both ways

---

[17] One common way to illustrate the difference between the two just described semantics in a non-extensional language is by observing that the so-called Barcan formulae (e.g., $(\forall x)\Box P(x) \supset \Box(\forall x)P(x)$ or $\Diamond(\exists x)P(x) \supset (\exists x)\Diamond P(x)$) are true in all Kripke models under the second, "fixed domain" semantics but false in some models under the first, "variable domain" semantics. A quantificational Kripke model is a sextuple (D, W, R, α, V, Π), where things are as in the propositional models of Section I with W, R, and α; D is a non-empty set (intuitively of individuals); V an assignment of extensions drawn from D to each pair predicate letter-world; and Π an assignment of extensions drawn from D to each world (intuitively Π is the restricting property resorted to in the "variable domain" semantics, but idle in the other semantics). Take $\Diamond(\exists x)P(x) \supset (\exists x)\Diamond P(x)$. Under the fixed domain semantics, if $\Diamond(\exists x)P(x)$ is true in α, there is a world w R-accessible from α such that some object o is in the extension of P in w; but then o is an object of the fixed domain D such that there is a world w R-accessible from α, such that o is in the extension of P in w; so $\Diamond(\exists x)P(x) \supset (\exists x)\Diamond P(x)$ is true in α regardless of the model, and hence true in every model under the fixed domain semantics. On the other hand, to construct a counterexample to $\Diamond(\exists x)P(x) \supset (\exists x)\Diamond P(x)$ under the variable domain semantics, let D = {1, 2}, W = {α, w}, R = W × W, V(P, α) = ∅ and V(P, w) = {2}; also, let Π assign {1} to α and {2} to w. In this model, $\Diamond(\exists x)P(x)$ is true and $(\exists x)\Diamond P(x)$ false in α.

[18] Under this conception of the quantifiers, it is reasonable to postulate that a quantificational sentence has not been given a content unless a non-empty domain for the quantifiers has been stipulated. (This postulate provides a way out of the exception to Kreisel's intuitive soundness premise, provided by sentences like '$(\exists x)(P(x) \vee \sim P(x))$', mentioned in an earlier note.) In the same way, it is reasonable to postulate that a sentence containing indexicals has not been given a content unless context determines some object(s) as the denotations of the indexicals under the context. See below for more on indexicals and the quantifiers.

the universal quantifier is an extensional quantifier in the sense above (or rather, in the case of the first way, it is extensional provided the property Π is extensional).

I think it is most natural to suppose that the universal quantifier behaves modally in the second way just described, i.e., that with it one quantifies in every world over all the objects of the domain that serves to interpret the quantifier. (Since there is no property Π to worry about, the universal quantifier understood in this way is a logical constant without qualification.) The reason why I think this is most natural requires us to review quickly some well-known ideas from the philosophy of language.

In standard treatments, a sentence containing indexicals becomes truth-evaluable in a circumstance or possible world only under a context of use, a context of use which provides a content for the indexicals and indirectly for the whole sentence under the context. If I now utter the sentence 'I am a philosopher', this sentence, given that I am the speaker of the context, acquires under the context the content that Mario is a philosopher. A content of this kind can be evaluated for truth or falsehood in the present and other possible circumstances. It will be true in those circumstances in which Mario is a philosopher, and false in those circumstances in which Mario is not a philosopher. A similar view is widely held about indexicals which, unlike 'I' or 'actually', are not pure, such as demonstratives like 'that'. If I say

(5) That is a philosopher,

my utterance will have acquired a content under the context if the context has provided a content for the word 'that'. This content cannot have been provided merely by the linguistic rules for 'that' and my uttering the word, unlike what happens in the case of 'I'—this is what it means that 'I' and 'actually' are pure. Something else, perhaps my action of pointing toward some person, or my intending that the word 'that' indicate a certain person, must have happened. I will suppose that this "something else" happens, and won't go into what it might consist in. Suppose that my word 'that', as a result of some feature of the context of my utterance, came to indicate Kripke. Then the content of my utterance of (5) was that Kripke is a philosopher. This content will be true at those worlds in which Kripke is a philosopher, and false at those worlds in which Kripke is not a philosopher.

This way of dealing with indexicals and their content is encouraged by certain intuitions about the *rigidity* and the *direct referentiality* of indexicals. Recall what Kripke (1972) called rigidity. A designator is rigid if it designates the same object in all possible circumstances or worlds. (Well, actually Kripke's notion of rigidity is a bit weaker than this; the one I just defined is what Nathan Salmon (1982) christened with the specialized name of 'obstinate rigidity'.) One way in which one checks that a designator is rigid is by considering what would have to happen in order for the content of sentences containing that designator to be true in different possible circumstances. Kripke asks us to consider (6) and (7):

(6) Aristotle was fond of dogs.
(7) The last great philosopher of antiquity was fond of dogs.

When we consider whether the content of (6) is true in other possible circumstances, we always ask ourselves whether the same person, Aristotle, is fond of dogs in those circumstances. So Aristotle must be the object designated by 'Aristotle' when (6) is evaluated at a possible circumstance, and 'Aristotle' is rigid. On the other hand, when we consider whether the content of (7) is true in other possible circumstances, we ask ourselves whether the person who is the greatest philosopher of antiquity in those circumstances is fond of dogs in those circumstances, and this need not be always the same person, in particular it need not be Aristotle. So different objects can be designated by 'The greatest philosopher of antiquity' when (7) is evaluated at other possible circumstances, and so 'The greatest philosopher of antiquity' is not rigid.

Similar considerations strongly suggest that indexicals taken under a context are rigid. Consider (5) again. When we ask ourselves whether the content of (5) under the context (or, in alternative terminology often taken to be equivalent, what I *said* with my utterance of (5)) is true in other possible circumstances, we always ask ourselves whether the same person, Kripke, is a philosopher in those circumstances. So Kripke is the object designated by 'that' under the context when (5) is evaluated at a possible circumstance, and so 'that' is rigid.

Direct referentiality is the name David Kaplan gave to what he took to be a related property of names and indexicals. Perhaps the best way of understanding this property is by noting than certain definite descriptions are rigid. One example is 'The ratio of the circumference of any circle to its diameter'. It designates the same object, the number π, in every circumstance. Consequently, when we consider whether the content of the sentence

The ratio of the circumference of any circle to its diameter is worshipped by the ancient Babylonians

is true at a world, we always ask ourselves whether the same object, π, is worshipped by the ancient Babylonians in those circumstances. But although the description 'The ratio of the circumference of any circle to its diameter' is rigid, it is not directly referential. It has a descriptive content by means of which, or through the mediation of which, the same object turns out to get determined in all circumstances as the designation of the description. A name and an indexical (under a context), on the other hand, are (on Kaplan's view) directly referential because their content is not such that it serves to "determine" the designated object; the content of a name or indexical (under a context) is directly the designated object; there is (in particular) no descriptive content conventionally associated with them which "determines" the designated object.

On reasonable assumptions about how to understand the classical first-order quantifiers, these would have properties analogous to rigidity and direct referentiality. These quantifiers are of course not designators in Kripke's sense (they are not singular terms), but there is a relatively clear sense in which they make implicit "reference" to a set of objects over which the quantifier variables range, or to the objects themselves as a plurality. Given a domain of objects for the variables, sentence (3) says approximately that every thing in *that domain* is either a non-president or a male, or that every thing among *those things* is either a non-president or a male.

(3) does not contain any descriptive element which helps determine the domain of objects over which its variables range. Rather, that domain or those objects are given

or fixed by the logician who uses a formal language containing (3) before (3) can be evaluated for truth and falsehood, in a way similar to the way in which something in the context provides a reference for 'that' in (5). Once the domain or its objects are fixed, (3) can be evaluated for truth and falsehood. But (3) itself contains no descriptive element which determines that domain or its objects.[19] So it would appear that in some sense the quantifier '(∀x)' in (3), under a choice of a domain, can be taken to make implicit reference to that domain or its objects *in a direct way*. We may say that '(∀x)' *directly ranges over* its domain or its objects.

This proposal about the quantifier '(∀x)' ranging directly over its domain is in harmony with the standard extensional semantics for it. This semantics requires only that a set of objects be given for the range of the quantifiers to be determined. No descriptive property characterizing or restricting the objects of the set is demanded in interpretation. (This is especially appropriate in model theory, since the model theorist clearly does not want to leave out of his considerations models whose domain is not picked out by any descriptive property.) Further, the standard satisfaction clause for a formula of the form '(∀x)φ' says that (∀x)φ is satisfied by a model and a valuation v of the variables with objects of the domain of the model if every valuation u of the variables with objects of the domain which differs from v at most at 'x' satisfies φ. In this clause there is no mention of any descriptive property required to hold of the objects which can be assigned to 'x' (by valuations which differ from v at most at 'x').

Direct referentiality implies rigidity in the case of proper designators (but not vice versa, as we saw). Must the quantifier '(∀x)' be taken to be "rigid" if we decide to understand it as ranging directly over its domain? In other words: suppose '(∀x)' ranges directly over its domain, and we are considering the question whether the content of (3), provided by a choice of a domain D for its variables, is true in other possible circumstances; must we adopt the view that we always have to ask ourselves whether each of *those same objects, the objects of D*, is either a non-president or a male (in those circumstances)? Or briefly put: is the domain D the range of quantification of '(∀x)' in all possible worlds? It appears so: assuming that (3), under a choice of a domain for its variables, makes implicit direct reference to that domain or its objects, and given that this domain or its objects are not determined by any descriptive element in the content of (3), it appears that the domain or its objects themselves must form part of the content of the quantifier. And then it appears that we must adopt the view that when we consider whether the content of (3) is true at a possible world, we always have to ask ourselves whether each of *those same objects, the objects of that same domain*, is either a non-president or a male (at that world). We thus have some powerful considerations in favor of the view that the first-order quantifiers behave modally so that the domain that serves to interpret the quantifiers is the domain over which one quantifies in every world.

---

[19] Recall that in (1936) Tarski adopted the convention that the domains of the models for first-order languages ought to be denoted by a non-logical predicate of those languages, which in some cases could express a descriptive property. He further required quantifications in those languages to be relativized to that predicate. So the present argument would not quite work under Tarski's (1936) conventions; it would work only under his later conventions.

An objection to this understanding of the classical first-order quantifiers as ranging directly and rigidly over their domain might run as follows: "It is true that the explicit textbook conventions about the first-order quantifiers don't say anything about whether they range directly over their domain or are 'rigid'. But a tacit assumption underlying what is explicitly said in textbooks is that the first-order quantifiers must be understood as being as close as possible to their correlates, or their closest correlates, in natural languages. Perhaps if we look at quantifiers in natural language we'll find that they are not 'rigid', or do not range directly over their domain."

Now, I'm not sure that what this objection calls the "tacit assumption" of formalization is in fact such. As noted above, formalization focuses on the precisification of truth-conditional content, and does not care about whether features relevant to modal character are determined in the process. But let's concede the "tacit assumption" for the sake of argument. Are the quantifiers of natural language not "rigid," or do they not range directly over their domain? Any answer to this question is bound to be complex and controversial. In particular, the answer is bound to be complex and controversial in the case of quantifiers which work as determiners, such as 'every', 'any', or 'some' (in its use as a determiner).

But the process of formalization leaves it settled that first-order quantifiers do not work grammatically as determiners. Unlike, say, 'every', '($\forall$x)' does not grammatically work as a determiner of a noun or a predicate that restricts the reach of quantification, and requires only a formula in order for a formula dominated by it to get formed. First-order quantifiers seem to be closer to pronominal uses of quantifiers in natural language, as in sentences like (8), (9), and (10):

(8) All are seen in the evening.
(9) Some are seen in the evening.
(10) All are, if presidents, males.

I think that a good case can probably be made that pronominal quantifiers range directly over their domain and are thus "rigid." The traditional terminology which calls them 'pronominal' even leads us *a priori* to expect this, since it is common in the treatments of indexicals that I've been relying on to consider pronouns as indexicals, hence as "directly referential."

Imagine that my student Mary and I are looking at close-up photographs of Mars, Venus, and Jupiter in an astronomy book. Then I utter (8). It may be useful to think of our intuitions about rigidity first. When considering whether the content of my utterance is true at a different possible circumstance, what will I consider? I think I will ask myself if each of these same objects, the objects in this domain containing Mars, Venus, and Jupiter, are seen in the evening. What I said with my utterance will be true at a possible circumstance just in case all of Mars, Venus, and Jupiter are seen in the evening in that circumstance. Something similar will happen in the case of (9). What I say if I utter (9) in the context just mentioned will be true in another possible circumstance just in case at least one among Mars, Venus, and Jupiter is seen in the evening in that circumstance. What I earlier called an "implicit" reference to a domain seems to be taking place also here, and it seems to be taking place in a "rigid" way.

Do pronominal quantifiers range directly over their domain? Presumably. The only way that occurs to me in which "direct ranging" might not be taking place here in the presence of rigidity would be that pronominal quantifiers were in some sense disguised quantifier determiners whose complement noun phrase picks out rigidly a set of objects, but has been elided. For example, on this view (8) would be elliptical for (11), or (12), or something similar:

(11) All objects identical with either Mars, Venus, or Jupiter are seen in the evening.
(12) All these are seen in the evening.

Arguably both 'objects identical with either Mars, Venus, or Jupiter' and 'these' pick out the set B rigidly. Now, even if we conceded that the content of some such sentence was the real content of my utterance of (8), this would be no objection to the view that there was "direct ranging" involved, since on the views we are relying on 'these' is directly referential. But I don't think one ought to concede even that. One basic problem with the suggestion is that there is no reason why my utterance of (8) should have the content of (11), or the content of (12), or the content of any one in particular of the many other options that we might think of. If there were in fact one noun phrase that I inadvertently elided when I uttered (8), presumably there ought to be some principle that determined what that noun phrase is and that allowed us to recover it; but there is no apparent principle that determines one such noun phrase as preferred.

Another, even more serious problem, lies in the fact that, even if there were such a principle determining a preferred noun phrase in some cases, it could hardly be seen as a principle determining indirectly the real content of my original utterance of (8) as the content of some other sentence. Presumably I can come to understand (8) before I understand 'objects identical with either Mars, Venus, or Jupiter', or 'these', or other noun phrases that are candidates for ellipsis on the objector's view. So, provided we grant the common idea that understanding is knowledge of content, one can know the content of my utterance of (8) before one knows the content of the sentences proposed by the objector; but on the objector's view this would be impossible. Together with the underdetermination argument of the preceding paragraph, this argument about independent intelligibility seriously undermines the ellipsis objection.[20]

Having thus argued for what I think is the most reasonable way of understanding the modal behavior of the first-order quantifiers, let me go back to the promised counterexamples to (T4). The counterexample that is more usefully presented first involves the further reasonable premise that a monadic predicate 'E' meaning "exists" is a logical constant. A primitive predicate with this intended meaning is taken as a logical constant in treatments of quantified modal logic and of intensional logic generally (see, e.g., David Kaplan's (1977), (1978) logic of demonstratives). Its intended meaning, a bit more explicitly, is given by the principle that it is to be satisfied by an object at a world if that object exists at that world. This is the

---

[20] Similar arguments against related syntactic ellipsis views of the quantifiers are provided by Stanley and Szabó (2000), though they don't consider the case of pronominal quantifiers.

common sense of existence according to which I exist in the current circumstance but I would not have existed if a certain spermatozoid and a certain egg had not met. It can be given a simple satisfaction clause in the definition of satisfaction in a classical quantificational model, very much like the clause for identity: a model M plus valuation v satisfies E(x) if v(x) exists.[21] Clearly 'E' is an extensional predicate in the semi-formal sense above. (In the sense that when t and $t\prime$ are terms with a shared denotation under a valuation v at a world w, E(t) is satisfied by v at w iff E($t\prime$) is satisfied by v at w.)

Is the predicate 'E' really a logical constant? As I just said, it has been taken to be such in intensional logic. But furthermore 'E' is certainly topic-neutral. Moreover, it (or its correlate in natural language) is widely applied across different areas of discourse, and thus satisfies the main criterion for logical constancy mentioned by defenders of "pragmatic" conceptions of this notion (see e.g. Hanson (1997), Warmbrōd (1999), Gómez-Torrente (2002)). Arguably, 'exists' is used in mathematics, for example; or at the very least, it is unclear that every appearance of the word as applied to mathematical objects is to be understood as an appearance of something like the first-order or a higher-order existential quantifier. Further, 'E' is a logical constant in the technical senses defined by Tarski, McGee, and Solomon Feferman. It is surely invariant under permutations of a model (Tarski (1966), Tarski and Givant (1987)) and even under bijections of models (with a domain of existing things), for its extension in any model (with a domain of existing things) is the full domain of the model; and it follows from its meaning that it is invariant under bijections of models (with a domain of existing things) (McGee (1996)); it is also invariant under homomorphisms of models (with a domain of existing things) in the sense recently defined by Feferman (1999). It is not a logical constant in the technical sense of Timothy McCarthy (1981)[22], since it is not invariant under bijections of models with a domain of existents plus possibly non-existing things; but this would seem a defect of McCarthy's proposal rather than a virtue.[23]

(It is worth mentioning that (T4) is easily refuted if one understands 'logical constant' not in a pretheoretical way as I do, but by means of any of the "permutationist" notions mentioned in the preceding paragraph (including McCarthy's), as I have argued in Gómez-Torrente (2002). Observe that a predicate 'H' meaning "is a married bachelor" has an empty extension in all possible worlds. Given this, it is invariant under bijections of all models without qualification; yet it is not a logical constant, provided there is any distinction at all between logical and non-logical constants. But suppose for the sake of argument that 'H' is a logical constant, as a "permutationist" characterization would have it. It follows that a sentence like '($\forall$x) $\sim$ H(x)' is a model-theoretic logical truth in the sense of (T4). But clearly this sentence is not a logical truth. Tarski (in (1936)) presumably understood the

---

[21] Note that 'E(x)' is not equivalent to 'AE(x)' under the somewhat standard model theory for 'A' used by Zalta, since an object may exist in a world without existing in the actual world.

[22] Nor in the related sense of Gila Sher (1991), (1996). Sher has emphasized to me that her domains contain non-existents. In this sense her proposal seems to me quite non-Tarskian.

[23] And a defect of the related proposal of Sher (1991), (1996) (see the preceding note).

notion of logical constancy pretheoretically, assuming at most that invariance under permutations (of actual domains) is a necessary condition on logical constancy. In what follows we will see that, under most conceptions of logical truth, (T4) is false even thus understood.)

Now consider the following quantificational formula[24]:

(13) $(\forall x) E(x)$.

Given only the reasonable premise that 'E' is a logical constant, this formula is a model-theoretic logical truth in the sense of (T4). Since there are no non-logical constants to worry about, a classical quantificational model for (13) is just a (set-sized) quantifier domain composed of existing things. And no matter what classical quantificational model for (13) we choose, (13) will be true in the model: no matter what actual (set-sized) quantifier domain composed of existing things we choose, every object in that domain will be an existing thing (in our world). This presupposes the usual understanding of quantificational model theory, according to which no non-existing objects form part of the domains of models.[25]

On the other hand, given only our earlier conclusion that the first-order universal quantifier appearing in a formula like (13) is "rigid," (13) can easily be interpreted so as to express a contingent content, and thus is not a logical truth under most conceptions of logical truth. Suppose that we specify the domain of the quantifier in (13) to be again the set {Franklin Roosevelt, Eleanor Roosevelt, George Bush, Laura Bush}. Under this interpretation, (13) is true in the actual world, but only contingently: in a possible world in which Franklin's parents had not met, (13) would not be true, because at least one of the people it quantifies over (they are the same four people in all possible worlds) would not exist in that world.[26] The same would happen with any domain containing a contingent existent.[27]

So given that 'E' is an extensional logical constant, and on the reasonable assumption we defended earlier, that the first-order universal quantifier is "rigid," (13) is

---

[24] I am taking the quantifier '$(\forall x)$' in (13) to be the classical first-order quantifier, and assuming that it behaves modally as in the "fixed domain" semantics described above—in short, that it is "rigid". For the purpose of giving a counterexample to (T4), however, one doesn't need to assume that the quantifier in (13) *is* the classical first-order quantifier. One may simply (and uncontroversially) assume that it is the "rigid", extensional, and logical quantifier described above, leaving aside the question whether it is to be identified with the usual first-order universal quantifier.

[25] Etchemendy's example (3) (and other examples of his), as well as most discussions of these issues, are based on this presupposition. If the presupposition is not made, the example does not clearly work. There may well be a non-existing person who is president of the United States and a female, even in our world—perhaps some fictional characters are non-existents with this property.

[26] For related reasons, Kaplan (1977) argued that 'I exist' is a (model-theoretic) logical truth (in his "logic of demonstratives") which is not necessary. For our purposes this is irrelevant, since 'I', unlike the quantifiers, is presumably a non-logical constant.

[27] Under some natural interpretations, (13) is even false; for example, if we take the domain of the quantifier to consist of "absolutely everything," (13) is false provided only that "something" (in the absolutely unrestricted sense) is not an actual existent. The same happens with any more restricted domain containing a non-existing object. However, I do not want to make my argument against (T4) rely on these interpretations, for the question whether it is possible to quantify over non-existing objects is substantially controversial. (Though I myself see no problem with this in the case of some mere *possibilia*.)

true in all classical quantificational models (that reinterpret only the constants other than extensional logical constants), is actually true in many domains for its quantifier, and, furthermore, for many of those domains it expresses a contingent truth. But under most pretheoretical conceptions of logical truth, a sentence that is a logical truth cannot be interpreted so as to express a non-necessary truth. It follows that under most pretheoretical conceptions of logical truth, thesis (T4), what I think was Tarski's thesis, is false.

The falsity of (T4) under most conceptions of logical truth does not depend on the fact that 'E' is a logical constant, for arguably other sentences analogous to (13), but not containing 'E', are model-theoretic logical truths in the sense of (T4) but can be interpreted so as to express contingent truths. A general lesson of (13) is that, for any monadic predicate F that is intuitively an extensional logical constant, applies to all existing things, but fails to apply to a thing in worlds where it doesn't exist, the sentence '$(\forall x)F(x)$' will be a model-theoretic logical truth in the sense of (T4), but it will not be a pretheoretical logical truth, under most conceptions of logical truth. It seems clear that other predicates with these properties exist. For example, a dyadic predicate 'T(x, y)' meaning "is a part of" is intuitively extensional, and has often been called a logical constant. Arguably the monadic predicate 'T(x, x)' applies to all existing things, but presumably fails to apply to a thing in worlds where it doesn't exist[28]. Another example with the same features as 'T(x, y)' may be a predicate 'S(x, y)' meaning "is simultaneous with" (in the sense of "it occupies the same stretch of time as"[29]); also arguably, 'S(x, x)' possesses the same relevant features as 'T(x, x)'. Given these assumptions, the sentences

$$(\forall x)T(x, x)$$

and

$$(\forall x)S(x, x)$$

will be further examples of sentences which are model-theoretic logical truths in the sense of (T4) but are not real logical truths under most conceptions of logical truth.

III

The examples in Section II are based on a natural assumption about the modal behavior of the classical first-order quantifiers, suggested by features of their classical extensional semantics. As I announced at the beginning, the examples in this final section

---

[28] The latter is again a frequent assumption about atomic predicates in modal logic. (Nevertheless, I doubt that it's a reasonable assumption for *all* such predicates. Elsewhere (Gómez-Torrente (2006)) I have argued that it may not be an intuitively compelling assumption about a number of atomic predicates whose application to an object at a world does not seem to depend on the object's existence. In my view, this is what happens with the predicate 'x = x' of self-identity, with the predicate 'x is a thing', and with predicates expressing a natural kind to which the object belongs essentially (such as 'x is a cat' or 'x is an electrical discharge').)

[29] Note that this predicate intuitively applies even to abstract, hence presumably eternal existents.

will be based on more speculative assumptions about the epistemology of the quantifiers, which are nevertheless also suggested by features of their classical extensional semantics. Just as the standard semantics of the classical quantifiers is not explicit about their modal behavior, it is also not explicit about some aspects of their semantics relevant to their epistemology. In this section I will explain how, given some plausible, but potentially more controversial ways of understanding the epistemology of the quantifiers, one obtains the result that some model-theoretic logical truths in the sense of (T4), and in fact even of (T1) and (T1(1)), are not *a priori*, and hence presumably not logically true.

We argued in Section II that the classical first-order quantifiers are most naturally understood as ranging directly over their domain and thus as "rigid" (though only the rigidity of the quantifier appearing in (13) was needed as an assumption for the argument that (13) is not logically true). The thesis that the quantifiers range directly over their domain suggests a familiar idea for finding model-theoretic logical truths in the sense of (T4) (and even of (T1) and (T1(1))) which are not *a priori* knowable. Before developing it in an explicit way, some clarifications.

What can it mean to say that a *sentence*, like (3), is *a priori* knowable? The question whether a sentence is *a priori* knowable does not seem to be a felicitous question unless it is made against the background of some suitable assumptions. A formula or even a sentence is not the sort of thing one can know or fail to know. Let's then ask the question about the *content* we gave to (3). Is it *a priori* knowable? Even this second question appears not to be proper. The reason is that it is unclear what has to happen for me to bear some psychological attitude toward the content of an expression containing directly referential or "directly ranging" expressions, like the content of (3). In fact, it is not clear that my bearing some psychological attitude toward something can be the same as my bearing some simple relation to what we have been calling a content.

The problem emerges in several ways. Think first of this famous example:

(14) Hesperus is Phosphorus.

Kripke noted that if 'Hesperus' and 'Phosphorus' are rigid, then the content of (14) is necessary given that it is true. He also noted that, intuitively, what we might call the (typical) attitudinal content of (14) is knowable only *a posteriori*. But if proper names are directly referential, (15) has the same content as (14), given the way we have been speaking of "content":

(15) Hesperus is Hesperus.

And yet the (typical) attitudinal content of (15) would seem knowable *a priori*. If we accept these intuitions at face value, it follows that the attitudinal contents of (14) and (15) must be different, even if their contents are the same. (Sometimes these attitudinal contents are called 'cognitive contents'.) But the case of (14) and (15) leaves open the possibility that the sentence is a relevant part of its attitudinal content; on this view, (14) and (15) might have different attitudinal contents simply because they are different sentences.

However, both Kripke (1979) and Kaplan (1977) have noted the possibility that different occurrences of the same sentence have the same content but different attitudinal contents. This can happen if different occurrences of the same directly referential word in the same sentence have the same content and yet differ in attitudinal content. For example, different occurrences of a demonstrative may share content but be accompanied with different demonstrations which carry distinct cognitive imports. Think of a weird but otherwise unobjectionable utterance of (16) where the utterance of the first 'that' is made on an evening, pointing to the Evening Star, and the rest of the utterance is made some later (much later) morning, pointing to the Morning Star:

(16) That is that.

This utterance of (16) has the same content as (14) and (15) (under the assumption that names and demonstratives are directly referential), and given our description of the conditions of the utterance, it appears that in some sense its attitudinal content is knowable only *a posteriori* (like that of (14)). On the other hand, if (16) is uttered quickly in the evening, with the speaker accompanying the two utterances of (16) with an action of continuously pointing to the Evening Star, it seems that the attitudinal content at stake can in some sense be known *a priori*. The two utterances of (16) would share sentence and content but would differ in attitudinal content.[30]

Here is another way in which the problem emerges, now illustrated directly for the case of the first-order quantifiers when these are taken to range directly over a domain. What can it mean to say that one knows (whether *a priori* or not) a quantificational sentence dominated by a quantifier '$(\forall x)$'? If we answer that it is to know its content, the content that is directly signified, there is still a problem. For in what sense can I be determinately said to know of *the objects* in the set, that all have a certain property? Suppose the domain of the variables is the set B = {Mars, Venus, Jupiter}, and that the predicate '$SIE(x)$' means "is seen in the evening", so that every utterance of

(17) $(\forall x)SIE(x)$

has roughly the content that all things in the *set* B are seen in the evening. One perfectly acceptable way of introducing B as the domain for the intended interpretation of (17) is for me to utter (17) while pointing to Mars, Venus, and Jupiter in the evening sky. If I do this it appears that in some sense the attitudinal content of my

---

[30] If proper names are directly referential, I suspect that (14) and (15) can have more than one attitudinal content, and, even in the case of (15), attitudinal contents that are *a posteriori*. I suspect that in this respect the case of proper names is more similar to that of demonstratives than often realized. It would seem perfectly possible that different tokens of a name like 'Hesperus' are associated by a speaker with different attitudinal contents, even if they are tokens with the same semantic content. (Compare the Paderewski cases brought out by Kripke (1979).) These tokens might appear in the same sentence. If (15) can have attitudinal contents that are *a posteriori*, even if its tokens of 'Hesperus' have the same semantic content (Venus), then we would have an argument that (15) is not a logical truth, under the assumption made below in the text that a logical truth cannot have *a posteriori* attitudinal contents. Yet the straightforward first-order formalization of (15), '$a = a$', is a model-theoretic logical truth. (Compare a related but different remark in Salmon (1986), p. 176, n. 5.)

utterance is not very informative. People can be assumed to have known it already. But if I utter (17) while pointing to close-up photographs of Mars, Venus, and Jupiter in the astronomy book (another perfectly acceptable way of introducing B as the domain of quantification), it seems that the attitudinal content at stake is informative, and people who trust me can learn a great deal from my utterance.

So if we accept all these intuitions at face value, it appears that sentences like (16) and (17) taken together with their contents are not (or not simply) what one is related to when one can be said to hold the relevant psychological attitudes, for example the attitude of *a priori* knowledge toward the attitudinal content of (an utterance of) (16) or the attitude of having learned toward the attitudinal content of (an utterance of) (17).

If a sentence and even a sentence taken together with a content for it are not determinately *a priori* or *a posteriori*, when should we call a sentence *a priori* (or *a posteriori*)? One possibility is to call a contentful sentence *a priori* when *some* of the attitudinal contents it may have are *a priori*. But this may lead to a trivialization of the notion of apriority for true sentences containing quantifiers. Think of me stipulating to my student John that the "rigid," directly ranged over domain of the quantifiers is "the set of planets seen in the evening in the coordinates such and such," and suppose with me, for the sake of argument, that this set is the set B above. I synthesize the application of this stipulation to formulae starting with a universal quantifier by means of

(18) $(\forall x)\varphi \equiv$ For every x which is a planet seen in the evening in the coordinates such and such, φ.

The two sides of this biconditional of course differ in content, but since (18) is a stipulation, John can perhaps be said to know its attitudinal content *a priori*. And then John can also perhaps be said to know *a priori* that

(17) $\qquad\qquad\qquad (\forall x)SIE(x),$

even if the content of (17) is contingent. (This is basically the argument used by Kripke in his famous example of the sentence "One meter is the length of the standard bar in Paris.") (17) would be a contentful sentence with at least one *a priori* attitudinal content. It seems thus that if logical truths are to have an epistemological property which distinguishes them from (17), this property ought not to be the property of having some *a priori* attitudinal content.

So let's opt for the other alternative that immediately suggests itself: let's agree to call a contentful sentence *a priori* when *all* the attitudinal contents it can have are *a priori*. (This principle is adopted by Salmon (1986).) Then we ought to accept this reasonable principle about interpreted logically true formulae: any attitudinal content they may come to have as a result of giving them whatever is needed to give them an attitudinal content ought to be an *a priori* knowable attitudinal content. If we find an interpreted model-theoretic logically true formula which is susceptible of having an attitudinal content that is knowable only *a posteriori*, it will follow from this principle that some model-theoretic logical truths are not *a priori*, and thus are not real logical truths under typical conceptions of logical truth.

Is (13), when one stipulates the range of its quantifier to be, e.g., {Franklin Roosevelt, Eleanor Roosevelt, George Bush, Laura Bush}, only susceptible of having an *a posteriori* knowable attitudinal content? One is tempted to say yes, since in a moderately intuitive sense one ought not to be able to know *a priori* that Franklin Roosevelt existed, and this would seem to be part of what one would have to know *a priori* in order to know *a priori* any attitudinal content expressed by (13) with its content. But, at any rate, it might seem clear that (13) with its content is susceptible of having at least *one* attitudinal content which is knowable only *a posteriori*. Think of someone uttering (13) while pointing to the portraits of Franklin Roosevelt, Eleanor Roosevelt, George Bush, and Laura Bush. Presumably whatever attitudinal content is involved in this situation is one I can only be said to be able to know *a posteriori*.

But here is a worry about the alleged aposteriority of an attitudinal content of (13). Perhaps in order for me to be able to quantify over some objects these objects must exist in my world. Perhaps this principle about quantifiability is knowable *a priori* by me. If this is so, then it is reasonable to think that I will be able to know *a priori* that any attitudinal content expressed by (13) with its content is true. Perhaps it is then also reasonable to think that I will be able to know *a priori* any attitudinal content expressed by (13) with its content. (13) with its content would then be one more alleged example of the contingent *a priori*.

I propose to forget about (13) in connection with apriority. Given our assumption that the first-order quantifiers are directly referential, in an utterance of a quantificational sentence containing two or more quantifiers it appears that we ought to find the same phenomenon as in (16). For definiteness, think of (19):

(19) $\qquad\qquad (\forall x)\text{SIE}(x) \supset (\forall x)\text{SIE}(x),$

where the quantifiers are interpreted as ranging again over the set B. (19) is a model-theoretic logical truth, even if 'SIE' is not a logical constant. Is (19) with its content knowable *a priori* or only *a posteriori*?

Again the answer would appear to be that that's not an appropriate question. Mere sentences containing expressions that signify their contents directly, or even sentences of this kind provided with a content, are not what one is related to when one can be said to know something. Before the question is asked, more needs to be said about the circumstances in which one comes into contact with (19) and its content.

Suppose that I utter (19) assertively in front of John. At the same time that I utter each token of the quantifier in (19), imagine that I point to Mars, Venus, and Jupiter in the evening sky. I think in this situation it can properly be said that John has not learnt anything that he could not have found out by himself *a priori*.

Now suppose that Mary also knows that the domain B is the range of the variables, and that I utter (19) assertively in front of her. At the same time that I utter my first token of the quantifier in (19), imagine that I point to Mars, Venus, and Jupiter in the evening sky, as before. But at the same time that I utter my second token of the quantifier, imagine that I point to the close-up photographs of Mars, Venus, and Jupiter in the astronomy book. I think that in this situation it can properly be said that Mary has learnt *a posteriori* something that she could not have found out by herself *a priori*.

It is certainly possible to introduce the domain of quantification for a first-order quantifier by any of the means we have mentioned, and so it is very reasonable to think that different tokens of a first-order quantifier having the same content differ in attitudinal content. What might raise more doubts is the claim that, once the domain has been introduced in some way, different tokens of a quantifier within the same sentence might have different attitudinal contents. Would it be more reasonable to think that, once the domain has been introduced in some way, some tacit principle prevents us from assigning different attitudinal contents to different tokens of a quantifier within the same sentence, so that something has gone wrong in my description of the attitudinal content of my utterance of (19) to Mary? I think not. The directly ranged over domain of quantification in the intended interpretation for a language often receives different descriptions from the person who introduces it, for example in a textbook or during a class. These descriptions often convey different descriptive information that gives rise to different attitudinal contents, and nothing seems to prevent these descriptions from being given in the time interval between the inscriptions of different tokens of a quantifier in the same sentence. The following situation seems perfectly natural: I am teaching a class and I stipulate that the domain of quantification of the sentence I am about to write on the blackboard is the set B above, which I introduce by pointing to three photographs of Mars, Venus, and Jupiter as tiny spots seen in the evening sky, but I don't call the planets by their names; then I write the antecedent of (19), but before writing the consequent I note that the objects in B are precisely the objects called 'Mars', 'Venus', and 'Jupiter' in some astronomy book containing close-up pictures of the planets, a book that my students know well; I then finish writing (19). It is most reasonable to think that the attitudinal content that the first token of '($\forall$x)' has for my students is different from the attitudinal content of the second token, and that (19) has for them an *a posteriori* attitudinal content.

The standard stipulations about the semantics of classical quantificational languages include the stipulation that all occurrences of a quantifier in a sentence are to range over the same initially fixed domain. This stipulation is respected in the examples. John, Mary, and my other students knew, just by knowing this and the other stipulations, that (19) had to be true, perhaps they can even be said to know this fact about the sentence *a priori*. But by means of my utterance of (19) Mary learnt something knowable only *a posteriori* while John did not learn anything he could not have learnt *a priori*. If we are not too fussy about using "word salads" mixing natural and formal language, such as are frequent in logic and mathematics texts, it even appears that we may use (19) embedded in attitudinal clauses to report what Mary learnt *a posteriori* and John could have learnt *a priori*. Imagine utterances of the following sentences, accompanied by actions of pointing similar to the ones I described earlier:

Mary learnt *a posteriori* that ($\forall$x)SIE(x) $\supset$ ($\forall$x)SIE(x).
John could have learnt *a priori* that ($\forall$x)SIE(x) $\supset$ ($\forall$x)SIE(x).

As noted in Section II, someone might embrace the idea that the first-order quantifiers must be understood as being as similar as possible to their closest correlates in natural languages, and object to the counterexample claiming that different utterances of a quantifier of natural language in the same sentence are not susceptible of having different attitudinal contents. However, just the opposite seems true. It appears that pronominal quantifiers have an epistemology essentially identical to that of indexicals. If I utter (8) or (9),

(8) All are seen in the evening,
(9) Some are seen in the evening,

neither the sentence nor the sentence taken together with its content seem appropriate objects for psychological attitudes. Something more is needed for an attitudinal content to arise. If I utter (8) to Mary while pointing to Mars, Venus, and Jupiter in the evening sky, intuitively she cannot be said to have learnt something she (we may assume) did not know. If I utter (8) to her while pointing to Mars, Venus, and Jupiter in the astronomy book, intuitively she can be said to have learnt something she (we may assume) did not know. Also, if I utter

(20) If all are seen in the evening then all are seen in the evening

(the sentence that would seem to provide the most accurate translation of (19) into English), accompanying the utterance of the first 'all' with an action of pointing to Mars, Venus, and Jupiter in the evening sky and the utterance of the second 'all' with an action of pointing to Mars, Venus, and Jupiter in the astronomy book, then Mary can be said to have learnt something she could not have learnt *a priori*.

Suppose one argued (as in the related ellipsis protest of Section II) that the two pronominal quantifiers in the relevant utterance of (20) are disguised quantifier determiners with two elided complement noun phrases that in this case have the same content but different attitudinal contents. It may be good to note that this by itself would be no objection to the claim that there are model-theoretic logical truths that are not *a priori*. Presumably those noun phrases would be syntactically identical, as in an utterance of

(21) If all these are seen in the evening, then all these are seen in the evening

(which we may imagine made accompanying the utterance of the first 'all these' with an action of pointing to Mars, Venus, and Jupiter in the evening sky and the utterance of the second 'all these' with an action of pointing to them in the astronomy book). But even if the relevant apparent utterance of (20) (and indirectly of (19)) is not really an utterance of (20) (or (19)) but is elliptical for (21), we still have reason to think that it is an utterance of a model-theoretic logical truth. After all, the sentence that it is supposed to be elliptical for, (21), may be translated back to formal language as a sentence of the form '$(\forall x)(P(x) \supset Q(x)) \supset (\forall x)(P(x) \supset Q(x))$', with antecedent and consequent having the same truth-conditional content. So this version of the ellipsis objection is no obstacle to the claim that there are model-theoretic logical truths that are not *a priori*.

A different version of the objection would be to claim that the two pronominal quantifiers in the relevant apparent utterance of (20) (though perhaps not in other utterances of (20)) are disguised quantifier determiners with two elided *syntactically different* complement noun phrases that have the same content but different attitudinal contents. For example, my described apparent utterance of (20) would on this view be elliptical for an utterance of (22), (23), or something similar:

(22) If all objects identical with either Mars, Venus, or Jupiter are seen in the evening, then all objects identical with either Mars, Phosphorus, or Jupiter are seen in the evening.

(23) If all these are seen in the evening, then all those are seen in the evening.

In this way, it might be argued, an appearance is created that (20) (and indirectly (19)) can have *a posteriori* attitudinal contents, though in reality this appearance is created by the fact that, in the circumstances of the relevant apparent utterance of (20), it is elliptical for (22), (23), or some other similar not logically true sentence, which can in fact have *a posteriori* attitudinal contents. However, even assuming for the sake of argument that the relevant apparent utterance of (20) is in fact elliptical for something, it's hard to see why it should be elliptical for an utterance of a sentence with two syntactically different complement noun phrases instead of a sentence with two occurrences of the same noun phrase. For example, an utterance of (21) is fully intelligible if I accompany the utterance of the first 'all these' with an action of pointing to Mars, Venus, and Jupiter in the evening sky and the utterance of the second 'all these' with an action of pointing to them in the astronomy book. Why should my utterance of (20) be elliptical for (23) and not for (21) (if it's elliptical for anything at all)?

In fact, however, I think it's reasonably clear that my utterance of (20) is not elliptical for any sentence of the sort, and that this is shown by underdetermination and independent intelligibility arguments analogous to those of Section II. There is no apparent principle determining that my apparent utterance of (20) should be elliptical for (21), or for (22), or for (23), or for any one in particular of the many other options that we might think of. And even if some such sentence was determined as preferred for some syntactic or psychological reason, it could hardly be seen as a required intermediate step in the determination of the real content of my utterance. Intuitively, this utterance (and presumably any felicitous utterance of (20)) is intelligible, and hence has a knowable content, independently of a previous knowledge of the meaning and hence the content of 'objects identical with either Mars, Phosphorus, or Jupiter', 'these', 'those', and other noun phrases that are candidates for ellipsis on the objector's view. Intuitively, we can easily imagine vicissitudes in which Mary does in fact not know the meaning of those noun phrases but my utterance of (20) is still perfectly intelligible to her, and makes her learn something she could not have learnt *a priori*. On the objector's view such independent intelligibility, and hence Mary's ability to learn something *a posteriori* with just an independent understanding of my utterance of (20), are impossible, against the intuitive view.

I conclude that (19) with its content is neither *a priori* nor *a posteriori* by itself. But (19) with its content, and any interpreted model-theoretic logical truth containing more than one occurrence of a quantifier, will have some attitudinal contents which are *a priori* and some which are *a posteriori*. If we accept the principle mentioned above, that all the attitudinal contents of an *a priori* sentence with its content must be *a priori*, then all the attitudinal contents of a logical truth with its content must be *a priori*. It follows that some model-theoretic logical truths in the sense of (T4), and even of (T1) and (T1(1)) (in fact many of them) are not real logical truths under typical conceptions of logical truth, which require logical truths to be *a priori*.

A reasonable hypothesis is, however, that all attitudinal contents of model-theoretic logical truths of unexpanded first-order quantificational languages in which the attitudinal contents of the different tokens of the quantifiers are all the same will be *a priori*. Thus a reasonable further constraint, that presumably turns (T1) and (T1(1)) into true theses, is the condition that all the attitudinal contents of the quantifiers must be the same.[31] Then the essence of the preceding considerations is that, if the reasonable (but not uncontroversial) assumptions of this section are accepted, some such constraint on the widely accepted (T1) and (T1(1)) is needed.

References

Bays, T. (2001) "On Tarski on Models", *Journal of Symbolic Logic*, vol. 66, pp. 1701–26.
Church, A. (1936) "An Unsolvable Problem of Elementary Number Theory", *American Journal of Mathematics*, vol. 58, pp. 345–63.
Cook, S. (1971) "The Complexity of Theorem-Proving Procedures", in M. A. Harrison, R. B. Banerji, and J. D. Ullman (eds.), *Proceedings of the third annual ACM symposium on Theory of computing*, ACM Press, New York, pp. 151–8.
—— (1991) "Computational Complexity of Higher Type Functions", in I. Satake (ed.), *Proceedings of the International Congress of Mathematicians in Kyoto*, Springer, New York, pp. 55–69.
Etchemendy, J. (1988) "Tarski on Truth and Logical Consequence", *Journal of Symbolic Logic*, vol. 53, pp. 51–79.
—— (1990) *The Concept of Logical Consequence*, Harvard University Press, Cambridge (Mass.).
Feferman, S. (1999) "Logic, Logics and Logicism", *Notre Dame Journal of Formal Logic*, vol. 40, pp. 31–54.
Gómez-Torrente, M. (1996) "Tarski on Logical Consequence", *Notre Dame Journal of Formal Logic*, vol. 37, pp. 125–51.
—— (1998/9) "Logical Truth and Tarskian Logical Truth", *Synthese*, vol. 117, pp. 375–408.
—— (2000) "A Note on Formality and Logical Consequence", *Journal of Philosophical Logic*, vol. 29, pp. 529–39.
—— (2002) "The Problem of Logical Constants", *Bulletin of Symbolic Logic*, vol. 8, pp. 1–37.
—— (2006) "Rigidity and Essentiality", *Mind*, vol. 115, pp. 227–59.
Hanson, W. (1997) "The Concept of Logical Consequence", *Philosophical Review*, vol. 106, pp. 365–409.
—— (2006) "Actuality, Necessity, and Logical Truth", *Philosophical Studies*, vol. 130, pp. 437–59.

---

[31] Perhaps a similar further constraint about the attitudinal contents of individual constants is also required, if individual constants are directly referential. Recall the comment about '$a = a$' in the preceding footnote.

Kaplan, D. (1977) "Demonstratives", in J. Almog, J. Perry, and H. Wettstein (eds.), *Themes from Kaplan*, Oxford University Press, New York, 1989, pp. 481–563.

Kaplan, D. (1978) "On the Logic of Demonstratives", *Journal of Philosophical Logic*, vol. 8, pp. 81–98.

Kreisel, G. (1967) "Informal Rigour and Completeness Proofs", in I. Lakatos (ed.), *Problems in the Philosophy of Mathematics*, North-Holland, Amsterdam, pp. 138–71.

Kripke, S. (1972) "Naming and Necessity", in D. Davidson and G. Harman (eds.), *Semantics of Natural Language*, Reidel, Dordrecht, pp. 253–355, 763–9.

—— (1979) "A Puzzle about Belief", in A. Margalit (ed.), *Meaning and Use*, Reidel, Dordrecht, pp. 239–83.

McCarthy, T. (1981) "The Idea of a Logical Constant", *Journal of Philosophy*, vol. 78, pp. 499–523.

McGee, V. (1992) "Two Problems with Tarski's Theory of Consequence", *Proceedings of the Aristotelian Society*, n.s., vol. 92, pp. 273–92.

—— (1996) "Logical Operations", *Journal of Philosophical Logic*, vol. 25, pp. 567–80.

Mancosu, P. (2006) "Tarski on Models and Logical Consequence", in J. Ferreirós and J. J. Gray (eds.), *The Architecture of Modern Mathematics*, Oxford University Press, Oxford, pp. 209–37.

Quine, W. V. (1954) "Quantification and the Empty Domain", *Journal of Symbolic Logic*, vol. 19, pp. 177–9.

Ray, G. (1996) "Logical Consequence: a Defense of Tarski", *Journal of Philosophical Logic*, vol. 25, pp. 617–77.

Salmon, N. (1982) *Reference and Essence*, Blackwell, Oxford.

—— (1986) *Frege's Puzzle*, MIT Press, Cambridge (Mass.).

Sher, G. (1991) *The Bounds of Logic*, MIT Press, Cambridge (Mass.).

—— (1996) "Did Tarski Commit 'Tarski's Fallacy'?", *Journal of Symbolic Logic*, vol. 61, pp. 653–86.

Soames, S. (1999) *Understanding Truth*, Oxford University Press, New York.

Stanley, J. and Z. G. Szabó (2000) "On Quantifier Domain Restriction", *Mind and Language*, vol. 15, pp. 219–61.

Tarski, A. (1935) "The Concept of Truth in Formalized Languages", in Tarski (1983), pp. 152–278.

—— (1936) "On the Concept of Logical Consequence", in Tarski (1983), pp. 409–20.

—— (1966) "What Are Logical Notions?", in J. Corcoran (ed.), *History and Philosophy of Logic*, vol. 7, 1986, pp. 143–54.

—— (1983) *Logic, Semantics, Metamathematics*, 2nd edn, Hackett, Indianapolis.

Tarski, A. and S. Givant (1987) *A Formalization of Set Theory without Variables*, American Mathematical Society, Providence, R.I.

Turing, A. (1936/7) "On Computable Numbers, with an Application to the Entscheidungs problem", *Proceedings of the London Mathematical Society*, ser. 2, vol. 42, pp. 230–65.

Warmbrōd, K. (1999) "Logical Constants", *Mind*, vol. 108, pp. 503–38.

Williamson, T. (2003) "Everything", in D. Zimmerman and J. Hawthorne (eds.), *Philosophical Perspectives 17: Language and Philosophical Linguistics*, Blackwell, Oxford, pp. 415–65.

Zalta, E. (1988) "Logical and Analytic Truths that are not Necessary", *Journal of Philosophy*, vol. 85, pp. 57–74.

14

Truth on a Tight Budget: Tarski and Nominalism

Peter Simons

Ku pamięci Czesława Lejewskiego (1913–2001r.)

TARSKI'S METHODOLOGICAL PLATONISM

It is a truth widely acknowledged that the single work from the Golden Age of Polish philosophy and logic that is most impressive in itself and most influential on subsequent philosophy is Alfred Tarski's monograph on the concept of truth in deductive science.[1] It is less well known that Tarski's doctoral supervisor Leśniewski was unhappy with the truth monograph. The reasons are not hard to find. Leśniewski was a convinced nominalist, and in his early years Tarski showed some signs of being willing to follow this position, but the truth chapter is saturated with references to and uses of abstract entities of many kinds. The metalogical description of languages chosen by Tarski opts to treat expressions as types or (what Tarski thinks is the same) as classes of equiform tokens. Leśniewski on the other hand always treated expressions as individual inscriptions. Tarski admits in a footnote that his treatment is an instance of "a widespread error which consists in identifying expressions of like shape"[2] and between definitions 17 and 18[3] explains at tortuous length why it would be very inconvenient to work with tokens. But it would be wrong to assume that Tarski embraced types merely for convenience. The assumption that a sentence has infinitely many consequences, which is an integral part of Tarski's metalogic,[4] is patently false for a language or theory whose expressions consist only of actual tokens; others of Tarski's results require types as well.

In his metalogic Tarski employs a certain amount of set theory. In other circles this would have raised no eyebrows, but Leśniewski was on record for his opposition to standard set theory[5] and had fallen out copiously and publicly with Sierpiński and

---

[1] Tarski 1983, 152–278.   [2] Ibid., 156 n.   [3] Ibid., 182–5.
[4] Ibid., 174–5, 182.
[5] Leśniewski 1992, 207 ff, which was first published in 1927. Because of the disagreement with Sierpiński and others, Leśniewski withdrew from the board of *Fundamenta mathematicae*.

other Polish mathematicians on the subject. Leśniewski considered that neither the null set nor singletons were intelligible. The merest use of set theory by Tarski meant he was offending against his teacher's sensibilities. It seems Tarski did not abandon Leśniewski's position without some pangs of conscience, but abandon it he did, because despite the admirable and indeed unprecedented exactness of his metalogical descriptions, Leśniewski's system constituted for Tarski a "thoroughly thankless object for metamathematical investigations."[6] Tarski's pragmatic or opportunistic acceptance of platonism—we might call it *methodological platonism*—may be seen as one of his first steps in emancipation from the cramping restrictions of Leśniewski and as part of Tarski's lifelong attempt to make logic appeal to mathematicians. So it may, but in private and to the end of his life Tarski continued to profess his philosophical preference for nominalism and against platonism understood literally.[7] There remained two hearts beating in that one breast. We may admit that platonism is convenient, especially for metalogic. But, to modify a saying of Queen Elizabeth, the word 'convenient' is not to be used with ontologists.

Further, when we look at the example Tarski actually gives of a theory of truth, it is not about individuals but the Boolean algebra of classes. Though these classes would usually be understood as sets, they can be given a nominalistically acceptable interpretation as a theory of pluralities rather than of sets, an interpretation which as a student of Leśniewski Tarski would have known. But other theories are less nominalistically innocuous. As Tarski makes clear, for deductive theories climbing somewhat higher in the type-theoretic hierarchy—which would indeed include first-order predicate calculus—the objects involved in the definition of truth will be of higher order than individuals or pluralities of individuals. Typically these are dealt with semantically as sets or functions, both of which are platonist conceptions.

So what if one puts aside the Papal finery of mathematical and linguistic platonism for the Franciscan hair shirt of nominalism? What does a theory of truth look like that is on the one hand nominalistically acceptable and on the other recognizably Tarskian?

Platonism of linguistic entities—including Tarski's truth-bearers, sentences—is of multiple value to logicians. It guarantees that such concepts as logical truth and falsity, validity, consequence, inconsistency, compatibility, etc. are well defined in a simply statable way which is independent of the vicissitudes and limitations of space, time, matter, and circumstance. It underwrites the objectivity of logic. This indeed was the principal motivation of such logical platonists as Bolzano, Frege, and Husserl. The ready assumption of the existence of linguistic types and of sets, relations, functions, and structures all expedite metalogic. This is first nature to mathematicians, for whom Tarski's initial scruples might seem fussy, quaint, boring, wrongheaded, or pathetic, depending on tolerance or perspective. It is not my intention here to persuade platonists that they are wrong but rather simply to invite everyone to a little

---

[6] This critical passage was left out of the English translation. I am grateful to Göran Sundholm for calling it to my attention. Sundholm has conjectured that Tarski's growing willingness to use set theory was probably bolstered by reading the important Zermelo 1930.

[7] On Tarski's private nominalism, see Suppes 1988, 81.

educative slumming in the messy but reassuringly real world of the nominalist. The purpose is of course not mere self-denial or puritanism for its own sake, but because I believe that, for the sake of mathematical effectiveness, Tarski's accommodation with platonism ends up compromising the best and true account of truth. The convenience of the mathematician or logician not being paramount, can we do better?

THE LIMITED GOALS OF A TARSKIAN THEORY

Tarski announces in his first sentence "This chapter is almost wholly devoted to a single problem—*the definition of truth*."[8] This may lead us to expect something in the grand historical tradition from Aristotle through Aquinas to Russell, and Tarski's reference to Aristotle's classic formulation reinforces this expectation. In fact, when the chapter is examined more closely, it turns out that the aim is modest and subject to considerable restrictions. Firstly, because of the semantic antinomies, no vernacular language or any other that is semantically open, i.e. that contains its own metalanguage, is treated. Languages containing vague expressions or indexical, context-dependent expressions are simply excluded from consideration. The languages for which Tarski sets out to define truth comprise a very restricted range: they all are formally precisely defined languages used in the deductive sciences, with exact and unambiguous conditions of grammaticality. Not only this: because of the limitation results that Tarski outlines in Section 5 and revises in Section 7, and which are the semantical analogues to Gödel's incompleteness results, truth will be definable for a language only within a metalanguage of higher order than the object language. Finally, and this is where many commentators on Tarski come unstuck, what he calls a *definition* of truth for a given language is not a definition in any intensional sense but simply a way of *delimiting* the true sentences from the false. This becomes patent when Tarski points out that in a finite language a mere list would suffice.[9] No list will give a traditional-style definition of truth, saying what truth is: it will simply tell which are the true sentences. Like his teachers Leśniewski and Kotarbiński, Tarski was deeply suspicious of intensional notions, and so the most he would be looking for was a way of delimiting truths. Nor is there any treatment of modal notions. In this he was of his time, and many would now take this as an unacceptable restriction. Modal notions might perhaps be susceptible to a non-platonist treatment, but the sort of scruples which prevent a nominalist from embracing abstract entities are sufficiently similar to those about possible worlds and other possibilities for it to be preferable to eschew modal assistance.

It is true that work on the theory of truth subsequent to Tarski has focused on removing Tarski's limitations as far as possible. Truth-theories have been proposed for vernacular languages, even such as permit the formulation of semantic antinomies, at the cost of relinquishing bivalence. Ways to deal with modality, vaguenesss, indexical, and contextual features of language have been proposed within the general vein of Tarski-inspired model theory. And semantic theories have been proposed which

[8] Tarski 1983, 152. [9] Ibid., 188.

attempt to define—as Tarski's expressly do not—concepts of intension, sense, or meaning. Without wishing to detract from the interest of such enterprises, let me say that I shall make no attempt in this chapter to go even one step in their direction. There are two reasons for this. Firstly, I think that Tarski's limited aims—despite the fact that they have been much derided in subsequent literature—were wise not only for their time but also in themselves. But secondly and more importantly, Tarski's very limited languages already pose sufficient of an obstacle for a nominalistically acceptable theory of truth, without bringing in all the additional complications. If semantical nominalism cannot even be made to look plausible here, what hope is there for it elsewhere?

METHODOLOGICAL RESOURCES

Tarski's teacher Leśniewski was perhaps the only person who fully carried through the nominalist ideal in prescribing how to actually *do* logic.[10] Leśniewski's aims were to formulate logical systems of sufficient expressive and deductive power to serve as antinomy-free foundations for mathematics. The obsessive and finicky detail with which this had to be carried out—and which is guaranteed to put off practically anyone not equally fanatical about nominalism—have been described elsewhere.[11] But as we saw at the outset, Leśniewski did not like Tarski's way of defining truth, and did not himself offer a nominalistically acceptable alternative. Perhaps, like Frege, he thought truth is too basic to be definable. Be that as it may, we need not follow him. Fortunately, for the purposes of giving a nominalist account of truth, we do not need to be as ambitious as Leśniewski was for logic: what we need to be able to say, for any given token sentence of a suitable language, is how it comes about that it is true, if it is true, or false if it is false. If that can, in principle, be done without resorting to platonist or modal realist resources, then Leśniewski's negative reaction to Tarski's theory of truth can be retrospectively justified. But what resources may we use, and what may we not use?

Truth-bearers: tokens, not propositions or types

Both propositions and sentence types are abstract entities (the latter of course much less controversial and mysterious than the former). But without abstract propositions or expression types, both equally anathematical to the nominalist, what should we take for our truth-bearers? While a proper account of meaning and truth, especially one which takes account of indexicality, should I believe take as primary truth-bearers our mental acts of understanding and meaning,[12] there are good reasons not to follow

---

[10] I have described Leśniewski's nominalism in logic and metalogic in Simons 2002. It is not clear to me to what extent the work of Leśniewski's contemporary Leon Chwistek, despite his own claims (Chwistek 1948, 22) constitute a successful logical nominalism.

[11] At even greater length in Luschei 1962, in much sketchier but more reader-friendly terms in Simons 2002.

[12] Cf. Simons 1992, 2003.

that line in this context, even though it is compatible with nominalism. Firstly, mental acts are private and difficult to evaluate intersubjectively. Secondly, and partly for that reason, they are relatively more mysterious and less well-understood and well-regulated than public tokens. There may perhaps be a grammar of thought, but if there is, we know precious little about it, and it would be unduly speculative as well as difficult to build a simple account of truth on something as complex. To take mental acts as truth-bearers is to revert (correctly) to the position of the Brentano school. The choice of publicly accessible items as truth-bearers was a conscious decision in the early history of the Lvov-Warsaw School, moving away from the mental truth-bearers, namely acts of judgement, of their Brentanian forbears.[13] Electing to follow the Polish choice here is a matter of expedient, not of ultimate principle.

Plurals, not sets

While sets are abstract entities, and so unavailable, we need not scruple to use plurally referring names and other plural terms. Talking of pluralities incurs no additional ontological commitment beyond that to the several individuals comprising the pluralities. When I say of the books on my shelf that they are valuable I am not committed to an additional item over and above the several books. Saying that they are valuable may mean that each one is valuable (severally), or it may mean that between them they are valuable (jointly). If it is the latter, we see that we can state truths with pluralities as logical subjects, and so justify the use of plurals pragmatically. Since some truths statable with plural subjects, such as predications of number, are theoretically important, there are considerable advantages to allowing plural terms into our vocabulary alongside singular and empty tems. This was in any case Leśniewski's practice in his logical system of Ontology.[14]

Extension, not meaning

There is a whole side to language and semantics that we are expressly avoiding. It has to do with sense or meaning. Tarski skirted around this too, as was standard among Poles then. His account of truth *presupposes* that the expressions for which truth is defined, and their parts, are *already* meaningful, whatever that consists in. There is no attempt to define meaning via truth or other extensional notions, as has been attempted since.[15] We should not *expect* a theory of truth to be of much assistance in accounting for meaning. Nor should we expect a theory of truth to give us any help in explaining how we are able to make true predications in a complex world. Consider for example the predicate 'is an electron' (ignore tense). It is believed that the known universe contains of the order of 10^{79} electrons. That's a lot of electrons. Clearly, someone who understands what 'electron' means cannot possibly have any kind of personal acquaintance with every electron in virtue of this understanding. Yet of any actual item in the universe, presumably it either is an electron or it isn't.

---

[13] Cf. Rojszczak 2005.   [14] See Lejewski 1958, Simons 1983.
[15] Misguidedly, in my view.

By having the noun 'electron' with associated criteria for application, any one of us is able to talk about 10^{79} or so of them, simply because the universe plays ball by providing all those things which are like that (as well as all those others which aren't). Now for the theory of truth we shall, like Tarski, consider only matters of extension, not intension, sense, or meaning. Meaning, content, sense, criteria for application, and so on are important topics in the philosophy of language, but we are simply not going into that issue. That means that the nominalistically available facts for giving our account of truth need not be finite or humanly graspable. The very possibility that the universe might be infinite makes this inevitable. Whether it is infinite or not, it is in fact very big and so a realistic account of how human beings learn and manipulate meaning must work with something other than extension.[16] In this regard the platonist account of extension is actually harmful, since it suggests there is a manipulable unit—the extension of an expression considered as a set—that we might be supposed to have in mind when we understand an expression. Not having sets at our disposal is actually conducive to a more adequate account of meaning than having them conveniently around.

No functions

Whether functions are primitive and *sui generis* as in Frege and Church, or defined via relations or sets, they are abstract, and so not allowed. So truth-functions, propositional functions and sentential functions are *a fortiori* inadmissible.

No satisfaction

Tarski famously defined truth via satisfaction. Satisfaction is a relation between a sentence (closed or open) and an infinite sequence of objects. A sequence is a mathematical object, best construed as a function from the natural numbers 1, 2, . . . into some set. As such it is abstract, so unavailable. Tarski's infinite sequences are a convenience only: in practice, and with a bit more effort, finite sequences can be used. But finite sequences are still either ordered multisets or functions, and so abstract. The effect of satisfaction will have to be obtained in another way.

No truth-values

Truth and Falsity, or the True and the False, are, taken literally, abstract objects. That indeed is how Frege conceived them. They cannot be concrete objects, which conceivably might not exist, but must be abstract. Frege thought they were "logical objects." To be fair, Tarski did not literally set out to define truth the abstract quality, or give a definition of the True, the abstract object. He set out to delimit the extension of predicates such as ' . . . is a true sentence of language L', for suitable L. We need do no more.

---

[16] As argued in Simons 1991.

No objects of higher type

The neutrally termed "objects" which in Tarski's account of truth are members of the satisfaction sequences are, with one exception, objects of order higher than individuals in a type hierarchy. The only exception would be a calculus of names, where the only variables are names, and the logical calculus contained only constants otherwise. Tarski considers the calculus of binary relations as another example. Relations, like properties or sets, are abstract. Further up the type hierarchy the objects are no less abstract. That is why the unifying notion of satisfaction is unavailable to a nominalist. The way in which expressions other than those from the category of name figure in the account of truth has to be different from the way Tarski portrays. Even the standard semantics for first-order predicate logic—a theory widely regarded as ontologically harmless—are up to their ears in commitment to abstract entities, and so must be done differently. The higher the order of the language in question, the more involved will be the way the expressions work. But at no stage will we be allowed to invoke, refer to, or quantify over higher order entities.

What can we use?

After this perhaps discouraging list of things we aren't allowed to do, let's consider what we can do. We can talk about concrete physical tokens of a language, actual inscriptions. Assuming the language is well-regulated in the manner Tarski requires, we can discern the parts of expressions, down to their smallest or atomic parts, tell to which syntactic category these parts belong and how they are syntactically combined to form complex expressions, and what the categories of these are. The categories are not sets: all we need to know is when two expressions are equicategorial. We can say which expressions are equiform to which other expressions. We can if need be actually point to particular expression tokens and illustrate how they are put together syntactically from examples. This is exactly how we proceed when learning a natural language anyway.

REQUIREMENTS ON THE LANGUAGE

It must be categorial

We require the language to be a *categorial* language in the following sense: every expression in it is *either* a sentence *or* a name *or* a functor expression *or* an operator expression. The notion of syntactic category for functor expressions is familiar. An expression is a functor of category $\alpha \langle \beta_1 \ldots \beta_n \rangle$ if, when combined according to the language's grammar with n expressions of categories $\beta_1, \ldots, \beta_n$ respectively, the resulting complex expression is of category α. A (token) functor need not in fact be so combined since it may be an argument of a functor of higher category, but were it to be so combined it would work in that way. Whether a functor is functioning as a functor or as an argument will be clear from the context of the sentence in which it occurs, which we assume to be well formed, since the question of truth does not arise for an ungrammatical complex.

An *operator* is a variable-binding expression, such as the quantifiers, description, and lambda operators in logic, the set abstraction operator in set theory, the generalized sum and product operators in arithmetic, the differential and integral operators in analysis, and so on. In use, an operator requires three elements. Firstly we have the operator expression itself. Then we have the sequence of variables which it binds. There may be one or more, and if there is more than one they may be of the same category or different categories. In general we shall allow variables to be of any category that a constant can have. Finally, the operator operates *on* an expression which we may call its *matrix*. For example in the case of the indefinite integral the matrix is the expression for the function over which is integrated, including its bound variable. For example, in the token displayed here—$\int \sin(x) \, dx$—the operator symbol is the integral sign, the binding variable is the second x, the matrix is the indefinite function expression, the sin(x), and the 'd' is a piece of punctuation prefacing the binding variable and separating it typographically from the matrix. Unlike functors, which take arguments (one or several) and produce an expression, adding one extra layer of grammatical complexity, an operator, by binding variables, can reach arbitrarily deep into the layers of grammatical complexity of its matrix. I have written elsewhere about how to provide a suitable extension of categorial grammar to cope with the additional complexities introduced by operators over and above functors.[17] It is absolutely necessary to be able to deal with operators as well as functors, since any logical language of any interest has them. Tarski's satisfaction idea is constructed expressly to be able to deal with variable binding such as that found in quantifiers and set abstraction. Since satisfaction is one of the things that gives nominalists most trouble, it is important to see a way clear to handling operators. I have previously outlined how this can be done in a platonist way, essentially by invoking functions and functional abstraction in the vein of Church,[18] so it is time to show how to bring the analysis down to earth.

We shall require all expressions involved *other than variables* to be meaningful. Being meaningful is not something that I shall attempt to give a theory of here, any more than Tarski did. That is not to say it should not be attempted, but it is far too big an issue to even scratch in an essay. We can excuse this lacuna by saying, as did Tarski, that the notion of truth only makes sense for sentences where the parts (other than variables) are meaningful.[19] Otherwise it does not arise at all. As to variables, these can be of two types: free or bound (in the context of the whole sentence). Both kinds of variables lack a meaning, that is, a determinate meaning of their category. Rather they "go along with" all meanings of that category, in an indeterminate way. Struggling to present the same idea, Frege says that variables *indicate* items of a particular category.[20] Free variables actually prevent a sentence from having a determinate meaning and hence being true or false, so we shall consider them later. Bound

---

[17] Simons 2006.   [18] Simons 2006, with some ideas taken from Simons 1985.
[19] Tarski 1983, 166 f. In this Tarski was following his teacher Leśniewski.
[20] Frege 1893, 31–2.

variables within the scope of an operator are different. They have two roles: they mark a place in a grammatical structure as being reached or operated on by the governing operator, the place being of suitable category, such as might have been occupied by a constant of that category. And they indicate by their form, or rather, by their equiformity or otherwise to other equicategorial token variables, how in evaluating a given matrix semantically the operator works on its matrix. This will become clearer below.

It is not as such part of the theory of truth how expressions get to be meaningful. All that matters is that, being meaningful of their category (which therefore excludes variables), they work in a certain way to determine truth. All meaningful and equicategorial expressions work in the same general sort of way, which is the semantic rationale behind their being of one category. And no two expressions of different categories work in the same way. The categories are semantically distinct in how they work as well as syntactically distinct in how they build grammatical complex expressions.

Every token expression is required to belong to one and only one category, and equiform tokens are required not only to be equicategorial but also to be equisignificant, or mean the same, within the context of any single sentence. This is simply the requirement of no ambiguity. For variables we can avoid problems of multiple occurrences within different scopes by insisting that a variable be bound only once (though there may of course be more than one equiform token of that type within the scope of the operator). A gain in clarity without loss of expressive power is incurred thereby.

If sets are not available to *be* extensions, we need other ways to deal extensively with expressions of all kinds, since our treatment is going to be extensional. To have a word for it, we shall say that expressions *signify* in a certain way, and that way constitutes what in platonist talk is called their extension or as we shall say, their *signification*.

HOW TO SIGNIFY

Sentences

A sentence contains only expressions with constant meaning, and, if it contains any variables, these are all bound. Apart from that, and depending on syntax, it may contain punctuation marks such as brackets, commas, dots, which are not meaningful but help to maintain an unambiguous syntactic structure. So a sentence is in line for having a truth-value, or, to avoid platonist talk, for being true or being false. From the point of view of giving an extensional account of truth, nothing else needs to be taken into account. In particular the sentence's meaning, and the state of affairs it describes, are not taken into account. In this chapter we shall consider only sentences which actually are either true or false, leaving truth-value gaps, vagueness, indeterminacy, and such like aside. This is a simplification over against natural languages but it is once again in Tarski's spirit. From the point of view of an account of truth it is not *that* a sentence is true that is important so much as *how* its being true depends on the way its parts signify. So for sentences, to signify is to be true or to be false, and these are different ways of signifying. But they cosignify, are coextensional, iff they are materially equivalent, so there are only two ways of signifying for sentences.

Adequacy

Tarski required a truth theory to be formally correct, i.e. consistent, and materially adequate.[21] The latter meant that the theory of truth for L expressed in a metalanguage for L, M(L), must entail every instance of the following schema (expressed in M(L)):

T X is a true sentence of L iff p

where 'X' stands in for an expression designating a sentence of L and 'p' stands for the metalinguistic translation of the sentence 'X' designates. Let us *designate* this translation here by 'P' (this is a schematic name of the metalinguistic sentence). 'Translation' simply means that every well-formed part of X corresponds one-to-one to a well-formed part of P with the same category and extension, from atomic parts through well-formed complex parts to the whole. Call such sentences as X and P *extensionally isomorphic*. They are alike in syntactic structure and their corresponding parts are coextensional throughout. Since we shall not be presenting a formal theory closed under logical consequence—which is after all a platonist conception—we shall merely require that when stating under what conditions a displayed token sentence of L is true, these must be materially equivalent to a sentence of our metalanguage extensionally isomorphic to the target sentence. T then drops out as a consequence since coextensionality for sentences *is* material equivalence. The correct way to understand T is not as a convention (despite Tarski's name 'Convention T'), but as a *requirement*. Our adequacy requirement is not the schema T itself but that every description of how a sentence signifies be compatible with its instance of T. Tarski required each T-instance to be derivable in the metatheory, which since he has an infinite language is beyond our reach.

Names

Sentences are not names. Sentences are true or false. Names denote. In standard predicate logic they all denote exactly one individual. In real life, and in more liberal logics, they may denote more than one individual, or no individuals at all. Whatever the extent of liberality in a language, we shall require nominal bivalence:

NB   for every individual x and every name n: n denotes x or n does not denote x

To say how a given name signifies, it is necessary and sufficient to say *which* individuals it denotes and which it does not. If we have a prior syntactic restriction to the effect that all the names are singular (as in standard logic) or singular-or-empty (as in free logic), this task is easy to carry out: we simply say which individual a name denotes, or if it is empty we can state that it denotes no individuals. The rest can go

---

[21] Tarski 1983, 187 f.

without saying. That is why denotation of names appears such a simple thing. If we say that 'Bertrand Russell' denotes that well-known philosopher and logician, we do not need to go on and say that the name does not denote anything or anyone else. So it is natural to say that philosopher *himself* is the name's extension. But this simplicity is deceptive. If a name is empty, there is no individual we can cite as what it denotes. Further, if we allow plural names, we may need a longer list of individuals it denotes. 'The authors of *Principia mathematica*' for instance (assuming it is a name, which I am happy to do) denotes Russell and also denotes Whitehead—and no one else. But the name 'the things that are not authors of *Principia mathematica*' denotes everything and everyone in the world *except* Russell and Whitehead. In that case to list everything it *does* denote would be a long, possibly an infinite task. In between, there are names like 'electron' (I assume, with traditional grammar, that this is a name) which denote a huge number of things but also do not denote a huge number of things. We are never going to be in a position to actually write down a complete list of the denotata and a complete list of the non-denotata, because the universe is too large.

But now—and here is the important point—*it does not matter* that we are unable to do so. All that matters is that *as a matter of fact* a particular name denotes certain of the individuals, and fails to denote all the rest. How it gets to do so is not our business, *qua* extensional theorists of truth. And our condition of coextensionality for names is simplicity itself:

NCoex   for any names n and m: n is coextensional with m iff for all individuals x: n denotes x iff m denotes x

If there are K individuals in the frame for being denoted, then there are K extensions if we are only allowed singular terms, $K + 1$ if we allow empty terms as well, and a whopping 2^K if we allow plural terms as well. Again, it does not matter that we may as a matter of practical necessity be unable to list all the objects a name denotes. What matters is that—however it gets to do so—it does denote them, and doesn't denote the rest. All the cases of its denoting or not denoting an object take the form of

Den n denotes x

being true for those objects x that it denotes, and false for those it doesn't. Thinking of standard semantics, it is tempting to gather all these cases up together, as a set (of the things denoted) or a conjunction (of the cases where it denotes). The temptation should be resisted. The former requires a platonistic object, the latter might require infinitely many metalinguistic names, indeed potentially a name for every object. We don't need either. It merely has to be the case that for any individual, either the name denotes it, or it does not, and names which denote the same individuals are coextensional. That's all. Note that since we require equiform expressions in a sentence to signify alike, i.e. be coextensional, any two equiform name tokens in a sentence are guaranteed to denote the same individuals. This is something that goes without saying in a semantics based on expression types, but in fact in everyday language it is not respected. For instance in American politics it is true if slightly perverse to say 'President George Bush is the father of President George Bush'.

Functors

A functor is an expression which combines with other expressions of given categories to yield a complex expression of given category. How the combining is carried out—by what grammatical means—we need not go into: there are numerous possibilities. If the input categories are $\beta_1 \ldots \beta_n$ and the output category α we designate the category of the functor as $\alpha \langle \beta_1 \ldots \beta_n \rangle$. The simplest cases of functors are sentential connectives, which have categories S⟨S⟩ (unary), S⟨SS⟩ (binary), and so on; function expressions, which have categories N⟨N⟩, N⟨NN⟩ and so on; and predicates, which have categories S⟨N⟩, S⟨NN⟩, and so on. But functors in general allow the output to be any simple or functor category, and likewise each of the inputs. For example a good way to categorize quantifier words such as 'every' is S⟨S⟨N⟩⟩⟨N⟩.

We saw how sentences signify—they are true or false—and how names signify—for each individual, they denote it or they do not. For both of theses categories there is a range of ways an expression can signify. Here is how a functor signifies. If its first argument signifies in a certain way, appropriate to its category β_1, and its second argument signifies in a certain way, likewise, and we run through its arguments in turn to the last β_n, and this signifies in a certain way, then the resulting complex signifies in a certain way appropriate to the category α. Each of these conditions has the form

If the first argument means thus, and . . . , and the last argument means thus, then the whole means thus.

The 'thus' stands in for a description of a way an argument or the resulting complex signifies. A functor signifies in a determinate way if each of these conditions is determined for every combination of significations for its arguments. The whole list of such conditionals, were it expressible, would indicate the extension of the functor. For small numbers they can actually be listed, or, more conveniently, tabulated. A standard truth-table is nothing other than a tabulation of such conditions for sentential connectives. With large domains of individuals, neither listing nor tabulation is practicable, but what matters is not that there be such a listing or tabulation, but that the conditions obtain which warrant it in principle. How they do so—how a functor gets to signify the way it does—is again not our business. All that matters is *that* they do. We can envisage the functor's signification as given by a virtual table, with input combinations on the left and output on the right, and one row per input combination. When a functor expression actually occurs in a token sentence, and works as a functor, i.e. takes arguments, then the way the complex of it and its arguments signify derives from the functor's total signification by simply selecting the condition given by the particular arguments' signifying as they do. Only one out of the many "lines" of the functor's virtual table applies: we "read off" the result. Knowing the functor signifies in a determinate way, we know in advance that there will be a determinate output way of signifying. So if the expression is 'son of' (mentioning both parents), the output value is a complex name denoting the sons of the persons listed, and denoting nothing if between them they have no sons. If the expression is 'loves' then the output value is a sentence which is true if the subject denotes someone who loves who or what the object denotes, and false otherwise. So the semantic effect of the application of a functor expression to a number of arguments is that of table lookup. Functors in general may have arguments and outputs of any category, and may be nested to arbitrary depth, but the way they signify and they way they form complexes with a given signification work in the same general way each time. If we have a sentence consisting solely of functors and atoms, then the whole sentence's truth-value, whether true or false, is determined by the ways in which each of the simple parts signifies and the way in which they combine together, directly or in stages, to yield the truth-value for the sentence.

Consider for example equations of the form $a + b = c + d$ over the natural numbers,[22] such as '$27 + 42 = 31 + 38$'. There are seven atomic parts to this sentence: four numerals (names of numbers), two tokens of the addition functor, of category N⟨NN⟩, and the equals predicate, of category S⟨NN⟩. We can compute the truth-value in any given case by looking up the value of the sum (in the infinite virtual table for addition) for each given pair of numbers, comparing the results, and evaluating the equation as true if the results are the same and false if they are not. Pocket calculators work on essentially this principle: inputting values for the atoms, performing computations tantamount to table lookup (but not actually being table lookup), storing intermediate results, and outputting a final value. In this case the intermediate results stored are two tokens of some sign representing the number 69, so the whole sentence gets evaluated as true, whereas the sentence '$2 + 5 = 7 + 3$' is evaluated as false.

It is to be stressed once more that the extensive—tabular—understanding of the signification of functors does not correspond to our way of understanding expressions for them, with perhaps a very few exceptions (the truth-table for the Sheffer stroke, perhaps?) But if a functor signifies in a determinate way—no matter how it gets to do this—then it *will* have a determinate output for any input: its extensional table will be complete.

Operators

Operators—variable-binding expressions—are more difficult to deal with semantically and were the main reason why Tarski introduced sequences and satisfaction. A side-effect of this was to give a way to assign a range of semantic values to open sentences, that is, sentences containing free variables. We shall attempt to cover the same ground in a nominalistically acceptable way.[23]

Whereas with a purely functorially constructed expression we can work inside out, or bottom up, with operators we have to work outside in, or top down. Suppose we are given an expression whose primary or outermost expression is an operator. There

---

[22] That the example is face-value platonistic is unimportant: it is selected only for ease of understanding.
[23] The basic idea of the syntax and semantics is in Simons 2006: we are here looking at how the semantics work nominalistically.

are three parts to how it works in this context. Firstly there is the operator itself: which one it is, how many variables of what categories it binds, what kind of tokens these variables are, what category of matrix it governs, and what category the result is. For example, the universal quantifier binding one nominal variable has a sentence as its result and a sentence as its matrix. It differs from the existential quantifier only in which operator of this kind it is: the two signify differently. It differs from a definite description operator in that the result of using the latter is not a sentence but a complex name. It differs from a universal quantifier binding a sentential variable in the category of variable it binds.

Let such an expression be of the form $Ov_1 \ldots v_n M$, where O is the operator itself, $v_1, \ldots, v_n$ are the variables (which may be one or more in number and if the latter, may be of one category or of several, but must all be typographically distinct), and M is the matrix. If α is the category of the result, $\beta_1, \ldots, \beta_n$ the categories of the variables, and μ the category of the matrix, we may notate the category of O itself by $\alpha \backslash \beta_1 \ldots \beta_n / [\mu]$. Thus a nominal universal quantifier of the sort discussed above has category S\N/[S], the description operator N\N/[S], and the sentential quantifier S\S/[S].

As was pointed out above, the characteristic feature of operators is that because they bind variables free within their matrix, these variables may be embedded to any finite structural depth within the matrix. There is thus no single way in which the variables combine with the rest of the matrix, unlike in the case of a functor. This affects how the operator determines the semantic value of the output. Assume that M contains no other free variables than those bound by O. In other words the remaining parts of M have determinate signification. How the variables affect the result depends on where they bind within M as well as what the rest of M is like. A binding variable matches bound tokens within the matrix: there may be more than one such token. For example in the quantifier sentence

$$\forall x[Fx \rightarrow \exists y[Rxy \wedge Ryx]]$$

both quantifiers bind their variables with more than one token within their respective matrices (F and R are here taken to be constants).

Suppose M is our matrix, of category μ, and $v_1, \ldots, v_n$ the variables occurring in it. In any given token the matrix will be given, and finite. Since it contains free variables, M will not itself have a determinate signification, but will have a range of significations which are determined by its structure, parts, and the occurrences of variables within it. So let $v_1, \ldots, v_n$ be assumed given temporary *fixed* significations, apt for their categories. For example if v_1 is nominal and v_2 is sentential, v_1 might denote Russell and Whitehead while v_2 is true. Note that for each equiform occurrence of a variable the temporary value it is given is the same for each token. With these significations fixed, since the rest of M is determinate, the result is determinate, apt for category μ. For other values of the variables, i.e. significations within their allotted ranges, M will typically yield a different value from the range for μ. For every combination of the values of the variables taken in turn, M will yield a determinate value, since the rest of M is determinate, and by fixing for each turn the values for the variables we get a fully determinate expression. The effect is as if we had tabulated the signification for a functor of category $\mu \langle \beta_1 \ldots \beta_n \rangle$. Given the effective signification of the matrix, the operator O then determines how the result of category α signifies. For example, a sentence of the form $\forall xM$ will be true if no matter how the variable x signifies, it yields a truth from M, and will be false if at least one way x can signify yields a falsehood from M. That is precisely how the universal quantifier signifies.[24]

We can imagine the semantics working as a three-part process: input a combination of values for the variables to M, get the result, and note it. Do this for all combinations of input values. This gives an effective extensional signification for the matrix M with variables worked out "on the fly." Then apply O to the result and use table lookup to see what value results. Of course in reality no such procedure is followed, but the logical order is as described. This means that from a semantic point of view, the distinctive work done by variable binding operators is done in the input-and-tabulate phase. This work is done in the same general way each time and so could in principle be done by a single family of operators which simply perform the input-and-tabulate task. These are λ operators. Normally λ is considered as a single operator in a type-free language, or else as a syncategorematic operator able to take different types of variables as required. But in a language with a categorial syntax λ should be considered as a family of operators working in analogous ways, each one of a fixed category according to the categories of its variables and matrix. Notationally nothing is lost by using just the one letter provided it is remembered this is an abbreviation. A λ taking variables of categories $\beta_1 \ldots \beta_n$ and operating on a matrix of category μ has category $\mu \langle \beta_1 \ldots \beta_n \rangle \backslash \beta_1 \ldots \beta_n / [\mu]$. It takes a matrix and some variables and yields the functor obtained from the matrix with those variables binding into those positions. We can then simulate the effect of *any* variable-binding operator by letting a suitable lambda do the binding and tabulating and then letting the resulting functor be the argument of a higher-order functor giving the result. For example we can replace the quantifier sentence

$$\forall x[Fx \rightarrow \exists y[Rxy \wedge Ryx]]$$

by a lambda-plus-functor sentence

$$\Pi(\lambda x[Fx \rightarrow \Sigma(\lambda y[Rxy \wedge Ryx])]).$$

The dispensability of other variable binders in favour of a pure abstractor was of course known to Church,[25] but it was also discovered independently by Ajdukiewicz.[26]

The example makes clear that complications must ensue when one operator is within the scope of another, that is, a token occurs literally within the other's matrix. What this involves from a semantic point of view is that the input-and-tabulate function of operators may be nested, with interim values and results being held in store

---

[24] Compare Tarski's account of the meaning of the quantifiers in his dissertation: 'for every (some) signification of the terms...'. (Tarski 1993, 1 n.). This formulation goes back to Leśniewski (1992, 203, 420).

[25] It is especially nicely illustrated in Church 1940.   [26] Ajdukiewicz 1935.

until the final evaluation. If this were genuine procedural evaluation, there is no limit to the complexity that could arise, and a real procedure would go on for ever if the domain of individuals is infinite, but once again the procedural talk simply is an expedient way of thinking about the logical order of dependence in which things are determined, for which mere numbers are of no account.

Any sentence with a quantifier-type operator as its outermost part will yield a truth-value according to the significations, variables, structures, and occurrences of variables within it, provided all the constant simple parts in the sentence are determinate in signification. How many input-and-tabulate complexes there are is determined by the number of variables bound overall. How complex the procedure is depends on how deeply the sentence's structure is nested and also how many arguments given functors have. But the ability to form complex structures of arbitrary length and depth ensures that in the semantics of operators we need to be prepared for any finite number of variables within an expression. That is why Tarski invented satisfaction, which copes with the complexity by a single platonistic expedient. Without the platonism, we have to be more roundabout and devious.

Another benefit of satisfaction is that it enables Tarski to deal semantically with expressions not having a single determinate signification because they contain free variables. The clue on how this can be dealt with is given by the way operators work. Each variable has a category, and for each category there is a range of ways an expression of that category can signify. For an open expression the range of significations is given by the outputs they yield for each combination of input significations for their variables. This range will in general be narrower than that of all the significations for the output category in question. In certain cases it will be confined to a single value. This applies for instance to the tautologies of propositional calculus, which contain free sentential variables, as to the valid formulas of first-order predicate calculus, which contain free predicate variables. It is usual to call such formulas true, or necessarily true, but this is strictly wrong. They have a unit range of significations, which is unaffected by variations of the significations of the variables.

CASE STUDIES

Consider a mini-universe of five people A B C D E and take the sentence

$$\forall x(\text{Adam loves } x \rightarrow \text{Ewa loves } x)$$

where
'Adam' denotes A and nothing else
'Ewa' denotes E and nothing else
and 'loves' has the following partial truth-table (other combinations do not matter)

Table 14.1.

First argument	Second argument	Result
A	A	true
A	B	true
A	C	false
A	D	false
A	E	true
E	A	true
E	B	true
E	C	true
E	D	false
E	E	true

Implication has the usual truth-table. The truth-table for the matrix of the quantifier tabulates as follows:

Table 14.2.

Value for x	Result for this value for '(Adam loves x → Ewa loves x)'
A	true
B	true
C	true
D	true
E	true

So by the meaning of the universal quantifier, which gives an output that is true iff all values of its matrix are true, the specimen sentence is true.

Next consider the following theorem of Leśniewski's Ontology, which states the transitivity of the 'is one of' or 'ε' functor:

$$\forall abc(a \, \varepsilon \, b \, \wedge \, b \, \varepsilon \, c \rightarrow a \, \varepsilon \, c)$$

Here the nominal variables *a b c* are not constrained to be singular. We can show that this is invariably true by a semantic *reductio* rather than running through combinations. We need to know how the functor 'ε' signifies which is as follows:

If '*a*' denotes just one object X and '*b*' denotes X too, whether or not it denotes other individuals besides, then '*a* ε *b*' is true; otherwise '*a* ε *b*' is false.

For the specimen sentence to be false, by the signification of the universal quantifier there should be values of *a* *b* and *c* such that the matrix is false, which means, by the signification of the implication and conjunction connectives, making both '*a* ε *b*' and '*b* ε *c*' true but '*a* ε *c*' false. Since '*a* ε *b*' is true, '*a*' must denote a single individual X. Hence if the consequent is false '*c*' cannot denote X. But since '*a* ε *b*' is true, '*b*' must denote X too, and since '*b* ε *c*' is also true we know '*c*' must also denote X, which contradicts the assumption that '*a* ε *c*' is false. Hence the matrix cannot be falsified so the universal sentence is true.

Consider the following sentential scheme, with two free sentential variables p and q:

$$\forall \phi (p \leftrightarrow (\phi(p) \leftrightarrow \phi(q)))$$

Here the variable 'ϕ' ranges over one-place sentential functors, of which (in an extensional bivalent logic) there are just four: the identity functor, negation, the tautological, and the contradictory. The identity functor applied to a sentence p always has the same truth-value as its argument, negation always has the opposite value, the tautological functor applied to a sentence is always true no matter what truth-value its argument has, and the contradictory functor always gives a false sentence no matter what truth-value its argument has. Let's look at the truth-table when the variable 'ϕ' is given the value of negation $\sim$:

Table 14.3.

p	q	$\sim p$	$\sim q$	$\sim p \leftrightarrow \sim q$	$(p \leftrightarrow (\sim p \leftrightarrow \sim q))$
true	true	false	false	true	true
true	false	false	true	false	false
false	true	true	false	false	true
false	false	true	true	true	false

Here the result is true only when q is true. So the universal cannot be true unless q is true. If we look at either the contradictory or the tautological functor we see it cannot give a true value if p is false, but does give a true value if p is true and q is true, as does the identity functor. Hence by the signification of the universal quantifier for sentential variables the truth-table for the whole open sentence is

Table 14.4.

p	q	$\forall \phi \ (p \leftrightarrow (\phi(p) \leftrightarrow \phi(q)))$
true	true	true
true	false	false
false	true	false
false	false	false

This truth-table is exactly that for conjunction, and enables conjunction to be defined as $\lambda pq[\forall \phi\, (p \leftrightarrow (\phi(p) \leftrightarrow \phi(q)))]$.[27]

The final example is from pre-school arithmetic. Part of what is behind the simple equation '2 + 2 = 4' can be expressed as

$$\forall ab(2(a) \wedge 2(b) \wedge a|b \rightarrow 4(a \oplus b))$$

Here the significations of the *predicates* '2' and '4'[28] and '|' are given by the conditions

'$2(a)$' is true iff 'a' denotes exactly two individuals
'$4(a)$' is true iff 'a' denotes exactly four individuals
'$a|b$' is true iff there is no individual that both 'a' and 'b' denote

while the nominal functor '$\oplus$' denotes all the individuals denoted by either 'a' or 'b' unless no individual is denoted by both names, in which case it denotes nothing. It is easy to show by a *reductio* that no instances of the matrix can make the conjunctive antecedent true and the consequent false.

IS THIS ALL GOOD NOMINALISM?

The guarded answer is: it appears so. The basic unit of evaluation is for names

  n denotes x

and its negation,

  n does not denote x

for sentences

  s is true

and its contrary

  s is false

and for functors

  if a_1 signifies thusly$_1$ ... and a_n signifies thusly$_n$ then $f(a_1 \ldots a_n)$ signifies thusly$_{n+1}$

where the schematic 'thusly's indicate that each argument and the whole signifies in a way appropriate to its category. As explained, since the number of individual facts contributing to such a 'thusly' is potentially large and not necessarily or even usually expressible, such conditions cannot be constrained to being expressed or expressible,

---

[27] The example has historical significance: the discovery of the truth-functional behaviour of the complex functor formed the centrepiece of Tarski's 1923 dissertation under Leśniewski, showing prototethic could be based on material equivalence and universal conjunction alone. See Tarski 1983, 1–23.

[28] That the numerical predicates are best interpreted as first-order and attaching to names is argued in Simons (2007).

but only to actually obtaining. For operators the way of signifying is determined not by the operator alone but by the form and signification of its matrix and the pattern of binding of its variables within the matrix, but the kinds of signification thereby determined are not different in sort from those found among functors: it is merely that they are determined in context rather than context-independently as is that of a functor.

It is obvious why the platonist way of doing things is so expedient. The nominalistically acceptable face of any given signification is usually a large number of highly distributed facts—and by 'facts' I simply mean that something is so, not that there is an ontological category of item, *facts*, which we need to invoke. Abstract entities (sets, functions, or whatever) in effect collect these distributed totalities together and offer them up as abstract unities which can be input and output as such by the semantics. The result is that whereas a nominalistic account gets more complicatedly distributive as we recede type-theoretically from names and sentences, for a platonist, while the types of arguments are of increasingly high order, the mode of complication is in each case the same: a step in the functional or set-theoretical hierarchy, since at each level the functor or operator takes inputs which are objects (concrete or abstract) and returns as value an object (concrete or abstract) determined suitably.

Nevertheless, expediency is not the same as truth. While this attempt to outline the way a nominalist can deal with truth cannot be expected to convince those who are already happy platonists or who think, with Church, that any semantic analysis *must* make use of abstract entities,[29] I hope it will bolster the confidence of convinced or wavering nominalists, and maybe even win one or two people over. Nominalism is a tough row to hoe, and the inconvenience of nominalism to semantics and the nominalistic inexpressibility of certain familiar semantic concepts because of their complex and distributed conditions of application mean an implicitly platonistic language will tend to remain the medium of choice for most work in semantics. But nominalism should not be committed to the impossible task of providing a paraphrase or translation for any and every platonistic proposition: it is an ontological position, not a methodological one. In this light, we might wish to re-evaluate Tarski's apparent schizophrenia—platonist at work, nominalist on reflection—as exactly embodying the human nominalist's dilemma. If God existed she would directly know what all the distributed significations are for every token expression, and would see how they combine to give the truth or falsity of the sentences in which they occur. There would be no need for the platonist expedients and shortcuts: God could easily be a nominalist.

References

Ajdukiewicz, K. (1935) Die syntaktische Konnexität. *Studia Philosophica* 1, 1–27. English translation: 'Syntactic Connexion', in S. McCall, (ed.), *Polish Logic* 1920–1939, Clarendon Press, Oxford, 1967, 207–31.
Church, A. (1940) A Formulation of the Simple Theory of Types. *The Journal of Symbolic Logic* 5, 56–68.

[29] Church 1951.

—— (1951) The Need for Abstract Entities in Semantic Analysis. *Proceedings of the American Academy of Arts and Sciences* 80, 100–13.
Chwistek, L. (1948) *The Limits of Science. Outline of Logic and of the Methodology of the Exact Sciences* (translated by H. C. Brodie and A. P. Coleman). London: Routledge and Kegan Paul. (Polish original 1935.)
Frege, G. (1893) *Die Grundgesetze der Arithmetik, begriffsschriftlich abgeleitet.* Jena: Pohle.
Lejewski, C. (1958) On Leśniewski's Ontology. *Ratio* 1, 150–76.
Leśniewski, S. (1992) *Collected Works.* 2 vols. Dordrecht: Kluwer.
Luschei, E. (1962) *The Logical Systems of Leśniewski.* Amsterdam: North-Holland.
Rojszczak, A. (2005) *From the Act of Judging to the Sentence: The Problem of Truth Bearers from Bolzano to Tarski.* Dordrecht: Kluwer.
Simons, P. (1983) On Understanding Leśniewski. *History and Philosophy of Logic* 3, 165–91.
—— (1985) A Semantics for Ontology. *Dialectica* 39, 193–216.
—— (1991) Inadequacies of Extension and Intension. In G. Schurz and G. Dorn, eds., *Advances in Scientific Philosophy.* Amsterdam: Rodopi, 1991, 393–414.
—— (1992) Verità atemporale senza portatori di verità atemporali. *Discipline filosofiche* 2 (1992), 33–47.
—— (2002) Reasoning on a Tight Budget: Leśniewski's Nominalistic Metalogic. *Erkenntnis* 56, 99–122.
—— (2003) Absolute Truth in a Changing World. In J. Hintikka, T. Czarnecki, K. Kijania-Placek, T. Placek, and A. Rojszczak, eds. *Philosophy and Logic. In Search of the Polish Tradition. Essays in Honour of Jan Woleński on the Occasion of his 60th Birthday.* Dordrecht: Kluwer, 37–54.
—— (2006) Languages with Variable-Binding Operators: Categorial Syntax and Combinatorial Semantics. In J. Jadacki and J. Paśniczek, eds., *The Lvov–Warsaw School. The New Generation.* Amsterdam: Rodopi.
—— (2007) What Numbers Really Are. In R. E. Auxier and L. E. Hahn, eds. *The Philosophy of Michael Dummett.* La Salle: Open Court 229–47.
Suppes, P. (1988) Philosophical Implications of Tarski's Work. *The Journal of Symbolic Logic* 53, 80–91.
Tarski, A. (1983) *Logic, Semantics, Metamathematics.* 2nd edn. Philadelphia: Hackett. (1st edn. Oxford: Clarendon, 1956.)
Zermelo, E. (1930) Über Grenzzahlen und Mengenbereiche. Neue Untersuchungen über die Grundlagen der Mengenlehre. *Fundamenta Mathematicae* 16, 29–47. English translation by M. Hallett: On Boundary Numbers and Domains of Sets: New Investigations in the Foundations of Set Theory. In W. Ewald, ed., *From Kant to Hilbert: A Source Book in the Foundations of Mathematics*, vol. 2. Oxford: Clarendon, 1996, 1208–33.

15

Alternative Logics and the Role of Truth in the Interpretation of Languages

Jody Azzouni

Theories of interpretation, in one notable tradition, require a truth-conditional analysis of the sentences of a language for their interpretation: necessary and sufficient conditions on when its sentences are true. In addition, the centrality of Tarskian approaches to truth force the imposition of the logic of the interpreting language on to the language to be interpreted. I show that there are reasons to break away from this perspective on interpretation in both these respects. We need an approach to interpretation that doesn't *a priori* require what the topic-neutral lexically primitive inferentially central terms[1] of a (to be interpreted) language are to be; and we need an approach that allows that sufficient (but not necessary) truth-conditions can suffice for interpretation. I first show that there is empirical pressure in the interpretation situation for sometimes regarding languages as containing specific logical idioms despite our inability to translate such directly to logical idioms (because of the antecedent resources in the interpreting language). Next, the role of the truth idiom in truth-conditional interpretations is analyzed, and it's shown that recursion clauses giving necessary and sufficient truth-conditions for sentences are sometimes trivial. Interpretation, however, requires nontrivial characterizations of truth conditions; but necessary *and* sufficient truth-conditions aren't to be had in a general setting if they are to be other than trivial. The chapter concludes with a sketch of an approach to interpretation that doesn't impose the logic of the interpreting language on the target language, and that provides not necessary *and* sufficient truth-conditions, but only sufficient ones.

One important subtheme of this chapter is an exploration of the relationship between ontology and truth conditions. It's an on-going claim both of this chapter, and of other work of mine, that there is less connective tissue between truth conditions and ontology than philosophers think. In particular, it's often claimed that Tarski's celebrated notion of satisfaction is not only the tool by which truth is to be defined

---

My thanks to Douglas Patterson for detailed helpful comments on *several* earlier versions of this. My thanks as well to Bradley Armour-Garb and Arnold Koslow for useful suggestions. Finally, I'm grateful to an anonymous referee for drawing my attention to an infelicity.

[1] I'll—hereafter—describe these as "logical" terms. See Section I for motivation for this.

(or characterized), but that it's the tool that reveals how truth relates to the world.[2] I will give reasons to doubt this.

I

Davidson's hope

An influential and famous late twentieth-century philosophical application of Tarski's theory of truth is Davidson's use of it in theories of interpretation—"theories of interpretation," so construed as theories giving truth conditions of sentences of languages, which in turn suffice for the understanding of those sentences. Davidson hoped that, in making truth the central primitive concept in theories of interpretation, he could both sideline the apparent significance of the informal notion of meaning, and provide evidential tractability for such theories by avoiding consideration of the "detailed" propositional attitudes of speakers.[3]

The role of translation, respectively, in Tarski's and in Davidson's approaches

A version of Tarski's approach, due to Kemeny, has become the standard model for what success in such a project looks like[4]; and indeed, provided a small number of

---

[2] Davidson (2005, p. 30) writes: "[Tarski's] construction makes it evident ... that, for a language with anything like the expressive power of a natural language, the class of true sentences cannot be characterized without introducing a relation like satisfaction, which connects words (singular terms, predicates) with objects."

It's notable that Tarski—to my knowledge—nowhere associates the notion of satisfaction with correspondence views of truth, and only treats that notion as relevant to ontology in an, at best, equivocal manner. (I discuss this further at the beginning of Section V.) In Tarski 1944, p. 342, he claims that his definition does "justice to the intuitions which adhere to the classical conception of truth," and he invokes what has come to be called Criterion T, but nothing about satisfaction whatsoever. So too, later in that article, when Tarski (p. 362) describes "the semantic conception" as neutral with respect to "any epistemological attitude we may [have]; we may remain naïve realists, critical realists or idealists, empiricists or metaphysicians . . . ," he again fails to mention the role—or absence of a role—of satisfaction for this issue. In the beginning of his 1932, p. 153, he raises the issue of the "classical" concept of truth: "true—corresponding with reality"; but it's striking that when he motivates his notion of satisfaction (p. 189), the problem described (that satisfaction is to solve) is that "in general, composite sentences are in no way compounds of simple *sentences*." Indeed, this problem motivates the "notion of the *satisfaction of a given sentential function by given object* . . . " "Object," however, stands for what can be characterized in the metalanguage (e.g., "classes of individuals") and Tarski doesn't even *allude* to the idea that satisfaction so construed shows how (formal) languages make contact with the world.

[3] Davidson 1984*a*, p. xiii, writes: "In the essays collected here I explore the idea that we would have an answer to the question [what is it for words to mean what they do?] if we knew how to construct a theory satisfying two demands: [the second] is that it would be verifiable without knowledge of the detailed propositional attitudes of the speaker. . . . The second condition aims to prevent smuggling into the foundations of the theory concepts too closely allied to the concept of meaning." Exactly how *meanings* are involved in a truth-conditional analysis of sentences is hotly disputed. I won't discuss the matter in much detail in this chapter; see, however, Section VII.

[4] See Etchemendy 1990, chapter 5, or Tarski 1932.

points that arise in Tarski's exposition of his own theory are overlooked, Tarski's approach *does* seem to yield paradigmatic theories of interpretation in the sense intended by Davidson. On Davidson's construal, the resulting interpretational theory is empirically sensitive both because it's finitely axiomatizable (which limits—to some extent—the underdetermination of such theories by their data), and because its Tarski-biconditional theorems are *directly* testable: sentences of the form: *A is true iff B*, where *A* is a structural description (in M, the language of the interpretational theory) *of* a sentence of the language O targeted by the interpretational theory, and *B* its translation in M, a translation induced by the (Tarskian) interpretational theory itself.[5] Indeed, such a theory is taken to reveal the truth conditions of each sentence in the language via the composition of the sentence by the application of a finite number of devices to elements drawn from a finite stock of vocabulary items.[6] In turn, and because of this, such a theory suffices for the understanding of the sentences of a language.[7]

By contrast, Tarski's approach to his theory of truth presupposes, *to begin with*, a translation of O into M.[8] (This is one of the small points to be overlooked that I mentioned earlier.) But Tarski's aims weren't Davidson's; and so, in the context of interpretation as Davidson understands it, one doesn't *start* with a translation, one *ends* with one: The theory itself yields a translation of all the sentences of the language O into the language M by virtue of its being a (modified) Tarskian theory of truth of O.[9]

---

[5] Delicacy regarding use and mention is required here, and in what follows. Having warned the reader, I'll continue to be studiously informal about it.

[6] "Compositional truth-conditions," aren't usually taken to require that the items—a sentence's truth conditions are recursively given in terms of—be visible syntactic subparts of that sentence. But as long as the resources are finite that the clauses of the interpretational theory avail themselves of, and the clauses themselves are recursive in the Tarskian sense, the requirement of compositionality is taken to be met—even if such clauses help themselves to items that aren't, strictly speaking, part of the sentences the truth conditions are of. None of the issues I raise in the rest of the chapter turn on demurrals of this "broadening" of the notion of compositionality.

[7] There are a number of wrinkles due to the idiolectical nature of languages—on Davidson's view—and to there being niceties about how "understanding" is to be taken. See, e.g., Davidson 2005. I have disagreements about these wrinkles, some of which emerge in the course of the exposition of this chapter—but none of the primary arguments of the chapter turn on any of this.

[8] Translation so utilized by Tarski needn't involve a particularly rich notion of "meaning." Co-referentiality, co-extensionality, etc., suffice. Davidson (1984a, p. xiv), contrasting his approach to Tarski's, writes: "One thing that only gradually dawned on me was that while Tarski intended to analyse the concept of truth by appealing (in Convention T) to the concept of meaning (in the guise of sameness of meaning or translation), I have the reverse in mind. I considered truth to be the central primitive concept, and hoped, by detailing truth's structure, to get at meaning." One of the issues that has become central in the current debates over deflationist notions of truth is whether deflationist truth presupposes a notion of meaning rich enough to infirm it for theories of interpretation. It will become clear that my version of deflationism neither presupposes a notion of sameness of meaning nor of translation; and so one of the more popular objections to the use of deflationist truth in theories of interpretation doesn't apply.

[9] *A qualification about the role of translation in theories of interpretation*: On standard views of how this approach extends to the context-sensitive utterances of a natural language—involving indexicals and tense—the resulting theory does *not* yield translations of (utterances of) the sentences of O in M. Rather, what a theory of the interpretation of (utterances of) the sentences of O in M gives

The interpretation, thus, of a language O in a language M seems shackled to the translation of O into M: A necessary condition on such an interpretation is the possibility of the extraction (from it) of a translation of the sentences of O into M. I'll later show that there are pairs of languages, the first amenable to interpretation in the second, although *not* so amenable to translation into the second. But I must first note another constraint that seems to emerge from Davidson's way of casting interpretation into a Tarskian mold.

The recursive interpretation of sentences

Tarski's approach, famously, involves a recursive definition (or axiomatization) of truth, and in the context of interpretation this is transformed into a recursive interpretative theory of the sentences of O. Imagine that the target language O has the connectives "&" and "¬," and that the language M has the connectives "and" and "not," which we regard as adequately capturing the meaning of "&" and "¬."[10] We can capture the truth conditions of the sentences of O containing "&" and "¬" by including *among* the recursion clauses of the interpretation theory in M:

(^) (A & B) is true if and only if A is true and B is true.
     ¬A is true if and only if not (A is true).

"*Among* the clauses," because there must be an explicit clause for every item that plays a role in the construction of the sentences of O—for every "recursive term" of O—except when such items are treated as purely syntactical (e.g., parentheses). The finite-number-of-clauses assumption places a powerful constraint on the theory: It forces both the nonlogical and logical vocabulary to be finitary, e.g., names, primitive predicates, quantifiers, and connectives.

A qualification

I've borrowed the phrase "recursive term" from Lepore and Ludwig (2001, p. 119). Recursive terms—at least as commonly presented—have three properties. (1) They are topic neutral: Their recursion clauses aren't restricted to subportions

---

are—more broadly construed—*truth conditions*: conditions under which utterances, containing such and such indexicals and tense indicators, uttered in such and such contexts at such and such times, and by so-and-so, are true. (Translations of the sentences of O into M are a special case—when context sensitivity is absent—of these characterizations.) Taking account of context sensitivity doesn't affect the arguments in this chapter, and so I frame what's forthcoming solely in terms of context-insensitive sentences. For evidence that my general approach to interpretation—via anaphorically unrestricted quantifiers—has the resources to handle context sensitivity, see Azzouni 2006, section 3.5.

[10] On the Davidsonian view, this is empirically confirmed by successful navigation among the speakers of O via the interpretations given by Tarski biconditionals: Our interpretation of "&" by "and" can be disconfirmed by a systematic failure (in our interpretation of the sentences of O) traceable to taking "&" as "and." See footnote 18 for explicit acknowledgement of the methodological idealizations involved here.

of O. (2) They are *lexically primitive*.[11] (3) Such items play a fundamental role in inference.[12] I'll usually describe such terms as "logical"; this is both because these properties are traditionally presumed to hold of logical vocabulary, and because this jargon suits the contrasts in the languages under discussion, e.g., ones based on classical logic as opposed to intuitionistic logic, etc.[13]

I mentioned a paragraph back that the finite-number-of-clauses assumption forces both the nonlogical and logical vocabulary to be finitary. But a more powerful constraint on possible interpretations of languages O seems also to emerge here: Successful interpretation requires imposing the logical idioms of M on O. Any sentence of O apparently containing a logical idiom must be translated to a sentence of M, and therefore to a sentence restricted in its logical resources to those available in M. The sample clauses (^) illustrate the point by their explicit recursion over sentences in O (containing "&" and "¬") in terms of sentences of M containing "and" and "not."

A necessary condition on successful interpretation is the possibility of translation?

Prima facie, this form—that a theory of interpretation must take—seems to rule out the *possibility* of interpretation if successful translation isn't possible.[14] To explore whether this is problematic (and how), presume an anthropologist who uses a language equivalent to a first-order classical language in its logical resources as her M, and imagine she is trying to interpret a language O.[15] Note first that alternative logics are pure mathematical objects defined into existence by sheer specification of their properties. Similarly, one can rest the notion of a "possible conceptual scheme" or "possible web of belief" on the basis of such logics—classical or otherwise.[16] In perfect analogy with—say—the existence of pure mathematical subject-areas, such as geometries, and to there being an empirical possibility that any such might best characterize the nature of space, it might seem that it should be an empirical possibility that the logic the language O is based on is different from what M is based on, involving—for example—branching quantifiers, intuitionistic connectives, or something

---

[11] My thanks to Douglas Patterson for stressing the importance of explicitly distinguishing between these first two properties.

[12] Lexical primitiveness and topic neutrality alone don't suffice for capturing the notion of a logical term: Imagine a pious community that allows a one-word blessing—grammatically structured like a one-place sentence-connective—at the beginning of any sentence.

[13] I'm not prepared, however, to argue that such constraints are *necessary* for the characterization of "logical term." That's a vexed question. In any case, the important work will be done by the forthcoming claim that the practice of interpretation should be open to the empirical possibility that there can be languages with different topic-neutral lexically primitive inferentially central terms—hereafter, "logical" terms—that can't be translated to one another.

[14] This claim is subject, of course, to the qualification noted in footnote 9.

[15] On my view, therefore, this anthropologist is *not* using English as the home language for her interpretational theories of other languages.

[16] Quine understands the revision of logic in just this way: a shift—at the heart of our web of belief—from one logic to another. See his 1951, where he considers specific examples of such—although superficially. A more searching engagement with alternatives is found in his 1986.

even more alien to M.[17] Furthermore, our anthropologist's *empirical* methods—so we might think—should be sensitive to this empirical possibility. But if her methods of interpretation require a theory along (modified) Tarskian lines, it seems that she *can't* take account of this empirical possibility. She must interpret the logic of any language O interpreted in M as having (possibly a subset of) the logical connectives of M.

II

Radical interpretation

There are two strategies seemingly available to deny the above analogy with the application of mathematics: Either deny the empirical possibility that the logic of any other group could (really) be different from ours (argue that our anthropologist *needn't be* live to the empirical possibility of a language based on a logic different from M's), or broaden the (apparent) theoretical resources available to M (argue that the description—apparently given in Section I by the example of (^) of the narrow form that an interpretation of O in M takes—is wrong). But before evaluating these strategies, a preliminary issue must be dealt with. We might describe the anthropologist's primary empirical instrument—in the "radical interpretation" context—to be, on the evidential basis of interactions with native speakers, the construction of a (modified) Tarskian theory of truth for O in M. What in the process of constructing such could reveal the possibility that O is based on an alternative logic?

Live empirical possibilities arise when our methods go awry in revealing ways. We suspect *additional* physical forces at work in the movement of certain objects when our description of the forces we take to be in effect doesn't yield correct predictions of their movements. (We may also doubt the laws we take to govern those forces—but that's a more drastic consideration after other moves have failed.) What corresponds to this methodological development in the radical-interpretation situation?

This: The anthropologist might find that all her provisional interpretations go wrong for a certain idiom. She manages—let's say—successful interpretations of a large class of sentences *without* that idiom. But—let's also say—that idiom is a recursive term that can be applied to any sentence (or, more generally, to any subsentential unit of the appropriate grammatical sort) to yield additional sentences; and let's suppose that all her provisional interpretations of sentences containing those particular terms go systematically awry.[18]

---

[17] See, e.g., Barwise and Feferman 1985, Gabbay and Guenthner 1986, or any of hundreds of other books.
[18] Several idealizations are implicit in this example. (1) Because of the holism of the interpretational situation—in the specific form of inferences—it will always be that data bearing on the anthropologist's interpretation of sentences without said idiom can mislead as well. (One can know that a rabbit is about, not because one sees the rabbit but because one sees other things from which the rabbit's presence follows.) This subtlety I take the anthropologist as able to handle—at least well enough so as to make empirical progress. (2) I'm presuming that the idiom in question is

The worry is that, despite the evident intelligence of the speakers of O, and the apparent significance of their noise-making, the anthropologist has no empirical way to distinguish two possibilities: (1) Despite appearances, there is *no* language O—speakers of "O" aren't communicating, at least, not by means of a language they *utter* the sentences of. Or, (2) O *is* a language, but one that eludes the empirical tools available to the anthropologist: an interpretational theory of O in M.

One might think that if noise-makers successfully coordinate their activities—or if, anyway, they are *human*—that's enough to support the hypothesis that they speak a language. Group-mind options should be ruled out; so too should innate genetic programming to behave in elaborately coordinated ways. *Perhaps*, but it would still be nice to have direct evidence that the sounds of a certain group of individuals indicate a language.

Direct immersion in a language

Luckily, there *is* learning a language *directly*. The anthropologist acquired her first language that way; why couldn't she do it again? And, having done so would seem to place her in the enviable epistemic position of being able to say with *some* authority, "this language is based on an alternative logic, and that's why my previous attempts to provide a (modified) Tarski truth-theory for it failed."[19]

Successful acquisition of a language provides, therefore, the empirical pressure needed: One can recognize that O is a language because one has acquired the ability to *speak* O; and this empirical pressure can operate independently—at least in principle—of whether or not an interpretation of O in M is possible. Further, given the acquisition of O, the anthropologist—it seems—is also in the position to establish, say, that the logic of O isn't the classical one of M: She can determine whether or not the logical idioms of O are intuitionistic, are branching quantifiers, etc. She can determine this, anyway, as well as anyone can determine such facts about a language she speaks.[20]

III

Let's turn to the various ways one might either try to deny the supposed empirical possibility of a language O that's uninterpretable in M (a possibility that I've been endeavoring to make space for), or—accepting that such *is* an empirical possibility—for broadening the description of the scope of interpretational theories to enable them to take account of it.

---

phonetically visible, and phonetically similar in all its instantiations. This needn't be; but again, the excised complicating details (in how the anthropologist should proceed) do affect the force of the argument.

[19] Quine (1986, p. 87), when writing of acquiring competence in intuitionism, deplores fuzzy explanation of its connectives: "One does as well to bypass these explanations and go straight to Heyting's axiomatization of intuitionistic logic, thus learning the logic by what language teachers call direct method rather than by translation."

[20] Recall the idealizations mentioned in footnote 18.

The flexibility of translation

Start with the claim that translation (and thus, interpretation) is a more supple tool than the simple connective-to-connective translations illustrated by (^) indicate. Consider intuitionism—a logic commonly described as having connectives different in "meaning" from classical ones, and a logic that, therefore, seems to elude the sort of interpretation exemplified by (^).[21] Still, there is the well-known recursive "translation" of first-order predicate intuitionism (FPI) into quantified S4 (QS4):

$$(\neg p)^* = \sim Bp^*,$$
$$(p \supset q)^* = Bp^* \to Bq^*,$$
$$(p \vee q)^* = Bp^* \vee Bq^*,$$
$$(p \wedge q)^* = p^* \bullet q^*,$$
$$(\forall xp)^* = \forall xp^*$$
$$(\exists xp)^* = \exists xBp^*.[22]$$

Although it looks like the translation operator * can be used to yield a good translation of FPI into QS4, it can be argued that—because of the logical operator B—it isn't an acceptable translation into the austere classical logic of M. After all, if we supplement the resources of M with an additional logical item (and treat the result as a successful translation into *M*), why not (more simply) supplement the logical resources of M with the intuitionistic idioms directly, and call *that* a successful translation?[23]

But there is a response: possible-world semantics[24]—itself couched in a classical first-order language with additional *nonlogical* vocabulary. Using a (Tarskian) theory of truth in a metalanguage M supplemented with this additional nonlogical

---

[21] An interpretational theory along the lines of (^) uses classical connectives (on the right side of the recursion clauses) to translate the intuitionistic connectives described (on the left side of the recursion clauses). How badly this skews the intuitionistic connectives is shown by the failure of the neat interdefinability of classical "&" and "¬" with "∨" and "¬"—and indeed, all the interdefinability results of the classical connectives—in intuitionism. On the other hand, there are sets of rules for the intuitionistic idioms identical with those for the classical idioms except for negation—in light of them, one might claim that only *negation* differs in meaning. But the holistic interaction of the logical idioms in the two settings, as well as the standard interpretations of the intuitionistic idioms weighs against this suggestion. There is no need to pursue this issue any further in this chapter.

[22] "B" is the necessity operator; the other connectives on the right sides of the clauses are classical; the ones on the left are intuitionistic. See Gödel 1933, p. 301. The full result is due to several logicians. See Troelstra 1986, 296–9.

[23] For the moment, I evade the important question this raises: Why not *always* treat an interpretation of a language O as only possible in a language M with (essentially) the same recursive terms as those in O? This issue will begin to be addressed in Section IV. I also set aside the important point that so introducing intuitionistic idioms into a setting where classical idioms already reside will cause an implicational collapse of the intuitionistic idioms into the classical ones. See Harris 1982: It's a melancholy fact of intellectual life that—in many cases—differing concepts cannot co-exist in the "same logical place." My thanks to Agustín Rayo for drawing my attention to this article—which illustrates this moral for a particular (logically clean) case. I should add that the theorem can be made especially transparent in the context of the approach taken to logical operators in Koslow 1992.

[24] See Kripke 1959, or, for the direct route, Kripke 1965.

vocabulary (to characterize possible worlds) will enable us to supply a translation of QS4 directly into M; composing that translation with the above one between FPI and QS4 results in a translation of FPI into M.[25]

The strategy is generalizable. Consider *any* (complete and sound) semantics S for a given alternative logic L that's available in a classical metalanguage M. S can be used to give a (modified) Tarski truth-theory in M for the sentences of a language O based on L only provided that M is supplemented with whatever nonlogical vocabulary (e.g., vocabulary referring to—and describing—properties of possible worlds, flows of time, proofs sequences, etc.) that S needs. The result (modulo the point made in footnote 9) is a translation of O into M. (The nonlogical vocabulary of O is presumed present in M.)

A point about this strategy

Any logic imposes a consequence-relation pattern on the sentences it governs. Successful translation into a classical language M needn't require a straightforward mapping of alien logical idioms to our logical idioms; and this is a good thing since the consequence-relation patterns the two sets of idioms impose on sentences isn't (in general) the same, or even close. The apparent trick to translation in such recalcitrant circumstances is to find a set of sentences (containing nonlogical idioms in addition to logical ones) *that have the same consequence-relation pattern as the alien sentences do*. According to this strategy, once we find one of those (and the numerous examples of semantic theories for alternative logics—all couched in classical metalanguages—shows how fruitful this approach is), we take the alternative logic as successfully translated to (indeed, interpreted in) our language.

Evaluating the value of induced translations as translations

Hereon, I'll describe these translations of alternative logics (and therefore of the languages O couched in those logics) as ones *induced* by classical metalanguage characterizations of their semantics, or (in short) as *induced translations*. The question naturally arises: Are induced translations *translations*? One objection to saying so is their failure to preserve significant properties of the logical terms of O. This failure shows up in three related ways. (1) The use of an induced translation to provide an interpretation of O in M fails to preserve the grammatical and inferential centrality

---

[25] Some may worry about the legitimacy of supplementing M with *nonlogical* vocabulary—in particular, they may be worried because of an antecedent worry about whether the distinction between logical and nonlogical vocabulary is being drawn too narrowly here. I'm setting aside these worries in favor of my opponent by giving (provisional) credence to the view that interpretation via an interpretational language M based on an austere classical logic is more flexible than it looks—thus granting the claim that the (possible) supplementation of M with nonlogical resources isn't problematic. For the record, I think it *is* problematic: A discussion of why, and what implications this has for theories of interpretation that require translations, involves substantial discussion of complex issues about the nature of names, kind terms, etc. I'm—for the most part—leaving this for future work. (See, however, pp. 43–5 of my 2006.)

of the logical terms of O so translated. This is because the sentences of O containing logical terms must be translated so that they contain classical logical connectives plus nonlogical predicates. Consider an interpretation of O in M that uses the translation operator * of FPI into a classical language via QS4, and where the translation of sentences without logical idioms is homophonic. The sentence "John is running and Peter is jumping," is translated straightforwardly into its classical cousin. Not so, alas, for "John is running or Peter is jumping." The sentence translating *this* is burdened with existential quantifiers and nonlogical predicates containing place-holders for possible worlds (or proxies thereof)—it ends up as something like: *(In every possible world John is running) or (in every possible world Peter is jumping).*

(2) Relatedly, as we see from the above example, these additional nonlogical predicates saddle the to-be-translated logical idioms of O with a special subject matter. And finally (3), because of the points just mentioned, any such translation can never be *onto* a classical conceptual scheme (no matter how rich—otherwise—the nonlogical vocabulary of O is).

Topic-neutral lexically primitive inferentially central idioms should be—all things being equal—translated to topic-neutral lexically primitive inferentially central idioms

Induced translations do preserve the topic-neutrality of the idioms so translated; it's the other two properties that are violated by such translations. It *is* sometimes reasonable to translate lexically primitive idioms by idioms that aren't primitive. "Robin-egg blue" may correspond to a lexically primitive term in another language. But if a term possesses the *three* properties I've singled out—if it's a logical term—then it's functioning in the language in a quite special way that translation should take account of. To translate it so that it ceases to be a logical term is—without good reason—empirically irresponsible because that amounts to the demotion of a logical term to a lexical complex of a quite different logical term *wedded to additional subject matter*.[26] A reasonable constraint on translation that follows from this is: logical terms should be directly translated to logical terms—unless there *really is* empirical evidence that such idioms are laden with an inarticulated subject matter. This constraint has empirical bite: We can recognize when it's not satisfied, and thus we can recognize how our purported translation fails to be a *good* translation insofar as it fails to satisfy this constraint. I'll now illustrate how this (defeasible) constraint operates.

---

[26] Because of the inferentially central role of the idioms so translated, there is a presumption of a widespread failure—among speakers of the target language—to fully analyze their "logic." Consider the possibility of such a charge directed towards ourselves: that the genuine logical terms of our language are actually intuitionistic, and that we have failed to recognize such because we have not teased out the hidden subject matter accompanying all our inferences that—only when kept tacit—makes them look classical. Such empirical presumption—if allowed in the practice of interpretation—fully deserves the charge of "logical imperialism." Notice this is quite different from the decision to *shift* from one logic to another, and to retroactively *reinterpret* one's previous inferences as laden with a specific subject matter.

Suppose we require of the translation extracted from an interpretation of a target language O only that it (for the most part) map truths to truths, that it be (relatively speaking) consequence-relation preserving, that it enables successful communication with speakers of O; and suppose an induced translation does all this. Given that the sentences we've translated O onto are precisely the truth conditions given by a semantic theory of a particular extensively studied alternative logic (intuitionism, say), however, why shouldn't we draw an entirely different conclusion? That we now have good empirical evidence that this (provisional) interpretation in M does *not* provide an adequate interpretation of O (because the logical idioms of O are interpreted by idioms that aren't logical). Furthermore (and interestingly), *by means of the tool of constructing a (modified) Tarskian truth theory of O*, we have facilitated *both* the empirical discovery that O is based on an alternative logic, *and* what that logic is.[27]

This empirical option is open to challenge. Why can't some languages be expressively richer than others? And if so, why can't it be that the "expressively richer" claim—about M in relation to O—is empirically confirmed by the success of interpretations that saddle purported logical terms of O with additional content that we can thus claim speakers of O have never explicitly expressed?

This is, of course, possible. But the mere "success" of an interpretation so imposing additional content on the logical idioms of O won't empirically confirm that interpretation because such interpretations—in general—are too easy to construct, and in ways conflicting in what's implied about the relative expressive richness of languages M and O. Recall there are also interpretations available of the classical predicate calculus in FPI.[28] So an easy (but ominous) symmetry in available interpretations (of O in M, and of M in O) arises if we so easily allow interpretations to saddle logical idioms with additional subject-matter.

In any case, if we so allow empirical confirmation, then we should so allow empirical disconfirmation: Success in saddling apparent logical idioms with a subject matter turns on that subject matter not being explicitly expressible by other sentences of O—this is how we motivate the claim that such idioms of O contain an implicit subject-matter systematically connected to how speakers infer that nevertheless speakers are unaware of. But we may find good evidence that the

---

[27] A methodological analogy: We initially apply a set of laws of motion to a type of moving object, and use the deviations of the motions of those objects to determine what the actual laws of motion governing those objects are. What it is about this strategy that makes tractable the discovery of laws is that it may not be easy to see by *direct examination of the movements of such objects* what laws of motion they obey—the pattern of movement may be hard to capture theoretically all at one go. But if we can approach the laws governing their motion *in stages*, by first developing a (simpler) set of laws that apply approximately, and then modifying the laws under the empirical impress of deviations from the approximating laws, we may (eventually) capture the actual laws. Similarly, we may find that once we have applied a theory of interpretation that imposes classical logic on O, we may see that we need only modify our interpretation of certain connectives (e.g., just "¬") to get a theory of interpretation that fits O much better. Applying "approximate" theories to empirical phenomena—this way—*structures* that phenomena so that its patterns may become more amenable to theoretical characterization.

[28] See Kleene 1952, §81.

nonlogical idioms utilized to provide our interpretation of these idioms are reasonably—and independently—taken as already present in the vocabulary of O; e.g., it may be reasonable to interpret other sentences of O as about "possible worlds." The claim, then, that certain topics aren't explicit in the language of O—that O is less expressive than M because of the absence of certain nonlogical terms that can be naturally interpreted by specific terms present in M—would in this way be empirically refuted.

"Empirically refuted," is too strong. There are always ways to redescribe (although perhaps baroquely) apparently recalcitrant empirical results in order to save an hypothesis. So there can be nothing fatal in insisting that an interpretation reveals apparent logical idioms of O to actually involve an implicit subject-matter; and this is because the empirical success of any interpretation is always a "more or less" affair; and so—at best—these points may be seen as infelicities in a candidate interpretation to be measured against deficits possessed by alternatives. In this particular case, the infelicity would be due to the imposition of a tacit subject-matter where—say—native speakers would deny one can be found. One may be reminded of Davidson's (1967, 1977) analysis of adverbial constructions quantifying over events, and conclude that the imposition of a tacit subject-matter to interpret a language is generally no big deal. There is a disanalogy, however. When these properties are sacrificed as a result, one is stipulating—because of the logic one's theory of interpretation is restricted to—that the logic of the target language is that of the home language; and perhaps that's too fundamental and drastic an assumption to build into one's theory of interpretation on the sheer grounds that interpretation—so construed—requires it.

In this respect interpretations should be like other scientific theories

The empirical pressure raised against the apparent requirement that interpretations of languages O must involve the imposition of the logic of the language M (that the interpretation occurs in) is due to taking seriously the idea that an interpretation of a language is on a par with scientific theories in general. One doesn't want to *stipulate* the empirical character of possible languages by *requiring* that *any* interpretation must be of such and such a form. And we especially don't want to do this if, as a result, other natural empirical routes to the nature of the phenomena—in this case, evidence of the logicality of certain idioms in the target language, or the evidence of bilinguals—are by definition of the methodology so imposed, stipulated as irrelevant. The fallibilism of empirical studies requires we not use our interpretational methodology—in this case, a (modified) Tarskian approach to interpretation—to *define* the potential range of the phenomena being studied.[29] One way this can be

---

[29] Some may protest, however, that more is at work in the supplying of an interpretation for a language than the (mere) empirical study of alien ways with words. The methods of interpretation utilizing a (modified) Tarskian approach include *rationality constraints*—circumscriptions on which logical idioms may or may not be used in a theory of interpretation; such constraints (on pain of imputing a hitherto unknown form of madness to alien speakers) *cannot* be open to empirical

illicitly done is to place premature necessary constraints—linked to our theories of interpretation—on what a language *is*. As with all empirical phenomena, one starts with an intuitive grasp of what's to be studied—languages in this case—and naturally includes as evidence for the presence of such languages the ability to acquire, and navigate in, them—to grasp them by means of direct immersion. It may be that our approach to interpretation eventually becomes robust enough that we can exclude certain examples of such "languages" on the grounds that interpretational theories for them are unavailable, and this despite our ability to so (pre-theoretically) "understand" them. But it's precisely here that ordinary scientific practice allows flexibility in what its terms are to mean: A generalization in the form such interpretational theories can take allows a concomitant broadening in what can be described as a "language."[30]

IV

Should interpretational theories have the same expressive resources as the languages they are to interpret?

One might concede that interpreting a language O is only possible in a language M that has the same logical resources as O does. Lepore and Ludwig (2001) seem to have this view: They argue that determining the logical form of sentences of a language O via the translation of such sentences into an "ideal" artificial language M creates problems best solved by requiring the recursive terms of O to be interpreted by recursion clauses in the interpretative theory, where such clauses utilize recursive terms that are translations of the recursive terms of O (as illustrated in (^)). Thus they require M to have (at least) the same logical resources as O. They particularly stress the superiority of this approach to natural-language idioms—such as quantified noun-phrases

---

refutation. Certainly the use of a (modified) Tarskian theory, to provide the interpretation of a language, has been seen in just this way: This is part of the motivation for coalesing interpretation—as so understood—with the supplying of the (only) means to the *understanding* of the sentences of the target language. Rationality constraints, however, must be established on independent grounds: They should not be established by mundane facts about how interpretations of languages in (other) languages can give rise to failures to preserve the logicality of idioms when (other) languages differ in their logical resources. Constraints on what's rational, thus established, are too dangerously a parochial matter of what approaches to interpretation we have managed to hit upon *to date*.

Davidson (2005, p. 63) writes: "The relations between beliefs play a decisive constitutive role; an interpreter cannot accept great or obvious deviations from his own standards of rationality without destroying the foundation of intelligibility on which interpretation rests. The possibility of understanding the speech or actions of an agent depends on the existence of a fundamentally rational pattern, a pattern that must, in general outline, be shared by all rational creatures. *We have no choice, then, but to project our own logic onto the language and beliefs of another*" (italics mine). This last assertion is in ominous tension with a construal of interpretational theories as *empirical*—something Davidson also commits himself to. Of course there is a challenge in the phrase "destroying the foundation of intelligibility on which interpretation rests..."—I respond to that challenge in Section VIII.

[30] My thanks to Douglas Patterson for raising issues in email that led to this paragraph.

or complex demonstratives—over attempts to capture the meaning of such idioms in an interpretative theory in a language M with only the resources, say, of first-order logic.[31]

Their approach might be supported—in our anthropological context—by the claim that all natural languages have the same logical resources: Thus an interpretative theory of a natural language O may occur in any other natural language M. Alternatively, if we recognize that whatever the universal constraints on natural languages are, it's unlikely they require the logical idioms of every natural language to be the same ones, one could instead bite the bullet by conceding that to interpret any natural language one has to acquire it first, and only then can one interpret that language (in a background language with the same logical terms).

There is an issue, however, that starts to surface when Lepore and Ludwig (2001, p. 118) extol truth-theoretic approaches because of their "ability to provide recursions needed to generate meaning specifications for object-language sentences from a finite base with no more ontological or logical resources than is required for a theory of reference." Meaning specifications for object-language sentences are to be supplied by truth-condition clauses—specifically the clauses of a (modified) Tarskian theory of truth for the language—and so the worry is this: If any manner of logical idiom—that appears in O—can also appear in M, why are we so sure that the resulting theory of truth, merely because it has the same *form* of recursion clauses, operates in the same way when different logical idioms are involved? What makes us sure that in any logical setting such specifications *always* supply equally desirable "truth conditions"?[32]

Indeed, if one set of recursion clauses employs intuitionistic connectives, and another set employs classical connectives, it might be thought that the resulting specifications (despite their being identical in form) involve such different logical resources that the specifications (despite the identical role of "true") can't be doing the same thing in both settings; specifically, they can't give informative "*truth* conditions" if they use *intuitionistic* connectives.

Merely pointing to the similar form of the recursion clauses in truth-theoretic approaches is therefore not entirely reassuring: One needs to look *closer*. To this end, I now take up the issue of the role of the truth idiom in (modified) Tarskian theories of interpretation: I consider how "truth," understood deflatedly, nevertheless suffices

---

[31] See Lepore and Ludwig 2001, especially pp. 126–8.

[32] What *are* truth conditions? Stalnaker (1998, p. 97) helpfully writes: "... think of the meaning of a sentence as a recipe for determining a truth value as a function of the facts. The recursive semantic structure of the sentence encodes such a procedure. One might identify the recipe with the truth conditions, since it spells out the procedure, or conditions, for determining whether the sentence is true. Here is a contrasting explanation: one might instead identify the truth conditions of a statement simply with the circumstances (the way things must be) for the proposition expressed to be true. ... Different recipes determined by statements with different constituent structure might end in the same place, no matter what the facts (as, for example, with statements of the forms $\sim(P \vee Q)$ and $(\sim P \wedge \sim Q)$). Such statements will have different truth conditions in one sense, but the same truth conditions in the other." This is an important distinction for what follows. I describe Stalnaker's first kind of truth conditions as *functional truth-conditions*. I describe his second kind of truth conditions as *circumstantial truth-conditions*.

for the role of truth—as it occurs in such theories. After doing so, I'll return to the issue just raised.

What is a deflationary view of truth?[33]

A "deflationary view of truth" is—following Quine (1986)—one embodying the presumption that the (only) point or function of the truth predicate is to facilitate the expression of blind truth-endorsements. The latter are expressions of assent to—or denial of—sets of sentences specified by descriptions or by names (e.g., "everything John said yesterday," "the theory of relativity," "Peano's axioms"). The value of this is especially evident when the sentences in question are large in number, or if one can't remember, or doesn't know, what specific sentences they are.

Several qualifications should be noted: A description of the function or point of the truth predicate is *not* a statement of the *meaning* of that predicate, or even of what's meant when one describes such a set of sentences as "true." So, it's *not* (necessarily) part of the view that to attribute truth to something is therefore to attribute assertoric force to that something. That the point of an idiom is to enable us to assert something, doesn't require that the meaning of that idiom therefore comes down to the (mere) attribution of assertoric force.[34]

Second, many—if not most—subscribers to a deflationist view of truth, although agreeing with Quine's claim about the function of "true," also believe the meaning of "true" is given by the set of (all the) sentences of the form, " 'S' is true iff S."[35] This second view can take the extreme form of claiming that " 'John is running,' is true" means the same as "John is running." There are motivations, no doubt, for such views—but no such synonymy claims are *required* by the functional characterization of the facilitating of blind endorsements. I take the functional characterization to be the heart of the deflationist view of truth, and the description of the meaning of truth as exhausted by—say—"T-sentences," to be an (unnecessary) accompaniment that many deflationists are also committed to.

Third, it's important to see *how* the truth predicate facilitates blind endorsements. What it does, and *all* it does is to—when coupled with a suitable quantifier—enable maneuverability between "use and mention." Instead of saying "John is running," by directly asserting "John is running," it enables *mention* of the sentence—by name or description—by instead, "That sentence on the blackboard is true," or "(1) is true," or even " 'John is running', is true." It's not, that is, a quantifier disguised in ordinary-language as a predicate—rather, it *is* a predicate that can be utilized in blind endorsements when, e.g., it's *coupled with* a suitable quantifier (one that ranges over sentences or propositions), as in "Everything John said is true." Being a predicate, of course, it may—*in addition to its functional role*—also turn out to be co-extensive with some *property* (of sentences, say); but the question of whether "true" is so co-extensive with

---

[33] Here I summarize a number of conclusions argued for in Part I of my 2006.
[34] I owe this crisp way of putting the point to Douglas Patterson.
[35] "S" and " 'S' " are cross-linked variables, so that when "S" is read as ranging over a sentence, " 'S' " is to be read as ranging over the quotation-name of that sentence.

a property, or if it is, what sort of property that would have to be, are questions entirely independent of its blind-endorsement role.

Truth deflationism, therefore, as I'm understanding it, doesn't *require* that "truth" *not* correspond to a property, nor that when a truth idiom is employed in interpretation, that it *mustn't* be compatible with there being some property holding of all true sentences. "Truth deflationism," as understood here, is neutral on these particular points; none of the arguments given in this chapter require anything more of "truth deflationism" than what I've described above.

Getting by without the truth predicate

It's important to realize that other devices—in principle—are available that can facilitate blind endorsements *without* a truth predicate. One fairly popular choice is sentential substitutional quantifiers, where sentential substitutional variables "x," "y," etc., stand stead for sentences, and predicates holding of sentences are regimented as *sentential* operators, so that "Written-on-the-blackboard(John is running)" is taken to be an application of a sentential operator—not a predicate—to the sentence "John is running." Sentential substitutional quantifiers, (Ex) and (Ax), introduced to bind substitutional variables, and taken to have substitution-classes of sentences, enable the expression of a blind endorsement like "Everything written on the blackboard is true," like so:

(#) (Ax)(Written-on-the-blackboard(x) ⊃ x).

Notice that the truth idiom has gone missing in (#): It's not needed because this way of expressing a blind endorsement of a set of sentences doesn't involve a shift from use to mention. The variable "x" appears in a use-role (in sentential position) everywhere—outside the quantifier itself—it appears in the sentence. An infelicity of this approach, however, is that we can't avail ourselves naturally of predicates applying to sentences (where "mention" occurs), but must always regiment them as sentential operators. This is awkward if we want to attribute something both to a set of sentences and to numbers (e.g., "Neither true sentences nor numbers are in space-time"): A predicate is required for so-attributing absence from space-time to numbers; but a sentence operator is needed to do the same for sentences. It can be even more sticky, say, if we want to claim that some sentences are numbers—and also attribute to them other properties that numbers are stated to have.

Anaphorically unrestricted quantifiers

A way around this problem is to introduce *anaphorically unrestricted quantifiers*.[36] Such quantifiers are not sentential substitutional quantifiers; their variables can appear in both "mention" positions (within the scope of a predicate) *and* in "use" (sentential) positions. To get an intuitive feel for such quantifiers, imagine a

---

[36] These were first introduced in my 2001. Reasons for preferring them to other approaches to the truth idiom—e.g., prosententialism—may be found in my 2006.

pronoun "it*" that, like pronouns generally, can be linked to noun phrases, but that nevertheless can also appear in sentential position. So, just as in ordinary English we can say, "John said something yesterday, and it's true," we can say (in Anaphorish: English supplemented with "it*"), "John said something yesterday, and it*."[37] Anaphorish is a language in which a truth predicate is redundant; and the dispensability of a truth predicate in Anaphorish proves that the truth predicate indeed has the functional role that Quine claimed for it. In anaphorically unrestricted quantifiers, a blind endorsement like "Everything written on the blackboard is true," goes over into: (Ax)(Written-on-the-blackboard$x \supset x$).[38] Here, unlike in (#), "Written-on-the-blackboard" is a straightforward predicate. The variable "x" following it occurs in nominal position, but a token of that same variable also occurs free-standing, in sentential position. Using "it*," this formalization corresponds to: If something is written on the blackboard, then *it**.[39]

V

The role of the truth predicate in truth-conditional interpretational theories

Recall (^) here repeated:

(^) $(A \& B)$ is true if and only if A is true and B is true.
    $\neg A$ is true if and only if not (A is true).

In such interpretational theories, taking truth "to be the central primitive concept" (Davidson 1984*a*, p. xiii), we can now ask: Is truth—so taken—playing *only* the functional role that Quine attributed to the truth idiom? If so, it can be replaced across the board in such theories by anaphorically unrestricted quantifiers. To illustrate this point in a convincing way, suppose we have a language O, based on the standard predicate calculus. Its vocabulary is: &, ¬, ∃, (,); the variables x, y, z, etc.; and suppose it has a single (interpreted) name: b, and a single (interpreted) one-place predicate: P. The formulas and sentences of O are defined and understood in the

---

[37] This may look like a cheap trick for getting around the use of a truth *predicate* by introducing a device that—as it were—buries the use/mention shifting role of the truth predicate within itself. Well, okay, it *is* a cheap trick that does exactly what it's just been accused of; but one should never underestimate the importance of cheap tricks—like zero, and like parentheses (for example). What's (philosophically) important about *this* cheap trick is that it relies on the actual role that the truth predicate has—as described by the truth deflationist—and replicates it honestly; this is why tricks, similar in spirit, can't be used to eliminate other predicates. That such trickery can replace the truth predicate across-the-board is a symptom that the deflationary role of the truth predicate is its only role.

[38] The notation: (Ax), now stands for the universal anaphorically unrestricted quantifier.

[39] For those with misgivings about such a strange-looking logical device, it should be reassuring to learn its metalogical properties differ only slightly from those of ordinary objectual quantifiers. The proof theory governing such items, for example, is virtually identical with that of the standard first-order predicate calculus. Furthermore, an intuitively plausible model theory is available in terms of which that proof theory can be shown to be complete. See my 2001.

Alternative Logics and Interpretation   407

standard way: For example, the quantifiers of O are taken to range over a domain D, the name refers to something in D, and the one-place predicate P holds of a subset of D. To replicate the work that satisfaction does on Tarski's approach, let D contain more than one element.

The language M, that's to house the theory interpreting O, contains much of the same vocabulary as O: &, ¬, ∃, (,), x, y, z. Apart from these items, it also contain b* and P*, which are—respectively—translations of b and P. In addition, its own additional and distinctive (anaphorically unrestricted) variables will be written as p, q, r, and it also contains the two-place equality predicate: =, apart from the other predicates it has that aren't in O. Furthermore, for each sentence S in O, it contains a canonical name, [S]. It has one last distinctive bit of nomenclature. It contains O-variable operators, i, j, k, to be applied to the variables of O like so: $i(x), j(y), k(x)$. The latter will be called *O-variable names* for a reason forthcoming in the next paragraph.

The domain of M is a two-sorted domain D + O*. O* is like O, except that its sentences are constructed from base sentences of the form P$i(x)$ and P*$i(x)$, for all O-variable names $i(x)$, in addition to the base sentence Pb. The quantifiers homophonic to those of O range over D. Further, P and P* hold of the same items of D; b and b* refer to the same item of D. But the canonical names of M refer to the sentences of O they are names of, the anaphorically unrestricted variables range over O*, and the O-variable operators map the variables of O to D. As on the Tarskian approach, for each item d in D, and for each variable x of O, there is an O-variable operator i such that $i(x) = d$. From the point of view of M, the O-variable operators map the variables of O to sets of names in M: $i(x)$, for particular i and x, functions indistinguishably—in M—from a name of an object in D.

For M to be sufficient to give an interpretational theory of O, it needs additional distinctive vocabulary that describes aspects of the syntax of O: C$_\&pqr$ (p is the concatanation of &, q, and r), N$_\neg pq$ (p is the concatanation of ¬ and q), Ip (p is the concatanation of P and an O-variable name), Jpq (q is the result of replacing all occurrences of P in p with P*), Ep (p begins with an existential quantifier), Vpq (p is the result of stripping an initial quantifier (∃x) from q, and replacing all tokens of the (now-)free variable (formerly) bound by that quantifier with tokens of an O-variable name).[40]

I now give the interpretational theory of O in full:

(♣) $(p)(p = [Pb] \rightarrow (p \leftrightarrow P^*b^*))$,
  $(p)(q)((Ip \& Jpq) \rightarrow (p \leftrightarrow q))$,
  $(p)(q)(r)(C_\& pqr \rightarrow (p \leftrightarrow (q \& r)))$,
  $(p)(q)(N_\neg pq \rightarrow (p \leftrightarrow \neg q))$,
  $(p)(q)(Ep \rightarrow (p \leftrightarrow (\exists q)(V(qp) \& q)))$.

Some of the simplicity in the presentation of (♣) is deceiving. The syntactic predicates adopted from the Tarskian approach are (usually) accompanied by further definitions based on the underlying grammar of O. But no important aspect of the spirit

---

[40] V(pq) is understood to apply to *any p* with an initial existential quantifier, and *any q*, where q is the result of the operation on p described above.

of interpretation—according to the (modified) Tarskian approach—has been violated by burying these details.[41] So too, Tarski's celebrated approach to truth via a detour through satisfaction is encapsulated by the supplementation of—in M—O-variable operators *i* that yield names—in M—when applied to the variables of O. But—nearly enough—this *is* the Tarskian satisfaction-strategy: Those who see satisfaction as an automatic indication of naked ontology at work—a characterization of the quantifiers of O, that is, that connects them *directly to the world*—should read their Tarski again. Connection to objects in the world there *may* be—this isn't in dispute. But such a connection occurs *only* via the antecedent linguistic resources of the "metalanguage" M—in particular, the antecedent ranging over of existents on the part of the quantifiers of M: something that isn't given by the satisfaction clauses for the sentences of O in M, and something that isn't, in any case, *required* of the quantifiers of M.[42]

Tarski's views on the relation of ontology to semantics

Tarski himself not only seems inclined towards an ontologically empty construal of his approach to truth, but even seems *impatient* with richer metaphysical readings of his techniques. In his 1932, p. 252, when describing concepts of what he calls "the *semantics of language*—i.e., such concepts as satisfaction, denoting, truth, definability, and so on," he writes:

A characteristic feature of the semantical concepts is that they give expression to certain relations between the expressions of language and the objects about which these expressions speak, or that by means of such relations they characterize certain classes of expressions or other objects.

But he adds, rather significantly: "We could also say (making use of the *suppositio materialis*) that these concepts serve to set up the correlation between the names of expressions and the expressions themselves." That is, Tarski seems to have left open the issue of what sort of ontology, if any, is imposed by the metalanguage by virtue of his approach to truth. Indeed, in his 1944, p. 345, he writes:

Semantics is a discipline which, speaking loosely, *deals with certain relations between expressions of a language and the objects (or "states of affairs") "referred to" by those expressions*. As typical examples of semantic concepts we may mention the concepts of *designation, satisfaction*, and *definition* . . .

The incautious reader, overlooking the scarequotes and the phrase, "loosely speaking" in the above quotation, may decide that serious ontologizing is afoot. But Tarski writes later in the article (p. 363) under the rubric "Alleged metaphysical elements in semantics," that

For some people metaphysics is a general theory of objects (ontology)—a discipline which is to be developed in a purely empirical way, and which differs from other empirical sciences

---

[41] e.g., it should be clear how predicates characterizing all of such sentences' "alphabetic" properties—locations of particular constants, predicates, etc., within the sentence—are available.
[42] See Azzouni 2004, chapter 3, for further discussion of this.

only by its generality. I do not know whether such a discipline actually exists... but I think that in any case metaphysics in this conception is not objectionable to anybody, *and has hardly any connections with semantics*. (Italics mine.)

Comparing the two approaches to recursion clauses

Despite the absence of a truth or satisfaction predicate from the recursions of (♣), it's clear that these recursion clauses—in Quine's memorable phrase—"chase truth up the tree of grammar" to the same extent as their cousin recursion-clauses stated in terms of satisfaction do.[43] That is, they seem to state the truth conditions of sentences of O in terms of sentential subunits of O, and—just as on the Tarskian approach—the free variables in such sentential subunits are interpreted as referring to items in the domain D[44]; the distinctive aspect of truth (and satisfaction) to transform what would otherwise be uses of sentences (or grammatical units of O with free variables substituted for names) into statements *about* those items is here captured by the use of an anaphorically unrestricted variable that (1) can stand stead directly for those sentences in sentential position, or (2) can appear in nominal position where something is attributable to such sentences, but also (3) where it can—from either position—be bound by a quantifier. Since there is no residue of a traditional interpretational theory of O—in terms of truth or satisfaction—that's missing from the unrestricted anaphoric quantifier approach to truth-conditional theories of interpretation, it has been established that the role of the truth predicate (*and* the role of its generalization, satisfaction) in such interpretational theories are only as devices that bridge the gap between use and mention. This result should be welcomed because satisfaction is often taken as possessing an ontological significance that goes beyond the resources of deflationist truth.[45]

---

[43] Why the qualifications here, and in what follows, about what these clauses and their cousins say? See Section VI.

[44] Recall footnote 6. Also, in the standard Tarskian approach, the variables themselves appear in the clauses—relative to an interpretation of them in D. The recursion clauses for quantified expressions of O then range over all possible interpretations. In this approach, the interpreted variable x is replaced by a name in M $i(x)$—defined on x—which indicates by virtue of what it names, the interpretation i assigns to x. The difference is nomenclatural. One may disagree with this last sentence. The Tarskian approach (sometimes) requires augmenting the target domain set-theoretically: mappings from variables to items in the domain, or ordered tuples of items in the domain, or whatnot. But the current approach (so the objection would go) bloats M with additional "names." This seems to give M excessive designative capacities. In response, notice that if we grant ourselves the capacity to so augment a domain of discourse to facilitate Tarskian satisfaction, why can't we (using similar set-theoretical tools) so augment the language of interpretation? In both cases, one quantifies over the augmented resources in order to give truth conditions for quantifier expressions. In neither case are specific items—individual ordered tuples, particular mappings of variables, or O-variable names—*specifically* utilized. For both approaches, these devices are only used insofar as they are quantified over *en masse*.

[45] But it's worth stressing again that no claim has been shown (or attempted) that because truth occurs only a device to bridge use and mention in interpretational theories, that therefore such theories can't be taken as compatible with the claim that "true" corresponds to a "substantial" property. As I indicated in Section IV, the entire discussion—in *this* chapter—is neutral on this.

VI

The role of deflationist truth in truth-conditional theories of interpretation

Although recasting interpretational theories in the idiom of anaphorically unrestricted quantifiers makes the use/mention role of the truth predicate in those theories transparent, it isn't as if that role couldn't have been seen in interpretational theories otherwise. Although satisfaction introduces some essential wrinkles, it's nevertheless clear that recursion on satisfaction otherwise inherits (a generalization of) the T-biconditional condition on truth: the equivalence of suitably chosen formulas S—of M—with statements of the form "S* is satisfied by everything"—also in M—but where S* are "structural-descriptive" names of sentences of O. It's this equivalence that allows the establishing of "formally identical" results in rather different logical contexts—and indeed, it's this property that Lepore and Ludwig (2001) rely on. Without it they could not guarantee that an interpretational theory in the Tarskian mold would be establishable in a context with recursive terms rather different from the classical ones that Tarski originally availed himself of.

An intuitionistic interpretational theory

Let's take the promised closer look at a nonclassical interpretational theory. Let O and M have the same recursive terms, and let these terms be classical except for an intuitionistic negation. That is to say, we don't assume the full body of intuitionistic lore with its (apparent) replacement of the notion of truth by (something in the neighborhood of) proof. We simply take it that O is a language lacking a classical negation, instead has an intuitionistic negation, and therefore that O is a language that excludes—in the reasoning that its speakers engage in—proofs utilizing "double negation elimination." Now, as Putnam (1978, p. 28) pointed out years ago, an interpretational theory of O in M of exactly the same form is available here as in the classical context.[46] In particular, ($^\wedge$i) looks rather like our earlier ($^\wedge$):

($^\wedge$i)   (A & B) is true if and only if A is true and B is true.
¬A is true if and only if not (A is true).[47]

In ($^\wedge$i), however, "not," and "¬,"—and most likely "and" and "&" as well[48]—are intuitionistic.

Putnam (1978, p. 28) draws—rather swiftly—the conclusion that the notion of "true" at play in clauses like ($^\wedge$i) is not the same one that's at play in clauses like ($^\wedge$); and indeed, this isn't an uncommon reaction. As alluded to earlier, intuitionism is often seen as best construed as replacing a classical notion of truth with something

---

[46] I don't know who *first* noticed this, but Kleene (1952, p. 501) makes a note of it to motivate his recursive realizability interpretation.

[47] It's worth noting that both the entire theory (♣)—and the standard version utilizing the Tarskian satisfaction—are available in this context.

[48] See footnote 21.

else.[49] But that conclusion —we're now in a position to see—is unwarranted; it's unwarranted, anyway, if it's to be drawn on the mere basis of the role of the truth predicate in truth-conditional semantic theories. The role of truth—of the use/mention shifting device—is exactly the same in both interpretational theories; what has changed is only the logical backdrop that use/mention shifting device plays against.

Do Tarskian recursion axioms express truth-conditional compositionality?

As we saw in Section III, the recursion clauses of interpretational theories (in languages M utilizing classical idioms) of languages O (that are based on nonclassical logics) can take fairly complicated forms. Using the translation operator * to give us such a truth theory for intuitionistic idioms characterizes, for example, intuitionistic negation in terms of classical quantifiers, and thus in terms of satisfaction. However, if recursive idioms of the same sort are utilized in interpretational truth theories—as in Lepore and Ludwig (2001)—these clauses all take a very simple form, essentially: T(C$p q$) iff CTpTq, where "T" is the truth predicate, and "C" ranges over various recursive terms.[50] Nevertheless, it's still thought that such clauses, provided the theory they occur in has the right properties, e.g., finite vocabulary, give compositional truth-conditions. Indeed, it's very natural to think of Tarskian recursion-clauses—in the context of such truth-conditional theories—as giving *nontrivial* truth-conditions, if only because of two aspects of the seductive form they take, first, that: p is true iff q; and second, because the left side of the recursion clause attributes a truth (or satisfaction) property to a sentential unit, and the right side attributes the same to certain designated subparts of that unit.[51]

So theories of this form are taken to tell us, *as they seem to say*, the conditions under which sentences—more broadly, meaningful sentential units—are true and false *on the basis of* the truth conditions for the sentential subunits that they are *composed* of (and, ultimately, also on the basis of the referential properties of other expressions). One sees the idea very clearly both from the sample recursion clauses of (^) and the full theory (♣): In both cases, one *seems to* describe the conditions under which a sentence is true in terms of the (recursively specified) conditions under which its various meaningful sentential subunits are true. In terms of Stalnaker's distinction (footnote 32), what's being supplied are taken to be their *functional* truth-conditions. We are supposedly being given—in terms of the structure of the sentences—how their truth conditions depend on the truth conditions of their sentential subparts. In this process, it's also presumed that we are simultaneously being given their *circumstantial* truth-conditions—by virtue, of course, of their functional truth-conditions.

---

[49] See, e.g., the essays in Dummett 1978.
[50] There are wrinkles introduced by the satisfaction clauses for quantifiers. But the point is robust under the changes these wrinkles induce. (e.g., the satisfaction clause of the O-sentence "Most A are B," will avail itself of the "Most" quantifier in M, and the distribution of the satisfaction predicate across it.)
[51] For this, and the paragraph below, recall the qualification in footnote 6.

I used the phrase "nontrivial" above, and an example of *trivial* truth-conditions should be given, and discussed. Here's one:

(†) "John is running," is true iff "John is running," is true.

No doubt (†) gives "truth conditions"—that is, necessary and sufficient conditions for when the sentence "John is running," is true. But, clearly, it does so only because necessary and sufficient conditions can trivially be given of something in terms of itself, although doing so obviously isn't informative. We can put the point in this other useful way: Although circumstantial truth-conditions have (presumably) been given; functional truth-conditions have *not* been given.[52] The above triviality charge can be made about a formulation of necessary and sufficient conditions for the truth of a sentence even if that formulation *isn't* a simple repetition of the truth of the sentence (as above). That a formulation of truth conditions is an easy transformation of the above trivial truth-condition can open that formulation to the charge of triviality, for example: "John is running," is true if and only if "John is running" is true or $2 + 2 = 4$.[53] Nevertheless, although it's easy to give examples of trivial necessary and sufficient conditions for the truth of a sentence, it's not as easy to characterize, in general, when a set of recursions giving truth conditions is trivial. We have to look carefully at the cases.

In doing so, it's helpful to note that we are not without pre-theoretical resources for recognizing if nontrivial truth-conditions have been given by a set of clauses—we can recognize, for example, when *functional* truth-conditions have *not* been given.[54] We can claim that nontrivial truth-conditions at least require functional truth-conditions for a sentence, and not the mere giving of circumstantial truth-conditions—because the latter are easily given by formulations such as (†). In the classical setting, in particular, one consideration that supports the view that the clauses (^) contribute functional (and thus nontrivial) truth-conditions for the sentences they apply to is that the classical connectives "&" and "¬" are "truth-functional"; sentences S of the form $(S_1 \& S_2)$ or $\neg S_1$ are true or false depending on—and only on—the truth values of their components. This is why a decision procedure—in the form of "truth tables"—is available for the sentential sublogic of the first-order predicate calculus.[55] So what the clauses (^) seem to tell us is exactly what we know in the case of these idioms: that the truth of $(S_1 \& S_2)$ does indeed depend on its subcomponents S_1 and S_2, *and* in the way that's *indicated* by the use of the idiom "&."

---

[52] This characterization of triviality is a preliminary one: I'll show, in Section VII, that in certain cases neither circumstantial nor functional truth-conditions are given by what are otherwise perfectly ordinary-looking recursion-clauses.

[53] Here again, notice, functional truth-conditions haven't been given although circumstantial truth-conditions have.

[54] "Pre-theoretical," here is in contrast to the *theory* of interpretation, when formulated—for example—in a (modified) Tarskian form.

[55] It's important to stress that functional truth-conditions aren't required to be *truth-functional* truth-conditions, where what's meant by the latter phrase is truth conditions in the form of a truth-table semantics. In particular, the truth conditions for languages based on intuitionism supplied by * aren't *truth-functional* truth-conditions in this sense, although they *do* supply functional truth-conditions, and ones housed in a classical first-order language.

But such is *not* true in the intuitionistic setting—where a formally identical recursion clause appears: It's simply not the case that the sentence (S_1 & S_2) has been given functional truth-conditions—ones that depend on the truth values of its components, S_1 and S_2.[56] So—*given that recursion clauses of exactly the same form are available in both cases*—whatever it is that this recursion clause is *actually* telling us, it can't be a functional characterization of when the sentence (S_1 & S_2) is true (or false) directly in terms of when its components are true (or false)—as the clause seems to indicate.

Similarly, the clause, ¬A is true if and only if not (A is true), seems to offer as a (functional) truth-condition for the negation of a sentence, the falsity of that sentence in a way at variance with how—in intuitionism—double negations of sentences aren't logically equivalent to those sentences. (Notice, in particular, that the variable "A" isn't restricted in its logical form: The negation of a sentence can be substituted for that variable.)

It may be thought that this objection overlooks the very different statements that are actually being made by (^) and (^i); the latter utilizes intuitionistic connectives on the right sides of the clauses. To say, using intuitionistic "and" that (S_1 & S_2) is true if and only if S_1 is true and S_2 is true, is to say something governed by the conditions of "and" *so understood*. In particular, since this "and" is intuitionistic, it does *not* connect the truth conditions of (S_1 & S_2) *simply* to those of S_1 and to those of S_2. The classical "and" does that precisely because it is truth-functional; the intuitionistic "and" doesn't.

This answer, however, raises the likelihood that the recursion clause is saying less than it seems to be saying: It *seems* to be connecting, rather directly, the truth conditions for (S_1 & S_2) to those for S_1 and to those for S_2, respectively. But now it's clear that the purported truth-conditional connection is held hostage to the idiom "and" that occurs on the right side of the recursion clause. In general, there need be no connection *whatsoever* between the truth conditions of the sentential unit on the right side of the recursion clause, and the sentential subunits on the left side that the recursion clause seems to describe those truth conditions in terms of.

Clauses indicating that truth or satisfaction predicates distribute over a recursive term don't necessarily give (nontrivial) functional truth-conditions

The lesson is simple: Appearances in this case (and as so often) can be deceiving. Davidson (1968, p. 94) has written:

What should we ask of an adequate account of the logical form of a sentence? Above all, I would say, such an account must lead us to see the semantic character of a sentence—its truth or falsity—*as owed to how it is composed,* by a finite number of applications of some of a finite

---

[56] This is why interpretational theories (in classical languages) of languages based on intuitionistic connectives must avail themselves of extra apparatus. In general, this is the cost of nontrivially capturing the (functional) truth-conditions of non-truth-functional idioms in a classical setting—at least when interpretation is restricted to Tarskian form.

number of devices that suffice for the language as a whole, out of elements drawn from a finite stock (the vocabulary) that suffices for the language as a whole. (Italics mine)

Although Lepore and Ludwig (2001, p. 117), after quoting this, describe it as not "precise" or "general" enough, nothing suggests they are abandoning the view that a "truth-compositional"—i.e., functional—description is being given of the recursive terms t of an arbitrary language O—even when the metalanguage M provides recursion clauses via terms t' with exactly the same metalogical properties as those in O.

Instead, I recommend this diagnosis: The interchangeability of a sentence (or, more generally, of a well-formed expression) with a truth-attribution to that sentence (or, more generally, a satisfaction-attribution) in the context of a recursive idiom suffices for a Tarskian recursion-clause; but the resulting clause does *not*, in general, contribute (nontrivial) functional truth-conditions for that sentence. This is because all the recursion clause actually says is that the truth predicate—or its generalization—distributes over the recursive idiom in question. And that is what allows the possibility of a charge of triviality. For $T(Cpq)$ iff $T(Cpq)$ is a trivial truth condition. Couple it with the *mere* property that equivalents can be substituted for equivalents, and—if nothing further is going on—we gain an equally trivial truth-condition: $T(Cpq)$ iff $CTpTq$. Indeed, numerous sorts of truth conditions are available using this property of truth: $T(Cpq)$ iff $TCTpTq$, $T(Cpq)$ iff $TCTTpq$, $T(Cpq)$ iff Cpq, and so on—not to mention what's available on the basis of the substitution of other equivalents for p and q (e.g., p & p). We need some way of singling out the truth-conditional formulations that aren't trivial—that at least provide functional truth-conditions.

In the classical case, the idioms in question *happen to be* truth functional. As a result, the standard Tarskian clauses also *happen to be* accompanied by (nontrivial) functional truth-conditions *in the form of truth functionality*.[57] However, that such clauses are accompanied by truth functionality—and are thus accompanied by a (nontrivial) functional contribution to truth conditions in the classical setting—is no credit to the expressive capacities of such interpretational theories in (modified) Tarskian form. It's not, in particular, an indication that such theories really *express*—even in the classical setting—(nontrivial) functional truth-conditions.[58]

---

[57] This is because it can be shown, not merely that $(A \vee B)$ is true iff (A is true $\vee$ B is true), but—and here *we really are closer to being given (nontrivial) functional truth-conditions*—that $(A \vee B)$ is true iff ((A is true & B is true) $\vee$ (A is true & $\neg(B$ is true)) $\vee$ ($\neg(A$ is true) & (B is true))). This result, notably, is *not* provable in, say, the intuitionistic context.

[58] Intersubstitutivity of an attribution of the truth idiom or its generalization to a sentential unit with that sentential unit itself is a very weak constraint—indeed, it's so weak that it can hold in languages in which the truth-conditions for a recursive idiom Cpq don't turn on p or q at all! When this happens, how is one supposed to avoid the charge that the truth condition $T(Cpq)$ iff $CTpTq$ is exactly as trivial (or informative) as the truth condition $T(Cpq)$ iff $T(Cpq)$?

It's worth noting that by "trivial," here, I mean only the clause's failure to contribute to the giving of functional truth conditions; I don't mean the claim (held by many) that Tarskian biconditionals are trivial by virtue of being analytic, or anything like that. (Recall my denial of any thesis of this sort in Section IV.)

Tarski's aims

It's worth noting that this denial that Tarskian-style interpretational theories *express* (nontrivial) functional truth-conditions doesn't negatively impact on Tarski's own aims for his theory of truth. Truth is recursively axiomatizable—and in certain contexts, definable—as long as (1) all sentences are ultimately composed of sentential subunits the attribution of truth or satisfaction to which can be replaced by something without such an attribution (the base clauses) and (2) truth or satisfaction attributions to more complex sentences or formulas can be "pushed inwards" across the recursive terms of those sentences or formulas. To point this out is just to point out a recipe that when applicable can be used to "recursively define away" *any* term that's applied to a set of sentences.[59] What I'm claiming, however, is that Davidson's attempt to reverse the Tarskian perspective on translation and truth faces a danger that Tarski's original approach doesn't face: Changing the recursive resources of the languages to which Tarski's approach applies has no impact on Tarski's approach (as long as it isn't technically blocked) since his approach doesn't require any more of recursive terms than that they allow truth to "distribute over" them—and so this is all Tarski's approach requires his recursion clauses to "say." But Davidson wants those clauses to say more: He wants them to express (nontrivial) functional truth-conditions of sentences they apply to; but this is something they are too weak to express. They don't express them in the classical situation, where (nontrivial) functional truth-conditions happen to fit the recursions as stated; and they certainly don't express them in nonclassical situations, where (nontrivial) functional truth-conditions *don't* fit the recursions as stated.[60]

What conditions need to be placed on a (modified) Tarskian theory so that it gives (nontrivial) functional truth-conditions—so that, in particular, the distribution of the truth predicate *across* a recursive idiom actually corresponds to a functional necessary and sufficient condition on the truth of a formula (in terms of

---

[59] Putting the matter this way is not, of course, to trivialize Tarski's achievement; although it does make more explicit the purely technical nature of that achievement.

[60] It's worth noting again how *seductive* the Tarskian clauses are. Recall: $(A \lor B)$ is true if and only if (A is true or B is true). Certainly this *sounds* like we're being told that a sentence is true if and only if either of its designated components are true; and that sounds like (nontrivial) functional "truth"-conditions (because the truth of the whole is being related to the truth of the parts). Of course, in the classical setting, as noted, this sentence implies the more extensive truth-functional characterization of disjunctions in terms of their truth tables—so that the truth and falsity of the whole *is* related to the truth or falsity of the parts. As my statement in the text (this footnote is appended to) indicates, I prefer to resist the idea that a sentences *says* everything it implies. But I guess one could claim that when a recursive idiom **C** is truth functional, then $T(\mathbf{C}pq)$ iff $\mathbf{C}TpTq$ does give (nontrivial) functional truth-conditions: I'd rather say this only when those (truth-functional) conditions are explicitly given, as illustrated in footnote 57. Certainly my requirement would prevent the misapprehension that $T(\mathbf{C}pq)$ iff $\mathbf{C}TpTq$ gives (nontrivial) functional truth-conditions of $\mathbf{C}pq$ (ones in terms of its parts) even when the role of **C** is utterly unexplicated. Notice the point: Although to generate the clause, Tp and Tq are substituted for p and q, respectively, and $T(\mathbf{C}pq)$ is replaced by $\mathbf{C}pq$, the right side doesn't give truth conditions of a whole in terms of the truth conditions of its parts, except in the trivial sense that the truth predicate has been distributed to those parts. But if the sentence $\mathbf{C}pq$ is true or false regardless of when p or q are true or false, then *in fact* the truth value of $\mathbf{C}pq$ does *not* turn on the truth values of p and q.

what the truth predicate has been distributed across the recursive idiom *to*)? My suspicion is that *the* condition that's required is that the recursive idiom in question be *truth-functional*. Then nothing remains unexplicated in the *recursion clause*: The application of the truth predicate to C*pq* really can be characterized as some (truth functional) condition on *p* and *q*. Otherwise we don't *know* that the structural interconnections among the sentences of O—that Davidson evidently requires of an interpretation[61]—actually coordinate with the recursion clause. Of course, in the case of a non-truth-functional idiom C, whether or not the truth conditions of the whole unit is or isn't explicable in terms of the items the truth predicate is applied to (on the right side of the clause) turns on what C actually does. The burden of argument is on anyone, who allows such idioms into the (right sides of such) clauses, to show exactly what they indicate by way of (nontrivial) functional truth-conditions.

VII

The lesson of Section VI is: Don't read too much into (specific) recursion clauses that may indicate only that the truth predicate distributes over recursive idioms; in particular, it's wrong to presume that in helping oneself to the same recursive idioms in the interpretational language M that appear in O, in order to write down particularly simple recursions of the form: T(Cpq) iff CTpTq, that one is even in the ballpark of giving (nontrivial) functional truth-conditions.

I have, therefore, engaged in a kind of "dilemma polemic." One point of Sections I–III is that (modified) Tarskian theories in (classical) languages M—when there is a good empirical case that logical idioms in languages O are nonclassical—provide only distorted interpretations of such languages. The point of Section VI is that the alternative of letting M help itself to the same recursive idioms as those in O is wrecked by the fact that recursion clauses express only the distributivity property of the truth predicate over the recursive idioms in question, and, prima facie, nothing genuine about functional truth-conditions at all; and this is despite such clauses looking rather like pure examples of *compositional* truth conditions—truth conditions of sentential units given in terms of their sentential subunits.

But the reader might protest that I have rhetorically (and unfairly) assimilated recursion clauses that don't give *functional* truth-conditions with those that are "trivial." This is unfair at least because *circumstantial* truth-conditions are important as well—and to some philosophers, perhaps more important than functional

---

[61] See, e.g., Davidson 2005, p. 62 where he explicitly contrasts interpretation with translation, writing, that "given a theory of truth for a speaker's language *L* stated in the interpreter's language *M*, it is fairly straightforward to produce a manual that translates (at least roughly) from *L* to *M*." He continues: "But the converse is false; there are many sentences we can translate without having any idea of how to incorporate them in a theory of truth. Demanding that a theory of interpretation satisfy the constraints of a theory of truth means that more structure than is needed for translation *must be made manifest*" (italics mine).

truth-conditions.⁶² It might be thought, therefore, that clauses such as (^i) although unable to give *functional* truth-conditions, can at least be credited with supplying *circumstantial* truth-conditions. In this way, it might also be hoped that the triviality charge directed towards such recursion clauses can be at least tamed to the extent that it applies only against their giving functional but not against their giving circumstantial truth-conditions. I now turn to cruelly dashing that hope.

In doing so, I argue that not only is truth-functionality a parochial characteristic of languages based on classical logic (perhaps no surprise there), but—indeed—even the possibility of nontrivial truth-conditions (necessary *and* sufficient conditions on the truth of sentences) *of either sort* are parochial features of languages based on classical logic. They are features, therefore, that needn't be present in languages *that we can nevertheless understand and learn*. For example, these parochial features are very likely absent from *natural languages*. The importance of this is that since such features have been central—at least in an important semantic tradition⁶³—to our view of what understanding sentences comes to, and what's required of interpreting sentences, a change in viewpoint is required if we are to get onto understanding and interpreting languages with nonclassical logical idioms.

Learnability constraints

Compositionality is often presented by Davidson, and by others as well, as a learnability constraint on languages that's linked to truth conditions: One recognizes the truth conditions of a sentence—the circumstances (to use Stalnaker's language) under which a sentence is true and false—in terms of the truth conditions (satisfaction conditions) of sentential subunits of that sentence.⁶⁴ Ultimately—on this view—one's capacity to learn a language is due to its sentences being the result of compositional processes of "some of a finite number of devices that suffice for the language as a whole, out of elements drawn from a finite stock (the vocabulary) that suffices for the language as a whole."

The circumstance result

"Circumstances"—though—is metaphor, and it must be unpacked. I'll first give a metaphysically loaded picture of what "circumstances" means, and then turn to metaphysically deflated ways of motivating what I'll describe as the "circumstances picture of truth conditions." Here's the metaphysically loaded line of thought: "Circumstances" are to be characterized in a language O—when they can be so-characterized at all—in terms of those sentences of O without logical terms, and negations of such (hereafter, *base sentences*). So, if the domain of discourse of the

---

⁶² For example, the more polemically minded reader may have noted that when I quoted Stalnaker's distinction in footnote 32, I failed to quote his very next sentence: "I will make only the weaker assumption that propositional contents have truth conditions in the second sense."
⁶³ See the opening paragraphs of Field 1994.
⁶⁴ Subject, as always, to the qualifications of footnote 6.

(classical) quantifiers of a language contains no unnamed objects, then possible "circumstances" can be described in O by (complete and consistent) lists of base sentences. Indeed, it can be shown (1) that any consistent set of classical sentences is compatible with at least one such list of sentences—although in general, consistent sets of classical sentences are compatible with many such lists; and it can also be shown (2) that any such complete and consistent list of base sentences determines a truth value for every sentence of O.[65] I'll call (1) and (2) *the circumstance result* (for classical logic). The circumstance result is what truth-functionality (and compositionality) provide in a classical setting.[66] Of course, these lists of base sentences can, in turn, be taken to correspond to various possible worlds where properties are held and not held by objects; and then one can take the picture as a "realist view" of how our sentences are "made true" in various circumstances by objects (and their properties) that such sentences describe. This is neat: correspondence truth, truth-functionality, and compositionality go nicely together.

Those who reject this metaphysical picture—who reject *global* correspondence assumptions—can assume instead that among our truths are many sentences that don't correspond to objects and their properties in this way.[67] Such a position—a *locally* anti-realist one—*isn't* incompatible with classical logic. It isn't even incompatible with objectual quantifiers since these, as noted in Section V, can occur with satisfaction conditions relative to a domain without anything that exists being presupposed. (That satisfaction involves a domain doesn't require that what's "in" that domain exist: The contents of a domain are stipulated relative to the—metalanguage—quantifiers of the language the semantic theory of those—object language—quantifiers is housed in.)

I'm not presupposing my rejection of the metaphysical correspondence view, as I've sketched it here; I *am* claiming that metaphysical correspondence assumptions shouldn't be *a priori* imposed on possible languages that are to be taken to be learnable and understandable.

A correspondence-free way of construing the circumstance result

This motivates, however, another—metaphysics free—way to understand the circumstance result: as a description of the "content" of sentences containing the (classical) logical connectives. Any statement containing logical connectives—according

---

[65] When O is expressively impoverished in relation to its domain of discourse—the base sentences characterizing sets of circumstances aren't in O—we can take such circumstances, metaphysically, as "ways" that domain of discourse could otherwise have been. In terms of this characterization of "ways of being," results similar to (1) and (2) follow, although not ones explicitly expressible in O itself. This way of putting the matter seems to require a realist metaphysical-construal of the impact of truth functionality and compositionality. I'm describing (1) and (2) as above to make clear issues about the value of truth functionality and compositionality when a realist correspondence metaphysics is missing.

[66] It generalizes only to certain other—highly restricted—logical settings.

[67] I have a view like this about abstracta. There are mathematical truths; but these truths don't correspond to states of affairs among objects—because there are no such objects. This isn't a *global* denial of correspondence: *Some* truths do correspond to states of affairs among objects.

to the circumstance result—corresponds to an intricate set of various (complete and consistent) lists of base sentences.⁶⁸ The meaning, therefore, of recursive terms can be taken—in the classical setting—as (recursively) characterizing various options in lists of base sentences corresponding to the sentences they occur in. There is no "content" to these recursive terms beyond what's given by the circumstance result.

This take on the circumstance result in turn motivates the view that non-truth-functional connectives—intuitionistic ones, for example—for which the circumstance result doesn't hold, consequently occur in sentences that have more content than the inference rules governing them *reveal*. Purely "expressive" connectives are ones the inference rules governing which allow the circumstance result; thus we interpret them as "expressing" (exactly and no more than) the circumstances that they code. Connectives with rules that don't allow the circumstance result I'll call *substantial* connectives: Their "content" isn't exhausted by any set of sets of circumstances.

This view of, for example, intuitionistic connectives makes natural attempts to interpret them by means of the induced translations described in Section III. Relative to such interpretations, the circumstance result holds, and the added content is explained in terms of the additional (nonlogical) terminology drafted for the purposes of this interpretation. But if we reject this approach, and similarly reject the relevance of the circumstance result (because of, say, a desire to respect the prima facie logicality of such connectives in the target language), then interpretation of such connectives must respect their substantiality.

One might worry that the supposed additional content of such substantial connectives "hangs in the air"; there is no way to manifest that content if we are barred from a description of circumstances they license. The response to this concern is to notice that in this respect, substantial connectives are (quite) similar to mathematical concepts—such as *counting number*—in that they are (sometimes necessarily) "incomplete": That is, the rules governing them are always open to supplementation. This—by the way—isn't an idle analogy between mathematical concepts and logical terms in imaginary languages. The generalized quantifiers of natural languages possess precisely the incompleteness properties I'm alluding to.⁶⁹ Connectives involving temporality may also be incomplete in this sense. So the response is this: The rules for substantial logical terms are open to subsequent supplementation; but at a time our understanding of how the semantic properties of the sentences containing such terms are affected by them is given by our understanding of what the rules—at a time—governing those terms are.

An expressive constraint on logical terms should *not* be imposed *a priori* on (possible) foreign languages O to be interpreted.⁷⁰ Therefore, I'm taking it that the

---

⁶⁸ So, for example, "Peter is jumping or John is jumping," codes the union of two (nondisjoint) sets of sets of complete and consistent sentences, ones in which "Peter is jumping," occurs, and ones in which "John is jumping," occurs.

⁶⁹ For details on this, see, e.g., Barwise and Feferman 1985.

⁷⁰ Brandom 2000, p. 67–9, may disagree. He claims that a necessary and sufficient condition for the avoidance of tonkish phenomena in Gentzen-style definitions of inferential roles for new connectives (see Prior 1960, Belnap 1962) is that such definitions "not license any inferences involving only old vocabulary that were not already licensed before the logical vocabulary was

constraint that an interpretation of a language supply nontrivial truth-conditions (nontrivial necessary *and* sufficient conditions on when the sentences of that language are true)—if supplying such is understood via the circumstance result—is the imposition of a parochial property of classical languages onto a much broader (and otherwise acceptable) class of languages that can be interpreted without such an imposition.

Notice that the standard philosophical understanding of truth conditions—in the classical setting—usually presupposes the circumstance result. Indeed, I hypothesize that it is thinking of truth conditions as circumstantial (rather than functional) that allows the thought that nonfunctional truth-conditions are nevertheless informative. But in the general setting—where the circumstance result doesn't hold—we are once again required to verify that our recursion clauses at least supply functional truth-conditions, on pain of them otherwise supplying nothing at all. Without functional truth-conditions, and without the circumstance result, the triviality charge towards, for example, clauses of the form (i^) surely strikes home. Notice that when semantic construals of modal logics or intuitionism—in classical settings—are given, that these succeed in supplying circumstantial truth-conditions: The base sentences couched in the nonlogical predicates—describing possible worlds, time flows, etc.—are drafted for a description of the "circumstances" under which the (modal or intuitionistic) sentences are to be true or false.

In deserting the circumstance result—something required for attempts at the interpretation of languages based on nonclassical logic that nevertheless preserve the logicality of their logical idioms—we need to determine what we are grasping when we understand such a language, and therefore, what an interpretation is supposed to characterize when one is given of such a language.

In a way—since I intend to offer what will amount to "fine-grained" functional truth-conditions—nothing particularly surprising is in the offing. Let's start with the following question: Are languages—e.g., ones based on intuitionistic connectives, or those involving various generalized quantifiers—otherwise learnable and understandable? It's hard to see why they wouldn't be, if only because the circumstance result really isn't relevant to this issue. What's relevant is how many terms are involved in a sentence, and whether one grasps the rules governing those terms. As long as there are only finitely many such terms, and as long as we grasp the finite number of rules (given at a time) that tell us what we can infer to and from those sentences, everything needed to guarantee both learnability and understanding is in place.[71]

---

introduced, that is, that the new rules provide an inferentially conservative extension of the original field of inferences." This condition isn't necessary in a general logical setting precisely because of the incompleteness phenomenon that governs substantial connectives. But, in any case, one should reject the "expressive" model of logical vocabulary—at least as an *a priori* constraint; in general, therefore, new connectives, and new rules supplementing rules for old connectives, can holistically affect inferences in a way that violates inferential conservativeness. There is no way—incidentally—to, *a priori*, rule out the possibility of inconsistency: One can only be careful.

[71] But if the rules governing a connective are conceded to be open-ended, don't we *to that extent* fail to understand that connective or what it contributes to the meaning of a sentence it appears in? Compare the case to our concept of "counting number." Any set of axioms governing such a notion

Licensing truth-preserving inferences

Here, therefore, is a picture—avoiding the circumstance result—that preserves the understandability and learnability of a language. We take logical terms to be devices by means of which (logical) rules—that determine truth-preserving (and falsity-avoiding) inferences—apply to sentences. We require—this is the learnability constraint—of any sentence that it be composed from a finite number of such logical terms, and that it otherwise involve a finite number of terms drawn from a finite lexicon.[72] We don't require—since in general it isn't true—that such rules provide circumstantial truth-conditions for the sentences they govern. That is, we don't require that the rules determine, of each sentence in each circumstance, whether that sentence is true or false.

Meaning and understanding without the circumstance result

This alternative picture takes the "meaning" (at a time) of a sentence to be, in general, composed of two elements: (1) the pattern of inferences to it from other sentences, and to other sentences from it, a pattern that arises because of rules that apply to that sentence via how it's composed of logical terms; and (2) a capacity to use that sentence in appropriate situations determined by how, in addition, it's composed of non-logical devices, such as predicates and names. Its individualized meaning, to the extent it has any at all, is due only to how these factors coalese into properties it has, and our understanding of its meaning comes *only* to our grasping the play of these two factors upon it.

Therefore: We understand such a language if we understand how its sentences are composed of logical elements and base vocabulary elements; if, further, we recognize the truth-preserving rules licensed by these logical elements; and if, further, we understand—to a degree—the appropriate application-conditions of sentences due to the base vocabulary elements. In giving a theory that describes all this, a (Tarskian) truth predicate *would be* of value if it didn't lead—as we saw in Sections I–III that such

---

must be incomplete. But (1), this doesn't mean that we fail to understand the concept, and the axioms governing that concept—at a time—nor (2), that there is some other—semantic—way to characterize the notion that gives us a "fuller understanding" of it. Any purported such (e.g., second-order) way of doing so buries the open-endedness of the notion elsewhere (in the logic, say). Incompleteness imposes a genuine—and unavoidable—semantic open-endedness to our concepts in the sense that there is no way (in general) to determine, *a priori*, the best way to extend the axioms governing such concepts. Any notion of "understanding" that requires either that our understanding of an open-ended notion is—somehow—defective because of that open-endedness, or that such an understanding must be supplemented in some way to circumvent that open-endedness, simply misconstrues what incompleteness entails. Directly applying this lesson to logical terms, here is the thought: Given the rules in play at a time, we can understand what a logical term is contributing to the inferential powers of the sentence; and should such rules be augmented, then we will understand how that logical term is now differently contributing to the inferential powers of the sentences it appears in (our understanding of how the rules for a logical term have been augmented will dovetail with our previous understanding of how the rules for that logical term operated).

[72] In practice, the lexicon—although finite at any time—is open-ended.

do—to translations and thus to interpretations of logical idioms that don't preserve their logicality. Precisely here anaphorically unrestricted quantifiers—because they are quantifiers—will be of value because they don't force interpretations that impose translations.[73]

VIII

I turn now to a sketch of the interpretation of a language O in a language M, where M has classical logical idioms, but O doesn't. This sketch will be—in many ways—that of a special case: It's not at all easy to provide interpretations once we allow M and O to differ—even slightly—in their logical and nonlogical resources, while honoring the strictures raised in Section III against empirically irresponsible translations. Before giving the sketch, I'll indicate some of the ways in which the exhibited example is special, and after giving it, I'll turn to a preliminary discussion of how far the tools on offer here can be generalized.

Appropriate circumstances

When a (modified) Tarskian theory in a language M is drafted for the interpretation of a language O, the interpretation of O's basic vocabulary (such as names and predicates) is handled by outright translation to corresponding terms of M—as the first clause of (♣) in Section V illustrates. The thought is that—in this way—our understanding of sentences containing only such terms of O is captured by an antecedent understanding of sentences of M containing only the translated terms. For the purposes of this chapter, I'm avoiding as much as possible any detailed answer to the question of how—or whether—basic vocabulary (names, predicates) not so translatable can nevertheless be interpreted. Despite this, we still face a special case of that issue now because even if the basic vocabulary of M and O is taken to be the same, the holistic impact of the surrounding logic affects when sentences composed only of such terms are true or false. In cases where we understand the language O to differ from the language M *only insofar* as the inference rules governing the logical terms of the target language are a strict subclass of those governing the logical terms of the interpretational language, we can still utilize our understanding of the basic terms of M to help provide an interpretation of those of O; the ♠-soundness conditions of the forthcoming interpretational theory (♠) illustrate this.

---

[73] Although anaphorically unrestricted quantifiers allow quantification into sentential position, those quantifiers don't (semantically) rely on substitutions—in this sense they are genuine quantifiers. Thus, they don't *use* the sentences of the language O that they range over. Tarskian truth predicates—and indeed, truth predicates in general—require Tarski biconditionals, of one form or another, to fix their logical properties. But these *do* use the sentences the truth predicate is of; and this means that the logic of the language the predicate *appears in* is imposed on the sentences the predicate *holds of.* This is the technical motivation behind the Davidsonian imperative that the logic of M be imposed on O (something anaphorically unrestricted quantifiers can avoid). For further details, see Azzouni 2006, section 3.4.

Quantifiers

The first-order (objectual) quantifiers, and the various generalized quantifiers, are similar to logical connectives insofar as they are governed by rules that license various inferences; but they also seem—semantically—to be analogical generalizations of names. Just as names, and predicates, are taken to be meaningful by virtue of how they are taken to refer to, and to hold of, items in a domain of discourse, so too, quantifiers are taken to be meaningful by virtue of operations on the domain that they correspond to. The sketch will exhibit the special case of the first-order quantifiers; and afterwards, I'll give indications of what interpretations of other types of quantifiers might look like.

Sound and valid arguments in O

An idealization is needed—but this idealization, like earlier ones,[74] can be successfully navigated empirically: The speakers of O engage in what are recognized to be *arguments* or *proofs*. The study of acceptable arguments—a study that crucially includes recognition of how speakers correct each other—enables the extraction of a set of rules (or *norms*) that govern *valid* inferences for speakers of O; furthermore, these rules induce—in principle—various classes of *sound inferences*; among these are the special cases of asserting what's true (a sound inference of one step).

(Modified) Tarskian theories, when utilized for (Davidsonian) interpretion, are taken to only implicitly identify validities by virtue of how the truth conditions of formulas are given in terms of other formulas. But, given the weakness of such theories (something exposed in Section VI), this isn't what's really going on; rather, the validities of O are identified by the imposition of the logic of M upon the syntax of O.[75] If such an imposition is to be avoided in an interpretation, the validities of O must be given by virtue of inference rules governing the recursive terms of O.

Soundness conditions

Inference rules for recursive terms provide the contributions of recursive terms to the "necessities" that a logic imposes on the language it governs: Given that any sentence A of such-and-such structural form is true then any sentence B of such-and-such structural form is true as well.[76] But a (modified) Tarskian theory of truth in the service of interpretation is officially a theory of *truth* for O, and that means that what it's taken to provide are (some of) what we might call soundness conditions for

---

[74] Recall footnote 18.

[75] That is, when we prove soundness and completeness of a formalism with respect to a set of (Tarskian) models, at that point in the proof where it's shown that a constructed (Henkin) model satisfies all the sentences of a particular complete and consistent set of such, the notion of satisfaction presupposes a classical truth-functional interpretation of the connectives. (This, of course, is unobjectionable in the contexts where such proofs are used.)

[76] I've put scarequotes around "necessities": What's supplied are general conditions on sentence-forms, not necessities in any metaphysical sense. Soundness conditions *that are due to the logic* are what such rules supply.

sentences of O: descriptions of the various ways that sentences of O might be true (or false). Such "conditions" involve more than the constraints supplied by inference rules—they include, most notably, "noninferential circumstances of appropriate application."[77] In a classical setting, Davidsonian interpretation seems to provide *all* of the soundness conditions—nearly enough[78]—by supplying translations of the sentences of O into those of M; the sentences of O thus inherit the soundness conditions of sentences of M. Furthermore, soundness conditions, in the classical setting, emerge in the neat form of the circumstance result.

As we've seen, however, when O contains substantial logical terms the circumstance result isn't to be had. Nevertheless, the appearance of all the soundness conditions for sentences being given can be stage-managed if, as in Lepore and Ludwig (2001), an interpretation is given that utilizes (nearly enough) translations of sentences from O into M (where M, that is, has the same logical terms as O). "Stage-managed," is the right word because substantial connectives—as described in Section VII—have genuinely open-ended soundness-conditions; therefore, the *appearance* of fully presented soundness-conditions is matched to an *appearance* of fully present soundness-conditions by a translation of a logical term in O to itself in M.

Where translation is eschewed, we must give—along with the inference rules governing logical terms—at least *some* soundness conditions. As we will see, finding soundness conditions, apart from the general rules governing logical terms, is harder and more controversial than finding those general rules themselves.

The sketch

Let's now consider a simple language O^I, along the lines of O in Section V, containing the vocabulary: &, ¬, ∃, (,), the variables x, y, z, and so on, as well as a single (interpreted) name: b, and a single (interpreted) one-place predicate: P. O^I is understood to be intuitionistic—in the sense of Section VI: All its logical terms—except for negation—obey classical inference rules. The language M is based on a classical logic, and has the vocabulary appropriate for that. In addition—as before—it contains b*, and P*, "translations" of b and P, and its own anaphorically unrestricted variables p, q, r, ranging over the sentences of O, the two-place equality predicate: =, as well as other nonlogical predicates indicated below. Also, for each sentence S in O^I, it contains a canonical name, [S], as well as the O^I-variable operators, i, j, k. The domain of discourse of O^I is D, and that of M is $D + O^{I*}$.[79] As before, M has additional distinctive vocabulary that describes aspects of the syntax of O^I. For the forthcoming ♠-Tarskian soundness conditions, I borrow "E" and "V" from the interpretational theory (♣) in Section V. For the ♠-inference rules, and for expository purposes, I invoke a proof predicate "PRσp," where σ is a variable ranging over sets of formulas of O, and p is an anaphorically unrestricted variable

---

[77] Brandom 1994, p. 21—when speaking specifically of the concept of *red*.
[78] Leaving aside, that is, wrinkles induced by context sensitivity.
[79] $D + O^{I*}$ is as defined in Section V.

Alternative Logics and Interpretation

(roughly: "there is a proof containing only the sentences of σ as premises, and the last line of which is p").

The interpretational theory of O in M, thus, divides neatly into two parts: the inference rules, structured by the (logical) principles governing the logical terms of O, and the other general soundness conditions.

♠-Tarskian soundness-conditions:

$$(p)(p = [Pb] \to (p \to P^*b^*)),$$
$$(p)(q)(Ep \to (p \to (\exists q)(V(qp)\&q))).$$

♠-inference rules:

$$(p)(\sigma)(PR\sigma p \to ((q)(q \in \sigma \to q) \to p)),$$
$$(p)((\sigma)PR\sigma p \to p).$$

Explication of the ♠-Tarskian soundness conditions

As I've mentioned already, in cases where the rules governing the logical idioms of O are a subclass of the rules governing the logical idioms of M, we can avail ourselves of the primitive terms of M to help provide interpretations of the primitive terms of O. In this case, our understanding is that the O-name b refers to the same thing that the M-name b* refers to, and that the O-predicate P holds of the same objects that the O-predicate P* holds of. Even more strongly, we assume—for the purposes of this sketch—that the quantifiers of O range over the same items that the quantifiers of M range over. That means that the factors contributing to the meaning of the quantifiers, the predicates, and the names of O "from below"—that is, *apart from* the contribution of the logical idioms themselves to the meanings of these terms, is the same; although the factors contributing to the meaning of these items by virtue of the logical idioms of O are strictly weaker than the corresponding factors in M. In constructing interpretational clauses for sentences containing such terms, we can therefore help ourselves, not to the biconditional clauses of the (modified) Tarskian approach for the quantifiers and the primitive expressions (for that would be to presume that these expressions of O were identical in their meaning to those of M) but only to the "only if" versions of such. That is, we understand *necessary* conditions on the truth of certain sentences (or on the satisfaction of certain formulas) via the truth (or satisfaction) of corresponding sentences (or formulas), respectively.

Consider a sentence, "Roses are red." Should it be clear that such is true for a speaker of O, it follows that the translation of such is true for speakers of M. Thus speakers of M will understand the sentence, "Roses are red" in O by their possession of a necessary condition on its truth. They don't possess sufficent conditions for its truth; but there are no such conditions to be had in cases where substantial logical terms are operative. Necessary and sufficient conditions would induce the circumstance result.[80]

---

[80] M, recall, has classical logical idioms.

A similar point holds for the quantifiers of O. If "Something is red," is true in O, then its translation to M is true as well. But, "Something is red," may be true in M without it necessarily being true in O. In both this case, and in the case of sentences of O like, "Roses are red," there is the impression of "extra content" in these sentences that eludes interpretation in M. In both cases this impression is due not to the interpretational theory in M falling short—failing to capture such content—but only to the fact that the logical idioms of O are substantial ones, and so the circumstance result doesn't follow.

Explication of the ♠-inference rules governing logical idioms of O

The predicate **PR** is a syntactic characterization of proofs (of a certain form); in this case, **PR** is defined in terms of the admissible proofs in the intuitionistic system as it has been described in Section V. That is, all the logical terms operate as classical ones (relative to the admissible steps in the construction of proofs) except for negation. Because of the holistic nature of how all such logical terms affect proofs, however, **PR** is best treated in a unified fashion. The ♠-inference rules also give necessary conditions not only on when sentences can be true, but also on when, given that certain antecedent sentences are true, others are true as well.[81]

Truth conditions?

I have described these conditions as soundness conditions—conditions on when certain arguments are sound (true results from true premises) and when, if certain sentences are true, certain other sentences must be true as well. Are these truth conditions? Well, in some sense, of course they are: they are conditions on the truth of the sentences of O. But, even restricting ourselves to functional truth-conditions, these are not "truth conditions" in the sense that necessary *and* sufficient conditions for the truth of the sentences of O have been given. But in the presence of substantial recursive idioms, no nontrivial necessary *and* sufficient conditions for the truth of such sentences exist.

Generalizations

It's easy to see that a proof predicate **PR** that places necessary conditions on the truth of sentences can be crafted for a wide range of languages where rather different logical systems place rather different necessary conditions on the truth of the sentences of those languages. But what about generalizations of the ♠-Tarskian soundness conditions; what about, that is, cases where the necessary conditions contributing to the truth of the sentences of a language O are *not* a subclass of the necessary conditions

---

[81] It's appropriate here to gesture in the direction of those who have done important work in inferentialist and non-truth conditional accounts of meaning, e.g., Gentzen, Prawitz, Dummett, Tennant. Unfortunately, reasons of space preclude any detailed engagement at this time.

contributing to the truth of the sentences of the interpretational language M? What sorts of supplementary soundness conditions are pertinent in this case?

The response is this—and it is an indication of future work: One must give a more careful analysis of exactly what the appropriate application-conditions of the base vocabulary of O are doing. If causation, for example, is relevant to the appropriate application-conditions of names and kind terms, than a description of how causation operates must be included among the soundness conditions. This raises irritating issues about normativity (encapsulated in the use, above, of the word "appropriate") but such cannot be avoided if interpretation is to go beyond the straitjacket of sheer translation as a tool for the understanding of alien languages.[82]

In cases where generalized quantifiers are present, but the domains of discourse of M and O are otherwise the same, the dilemma this chapter has offered for non-truth-functional logical terms, seems to arise. My suggestion is that such quantifiers have necessary soundness conditions along the lines of the second clause of the ♠-Tarskian soundness-conditions above, but corresponding to the semantic conditions for such quantifiers; the inference rules they licence are, in general, open-ended.

But what about languages where the domain of discourse is not the same as the interpretational language of M? Here, as with the predicates and constants, an explicit analysis of the soundness conditions for quantified sentences must be given—one that goes beyond the simple adoption—in M—of satisfaction in a domain constructed from that of O.

CONCLUSION

Davidsonian and neo-Davidsonian approaches to interpretation put truth center-stage. In doing so, they also place soundness above validity in what's central to interpreting and understanding another language. I've argued that such views enshrine certain parochial features of classical languages—namely those encapsulated in the circumstance result. If we give that up, we also give up the centrality of (nontrivial) necessary *and* sufficient conditions on the truth of sentences in our understanding of such sentences. What's left is, I hope, *sufficient* for such understanding—despite its only being *necessary* for truth.[83] To interpret a sentence in a language is to recognize how its composition indicates (1) that sentence's inferential powers: what it implies and what implies it; and (2) other necessary conditions on when it's true and when it's not, due to the basic vocabulary it contains.

---

[82] For a depiction of what such conditions on kind terms are supposed to look like (on my view) see Part IV of my 2000.

[83] My succumbing to word-play here shouldn't be allowed to breed misunderstanding. Giving up necessary *and* sufficient truth conditions nevertheless allows the supplying of sufficient conditions, necessary conditions, and even—in certain circumstances—both necessary conditions and sufficient conditions that *don't* amount to necessary *and* sufficient conditions. "Certain circumstances," of course, refers to details about the logics of the languages O and M, in question, as well as facts about the noninferential circumstances of appropriate application.

Davidsonian and neo-Davidsonian approaches also makes translation a necessary condition on interpretation:[84] A language is interpreted by providing translations for all its significant vocabulary items, and therefore, for all its sentences. Although I have faulted the general presumption that (modified) Tarskian interpretational theories provide nontrivial truth-conditions, I haven't faulted the claim that such theories—when modified to give nontrivial truth-conditions of a language—will, in doing so, provide both an interpretation of that language, and an understanding of it. I have claimed, however, that nevertheless the result is useful only in quite restricted cases: when the translation it induces is fair to the properties of the target language. When the logic a language is based on differs from that of the language the interpretational theory occurs in, such an interpretational theory distorts the target language in empirically recognizable ways. But I haven't denied that there may be cases in which that's the best we can do. There is no likelihood that given any language M and any language O, an interpretation of O in M is available that doesn't—objectively speaking—distort the semantic properties of the sentences of O. I hope I've shown this: that in certain cases, where translation is not available, or where a translation of the sentences of a language O into M would distort the meaning of those sentences, nevertheless an interpretation is available. I have not, however, tried to characterize the necessary and sufficient conditions (on arbitrary O and M) that allow such interpretations.

References

Azzouni, Jody (2000) *Knowledge and Reference in Empirical Science*. London: Routledge.
—— (2001) Truth via anaphorically unrestricted quantifiers. *Journal of Philosophical Logic* 30: 329–54.
—— (2004) *Deflating Existential Consequence: A Case for Nominalism*. Oxford: Oxford University Press.
—— (2006) *Tracking Reason: Proof, Consequence, and Truth*. Oxford: Oxford University Press.
Barwise, Jon and Soloman Feferman (eds.) (1985) *Model-Theoretic Logics*. New York: Springer-Verlag.
Belnap, Nuel D. (1962) Tonk, plonk, and plink. *Analysis* 22: 130–4.
Brandom, Robert B. (1994) *Making it Explicit: Reasoning, Representing and Discursive Commitment*. Harvard: Harvard University Press.
—— (2000) *Articulating Reasons: An Introduction to Inferentialism*. Harvard: Harvard University Press.
Davidson, Donald (1967) Causal relations. In Davidson 1984*b*: 149–62.
—— (1968) On saying that. In Davidson 1984*a*: 93–108.
—— (1977) The method of truth in metaphysics. In Davidson 1984*a*: 199–214.
—— (1984*a*) *Inquiries into Truth and Interpretation*. Oxford: Oxford University Press.
—— (1984*b*) *Essays on Actions and Events*. Oxford: Oxford University Press.

---

[84] Setting aside, of course, issues raised by contextuality.

—— (2005) *Truth and Predication*. Harvard: Harvard University Press.

Dummett, Michael (1978) *Truth and Other Enigmas*. Harvard: Harvard University Press.

Etchemendy, John (1990) *The Concept of Logical Consequence*. Harvard: Harvard University Press.

Field, Hartry (1994) Deflationist views of meaning and content. In Simon Blackburn and Keith Simmons (eds.) *Truth* (1999). Oxford: Oxford University Press, 351–91.

Gabbay, D. and F. Guenthner (eds.) (1986) *Handbook of Philosophical Logic. Volume III: Alternatives in Classical Logic*. Dordrecht: D. Reidel Publishing Company.

Gödel, Kurt (1933) An interpretation of the intuitionistic propositional calculus. In (ed. Feferman, et al.) *Kurt Gödel: Collected Works*, vol. 1. Oxford: Oxford University Press, pp. 301–3.

Harris, J. H. (1982) What's so logical about the 'logical' axioms? *Studia Logica*, 41: 159–71.

Kleene, Stephen C. (1952) Introduction to metamathematics. Amsterdam: North-Holland.

Koslow, Arnold (1992) *A Structuralist Theory of Logic*. Cambridge: Cambridge University Press.

Kripke, Saul (1959) A completeness theorem in modal logic. *The Journal of Symbolic Logic* 24, 1–14.

—— (1965) Semantical analysis of intuitionistic logic. In J. N. Crossley and M. A. E. Dummett (eds.) *Formal Systems and Recursive functions*. Amsterdam: North-Holland, 92–130.

Lepore, Ernest and Kirk Ludwig (2001) What is logical form? In Peter Kotatko, Peter Pagin and Gabriel Segal (eds.) *Interpreting Davidson*. Stanford, California: CSLI Publications, 111–42.

Prior, A. N. (1960) The runabout inference ticket. *Analysis* 21: 38–9.

Putnam, Hilary (1978) *Meaning and the Moral Sciences*. London: Routledge and Kegan Paul.

Quine, W. V. (1951) Two dogmas of empiricism. *From a Logical Point of View*. Cambridge, Mass.: Harvard University Press (1980), 20–46.

—— (1986) *Philosophy of logic*, 2nd edn. Cambridge, Mass.: Harvard University Press.

Stalnaker, Robert (1998) What might nonconceptual content be? In York H. Gunther (ed.) *Essays on Nonconceptual Content*. Cambridge, Mass.: MIT Press (2003), 95–106.

Tarski, Alfred (1932) The concept of truth in formalized languages. In J. Corcoran (ed.), *Logic, Semantics, Metamathematics*. Indianapolis, IN: Hackett Publishing Company, Inc. (1983), 152–278.

—— (1944) The semantic conception of truth, *Philosophy and Phenomenological Research*, 4, 341–75.

Troelstra, A. S. (1986) Introductory note to 1933f. In S. Feferman, et al. (eds.). *Kurt Gödel: Collected works*, vol. 1. Oxford: Oxford University Press, 296–9.

Index

abstract entity *see* object, abstract
adequate usage 161 *see also* linguistic usage
Ajdukiewicz, Kazimierz 61ff
American postulate theorists 4, 49, 76
analytic/synthetic distinction 297
Aristotle
   on truth 23–4, 25
   on definition 114
arithmetic 241
attitudinal content 360
axiom
   as definition of primitive term or concept 99, 160, 161 n. 6, *see also* truth in a structure
axiom system 63ff, 76
   as definition of a structure 126–7
   see also deductive science
Ayer, A. J. 193

Banach-Tarski Paradox 48
Barcan Formula 351 n. 17
Bernays, Paul 63–4
bivalence 208
Bolzano, Bernard 14, 22, 61, 295
Boole, George 85
Brentano, Franz 196
Burley, Walter 98

calculus ratiocinator 57
Cantor, Georg 74
Carnap, Rudolf 87, 112, 113, 160, 192ff, 217–19, 226, 247, 253, 318
categorial grammar 375ff
categoricity 163–5, 181
characteristica universalis 45
Church, Alonzo 96, 340
Chwistek, Leon 37, 46, 372 n. 10,
completeness of a theory with respect to its specific terms 164, 187
completeness theorem 275, 331
concept 113, 149, 166–9
conceptual analysis 6, 9, 72, 74, 112, 113, 192, 294
contentual vs. formal axiomatics 63–7
context-sensitivity 150, 392 n. 9
continuum hypothesis 276–7, 321, 328, 329 n. 22,
convention 153–5
convention T 6, 133–4, 140ff, 173, 249

correspondence *see* truth, correspondence theory
Czeżowski, Tadeusz 26

Davidson, Donald 178, 247, 391, 413, 416 n. 61
Dedekind, Richard 76
decidability 331
deductive science, theory 64, 80, 96, 112, 117, 126, 128
   practising, peforming 97–8, 108
definition 5, 94, 126–7
   Ajdukiewicz on 63
   closed form 106
   eliminability 100, 104
   establishing vs. reporting on meaning 100, 154, 257
   evaluation of 170
   extending a theory vs. abbreviating its expressions 5, 102, 159, 175
   extensional correctness 114, 173
   explicit 116, 119
   formal 157–9 *see also* definition, explicit
   formal correctness 115–7, 138
   inconsistent 116
   implicit 116 *see also* axiom, as definition of primitive term
   in Warsaw topology 74
   Kotarbiński on 99
   Leśniewski on 104
   mathematical and metamathematical 105–111, 124
   non-creativity 104–5, 115, 127
   of a set 81, 108–111, 162–3
   of semantic notions 120, 125, 138
   of truth *see* truth, defintion of
   Padoa's method 104–8, 124, 125, 127, 163
   possible 116, 119, 159
   relativization to a theory 140–1, 159, 165–6
   relativization to a set of contexts 175
   semantic 162–6 *see also* Kotarbiński
   Suppes' criteria 104
   syntactic and semantic conditions 106–8, 125
definability
   see definition
De Morgan, Augustus 85
direct reference 353
domain, cardinality of *see* universe, size of

elucidation 200
empirically establishing something 98, 111
empiricism 195, 216
   logical 197, 216, 233, 235
equivalence of the form T *see* T-sentence
essentially richer language 133 n.1 *see also* definition; truth, indefinability of
existence 62
existence predicate 356–9
explication 112, 113
excluded middle 294
extensionality 175
   axiom of 96, 114

Field, Hartry 9, 84 n. 21, 241–4, 254–5
finitism 227, 230, 273 *see also* Hilbert, David
formalism 63ff, 83, 161 n. 7, 169
formally correct *see* definition, formal correctness
foundationalism 308–12
Fraenkel, Abraham 164–5
Frege, Gottlob 85, 96, 97, 112, 127, 169, 305, 318
functor 380
Fundamenta Mathematicae 51

geometry 76
Gödel, Kurt 11, 76, 91, 125, 296, 329
Goodman, Nelson 226, 232–3

Hempel, Karl 192, 203–4, 226
Hilbert, David 49, 63–4, 76, 84, 96, 124, 127, 160, 169, 227, 230
Hintikka, Jaakko 58
Husserl, Edmund 85, 186

identity predicate 279, 336–7
implicit definition 161 n. 7; *see also* axiom, axiom system, truth in a structure
incompleteness 276, 312, 420 n. 71
inconsistent language 83, 161
indefinability of truth *see* truth, indefinability of
indexicals 150, 281, 352
inductive 98, 122 *see also* empirically establishing something
infinity 236
   axiom of 85, 273, 328
intensionality 176, 347, 371
interpretation 390ff
intuitive, intuition 96, 125
   and material adequacy 103, 114
   see also Leśniewski, intuitionistic formalism

Janiszewski, Zygmunt 73

Kaplan, David 288, 353, 356, 361
Kant, Immanuel 270, 305
Knaster, Bronislaw 74, 128
knowledge, causal theory 243
Kokoszynska, Maria 195, 206, 212–15
Kotarbiński, Tadeusz 12, 99–100, 114, 120, 121, 162 n. 9 196, 210, 227, 237, 371
Kreisel, Georg 275 n. 7, 341 n. 5
Kripke, Saul 280, 282, 352, 361
Kuratowski, Kazimierz 47, 73, 109, 112, 124, 125

Langford, Cooper 4, 49, 77, 164
language
   as including a theory 84, 161
   calculus 231
   conditions for intelligibility 227–33
   interpreted 250 *see also* meaning, intuitive
   of the calculus of classes 134
   of the general theory of classes 138, 179
   related to reality 199–200, 204, 208, 214–16, 251
Lebesque, Henri-Leon 112
Leśniewski, Stanislaw 45ff, 102–5, 115, 120, 128, 369, 371
   intuitionistic formalism 53, 103, 112, 119, 169, 250 n. 9
   on directives 58, 102
   on metalanguage 55
   on names 54
   on protothetic 47
   on semantics 59–60
   on semantic antinomies 53
   on semantic categories 85, 186
   on set theory 48
   practicing a deductive theory 97
liar paradox 32, 83, 188
Łukasiewicz on 25
lingua characteristica 57
linguistic usage 135
logic
   and mathematics 307, 316, 318–21
   and set theory 139, 320 *see also* logicism
   as formal 307–8, 315, 322
   as maximally general 302, 305–7
   as normative 317
   as universal language 85; *see also calculus ratiocinator, characteristica universalis*
   first vs. higher order 119, 161, 165 n. 12, 187, 228, 234–5, 276, 321, 357
   see also type theory
   intuitionistic 397ff, 410
   invariance under homomorphism 325, 334–8
   invariance under isomorphism 302, 305ff, 315, 320

many valued 25, 50
non-classical 394
ontological commitments of 328
topic neutrality 306, 357, 398–401, 419
logical consequence 3, 14, 61–2, 263ff, 398
  epistemic aspects 266, 268, 313, 317
  intuitive concept of 265
  modal aspects 266, 270–1, 315
  reduction to truth plus generalization 266ff
  semantic vs. syntactic conception 264
  truth preservation 267–8
  vs. analytic consequence 279, 297
logical constant 266, 269, 278, 282ff, 293, 295–6, 314–16, 344, 393–4
  see also logical notion
logical notion 78, 300ff, 347
logical truth 263
  a priority 341, 362ff
  and arithmetic truth 274 see also logicism; universe, size of
  as truth of associated generalization 25–26, 32, 265, 286
  necessity 341, 346, 348
logicism 17, 49, 52, 75, 78, 160 n. 2, 296, 318, 320
Löwenheim, Leopold 49
Löwenheim-Skolem Theorem 49, 79, 329–331
Łukasiewicz, Jan 24, 25, 96, 112, 120, 196
  liar paradox 25
  on probability 25
  on definition 115

material adequacy 7, 86, 103, 114–5, 138, 173, 279, 378
Mazurkiewicz, Stefan 73
meaning 38, 373, 421
  contingent vs. necessary property of expressions 174–8, 248, 251
  determined by sentences held true 158, 160, 174–78
  of primitive expressions of a language 84, 95, 99, 105–8, 113, 158
  intuitive 114–5, 173, 166–9, 184, 187, 189, 232
Menger, Karl 74
metalanguage 53, 94, 134, 138, 170, 199–200
metamathematics 49, 59, 67, 80–1, 96, 112, 118
metatheory 94, 115, 138 see also metalanguage
metametalanguage 170 see also metalanguage
metametatheory 123 see also metatheory
metaphysics 194ff, 233
modal objection to Tarskian semantics 247ff
model theory 59, 76, 79, 88–90, 108, 118, 127, 228, 251, 264, 285ff, 341ff

see also truth in a structure
morphology of language 138

Naess, Arne 196, 212, 216
name 378ff
  see also structural-descriptive name
necessity 174–6
Neurath, Otto 192ff
nominalism 12, 27, 34–8, 78, 226, 229, 232–3, 243–5, 369, 387
  indispensibility argument 244
  materialistic taint 229 n. 7
notion 168
number 240–3

object
  abstract 12, 160 n. 3, 226, 375
  physical 229–30
object language 134
operator 376, 381–4

Padoa, Alessandro 100, 105–8, 120, 124, 163
pansomatism 12, 27
paradox
  semantic 81
  set-theoretic 81
  see also liar paradox, Leśniewski on semantic antinomies
Peano, Giuseppe 74, 96, 112
platonism 12, 226, 234, 318, 370, 388
plural 373
Popper, Karl 87, 196, 247
Presburger, Mojzesz 77, 122
Principia Mathematica 139, 186, 300
protothetic see Leśniewski, Stanisław
Putnam, Hilary 174–8, 247–8, 253

quantifier elimination 49, 76–7, 110, 121–2, 123
quantifier, semantics of 350ff
Quine, W. v. O. 89, 226, 232–3, 234–5, 277, 295, 297, 404, 409

reference, causal theory 243
reism 13, 27, 227, 237
rigid designator 352
Russell, Bertrand 24, 85, 89, 97, 169, 226, 233, 267, 273, 295

satisfaction 121–2, 391, 408 n 2 see also definition, of a set; truth, definition of
  Ajdukiewicz on 63
Schlick, Moritz 202–205
Schröder, Ernst 85
science
  classical ideal 45–6, 57ff

science (cont.)
  Neurath's conception 193, 200–2, 204–5
  Schlick's conception 202
self-reference 32
semantic closure 210, 258
semantics
  compositional 8, 124, 127–8, 392, 417
  definition vs. theory 259
  descriptive vs. theoretical or pure 257–9
  ontological commitment 408
  possible worlds 280, 291, 397
semantical category 52, 186, 307
set theory 73–4, 117, 226, 234, 369
Sierpinski, Waclaw 48, 73, 369
situation 311
Skolem, Thoralf 76, 110, 123
soundness theorem 275
structural-descriptive name 134
structuralism 307, 318, 327
subjective certainty 98
Sundholm Göran 54, 56, 86
Suppes, Patrick 104, 225
Szmielew, Wanda 90 n. 27

T-schema 32, 86, 121, 135
  see also T-sentence
T-sentence 26, 135, 247, 250, 392, 415
  as determining meaning of "true" 158–9, 171–2, 179, 188, 404
  implication by truth definition 140, 171, 212, 248
topology in Warsaw 73–4
translation 135, 173, 249, 254, 378, 392
truth
  adverbial theory 27
  axiomatic theory of 181–2, 260
  Aristotle see Aristotle, on truth
  Aristotelian conception of 31, 155, 171, 173, 178, 187, 371
  bearers of 11ff, 34–6, 137, 310 369, 372
  classical conception of see truth, Aristotelian conception of
  coherence theory 155, 193, 203, 206, 213; see also science, Neurath's conception
  concept of 148ff, 166–9, 189, 218
  condition 173, 403 n. 32, 411ff, 426 see also T-sentence
  context sensitivity 150ff
  correspondence theory 24, 29, 30, 193, 200, 204, 206, 309, 418
  definition of 8–9, 136–7, 140, 172
  deflationary theory 29, 404, 410
  disquotational theory 29
  eternality 23 see also bivalence
  function 301
  indefinability of 2, 11, 86, 91, 125, 138, 179–89
  in a structure 32–3, 75–6, 79, 83–4, 87, 126, 127, 251
  minimalism 29
  nihilistic theory 26, 29, 33
  partial definition of 135
  pragmatist conception 155 see also truth, utilitarian conception
  prosentential theory 28 n.16
  redundancy theory of 29, 218
  relativization to a language 141ff
  scheme see T-schema
  semantic conception 121, 172, 217
  utilitarian conception 29, 155
Turing, Alan 72, 103, 340
Twardowski, Kazimierz 22–6, 196
  on relativism 22–4
type theory 52, 55, 77–9, 85, 124, 185, 300, 370

Unity of Science Congress, Paris, 1935 88, 125, 193, 196–99
universality
  of language
  of logic 85
universals 235 see also, logic, first vs. higher order; object, abstract; type theory
universe
  size of 236–8, 242–3, 272–4, 329, 332–4, 349
unrestricted quantification 291, 349

validity see logical consequence
Vienna Circle 88, 192ff

Waismann, Friedrich 199
Weierstrass, Karl 74
Weyl, Hermann 82 n. 18, 96
Wittgenstein, Ludwig 198, 199–200, 209, 215, 233, 235–6, 267, 273, 296

Zawirski, Zygmunt 26
Zermelo, Ernst 52, 78, 128